MODERN DEVICES

MODERN DEVICES

The Simple Physics of Sophisticated Technology

CHARLES L. JOSEPH
Department of Physics & Astronomy
Rutgers University

and

SANTIAGO BERNAL
Institute for Research in Electronics and Applied Physics
University of Maryland

With contributions by

TIMOTHY KOETH
Institute for Research in Electronics and Applied Physics
University of Maryland

WILEY

Published by John Wiley & Sons, Inc., Hoboken, New Jersey
Published simultaneously in Canada

Library of Congress Cataloging-in-Publication Data:

Names: Joseph, Charles Lynn, author. | Bernal, Santiago, author.
Title: Modern devices : the simple physics of sophisticated technology / by Charles L. Joseph and Santiago Bernal.
Description: Hoboken, New Jersey : John Wiley & Sons, 2016. | Includes bibliographical references and index.
Identifiers: LCCN 2015044568 (print) | LCCN 2015050864 (ebook) | ISBN 9780470900437 (cloth) | ISBN 9781119011835 (pdf) | ISBN 9781119011828 (epub)
Subjects: LCSH: Manufactures–Technological innovations. | Electronic apparatus and appliances–Technological innovations. | Industrial equipment–Technological innovations. | Optical instruments–Technological innovations. | Technology. | Physics.
Classification: LCC TS145 .J67 2016 (print) | LCC TS145 (ebook) | DDC 670–dc23
LC record available at http://lccn.loc.gov/2015044568

Set in 10/12pt Times by SPi Global, Pondicherry, India

Printed in the United States of America

10 9 8 7 6 5 4 3 2 1

CONTENTS

Preface **xi**

About the Companion Website **xv**

1 Principles of Physics and the Relevance to Modern Technologies **1**

1.1 CM, EM, and QM: The Backbone of Physics 3
1.2 Photonics and Electronics 5

2 Everyday Home Appliances **9**

2.1 The Air Conditioner 10
2.2 Microwave Ovens 18
2.3 Smoke Detectors 25
2.4 Compact Discs, Digital Versatile Discs, and Blu-Ray Discs 27
2.5 Photocopiers and Fax Machines 37

3 Devices Encountered in Modern Life **43**

3.1 Metal Detectors for Airports and Traffic Lights 43
3.2 Barcode Scanners, Quick Response Codes, and Radio-Frequency Identification Readers 47
3.3 Global Positioning 53
3.4 Transportation Technologies 57
3.4.1 Internal Combustion Engines versus Electric Motors 57
3.4.2 Alternative Fuels 58
3.4.3 Speed Radar Guns 60
3.4.4 High-Speed Rail 67

4 Vacuum Systems: Enabling High-Tech Industries **69**

4.1 Vacuum Chamber Technology 70
4.2 Physics of Some Vacuum Gauges 76

4.3 Low Vacuum via Venturi, Mechanical, or Sorption Pumps 78
4.4 HV via Diffusion, Turbomolecular, or Cryogenic Pumps 80
4.5 UHV via Ion Pumps 84

5 Cleanrooms, an Enabling Technology 87

6 Solid-State Electronics 91

6.1 Conducting, Semiconducting, and Insulating Materials 95
6.2 Resistors, Capacitors, and Inductors 101
6.3 Diodes and Transistors 110
6.4 FET, JFET, MOSFET, CMOS, and TTL 119
6.5 Summary 124

7 High-Tech Semiconductor Fabrication 127

7.1 Thin Films 127
7.2 Thin-Film Deposition Methods 132
7.3 High-Purity Crystals via MBE 138
7.4 Photolithography and Etch Techniques 141
7.5 In Situ and Intermediate-Stage Tests 145
7.6 Device Structures and IC Packaging 152

8 Materials Science—Invaluable High-Tech Contributions 155

8.1 The Use of Composite Materials 156
8.2 Thin-Film Multilayers 157
8.3 Nanotechnology 158

9 Light Sources 161

9.1 Incandescent Lamps 166
9.2 Gas Discharge Lamps 168
9.3 Fluorescent Lamps 171
9.4 Light Emitting Diodes 174
9.5 X-Ray Sources 175
9.6 Lasers 177
9.7 Synchrotron Light Sources 180
9.8 Summary of Light Sources 180

10 Some Basic Physics of Optical Systems 183

10.1 Refractive and Reflective Optics and Their Uses 184
10.2 Polarization and Birefringence 188
10.2.1 Law of Malus and Brewster's Angle 188
10.2.2 Dichroism and Birefringence 190
10.2.3 Retarder Plates and Circular Polarization 192

10.3 Diffraction 194
10.3.1 Huygens' Principle and Diffraction from a Single Slit 194
10.3.2 Fresnel Zone Plate 196
10.3.3 Diffraction Gratings 198
10.4 Holography 200
10.4.1 Basic (Absorption) Holography 200
10.4.2 Temporal and Spatial Coherence 202
10.4.3 Other Methods of Holography and Applications 203
10.5 Primary Aberrations 205

11 Optical Couplers Including Optical Fibers 217

11.1 Optical Fibers and Hollow Waveguides 218
11.2 Couplers for Long Distances 223
11.3 Optical Couplers as a Means of Electronic Isolation 228

12 Spectrographs: Reading the "Bar Code" of Nature 231

12.1 Prisms, Ruled Gratings, and Holographic Gratings 240
12.2 Long-Slit Spectrographs 248
12.3 Integral Field Unit and Fabry–Pérot 249
12.4 Echelle Spectrographs 254
12.5 Raman Spectrographs 255

13 Optical and Electron Microscopy 259

13.1 Optical Microscopes 260
13.1.1 The Magnifier 260
13.1.2 The Compound Microscope 261
13.1.3 Numerical Aperture, Resolution, and Depth of Field 262
13.1.4 Alternative Methods of Optical Microscopy 265
13.2 The Transmission Electron Microscope 266
13.3 Electron–Matter Interactions 271
13.4 Bragg's Diffraction 273
13.5 Scanning Probe Microscopes 275

14 Photoelectric Image Sensors 277

14.1 Solid-State Visible Wavelength Sensors 280
14.2 Photoemissive Devices for UV and X-Rays 284
14.3 Infrared "Thermal" Sensors and Night Vision Sensors 287

15 Image Display Systems 291

15.1 The Human Visual System 293
15.2 Who Invented Television? 300
15.3 Traditional and High-Definition TV Display Formats 301
15.4 Cathode Ray Tubes 306

15.5 Liquid Crystal Displays 308
15.6 Plasma Displays 310
15.7 Digital Micro-Mirror Devices 311
15.8 Touch Screens 314
15.9 Electrophoretic Displays 315
15.10 Near-Eye Displays, Augmented Reality, and Virtual Reality 317
15.11 Stereoscopic, Autostereoscopic, and Holographic 3D Displays 319

16 Spacecraft Systems 325

16.1 Operating in Space: An Overview 326
16.2 Attitude Control System 330
16.3 Spacecraft Power 337
16.4 Thermal and Other Environmental Control 339
16.5 Command, Control, and Telemetry 341
16.6 Launch, Propulsion, Station Keeping, and Deorbit 345

17 Astronomical and Planetary Observatories 353

17.1 Telescope Designs 354
17.2 Very Large, Ultra-Lightweight or Segmented Mirrors 358
17.3 Adaptive Optics and Active Optics 362
17.4 Space Observatories 365
17.5 Planetary Probes 372

18 Telecommunications 377

18.1 Physical Connections: Phone Lines, Coaxial Cable, and Fiber Optics 378
18.2 Analog Free-Space Channels: TV, Radio, Microwave Connections 384
18.3 Digitally Modulated Free-Space Channels 390
18.4 The Network, Multiplexing, and Data Compression 392

19 Physics of Instruments for Biology and Medicine 397

19.1 Imaging Instruments 397
19.1.1 CT Scanners 398
19.1.2 Magnetic Resonance Imaging 398
19.1.3 Ultrasonography and Ultrasonic Lithotripsy 408
19.2 Minimally Invasive Probes and Surgery 410
19.3 Laser Technologies 411
19.4 Miscellaneous Electronic Devices 415

20 A-Bombs, H-Bombs, and Radioactivity **419**

20.1 Alpha, Beta, and Gamma Ray Radiation 421
20.2 A-Bombs, H-Bombs, and Dirty Bombs 423
20.3 Radiation Safety, Detection, and Protection 428
20.4 Industrial and Medical Applications 431

21 Power Generation **433**

21.1 Principles of Electric Generators 434
21.2 Power Storage and Power Content of Fuels 435
21.3 The Power Grid 439

22 Particle Accelerators—Atom and Particle Smashers **443**

22.1 Lorentz Force, Deflection, and Focusing 446
22.2 Beam Generation, Manipulation, and Characterization 448
22.3 DC Accelerators 450
22.4 RF Linear Accelerators 450
22.4.1 Motivation and History 450
22.4.2 Linac Components and Operation 452
22.4.3 Beam Bunch Stability and RF Bucket 454
22.4.4 Power Budget and Linac Applications 454
22.5 Cyclotrons 456
22.6 Synchrotron Radiation and Light Sources 462
22.6.1 Dipole Radiation and Larmor's Formula 462
22.6.2 Wigglers and Undulators 464
22.6.3 First-to-Fourth Generations of Light Sources and Applications of SR 466
22.6.4 Free-Electron Lasers 468

23 Jet Engines, Stratospheric Balloons, and Airships **471**

23.1 Ramjets, Turbojets, and Turbofan Jets 474
23.2 Stratospheric Balloons 476
23.3 Future Airships 484

Appendix A Statistics and Error Analysis **489**

Bibliography **497**

Index **503**

PREFACE

We all encounter sophisticated technology in many aspects of our daily lives. There are, of course, items that readily come to mind such as smart phones, tablets, personal computers, and the like. Technology, however, has also invaded every category of employment in the workplace. It is no longer unusual to see family farmers or construction workers using lasers to make precision measurements. Farmers in the not-too-distant future may commonly harvest crops at night as well as during the day. This flexibility will increase their efficiency. For instance, farmers would be able to share expensive field equipment with one farmer working in the night shift. Alternatively, they would have the option if necessary to conduct a burst of activities, performing critical farming operations in advance of say a pending change in weather. The night farming is made possible by the use of global positioning, which enables the farmer to maneuver his large equipment exactly (within inches) to plow or harvest during the darkest nights. He will also be able to apply minimal amounts of water, insecticides, and fertilizers only to those portions of his field requiring it based on satellite imaging.

In addition, many of us will be expected to become familiar with many more devices than we current are. Not only is there an ever-increasing array of new products to increase our productivity or to enhance our leisure, many of us will switch careers several times throughout our lives. Each time, we are likely to encounter a new, uniquely, and highly specialized set of tools and devices. Even those of us who stay with a single company are likely to encounter a diversity of technological gadgets. There has been, for example, a long tradition in the aerospace industry for senior engineers to evolve toward and eventually become managers where they then frequently find themselves making decisions about completely unfamiliar technologies. As more and more businesses incorporate technologically sophisticated instruments, the general work force will increasingly resemble aerospace companies in some respects. Their personnel will have to learn to deal effectively with new, unfamiliar technologies, repeatedly.

Moreover, many individuals will have to make decisions in response to new technological advances. Such individuals include financial investment advisors or economists, who will have to assess the potential payoff from costly and sometimes risky investments in emerging technologies. Government regulators will require a grasp of the benefits and downsides to new technological solutions to say, cleaning up the environment. Local government officials will have to assess the impact that the introduction of sophisticated technologies can have on a community. For that matter, political leaders will be making decisions on which high-tech systems should be deployed by the military as well as by other agencies. Many new systems will be expensive, and we as a society will have to weigh the effectiveness of these systems against the costs and then balance the new capability against other social needs. Even lawyers and judges will have to litigate more issues associated with technology, both from complainants with grievances as well as law suits over intellectual property rights.

A wide variety of individuals need to understand the basic concepts and the corresponding limitations behind various technologies. They need information that articulates the inherent strengths and limitations of these technologies, which can be obtained from an understanding of the physics of these modern devices. While technology is constantly evolving and improving, the basic underlying physics evolves very slowly, much longer than a human lifetime. There are relatively few physics principles, and fortunately each can be applied to numerous applications. This is the advantage of understanding the physics behind technologies. Rather than memorizing thousands of seemingly unrelated facts and data, one can learn a relatively few number of physical principles and then readily understand how many unfamiliar devices work. A better understanding of the physics will also aid in the interpretation of results, especially in identifying spurious or absurd measurements.

The current book is designed for professionals with a bachelors or higher degree who require a basic understanding of the operation of complex technologies. In the current "information glut" created by the Internet, there are numerous websites established to provide technical backgrounds on almost any subject. However, these web pages vary substantially from one site to another in the level of sophistication and the technical accuracy. Moreover, these sites often contain unclear or misleading physics. Some excellent websites at universities (e.g., at the University of Colorado and at Georgia State University) provide superb backgrounds over a wide range of physics topics but do not focus specifically on the common recurring physical principles behind sophisticated modern devices. This book fills this need at a consistent standard and is also suitable as textbook for an upper-level undergraduate college physics course for nonmajors. We describe the basic physics behind a large number of devices encountered in everyday life. There are many more topics than can realistically be taught in a two semester course. This encyclopedia style format is deliberate. Each device or topic is written essentially as a stand-alone article, allowing individuals or instructors to select the topics most suitable for their interests. Instructors at Rutgers University teach a non-majors course: The Physics of Modern Devices and have their students prepare 15-minute presentations on a topic of their choosing. The present text serves as a starting point for further exploration of other topics as well as serves as a useful reference book to professionals.

The approach is to present physics principles in an interesting context and in a non-threatening highly descriptive manner. It is thus an excellent resource for students who feel somewhat intimidated by physics and mathematics in general. It is assumed, however, that the student has taken an introductory physics course, college algebra, and had at least a minimal exposure to calculus.

This text contains many diagrams and sketches as well as graphs to help the student visualize the physics. There are also equations and mathematical analysis where appropriate. Both approaches enhance the reader's overall understanding and appreciation of the subject. The goal is to bring the reader quickly up to speed on the essential issues leading to the inherent strengths and limitations of various devices and associated technologies. Where possible, alternative technologies and any ongoing efforts to mitigate existing shortcomings are discussed. Each topic is intended to provide the reader with the insights necessary to ask intelligent questions on the relevant topics and to gain a greater appreciation of the basic physics principles that impact the operation of various modern equipment. It is hoped that the student gains an appreciation that no single device is optimal for all applications.

Charles L. Joseph
New York, USA

Santiago Bernal
Maryland, USA

ABOUT THE COMPANION WEBSITE

This book is accompanied by a companion website:

www.wiley.com/go/joseph/moderndevices

The website includes:

- PowerPoint Slides

1

PRINCIPLES OF PHYSICS AND THE RELEVANCE TO MODERN TECHNOLOGIES

The basic motivation that science, the scientific method, and scientific reasoning should be mastered by an increasingly large fraction of our population can be seen in Figure 1.1, which shows the volume of an individual's knowledge and understanding compared to the collective, comprehensive volume of all human experience. The gray areas represent the fragments grasped by an individual with some areas being connected (i.e., related) through various mental paths. Gray blobs that are clustered represent the formation of expertise in some field. Most of the volume (white area) is empty, indicating those topics where the individual is uninformed. As the figure depicts, the overall volume of knowledge and understanding is increasing rapidly with time. While the individual continues to grow and learn, adding more fragments as well as enhancing his/her expertise in some fields (larger, more concentrated gray area clusters), it is difficult to keep pace with all that one ought to understand. This task becomes virtually impossible if one relies solely on the incorporation of more factual knowledge, especially in a world that is increasingly becoming more reliant on technologies. A human being has a limited amount of memory that can be accessed with any reliability. The person who develops and incorporates scientific cognitive skills has a significant advantage since there are relatively few concepts underlying the physics behind all science and technology. Each fundamental theory can be applied to numerous applications, providing shortcuts to acquiring an understanding of new, unfamiliar equipment. The laws of physics are unchanging, and after basic concepts have been established, these evolve slowly on timescales of centuries. Basic scientific cognitive skills provide the individual with more mental tools, and he/she can exploit the observed commonalities between recognized and unfamiliar technologies.

All modern technologies are the exploitation of one or at most a few basic laws of physics. Insights into these governing principles illuminate simultaneously the intrinsic

Modern Devices: The Simple Physics of Sophisticated Technology, First Edition.
Charles L. Joseph and Santiago Bernal.

Companion website: www.wiley.com/go/joseph/moderndevices

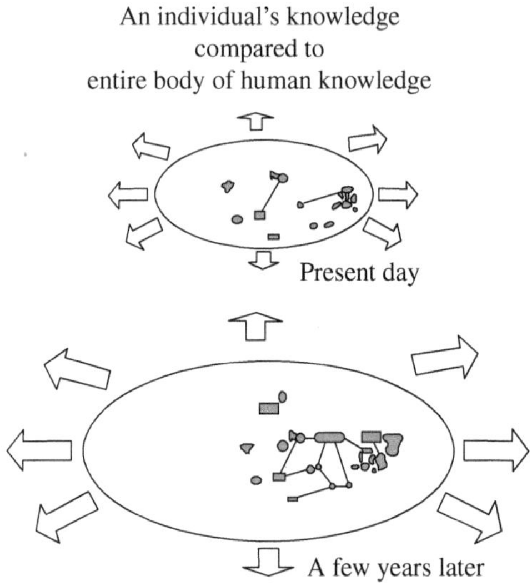

FIGURE 1.1 In the ever-expanding body of human knowledge, it is difficult for an individual to keep pace by only absorbing factual information. Gray areas represent small fragments of an individual's knowledge compared to all of the available data. Some of these fragments are connected (shown as lines) via various means (e.g., factual, cognitive, and reasoning).

operation as well as the inherent strengths and limitations of any apparatus or piece of equipment. Once optimized, there are only two ways to enhance the performance further. First, one performance parameter can often be enhance within limits at the expense of another. For example, power and speed in many electronics devices can be sacrificed against each other. Computing speed can be increased, but only at the expense of needing more power. Increased power consumption normally carries the penalties of greater cooling requirements, greater mass, and greater volume. Second, the only other way to enhance the performance of a device that has already been optimized is to switch to a totally different technology, one obeying a different set of physical laws.

The mastery of the underlying physics of modern equipment is satisfying, giving the student added insights into the equipment used throughout their careers. However, acquiring these cognitive skills does require some serious effort. It is important to bear in mind that in the early stages of learning physics, the individual has to absorb each rudimentary concept through the process of solving a number of similar problems. This learning process is similar in nature to a student learning a musical instrument, who must repetitively practice his or her scales and perform other repetitive exercises prior to the thrill and enjoyment of performing. The same is true of an individual taking up a new sport activity. He/she cannot expect to become a star without first receiving instruction on various techniques and plenty of practice. While rudimentary training cannot be avoided if the individual is to gain a solid understanding, the approach of the current text seeks to provide the motivational framework necessary to entice the student. The acquisition of new knowledge and new reasoning skills has to be a life-long endeavor, if one wants to rise above the crowd.

The current text contains a series of boxes titled *Intro Physics Flashback* to assist the individual identify the appropriate concepts from his/her freshman physics course. Individuals with strong backgrounds in high school or freshman-level college can ignore these Flashbacks. Throughout the textbook, the student is advised to search for recurring principles and to organize his or her thoughts according to a hierarchy of importance. Merely identifying the appropriate equations to solve a problem is simply substituting one factual database for another, a list of equations instead of a list of facts. Such an approach leaves the student unable to recognize the underlying physics for an unfamiliar device.

Moreover, the student is encouraged to step back on a regular basis and contemplate the reasonableness of his or her assumptions, measurements, or conclusions. Always ask: Is this statement consistent with other facts and knowledge? How does my answer compare with other information? It is of great assistance in answering these types of questions if the individual has at his or her finger tips a few benchmark numbers. For example, it is not uncommon for students to calculate the mass of a subatomic particle to be more massive than that of the Earth. The individual who knows one or more crude benchmark values, say the mass of a proton (10^{-27} kg) or of the Earth (6×10^{24} kg), easily recognizes if his calculation is amiss or the significance of someone else's presentation of facts. It is important to memorize or if necessary look up benchmark values for everything. For instance, what value constitutes a large amount of electrical charge? At what maximum voltage will there likely be a breakdown, leading to a discharge? Is this value the same for different environments (e.g., using an insulator or operating in a vacuum)? What is the smallest amount of electrical current that can be reliably measured? Incorporating benchmark numbers dramatically assists a researcher to identify spurious or suspicious measurements and to perform consistency checks on his calculations. Many investigators refer to this mental process as performing sanity checks.

1.1 CM, EM, AND QM: THE BACKBONE OF PHYSICS

Classical mechanics (CM), quantum mechanics (QM), and electromagnetism (EM) are topical areas that form the backbone of most physics knowledge and reasoning. CM deals with objects, how the objects respond to forces and changes in gravitational potential energy, while electromagnetism involves electric charge and the response of these charges to electric and magnetic fields, all of which may vary over time. QM came into its own in the early part of the twentieth century. QM is the physics of atoms and subatomic particles as well as the discrete quantization of energy. There are, of course, important other physics disciplines such as optics and more exotic topics such as *relativity*. The latter deals with the strange properties that objects or particles exhibit when moving close to the speed of light. While a global positioning system (GPS), for example, has to take into account the effects of general relativity to function properly, the basic concepts of a GPS can be understood in a simple Newtonian environment with *relativity* being a small correction factor.

Most technologies are essentially a component of one or more of these three backbone areas of physics. For example, optics is an application of EM, dealing with the transportation, absorption, or reflection of EM waves (most notably visible light) interacting with various materials. Electronics, magnetism, and electricity also fall under the EM umbrella. Most everyday experiences and the operation of devices can be shown to be specific applications of CM, QM, or EM. In turn, each of these topic areas can be reduced essentially to a small number of equations, embodying virtually a complete description of all natural phenomena. The physicist, who generally has a fondness for elegance, tends to prefer thinking in abstract, broad-brushed generalizations that describe a wide range of observed attributes. Unfortunately, physics classes have been taught historically in these abstract terms, leaving many students with the impression that physics has little relevance to their everyday life experiences.

For instance, a simple pulley taught seemingly laboriously in an introductory physics class might seem blasé to the student. He or she might think it is some archaic tool used only by their grandfathers' and earlier generations, a relic of the past that is only used in very old antiquated equipment that should have been replaced decades ago. In fact, pulleys continue to be the best choice for many new applications. A set of pulleys is still the most effective method used by hospitals to apply traction for certain types of skeletal injuries. Pulleys are crucial for supplying very precise amounts of pull in accurate directions. As a result, pulleys are used in the most advanced prosthetics (i.e., artificial limbs). Figure 1.2 shows several examples where pulleys continue to be employed as the most effective tool.

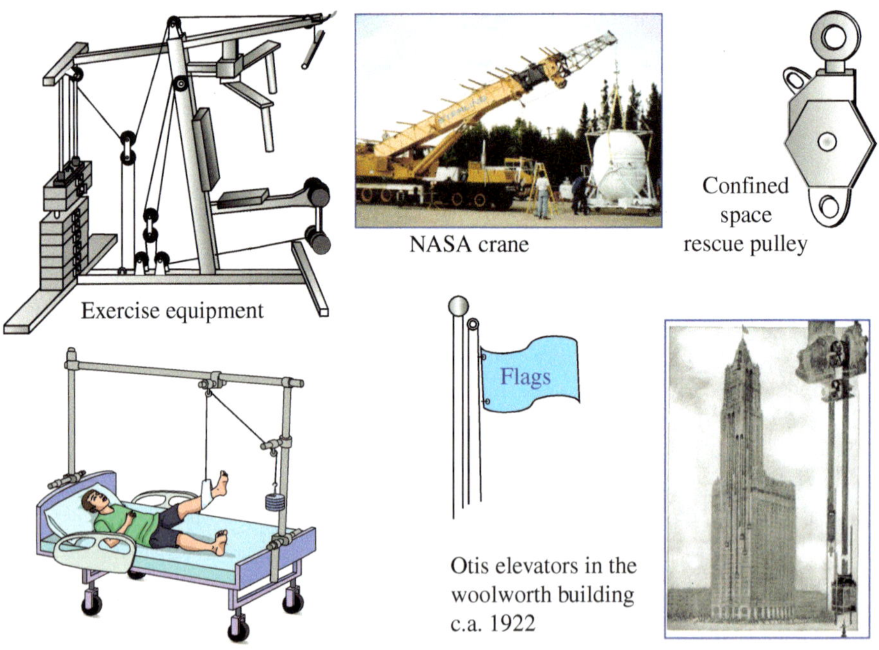

FIGURE 1.2 A few of the many uses of modern day pulleys.

Obviously, pulleys are used in many more applications than just those shown in Figure 1.2. Likewise, various other principles of *classical mechanics* are at the heart of various technologies used in many engineering, biological, and medical specialties. For example, Leonardo da Vinci, the quintessential Renaissance man, deduced that eddy currents in the blood flow, created by structures in the main aorta artery, significantly assist the heart valves to close. An eddy is a circular current or vortex often seen in fluids and gas flows. It is a classical mechanics problem associated with instabilities introduced at a boundary between a moving fluid or gas and a solid object restricting its flow. A heart surgeon must maintain or repair these structures in the main aorta artery to insure proper mechanical functioning of the heart valves. It is not enough simply to clear any clogged aorta. This basic physics concept is essential to understanding how a heart functions and cannot be ignored.

1.2 PHOTONICS AND ELECTRONICS

Except for mechanical components, most modern technologies rely on electronics, photonics, or some combination of the two. (Both subjects fall under the EM umbrella.) Moreover, virtually every mechanical device has either electronic or photonic components. For example, milling machines and lathes found in advanced machine shops use inputs from computer assisted design (CAD) systems or a series of machining steps that can be programmed on the spot by the machinist. Electronics and photonics are so pervasive in modern technologies that these topics make an excellent starting point for the study of the physics of modern devices.

Electronics are instruments that manipulate and sense the electron, while photonic devices are those that exploit the properties of the photon, a quantum of light. *Note*: all matter and energy exhibit a particle–wave duality over very small atomic or subatomic length scales, sometimes behaving like discrete highly localized entities (particle-like) and in other circumstances, behaving like waves that spread out or interfere with one another. Electronic equipment reigned supreme at the forefront of advanced technology throughout most of the twentieth century. Although photonics have been around almost as long, these instruments really came into their own during the 1980s and by the beginning of the twenty-first century, many apparatuses were actually electronic-photonic hybrids. Examples of traditional electronic devices are the radio or the telephone. Fiber optics and lasers are members of photonics.

The electron is a Fermi particle, also known as a *fermion*, which cannot share the exact same physical location and spin at exactly the same time as another electron. This principle is known as the *Pauli Exclusion Principle*. Two or more electrons approaching each other are repelled or scattered by any other electrons. In contrast, the photon, being a *boson*, can be superimposed with many others simultaneously. Light beams consisting of a stream of photons can pass through multiple other photon streams, each emerging undisturbed by the presence of the others. *Note*: there will be interference both constructive and destructive over the volumes where the beams overlap, but these waves emerge outside of the overlapped regions unaffected by the presence of the others. The primary difference between the two particles reflects the distinction between photonic and electronic devices.

Photonics inherently enable much higher degrees of multiplexing. For instance, a wire carrying a telephone conversation can be multiplexed with several other conversations simultaneously, using time-division multiplexing (TDM). The process, which will be discussed in Chapter 18, makes use of the fact that a large fraction of each conversation is dead time with no information being transmitted. The multiplexer divides two or more signals into recurrent timeslots. Each conversation is compressed and put into its assigned TDM frame, the composite signal in the form of a train of these frames is sent over the line, and then unsorted at the other end. If the calls are carried on a fiber optic cable, approximately 30 times more data can be sent. Several laser beams, each with a distinct wavelength (color), can be sent down the fiber at the same time. The information contained in a conversation consists of the modulation of the intensity of one of these colors and each separate wavelength can be multiplexed using TDM.

As noted, many devices are actually hybrids. Many high-end computers, for example, have internal optical couplers to transfer data. Similar optical couplers are used on spacecraft to prevent electrical shorts in one subsystem from rendering the others useless. Consumer computers have CD-ROM or DVD devices where the data are stored photonically via a modulated laser beam burning information onto a plastic disk. Retail checkouts use optical scanners to read universal product bar codes (UPC), then use electronics for the rest of the transaction.

Finally, let us consider some benchmark numbers regarding electronics and photonics instruments. These will serve as conceptual aids throughout this text. Most electronic circuits inside a complex integrated circuit (IC) are measured in microns (μm, or millionths of a meter). IC development since the start of the twenty-first century has been trending toward nanoscale, measured in billionths of a meter. Atoms are typically 0.1 nm, indicating nanotechnology is equally well measured in terms of tens of atoms. Figures 1.3 and 1.4 provide some reference scale lengths of common items as well as scales of visible light and subatomic particles compared to the size of an atom. From these figures, we might infer that one of the thousands of electrical components in an IC chip is the same size as the separate ridges of a fingerprint. That

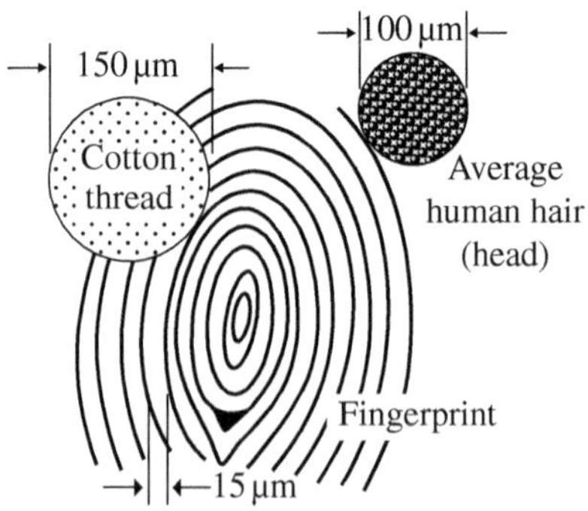

FIGURE 1.3 Benchmark sizes of fingerprint ridges, a cotton thread, and a typical hair from a human head, showing the relative scale of objects that can be seen by the eye.

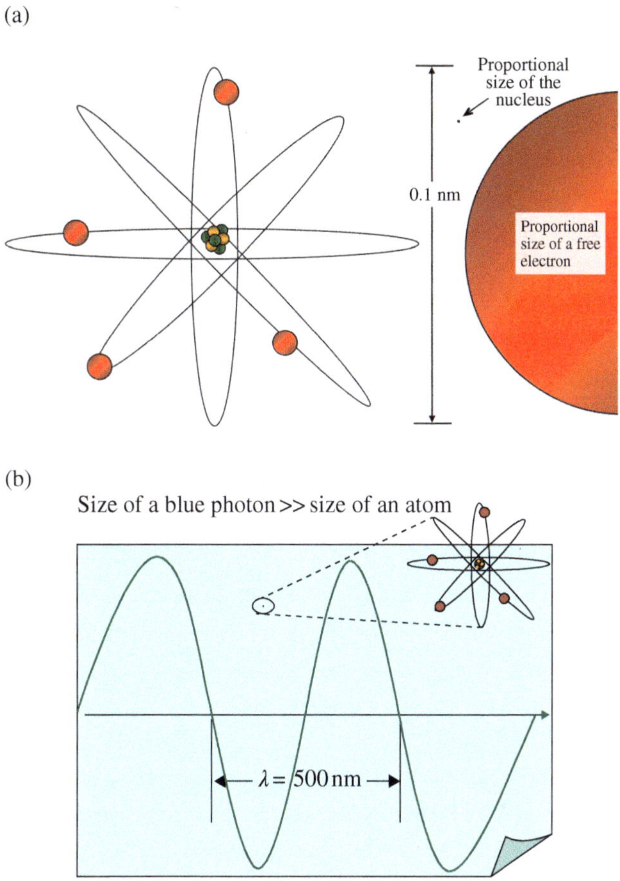

FIGURE 1.4 (a) It is often useful to conceive of an atom as a planetary model with a central nucleus, surrounded by orbiting electrons. The size of a free electron (as determined from its de Boglie wavelength) is approximately the size of the atom itself. The size of the nucleus is so small that it cannot be drawn to scale. (b) Pictured is a single photon ($\lambda = 500\,\text{nm}$) scaled equivalently to half of an 8.5″ × 11″ sheet of paper. Also drawn to scale is the size of an atom, which is 1/5000 times smaller (the dot).

same IC component is approximately 1/5 as big as the width of an average human hair. Figure 1.4a and b shows the relative sizes of an atom, a free electron, and a visible wavelength photon, all compared to the size of an atom. (*Note*: the sizes of atoms of all the elements are approximately the same.) It is fascinating that a photon, which is roughly 5000 times larger than an atom, can be absorbed and reemitted by it. As noted, one can also infer from the figure that nanotechnologies must be working with structures that may be as small as 10 atoms across. The polished smoothness of a high-quality optic is typically measured in 1/10th or 1/100th of a wavelength, suggesting a few 10's of atoms in roughness. Optical surfaces having a smoothness of 1 atom have been demonstrated.

INTERESTING TIDBIT TB1.1

Many Americans find it difficult to switch from Old English units to metric ones (e.g., Fahrenheit to Celsius temperature scale). One major obstacle is their lack of benchmark numbers in the unfamiliar system. All that a person needs to do is to establish his/her comfort range on the new scale, adjusted to the nearest 5-degree range. In other words, if the person likes warm outdoor temperatures (say 80–90°F), then their ideal temperature range on the Celsius scale might be: 25–30°C. This Celsius range actually equates to 77–86°F. If the individual likes it even warmer, then 30–35°C (86–95°F) is the appropriate range. A better, more complete appreciations is had by adding a couple of extra benchmark numbers, 40°C (104°F)—starting to be life threateningly hot, and −10°C (14°F) a very cold winter day in the most northern states of the continental US. Now the person has a complete and easy to recall mental map of the Celsius scale: −10 very cold winter, 0 freezing, 25–30 nice outside, 40 very hot, and 100 boiling temperature. Once a person has internalized their own set of benchmark numbers, it becomes easy for them to appreciate immediately any temperature expressed in Celsius without having to translate back to the Fahrenheit equivalent.

2

EVERYDAY HOME APPLIANCES

As noted, technology has invaded virtually every part of industrialized societies, and the physics behind home appliances represents excellent opportunities to showcase the very same physical principles that are commonly found in the most advanced industrial and government facilities. While significantly improving lifestyles and quality of life, these technologies usually carry some downsides. For instance, many students and the public in general might be surprised to learn the variety and total amounts of hazardous chemicals, including sources of radioactive materials, they readily bring into their homes. All rechargeable batteries in cell phones, laptop computers, electric tooth brushes, and in other appliances must be recycled separately since these contain hazardous chemicals. Energy-saving compact florescent light bulbs contain small amounts of mercury that should be recycled at special facilities. Moreover, most household smoke detectors carry small amounts of radioactive materials.

As an economic driving force, the private sector continues to have a major impact on the further development of various state-of-the-art technologies. Devices specifically designed for personal use represent a sizable fraction of the total technology produced as well as the total energy consumed in their operation. Throughout much of the twentieth century, equipment developed for the US Department of Defense was often exceedingly expensive and very advanced technologically compared to commercial devices. Today, the military uses a much higher percentage of commercially available off-the-shelf (COTS) gear, since the reliability and versatility of a significant fraction of COTS technologies have become comparable to those produced specifically for the DOD.

Modern Devices: The Simple Physics of Sophisticated Technology, First Edition.
Charles L. Joseph and Santiago Bernal.

Companion website: www.wiley.com/go/joseph/moderndevices

2.1 THE AIR CONDITIONER

An air conditioner (A/C), a heat pump, a refrigerator, dehumidifiers, or for that matter, certain types of vacuum pumps all operate on some variant of the *Equation of State* of a fluid. The simplest equation of state is that of an ideal gas (Eq. 2.1). While seldom adequate to describe in detail the behavior of a given fluid (gas or gas plus liquid) in a sealed container, it nevertheless demonstrates heuristically the generalized behavior of most fluids, relating relative changes in temperature, pressure, or volume to each other. For example, increase the pressure of the fluid and its temperature will rise. Decrease the pressure and the temperature falls. Air conditioners, heat pumps, and refrigerators all make use of this fundamental property of fluids.

$$PV = \text{constant} * T \tag{2.1}$$

Figure 2.1 contains a schematic drawing of a window-mounted air conditioner. The compressor is the heart of the process. It is responsible for establishing and maintaining a cyclical flow of refrigerant (fluid) under conditions that are capable of exchanging heat. While there are several types of compressors, we consider here a piston inside a cylinder. The cycle starts when the piston reduces the volume of refrigerant inside its cylinder, causing the pressure and temperature of the fluid to increase. The elevated temperature must exceed that of the ambient, outside air or no heat can be lost to this warm environment. The gas snakes through a long pipe called the condenser, allowing it adequate time to lose heat to the surrounding environment and to condense to a saturated liquid state, holding as much thermal energy as it can without boiling.

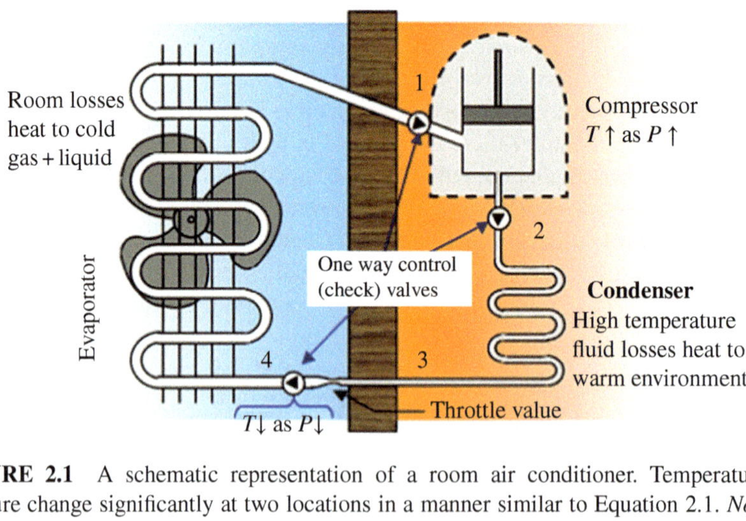

FIGURE 2.1 A schematic representation of a room air conditioner. Temperature and pressure change significantly at two locations in a manner similar to Equation 2.1. *Note*: the four locations. These denote the thermodynamic positions of the refrigerant on the graphs that follow.

After the fluid has lost its heat to the outside environment, it is primarily in the liquid phase. Next, the liquid passes through a restriction called a throttle valve, which impedes the flow and assists the tubing shown on the right-hand side of the figure to maintain a higher pressure than the fluid in the left-side pipe. There typically are two types of throttle valves: capillary tube and thermostatic expansion valves. Capillary throttle valves have interior diameters of 0.5–2.3 mm (0.02–0.09 inches). Thermostatic expansion valves commonly are solenoid controlled and most often used for medium-to-large central air conditioning systems. These work automatically and are not influenced by the ambient temperature. Both types of throttle valves adiabatically flash evaporate somewhat less than half of the refrigerant, causing its temperature to drop abruptly and leaving the refrigerant in a saturated liquid stream, a process also known as auto-refrigeration. The left-side pipe normally has the larger diameter of the two, facilitating a larger drop in pressure and a more significant decrease in temperature. The refrigerant is now cold and enters the evaporator coil where it is able to absorb heat from the interior of the building to evaporate the remaining liquid. For a saturated fluid, heat can only be absorbed through further evaporation of the refrigerant. The cycle is completed when the piston moves up in Figure 2.1, allowing the refrigerant to flow back into the compressor.

Two or three check valves insure the refrigerant flows only clockwise through the system pictured without any backward streaming. Metal fins are normally attached to both the condenser and evaporator coils (the snaked portions of the pipe). These fins, only shown on the left side of the figure, are good conductors of heat, facilitating faster heat exchanges. Fans are also incorporated to increase convective heat exchange. In addition, air conditioners are designed so that the changes in refrigerant temperature and the flow speed of the refrigerant through the evaporator stage result in cooled room air with a relative humidity between 35 and 50%. (Air that is 20°C [~70°F] with a relative humidity of 90% is often uncomfortable.) There are two competing factors at play in achieving the optimal range of relative humidity. A simple drop in room temperature increases the relative humidity of the room air. However, moisture from the interior room air will condense on sufficiently cold evaporator coils, much as a glass of ice tea will form water droplets on the outside of the glass. This condensation dehumidifies the room air and the excess must be drained away. In most window air conditioners, the unwanted water collects at the bottom of the A/C unit, which may be tipped slightly so that the water runs off from the outside portion of the unit.

The earlier discussion is heuristic. Thermodynamics must be used to obtain a detailed, quantitative understanding of refrigerators and air conditioners, especially since refrigerants operate primarily via phase transitions. (See Intro Physics Flashback FB2.1 for a refresher background on phase changes.) An air conditioner or refrigerator is a device that causes heat to flow against its natural direction. The laws of thermodynamics tell us that heat naturally flows from warmer objects to cooler objects, tending to equilibrate the temperature. The air conditioner or heat pump reverses the natural flow of heat by applying external work to its refrigerant, making it cold compared to the cool volume and hot to the warm one as depicted in Figure 2.2.

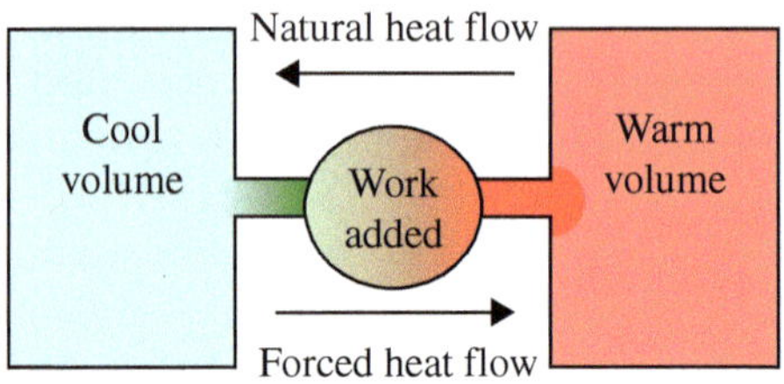

FIGURE 2.2 An air conditioner must add work to extract heat from the cold environment and move it to the warmer one.

Mathematically, the process is expressed in Equations 2.2, 2.3, and 2.4.

$$-Q_c = \frac{W \cdot (T_c)}{(T_W - T_c)} = W \cdot K, \tag{2.2}$$

where Q_c is the heat removed from the cool volume, K is the coefficient of performance, W is the external work, and T_W and T_c are the temperatures of the warm and cool volumes, respectively. Thus, the heat supplied to the warm or hot environment, Q_W, is as follows:

$$Q_W = -Q_c + W \tag{2.3}$$

In the ideal case where the process is completely reversible, the heat engine describes the Carnot Cycle and has the theoretically maximum efficiency. It is given by

$$\text{Efficiency} = \frac{(T_W - T_c)}{T_W}, \tag{2.4}$$

where the temperatures are in Kelvin (T(K)=T(°C) −273.15), the natural unit of temperature.

The thermodynamics of an air conditioner is depicted graphically in Figure 2.3. Starting again with the compressor, the refrigerant enters as a saturated vapor, a gas holding as much fluid as it can without condensing. The compressor takes the fluid from point 1 to 2 on the plot by compressing it to a high-pressure superheated fluid, which is above its boiling point but not boiling. (A superheated fluid is one that is in a metastable condition, which normally occurs for a pure, homogeneous substance in an exceptionally clean container to avoid nucleation sites that create bubbles. Certain refrigerants such as R-410a, which is a 50:50 nearly azeotropic blend of R-32 and R-125 refrigerants, have properties very close to that of a homogeneous substance.) In the condenser, the superheated refrigerant gas first cools along path 2 to 2a and then the vapor loses more heat during its phase transition to a liquid (path 2a to 3). *Note*: the isobar curves (thin solid lines) in Figure 2.3, which correspond to the phase transition curve in Intro Physics Flashback FB2.1 that fall off at the left, are flat in the center, and rise at the right of the figure. As it leaves the condenser, the refrigerant is essentially completely in a liquid state and is again in a saturated state, holding as much thermal energy as possible without boiling.

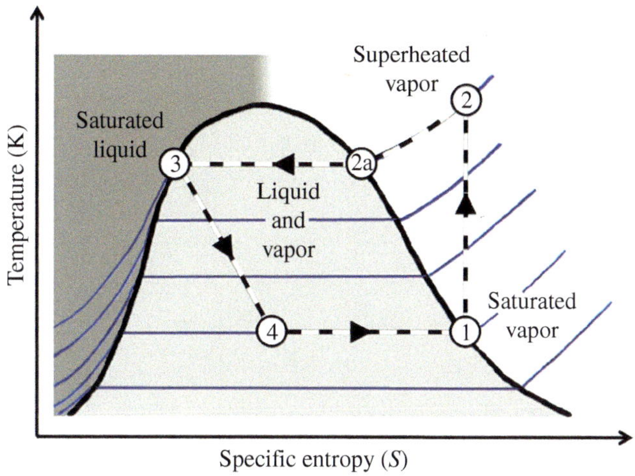

FIGURE 2.3 The thermodynamic path of taken by a refrigerator or air conditioner. Compression of vapor occurs along the path from point 1 to 2. Superheated vapor is removed in the condenser along 2–2a. Vapor to liquid in the condenser is along path 2a–3. From 3 to 4, the liquid flashes into vapor + liquid in expansion valve. The two-phase fluid converts completely to vapor in evaporator along path 4 to 1. Source: Adapted from graphic authored by Kennan Pepper.

Path 3 to 4 occurs abruptly in the expansion valve, being adiabatic since there is no time for external heat to be exchanged with the fluid. *Note*: this flash evaporation process (also known as auto-refrigeration) moves the refrigerant off the boundary between the liquid phase and the liquid-plus-vapor states. Approximately half of the liquid is flashed vaporized, causing the large drop in the temperature of the refrigerant. It then enters the evaporator stage where most of the remaining liquid refrigerant is transformed into vapor. This phase transition occurs along a lower isobar as seen on the plot (path 4 to 1). The temperature for the phase transformation of the refrigerant is well below that of the comfort level of an interior building, allowing room air to supply the necessary heat for this portion of the path. The cycle is now complete with the refrigerant being drawn back into the compressor.

The design of any air conditioner, heat pump, or refrigerator depends critically on the refrigerant to be used. There are a dozen commonly used refrigerants that fall under the umbrella name of "Freon." It is important to emphasize that no single refrigerant is suitable both for air conditioners and for refrigerators. Other refrigerants that are and have been used (especially in the past) include ammonia, sulfur dioxide, and highly purified propane. There are five desirable properties for a refrigerant: (i) its efficiency as a heat transport material, (ii) it should be non-flammable, (iii) it should not be corrosive to A/C components, (iv) it should represent a limited toxicity threat to humans and other animals, and (v) it should be safe for the ozone and atmosphere. None has all of these desirable features. Ammonia, for instance, is toxic and its use is presently limited to large packaging plants, ice plants, and large cold storage facilities.

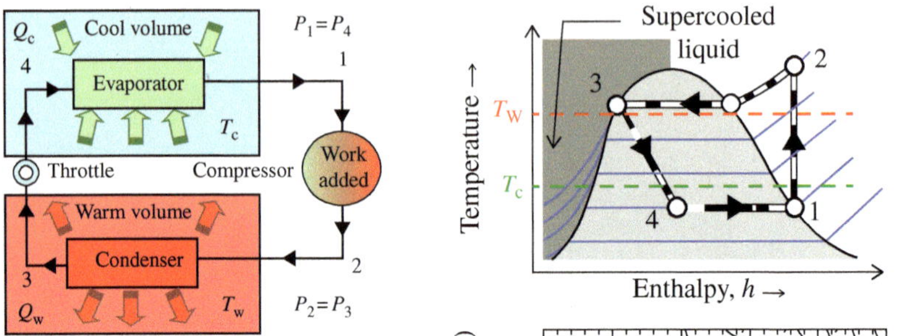

P–T chart

T(°C)	P of R-22 (psi)	P of R-410a (psi)
30	158.4	258.0
35	182.0	295.0
40	207.2	334.6

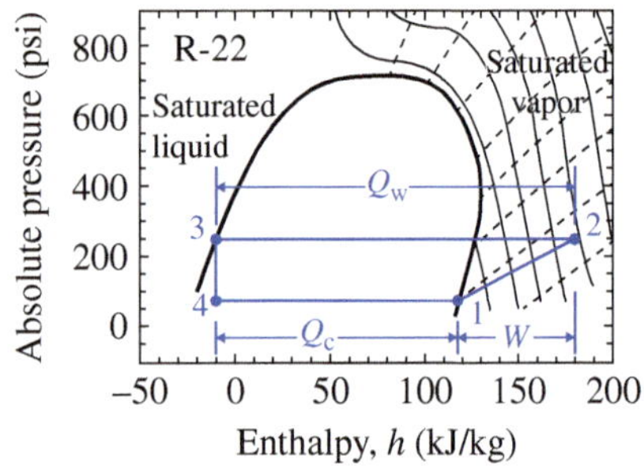

FIGURE 2.4 Top left: the four stages of an A/C corresponding to those in Figure 2.1. Top right: the thermodynamic curve given in Figure 2.2 for reference. Bottom left: an excerpt from pressure–temperature table used by A/C technicians and engineers. Bottom right: an example graphic used by A/C engineers.

To see the critical role of the refrigerant in A/C or refrigerator design, consider the differences between a residential window unit that uses the standard R-22 (currently being phased out) and a unit created for R-410a. The latter requires pressures that are about 1.6 times higher than those used for R-22. New A/C units require internal plumbing that can withstand the higher pressures. (One must not replace the R-22 refrigerant with R-410a in an older model A/C since the unit would not function well and would soon fail thereafter. As it is, technicians using the correct refrigerant have to be careful not to overcharge the system fluid; an A/C with too much refrigerant will overload the compressor, damage the throttle valve, and dramatically increase the chance of a leak.) For R-22 refrigerants, head pressures (HPs) exiting the compressor are typically 200–250 psi (pounds per square inch) and the piston suction line pressure (LP) is 65–70 psi. The corresponding numbers for R-410a are 330–430 psi HP and approximately 110 psi LP, respectively. Obviously, the cooling effectiveness of the refrigerant depends on the interior and exterior temperatures, which vary from day to day, and air conditioner designs vary according to one of nine temperature zones in the United States.

Knowing the pressures and temperatures, one can estimate the enthalpy and the cooling efficiency of an air conditioner. More information is required than can be gleaned from Figure 2.3 alone, especially since numerical scales are not provided. In practice, engineers use tables or graphical information of the type shown in Figure 2.4 to design an A/C unit. Technicians and repair personnel use a pressure versus temperature (*P–T*) chart to determine the correct amounts of refrigerant that an A/C unit should have. An excerpt from a *P–T* chart for two refrigerants is given at the lower left of Figure 2.4. In charging or recharging an A/C unit with refrigerant, the technician notes the external temperature surrounding the compressor. Then, he or she notes the corresponding pressure from the *P–T* chart for that particular refrigerant. If the internal pressure measurement is too low, additional refrigerant is added. Fluid should be removed if the internal value is too high.

The panel at the top left retraces thermodynamically the same cycle shown in Figure 2.1, with corresponding locations 1 through 4 denoted in both figures. As before, the compressor takes the fluid from point 1 to 2 and the throttle valve from 3 to 4. Inside both the condenser and the evaporator, the pressures remain constant so that $P_2 = P_3$ and $P_4 = P_1$, respectively. Q_C gives the energy lost from the cool indoor air and the energy imparted to the warm outside air is Q_W. The *T* versus *h* curve of Figure 2.3 is provided for reference in the upper right, along with T_W (the outdoor temperature) and T_C (the indoor temperature) for comparison. The pressure–enthalpy graph at the bottom right of Figure 2.4 provides the key information that is unique to each type of refrigerant. *Note*: engineers work with graphical data that are far more detailed and precise than are provided here. There are a number of sources where tables and graphical information for the various refrigerants can be found, including the American Society of Heating, Refrigerating, and Air-conditioning Engineers (ASHRAE). Detailed *P* versus *h* plots contain numerous contours of constant temperature (thin solid lines) as well as contours of constant entropy (thin dashed lines), each labeled with its numerical value. Also plotted are the pressure and enthalpy values of the R-22 refrigerant at the corresponding locations throughout the cycle. The corresponding mathematical relationships are as follows:

$$W = h_1 - h_2 = Q_W + Q_C \left(\text{kJ / kg}\right) \tag{2.5}$$

$$Q_C = h_1 - h_4 \tag{2.6}$$

$$Q_W = h_3 - h_2 \tag{2.7}$$

In the United States, most heat pumps, A/C units, and refrigerators have capacities and mathematical relationships given in English units. For the heating capacity, these are expressed in British Thermal Units (BTUs), the heat required to raise the temperature of 1 lb of water one degree Fahrenheit. In the metric system, the corresponding value is the amount of energy to raise 1 kg of H_2O by 1 °C and the enthalpy of a refrigerant is given in kiloJoule per kilogram. (For reference, 1 BTU equals 1055 J and 1 BTU/lb°F equals 1289 J/kg°C.)

Finally, the energy efficiency rating (EER) of an air conditioner is its enthalpy rating over its power usage. For example, a 1,200 W A/C that produces 10,000 BTU worth of cooling has an EER of 8.3 (10,000 BTU/1,200 W). Most newly manufactured A/C units in the US have an EER over 9 with some central units achieving an EER of 13. Actually, the use of EER values is strange since it is an Imperial number divided by a Standard International (SI) value and since the EER, being sensitive to changes in the exterior temperature, varies from day to day. The situation is improved somewhat by using seasonally adjusted EER (SEER) that is adjusted for each temperate zone, which has units of BTU/watt-hour. A more natural parameter is the coefficient of performance (COP), which is a measure of system efficiency and COP is unitless number since it is a ratio of Joules divided by Joules. For A/C units, $COP_{cooling} = T_C/(T_W - T_C)$, while for heat pumps, $COP_{heating} = T_W/(T_W - T_C)$. In both cases, the COP varies with the fluctuating external temperature as does the EER.

INTRO PHYSICS FLASHBACK FB2.1

Phase Transitions

Recall some basic thermodynamics related to phase transitions. Consider the amount of heat or energy required to raise slowly some H_2O from a temperature below its freezing point to a temperature above its boiling point, subsequently converting ice into water and then into steam vapor. The process, shown graphically in Figure FB2.1, shows the temperature of the ice rises as heat is supplied to the solid phase of H_2O until it reaches 0°C (32°F). (The entire process is done at a constant pressure.) At that point, a substantial amount of heat must be added to convert the ice into the liquid phase. This is depicted by a horizontal line, indicating the temperature remains at its freezing temperature during the phase transition. Once all of the ice has melted, the temperature once more rises as heat is added to the water. The temperature again ceases to climb once the boiling point is achieved, corresponding to the conversion of water into the gaseous phase. *Note*: this horizontal line is much longer than the solid-to-liquid phase transition, indicating the liquid-to-gas phase transition requires significantly more energy. Air conditioners, refrigerators, and heat pumps all operate over this portion of the graph, but for special fluids referred to as refrigerants rather than water.

The process is completely reversible in that removing the exact amounts of heat from the H_2O corresponds to moving along the curve in Figure FB2.1 from

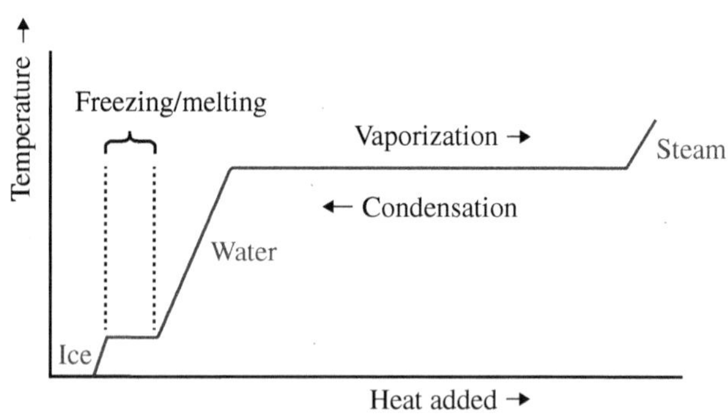

FIGURE FB2.1 Phase transition diagram for H_2O, specifically temperature versus energy for a single isobar. In a closed system such as an air conditioner where the pressure is also a variable, there is a family of curves representing various isobars.

right to left. The heat required to convert from one phase to another is called the *latent heat*, which is the energy that must be added or removed from 1 kg of the substance to convert its phase. For H_2O, the latent heat of fusion (solid/liquid transition) is 3.35×10^5 J/kg and the latent heat of vaporization (liquid/gas transition) is 2.26×10^6 J/kg.

INTERESTING TIDBIT TB2.1

All CFC refrigerants damage the ozone layer. The US EPA has mandated that residential air conditioners manufactured after January 1, 2010, must use R-410a instead of R-22. R-410a does not contain a Cl atom in its molecular makeup, the main culprit in ozone destruction. A CFC molecule takes approximately 15 years to find its way to the stratospheric ozone layer where the Cl atom now dissociated from the rest of the molecule continues to encounter and break apart O_3 molecules for up to approximately 200 years. Regardless, all refrigerants including R-410a, have high global warming potentials (GWPs), greenhouse gases that are 1500–1800 times worse by weight than CO_2 emission.

COMPREHENSION VERIFICATION CV2.1

A steam engine takes steam from the boiler at 200°C and exhausts it directly into the air at its boiling point temperature. What is the theoretically maximum efficiency? What would you expect for a real efficiency? Explain.

Answer: from Equation 2.4,

$$\text{Efficiency} = \frac{473 - 373\ \text{K}}{473\ \text{K}} = 0.21 \text{ or } 21\% \qquad \text{(CV2.1)}$$

Actual efficiencies are always lower than the theoretical maximum due to losses from friction, turbulence, or other inefficiencies. Real-life efficiencies tend to be 60–85% of the theoretical values. For instance, the theoretical efficiency for an automobile with an internal combustion engine currently is approximately 56%, but practical considerations reduce this to approximately 25%. These numbers are for cars built since the 1990s and for the efficiency of moving a half-ton auto. If one considers the efficiency of moving the driver, the efficiency is only approximately 2%. For cars made in the mid-twentieth century (1950s and 1960s), those efficiencies were about half that obtained at the start of the twenty-first century.

2.2 MICROWAVE OVENS

Microwave ovens provide the opportunity to showcase several physical phenomena. There are three essential components to a microwave oven, consisting of a source of microwave radiation called a magnetron, a waveguide, and a cooking chamber. Figure 2.5 shows these basic components. These ovens operate at 2.45 GHz (electromagnetic radiation with a wavelength of $\lambda = 12.24\,\text{cm}$), although some high-powered commercial microwave ovens operate at 9.15 GHz. Microwave ovens operate by excitation of dipole molecules, primarily H_2O and, to a lesser extent, fat and complex sugars. The process is known as dielectric heating.

Water is the principal dipole molecule excited in microwave cooking. As depicted in Figure 2.6, the water molecule consists of an oxygen atom and two attached hydrogen atoms with a 104.5-degree separation. The point where the two dashed lines intersect denotes the center of mass of the H_2O molecule. The oxygen atom has a greater affinity for the covalence electrons than do the hydrogen atoms. Thus, the water molecule as oriented in Figure 2.6 has an internal electric field pointing up. By symmetry, there is no net electric field left to right. Two or more water molecules tend to orient to adjacent molecules similar to the way bar magnets do as shown at the right.

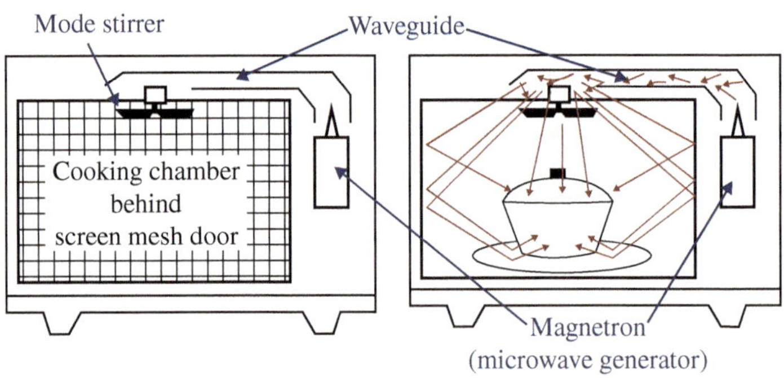

FIGURE 2.5 The basic components of a microwave oven.

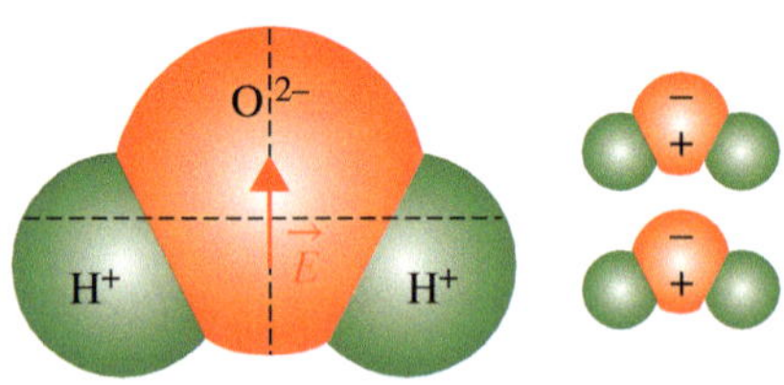

FIGURE 2.6 The water molecule, consisting of one oxygen atom (red) and two hydrogen atoms (green). The molecule is a dipole with an internal electric field pointing up in this example. One molecule tends to attract weakly another with the same orientation (right).

The molecule experiences a torque in the presence of an external electric field. In Figure 2.7, the molecules in the left panel are randomly oriented when there is no electric field. If an electric field is applied pointing up, the molecules orient preferentially in one direction and in the opposite direction if the field points down as shown in the center and right groupings. The negatively charged O atoms (red) seek to climb the electric potential, while the positively charged H ions (green) slide down it. *Note*: there is still some randomness among the H_2O molecules; not all molecules align exactly. If the electric field can be made to oscillate back and forth, the water molecules respond by flipping their orientations, matching the oscillating $\vec{E}$ field. A temperature increase is nothing more than an increase in the internal motions of the molecules contained in a material. In the case of microwave ovens, the water molecules inside the food become hot and through collisions with adjacent molecules some of which are not H_2O, agitate all molecules causing rest of the food to cook. As noted, microwaves can directly impart internal motions to dipole molecules other than water, but normally with far less efficiency.

There is a general misconception that microwave ovens cook from the inside out. In reality, most microwaves are absorbed in a short distance (typically 2.5 cm) from the surface of the food. For many small prepackaged frozen dinners, the microwaves do penetrate almost to the center. However, microwave cooking of say a 10–20 lb piece of meat is only somewhat more penetrating than heating with a conventional convection oven, which transfers energy at the surface. The principal advantage of a microwave oven is the fact that virtually all of the energy is deposited into the food and very little goes into heating the oven walls. The penetration depth depends on the water content of the food being prepared with a higher H_2O content leading to a shorter mean free path. The microwaves enter the cooking chamber as seen in Figure 2.5 and are reflected off the walls of the chamber. A wire mesh on the door of the oven also reflects microwaves, preventing these from escaping the oven and potentially harming humans. *Note*: if a metal mesh has gaps smaller than the wavelength of the electromagnetic radiation striking it, then the waves will be reflected

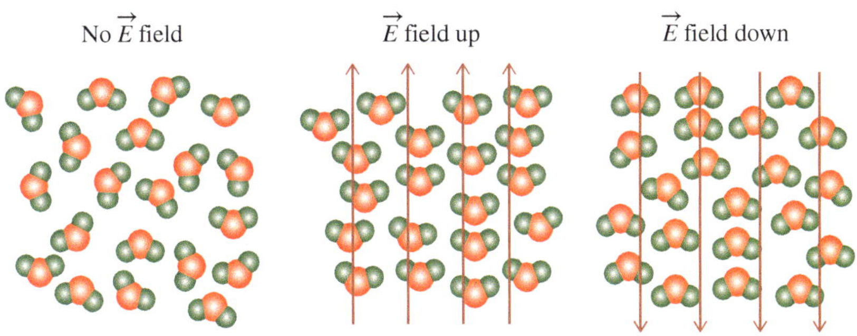

FIGURE 2.7 The schematic representation of the response of water molecules to an electric field. The H_2O molecules start flipping and spinning if the $\vec{E}$ field continually changes its orientation. The resulting higher internal motions correspond to an increased temperature.

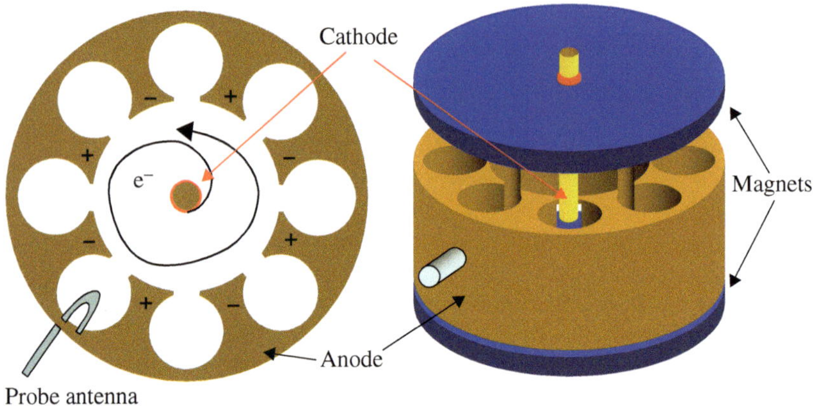

FIGURE 2.8 Cross sections of a magnetron, a 2D (left) and a 3D (right) rendering.

rather than be transmitted. As shown at the right, the reflection of the microwaves causes heating from all sides. To improve the uniformity of the heating further, many microwave ovens have a rotating platter and a mode-stirrer fan to scatter the microwave radiation field more randomly.

As noted, the magnetron is the source of the microwaves. A magnetron, as depicted in Figure 2.8, consists of a cylinder of a metallic conductor with an even number (typically 6–8) cavities arranged symmetrically around a central cavity. The entire cavity is under vacuum. There is a thin metallic rod at the center, running most of the length of the cavity. The center rod is operated hot and is voltage biased negatively at a few thousand volts to form a cathode. At the top and bottom of the magnetron cylinder are two short, but strong, bar magnets that create a uniform magnetic field ($\vec{B}$ field) running the length of the cylinder. A probe antenna, shown as protrusion out the side of the cylinder, picks up the generated microwaves and sends these to the waveguide, which delivers the microwaves to the cooking chamber. *Note*: the top magnet in Figure 2.8 is shown elevated above its real position to reveal the internal three-dimensional structure of the magnetron.

A two-dimensional rendering of the magnetron is shown at the left. The creation of microwaves begins with electrons being boiled off the hot, negatively biased cathode and accelerating radially outwardly toward the anode along an $\vec{E}$-field line. The magnetic $\vec{B}$ field, which is parallel to the axis of the cylinder, causes the electrons to move circularly. The combined forces of the $\vec{E}$ plus $\vec{B}$ fields cause the electrons to spiral outwardly. (See Intro Physics Flashback FB2.2 for the underlying physics.) The spiral trajectory of an individual electron, however, is distorted by the anode geometry and its response to the electric field of the free electron. As shown on the left of Figure 2.8, the spiral trajectory of a single free electron is distorted and the $\vec{E}$ field from this electron causes electrons inside the outer anode structure to move, temporarily setting up oppositely charged electrical poles in the anode. These momentarily created poles in the outer anode further distort the trajectories of any free electrons, establishing positive feedback.

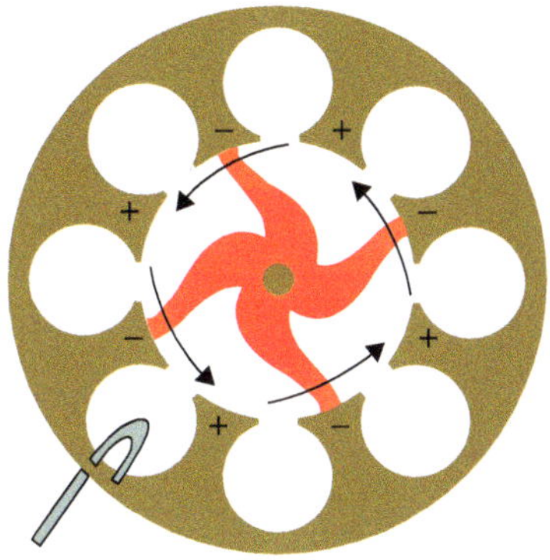

FIGURE 2.9 A cross section of a magnetron, showing the approximate shape (red) where the electrons are located at one instance in time. This cloud of electrons rotates counter clockwise as shown.

The interaction between many free electrons, streaming out from the cathode, and the movement of electrons inside the anode establish a reinforcing feedback mechanism known as a resonance. The free electrons become bunched into the spiral-shaped groups as depicted in red in Figure 2.9, sweeping around counterclockwise. Electrons are continually being emitted by the central cathode and collected at the outer anode. The electrons inside the anode slosh (oscillate) back and forth between adjacent poles in sync with the streaming electrons from cathode. The circular motions of electrons inside the anode cavities enhance alternating magnetic fields. Oscillating electric-magnetic fields are known as electromagnetic (EM) waves or EM radiation. (EM radiation is discussed in greater detail in Chapter 9.) In the case of magnetrons inside the oven, the frequencies produce EM microwave radiation.

Each outer cavity becomes an LRC (inductor, L, resistor, R, capacitor, C) circuit, which has been tuned to have a resonate frequency of 2.45 GHz. (LRC circuits including resonances are discussed in much greater detail in Chapter 6.) Depicted in Figure 2.10 is an LRC circuit and the corresponding components from one of the cavities. As observed in Intro Physics Flashback FB2.2, any net charge in a conductor only exists at the surface so the electrons are confined to move in a circular path just as in an inductor consisting of looped wire. There is a small capacitor formed across the volume where the outer cavities connect to the central one. There is also a very small amount of resistance in both diagrams either through the wire (left) or as the electrons move back and forth through outer portion of the magnetron. The inductor in LRC circuits resists changes to the flow of current, creating inertia in its flow. The capacitor stores the potential energy of the system, absorbing the transfer of charge until it becomes oppositely biased. The resister dissipates (removes) energy from the system to prevent the positive feedback from the LC portion of the circuit from creating a run away system. Resonance is established, which feeds most of the energy into a particular frequency, when the dissipative losses balance the injection of new energy from the cathode.

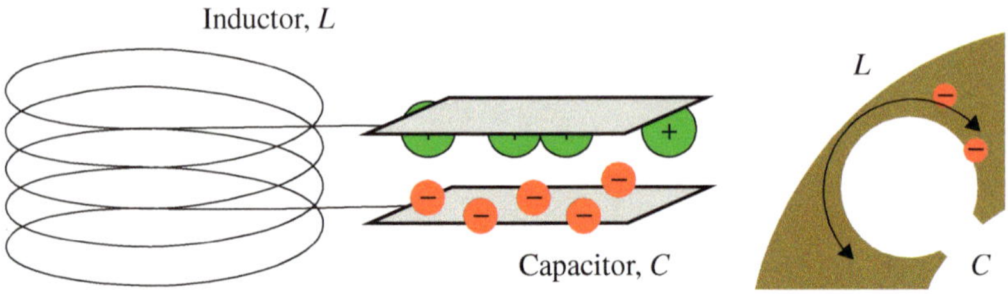

FIGURE 2.10 An inductor-capacitor (LC) oscillating circuit (left). The voltage across the capacitor will induce current through the inductor, but the coil inductor provides "inertia" to any generated current. Current will continue to flow until the voltage across the capacitor becomes oppositely charged to that shown in the figure. One cavity of the magnetron is shown (right), which forms a small LC circuit as electrons oscillate around the evacuated cavity. In this case, the inductor is less than one complete loop.

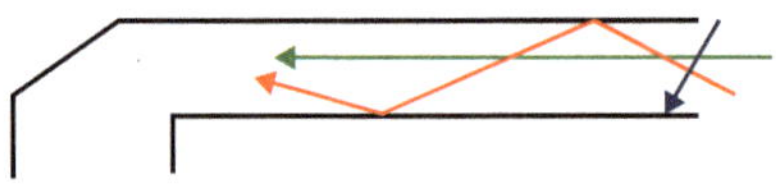

FIGURE 2.11 A waveguide. Waves traveling down the axis of the guide (green vector) travel unimpeded. Waves entering with sufficiently low incident angle (red vector) reflect off the sides of the guide, eventually exiting the other end. If the wave's ray is too steep (blue), it will be absorbed or escape the waveguide. The critical angle is wavelength dependent.

The final key component of a microwave oven is the waveguide to deliver the microwaves created by the magnetron to the cooking chamber. The concept of a waveguide is quite simple. An outer boundary that reflects any EM waves that strike its edge is required for a waveguide. One of the most familiar waveguides is a fiber optic used to transfer visible light over large distances with minimal losses. Fiber optics will be discussed in Chapter 11 on optical couplers. For microwaves, which are short radio waves, a waveguide consists of a simple hollow metal shell that can have a circular or rectangular cross section. A profile of a small section of a waveguide is provided in Figure 2.11. Vectors show the various paths taken down the waveguide. The green arrow is nearly aligned with the guide and the wave travels unimpeded down it. Some waves as depicted by the red arrow enter the waveguide at an angle. These bounce from side to side as these travel down guide. If the angle of a wave in the guide is too steep (e.g., the blue vector), it will be absorbed or lost to the waveguide. One can infer that there is a minimum rate of curvature for which a waveguide can change the direction of the EM waves. (Also see the discussion of fiber optic waveguides in Chapter 11.) There are several methods that can be used to change the direction of EM waves abruptly with minimal losses. Reflection off the far side (left side of Fig. 2.11) of the waveguide in microwave ovens is practical, especially since the short path into the cooking chamber allows most of the waves to leave the guide without any further reflections.

INTRO PHYSICS FLASHBACK FB2.2

Electromagnetism

To appreciate fully the operation of a magnetron, it is important to recall some basic concepts in electromagnetism. First, electrons in a conductor move to cancel out any internal $\vec{E}$ fields. As a consequence, all net charge resides on the surface of a conductor. Consider a piece of copper as shaped in Figure FB2.2. If some external electrons are brought into proximity near the center leg, a fraction of the electrons in the conduction band move away from the center leg, making it positively charged and the other two legs negative. By symmetry, half of the electrons go left and the other half go right. If the external electrons are then moved beneath the right leg, once again a charge shift occurs in the conductor. This time, however, the middle leg has more of the negative charge than does the left leg. While the left leg is negative, it is more positive than the center one.

Second, the motion of an electron in a uniform magnetic field with no external electric fields is circular and perpendicular to the magnetic field, $\vec{B}$. As shown in Figure FB2.3, the right-hand rule (RHR) for vectors cross multiplication applies. (See Section 22.1 for a reminder of the RHR.) At any instance in time, the centripetal acceleration vector is the negative of the velocity vector crossed into the $\vec{B}$ field vector (into the page). If a pair of concentric conductors is biased to create a radial $\vec{E}$ field, the motion of the electron is transformed from circular into a spiral as shown on the right side of Figure FB2.3.

Finally, electrons constrained to move in circular paths, create a magnetic field. We all learned in grade school that you could create an electromagnet by

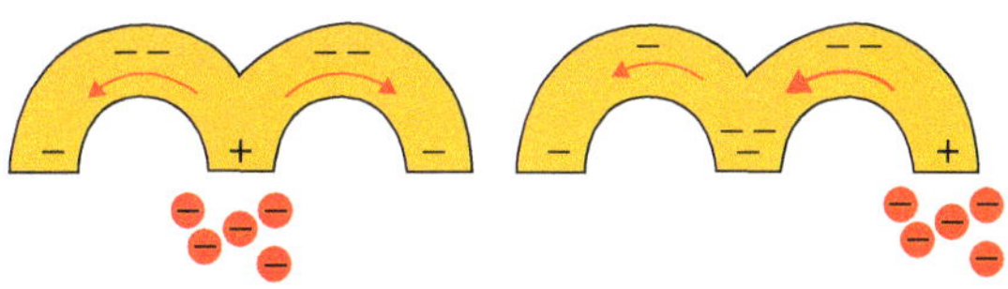

FIGURE FB2.2 Charge shifts inside a conductor.

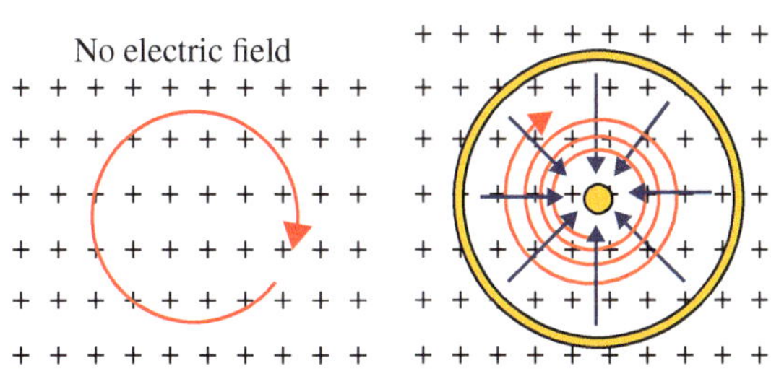

FIGURE FB2.3 Motion of an electron in a uniform magnetic field (left) and in the same magnetic field with a radial electric field (right).

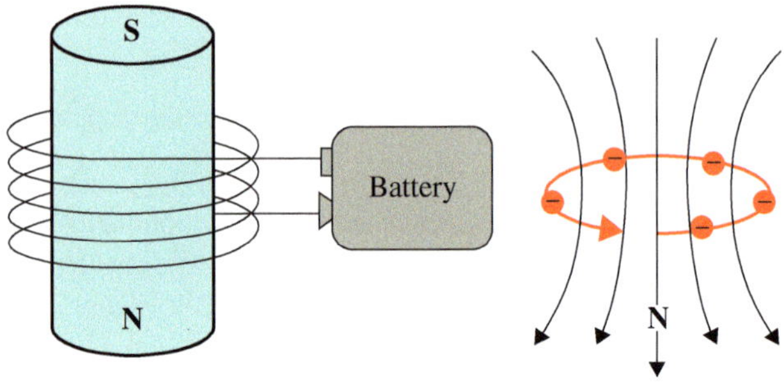

FIGURE FB2.4 Circular currents create magnetic fields and even a single electron forced to move in a circular path creates a magnetic dipole.

wrapping a coil of wire around an iron rod and hooking it to a battery. The left-hand side of Figure FB2.4 shows a schematic representation of an electromagnet. If the current is strong enough, one can then detach the battery and the iron core remains magnetized. In fact, a magnetic field can be created without the central iron piece since the electric charge moving in a circular path through the wire was responsible for creating the initial $\vec{B}$ field. The iron rod only enhances the uniformity of the magnetic field, extending the length of the dipole. A current moving through a single loop or even a partial loop will generate a weak magnetic field, albeit one with dipole that quickly diverges as depicted on the right. *Note*: the RHR applies to the motion of a positive charge. Electrons, being negative and the primary carrier, move in the opposite direction. Thus, the magnetic field, which points from South to North poles, is downward as depicted in Figure FB2.4.

Note the symmetry between the electric and magnetic fields and the interaction with moving charge. A uniform $\vec{B}$ field causes an electron to spiral circularly orthogonal to the magnetic field lines. An electrical charge moving circularly is constantly undergoing a centripetal acceleration, which creates an opposing $\vec{B}$ field. This is the underlying physics behind an electronic inductor, one of three primary types of components of any electronics circuitry. The symmetry between moving charges, and the $\vec{E}$ and $\vec{B}$ fields also have the consequence that light waves or photons consists of simultaneously oscillating electric and magnetic fields.

COMPREHENSION VERIFICATION CV2.2

Question: Why do magnetrons have an even number of cavities around the central one?

Answer: The electrons moving back and forth inside the anode require an even number of poles. At any instance in time, there has to be one positively charged pole for every negatively charged one.

INTERESTING TIDBIT TB2.2

The element americium (atomic number 95) was discovered in 1945 during the Manhattan Project in the United States. The first sample of americium was produced by bombarding plutonium with neutrons in a nuclear reactor under the football stadium at the University of Chicago—World Nuclear Association

2.3 SMOKE DETECTORS

There are two principal types of smoke detectors: those employing radioactive material and those using a light beam and sensor. These are normally referred to as ionization and photoelectric smoke detectors, respectively. By far, the most common smoke detectors use the radioactive release of alpha particles to ionize the air inside the detector, setting up a small electrical current as shown in Figure 2.12. Alpha particles consist of two protons and two neutrons, equivalent to He^{++}, a doubly ionized helium atom. The kinetic energy of alpha particles is around 5 MeV, enough energy to ionize approximately 500,000 gas atoms for each alpha particle emitted. Inside the ionization chamber of the smoke detector, the free electrons flow toward the positively charged plate while the ionized molecules drift slowly toward the other plate. When smoke and its accompanying small particulates enter this ionization chamber, the electrical current is quickly and dramatically suppressed since the fine particulates in the smoke absorb many of the alpha particles. These fine particulates also attract and absorb many of the residual free electrons that were generated by alpha particle collisions with air molecules. The integrated circuit (IC) in the system is designed to detect any interruption in the electrical current established inside the ionization chamber and to trigger the sounding of the alarm.

The source of alpha particles is a sample of Americium oxide (AmO_2), which is sold to smoke detector vendors by the US Atomic Energy Commission. Americium 241

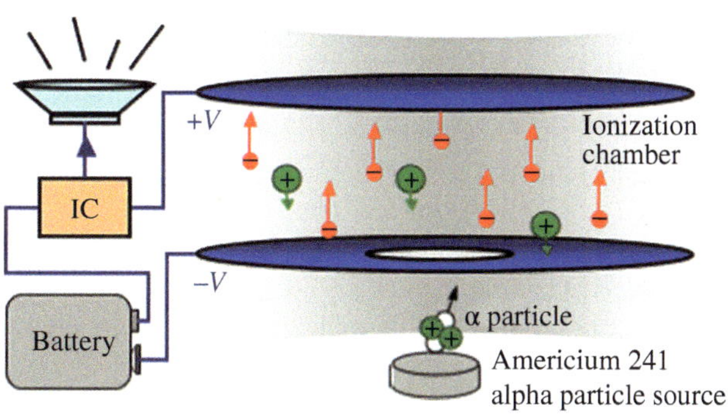

FIGURE 2.12 The common configuration of a household smoke detector. Alpha particles collide with gas atoms to create an electrical current, which is interrupted in the presence of smoke particles.

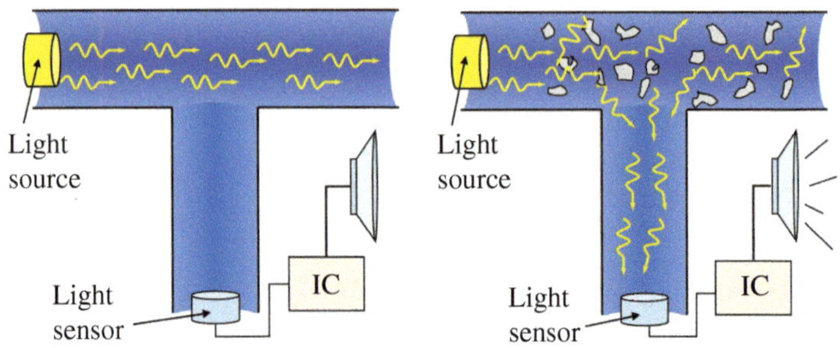

FIGURE 2.13 Photoelectric smoke detectors that are most often commonly found in public buildings. Any smoke particulates entering the detector scatter light out of the beam in all directions, including toward the photosensor and triggering the alarm.

with a half-life of 432 years is a by-product of the decay of plutonium-241 in nuclear reactors used for power generation. Typically, smoke detectors contain approximately 1 μCi (microcurie) source, creating about 37,000 alpha particles per second. Fortunately, alpha particles are easily and very effectively stopped by the smallest amount of shielding material surrounding the source. Alpha particles cannot penetrate more than a few centimeters of air and can be stopped by a thin metallic foil. The radioactive Americium contained in smoke detectors represents no threat to the occupants in the home. Naturally occurring as well as anthropogenic radioactive sources, the relative hazards, and potential doses are discussed more thoroughly in Chapter 20.

The second type of smoke detector, the photoelectric, uses a light beam and a light sensor positioned at a right angle. See Figure 2.13 for a schematic representation of the configuration. A light beam typically from a light-emitting diode (LED) passes straight through this type of smoke detector unhindered in the absence of smoke. The light sensor resides in the dark with virtually no light striking it. When the fine particulates in smoke enter the detector, light gets scattered in all directions, some of which is scattered toward the photodiode. The scattered beam causes both a jump in voltage and an electrical current to start flowing in the sensor. In other words, the light sensor is a device that converts light into an electrical signal, giving it the name of photoelectric. This sensor is attached to an IC that triggers the alarm when photons hit the photodiode.

There are very minor differences in the performance between these two classes of smoke detectors. The obscuration trigger levels are 2.6–5.0% observed per meter for ionization devices and 6.5–13.0% obs/m for photoelectric. According to studies conducted by the National Institute of Standards and Technology (NIST), both types of smoke detectors have very similar sensitivity, are rapidly activated by trace amounts of smoke, and are highly reliable. While ionization devices are somewhat faster at detecting smoldering (incomplete) combustion, photoelectric smoke detectors are slightly better for flaming hot (complete combustion) fires. The ionization type is the least expensive to fabricate, which is the main reason it is by far the most commonly used in homes. The more-expensive photoelectric devices, however, are more popular as commercial smoke detectors since these are more easily wired up to fire alarm panels or to central monitoring systems.

COMPREHENSION VERIFICATION CV2.3

Estimate the current generated in a home smoke detector from alpha particles. Use the following two benchmark numbers to assess the required sophistication of its electronics:

Benchmark 1: 1 pA (1×10^{-12} A), the smallest current that can be measured without extraordinary measures. Benchmark 2: 1 fA (1×10^{-15} A), the smallest current that can be reliably measured in a controlled laboratory environment.

Answer: The typical radioactive source in a smoke detector is 1 μCi that emits 37,000 alpha particles/s. Each alpha particle ionizes approximately 500,000 atoms, creating 500,000 electrons. Thus, the current is $37{,}000 \times 500{,}000 = 1.85 \times 10^{10}$ electrons/s. Each electron has a charge of 1.60×10^{-19} Coulombs so the current is (1.85×10^{10} electrons/s) $\times 1.6 \times 10^{-19}$ C or approximately 3×10^{-9} A. Our answer of 3 nA is 3000 times higher than Benchmark 1 and is easily measured with inexpensive electronics.

2.4 COMPACT DISCS, DIGITAL VERSATILE DISCS, AND BLU-RAY DISCS

Compact discs (CDs), digital versatile discs (DVDs), and Blu-ray discs (BDs) are all data-storage media that use optical interference as a means of retrieving information. The basic physics for all three is given in the Intro Physics Flashback FB2.3. A light beam is separated into two rays: one used as a reference and the other sent to the optical disc. Inside the optical disc is a reflecting layer referred to as a land. Shallow, flat-bottomed pits are impressed into the land surface, moving the reflecting planes by a discrete amount sufficient to change the phase of the reflected beam. When the beam returns from a land, it is in phase with the reference beam and out of phase from a pit. The instantaneous intensity will correspond to a bright or a faint signal, respectively.

The focused laser beam encounters several other optical surfaces of the disc besides the one of interest, especially if it is a dual layer. Each of these planes reflects a portion and scatters small percentages of the beam. Fortunately, the interference signal dominates and the undesirable reflective contributions add analog noise that has no impact on the digital signal. The DVD or other players use the bright-to-faint or the faint-to-bright transitions to denote "1's" in the bit stream. In comparison, a "0" is registered any time the light intensity remains constant during the sampling time. In other words, a bright-bright or a faint-faint sampling is a zero.

The standard size for all three types is a disc with a 120 mm diameter and 1.2 mm thick. CDs were developed first, followed by DVDs, and BDs being the most recent and most advanced. Each new generation of disc player is backward compatible, meaning a BD device can also read a DVD or a CD. However, a CD player does not have sufficient resolution to decrypt either a DVD or BD. The principal difference between these devices is the wavelength of the light used to record and play back the information. CD, DVD, and BD players use near-infrared

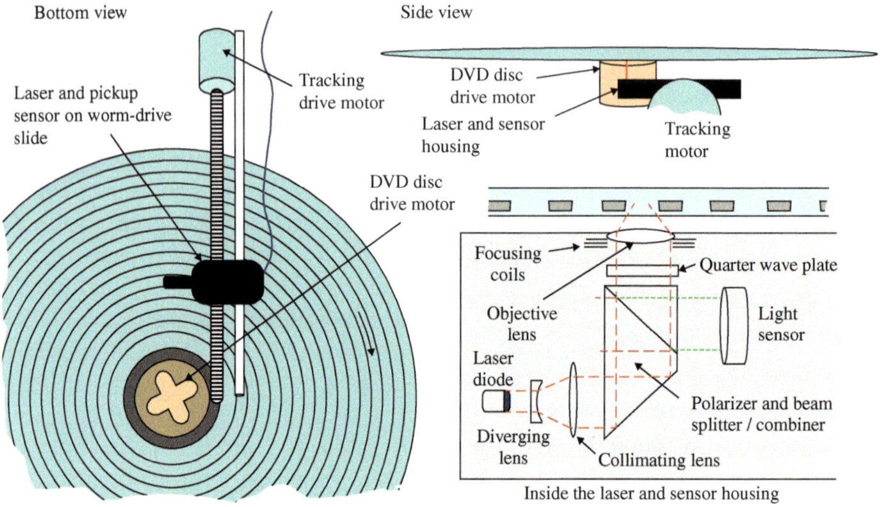

FIGURE 2.14 A schematic representation of a DVD player from several vantage points. The bottom view (left side) is looking up from underneath the disk. The side view (top right) shows the components as seen looking edge on to the disk. An expanded detail of the laser and sensor housing is shown (bottom right).

light (wavelength: $\lambda = 834\,\text{nm}$), red light ($\lambda = 630$–$650\,\text{nm}$), and violet light ($\lambda = 405\,\text{nm}$), respectively. The data storage density, and correspondingly the total amount of information that can be saved on one disc, depends on the inverse square of the wavelength ($1/\lambda^2$). A detailed description of the DVD player is given below since it is a good representation for all three devices.

Figure 2.14 schematically depicts the inner workings of a DVD player, while Figure 2.15 is an interior picture of a drive. A laser and photosensor are mounted in a housing attached to a worm drive. The disc rotates above this housing as shown in the figure. Each track is actually a tight spiral, running from the inner hub to the outer radius or visa versa. The length of a track segment contained in one disc revolution is $2\pi r$, where r is the radius of that segment from the axis of rotation. To insure a constant bit stream, early players continuously varied the speeds of its mechanical mechanisms. Both the angular rotation rate of the disc and the linear speed of the housing outward were reduced as the laser beam moved from the innermost to the outermost portion of the track. Later DVD players still vary the radial, linear motion of the laser continuously, but use a set of constant discrete angular rotation speeds. The irregular data rate is absorbed in a first-in first-out (FIFO) buffer, which allows the player to stream the data out at a constant rate.

The laser plus sensor housing is the heart of the DVD player, schematically represented in at the lower right of Figure 2.14. A laser diode provides a monochromatic and coherent light ray, which first passes through a concave lens to broaden the beam diameter. Next, it encounters a collimating lens to make all rays parallel again (i.e., the beam size is neither diverging nor converging). The collimated, broad beam goes

FIGURE 2.15 The internal components of a DVD drive. Source: Slowacki, https://upload.wikimedia.org/wikipedia/en/e/eb/Dismdvd.jpg. Public Domain.

through a splitting and recombining optic, which sends part of the light to the light sensor and the rest to an objective lens just below the disc. This particular set of optics leaves each beam polarized orthogonally to each other. The objective lens focuses the transmitted light beam onto one of the disc's reflecting surfaces, which has to be capable of returning the beam either in phase with the reference or with a phase shift between 0.4 and 0.6 of the wavelength. Photons traveling to the DVD and back pass through a quarter-wave plate twice, which alters the direction of polarization by $\pi/2$, orienting its polarization vector parallel to the reference beam and causing total reflection toward the light sensor when encountering the prism for the second time. (See Intro Physics Flashback FB2.3 for additional details on the principles of interference, wave plates, and polarization.)

Refer to Figure 2.16 to understand how a DVD encodes interferometric data. The disc is made of a polycarbonate substrate, which may have as many as two reflecting surfaces incorporated into each side. The data capacity is approximately 4.5 GB per reflecting interference layer. The deepest surface is aluminum, while the second, shallow layer (if the disc has one) is a thin film of gold, which transmits approximately half and reflects the rest of the light beam. If the disc is a double layer, the laser has to be able to focus either on the deep or shallow layer. These two cases are depicted in Figure 2.17, where the laser beam has been focused on the deep reflecting layer (center image) and on the shallow plane (bottom). *Note*: the focused beam has

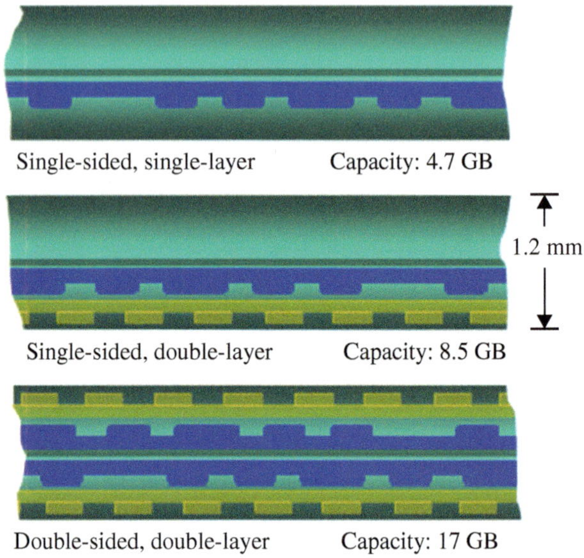

FIGURE 2.16 Optical disc storage devices come in one or two layers on a side. The storage capacities for DVDs are shown in the figure.

a minimum diameter before it diverges again. Also, the intensity of the beam is concentrated in the center and falls off at its edges. Regardless of which surface is being read, some of the photons are reflected from the out-of-focus layer. Fortunately, the unfocused portion of the beam reflects from a much broader area, encompassing several lands and pits simultaneously. This out-of-focus plane contributes about as many in-phase as out-of-phase photons, adding to the noise but not altering the digital state of the beam (Fig. 2.17).

DVDs, CDs, or BDs that already have recorded content usually are manufactured using a stamping method. The underlying reflecting surface (land) has flat-bottomed indentations (pits) pressed into it. *Note*: that pit surfaces are actually closer than the lands are to the incoming beam, since the disc is read from the underside. Pits are pressed into this reflecting surface prior to it being incorporated into the multilayer disc. As a wave enters a DVD, its speed slows and the wavelength shifts from 650 to 440 nm due to the index of refraction of the polycarbonate substrate. Thus, the depth of the pits has to be 88–130 nm to cause a 0.4–0.6 phase shift during the round-trip path of the beam. Once the reflected beam leaves the polycarbonate disc, it resumes its original speed and wavelength.

The minimum pit lengths are 0.834 μm for CDs and 0.4 μm for DVDs. The track pitch (the separation between adjacent track segments of pits) is 1.6 and 0.74 μm for CDs and DVDs, respectively. Figure 2.18 is a picture of a DVD taken with an electron microscope, showing the pits impressed into the substrate. Overlaid is a set of

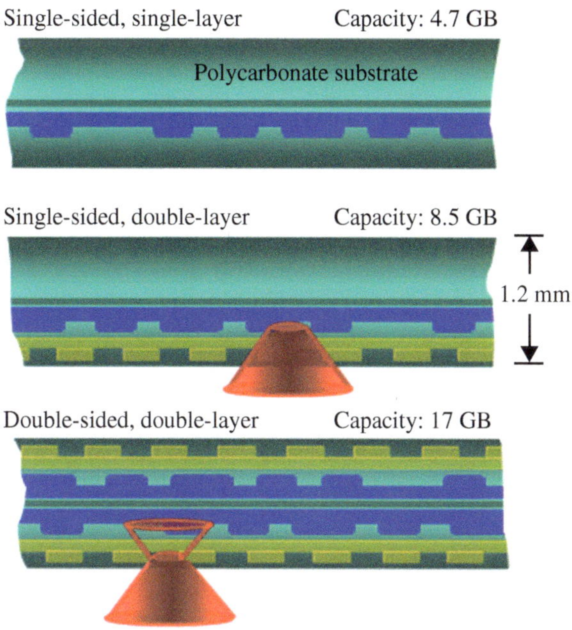

FIGURE 2.17 Same as Figure 2.16 but showing the focal depths. The center disc is depicted with the laser beam focused on the deeper layer, while the bottom disc shows the beam focused on the shallower semireflective layer. The out-of-focus portion of the beam reflects off several in-phase and out-of-phase surfaces.

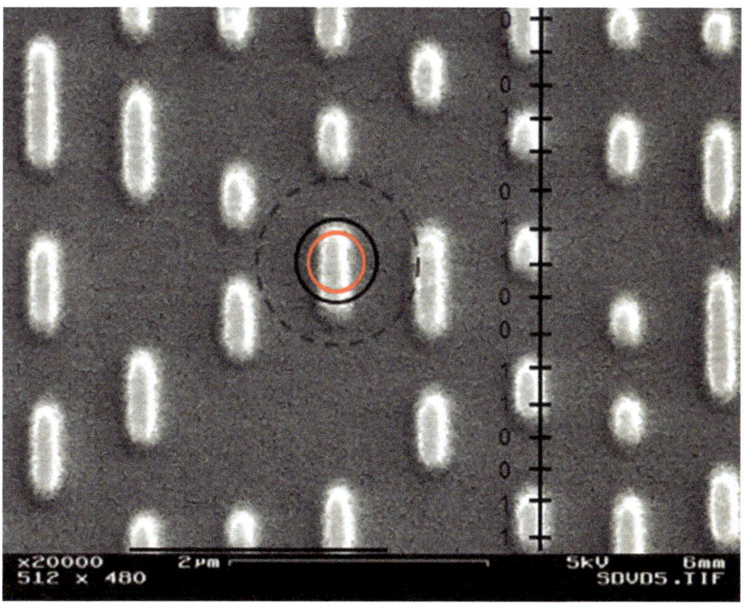

FIGURE 2.18 An image of the pits over a portion of several tracks of a DVD as seen under an electron microscope. Over plotted are three circles showing the 67, 87.5, and 95.5% energy encirclements of the DVD laser beam. Source: Image courtesy of University of Rochester: URnano.

circles. The innermost one represents the laser beam size associated with a 2/3 energy encirclement. The full width at half max (FWHM) of the beam (second solid circle) shows the ring where the beam intensity has fallen to half of its central value. Slightly more than 3/4 of the total energy is encircled within the FWHM radius. The dashed line shows the area of 99% encirclement. When the beam is over a pit, approximately 70% of the focused light is shifted to destructive interference with another 6–8% lost due to scattering from the pit edges. (That still leaves 23% of the focused beam falling on land area and undergoing constructive interference.) In contrast, the entire focused portion of the beam is constructively interfering when the beam is over a land. *Note*: these percentages apply only to the portion of the beam interacting with the read surface. If the disc is a double layer, half of the beam is reflected by the out-of-focus surface, regardless of which layer is being read.

For blank CD-R, DVD-R, or BD-R that can record information only once, there is a separate layer that consists of a light-activated dye. Exposure to an intense laser beam changes its opacity at those locations inside the disc. The process is permanent and the disc cannot be erased. Discs that are "re-writable" (e.g. DVD-RW) use a different procedure. These discs contain recording layers that transition from amorphous to crystalline material at locations heated via exposure to intense light. The recording layer deforms itself along with the reflecting and dielectric layers, at those places to form pits. A cross section of a DVD-RW is shown schematically in Figure 2.19. These recording layers can be erased, returning the material to back to its amorphous state via laser reheating. DVD recorder/players typically use a 5-mW laser to read data, but need a powerful 100–400 mW laser to record or erase data. The read laser has to be sufficiently weak not to alter the data content during the read

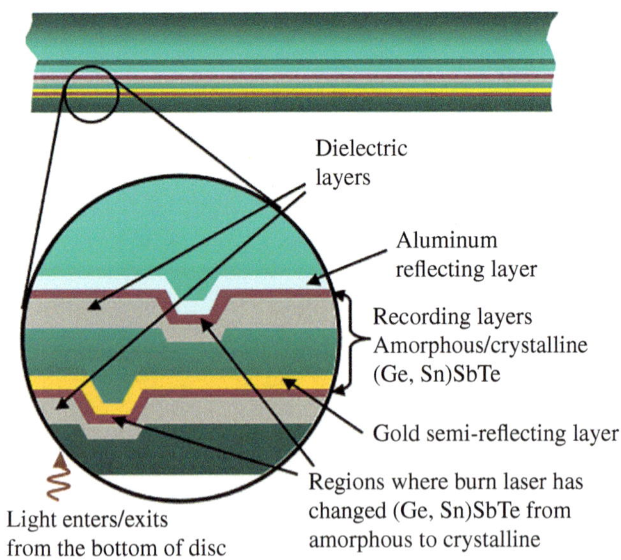

FIGURE 2.19 The cross section of a DVD-RW, showing two pits that have been "burned."

process. A fully read/write recorder has to produce a laser beam with three separate intensities. To erase a previously recorded pit, the laser has to be able to heat the (Ge, Sn)SbTe material of the pit above its melting temperature (600°C). To create a new pit, the material has to be heated to a temperature somewhat below its melting temperature.

Finally, keeping the laser housing properly aligned with respect to the track of the disc is a major challenge. There are a number of methods used to maintain this correct positioning. One of the simplest procedures is to separate the read beam into three components, two of which are offset slightly with respect to the central one. If the position of the laser housing starts to get ahead of the correct location, the digital signal from one of the auxiliary beams becomes stronger at the expense of the other two. If the position starts to lag the proper location, the other auxiliary beam starts to become the dominant digital signal. This feedback information allows the recorder/player to make real-time adjustments to its tracking. Discs have angular rotation speed between 200 and 500 rpm (revolutions per minute). It is nearly impossible or at best very expensive to be able to make focus adjustments to accommodate any wobble of the disc. Fortunately, the rotation rates are sufficient to help seat the disc correctly on the drive hub. DVD disc start to become unstable at 10,000 rpm, potentially leading plastic creep and the disc shattering.

INTERESTING TIDBIT TB2.3

Ever larger optical storage capacities have been obtained by moving to shorter and shorter wavelengths. Unfortunately, most materials that are transparent in the visible become opaque at ultraviolet wavelengths. The polycarbonate of optical discs is a particularly good UV-blocking material for $\lambda < 400$ nm. Greater data storage density for the next generation disc will require other methods or other substrate materials.

INTRO PHYSICS FLASHBACK FB2.3

Wave Interference

CDs, DVDs, and BDs are all photonic devices that encode information optically, using interference as its primary method. Interference is a property of waves that can enhance or destroy the amplitude of the combined wave. We will concern ourselves here only with the special case where a single monochromatic wave is separated into two beams and then those two beam recombined, traveling parallel to each other. If the two beams arrive in phase (i.e., if the crests and the troughs of the two are exactly aligned), the maximum signal strength is observed. However, if the crests of one beam exactly align with the troughs of the other and visa versa, the two beams are 180° (or equivalently π radians) out of phase and the effective signal is absent (no signal). These two conditions are shown graphically in Figure FB2.5.

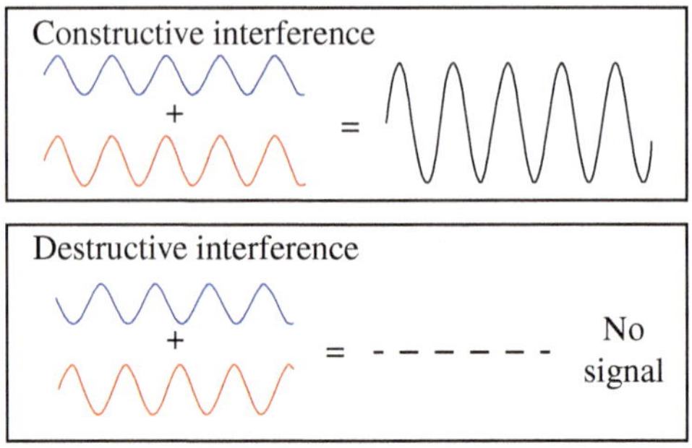

FIGURE FB2.5 A graphical representation of perfect constructive and destructive interference.

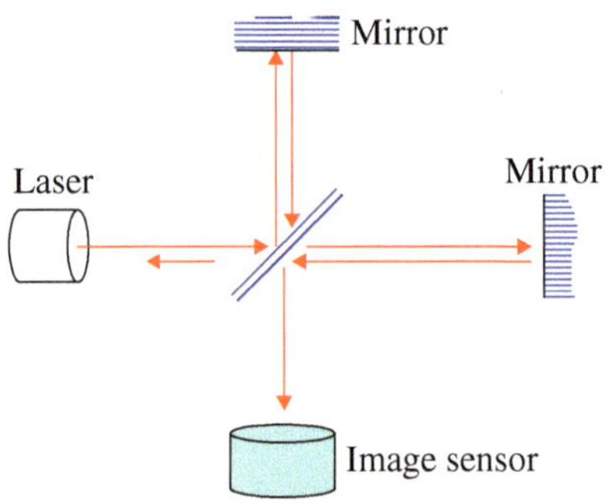

FIGURE FB2.6 Schematic representation of a Michelson interferometer.

One of the earliest and mostly widely used methods of separating and recombining two beams is the Michelson interferometer (Fig. FB2.6). The optical paths used in CDs, DVDs, or BDs are similar to those found in a Michelson interferometer. A laser emits a monochromatic beam of light that first encounters a partially silvered mirror that reflects some of the light toward the top mirror and transmits the rest straight through to a second mirror. Each of these mirrors reflects its beam back to the center, where in each case some of the light passes and the rest reflects. The top mirror reflects toward the laser and transmits toward the image sensor, while the mirror on the right side of Figure FB2.6 does the opposite. The system is inefficient in that nearly half of the light returns to the laser and is lost. However, each of the paths to/from each mirror contributes approximately 1/4 of the original beam intensity to participate in the interference measurements. Normally, one of the two mirrors in a Michelson interferometer is fixed and the other can be moved precisely along the path of its beam. When the distance of the adjustable mirror returns a beam at the image sensor that is in phase with the beam from the fixed mirror, the signal is strong. Additional movement of the adjustable mirror in the same direction by a distance equivalent to a half wavelength, then results in a weak signal due to destructive interference. (Continuous motion of the adjustable mirror in one direction causes a cyclic bright and dark signal at the image sensor

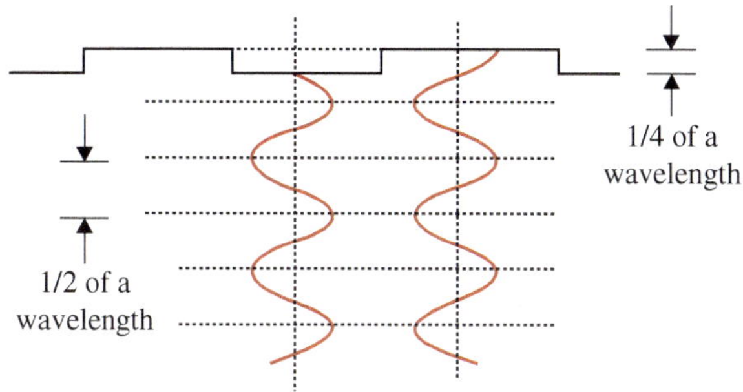

FIGURE FB2.7 Conceptually, the pits and lands inside a DVD provide two reflecting surfaces equivalent to the two positions of the adjustable mirror in a Michelson interferometer.

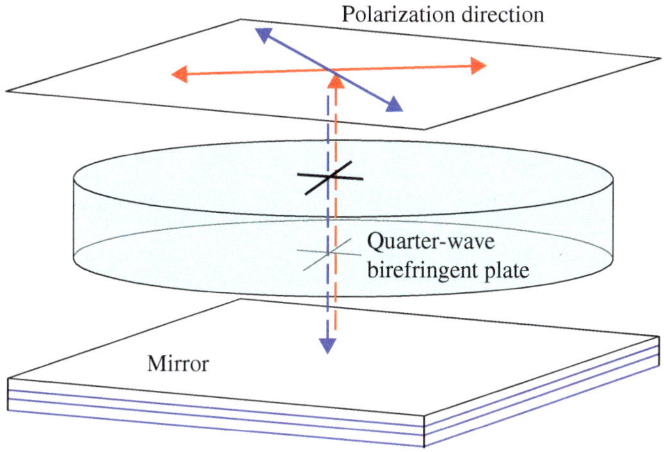

FIGURE FB2.8 A quarter-wave plate made of a birefrigent crystal will shift the polarization direction by 90° after a polarized wave passes through it twice. The proper rotation of polarization only works for specific wavelengths.

as the two signals repeatedly constructively and destructively interfere.) Distances can be measured very precisely using this type of interferometer.

Each track on CDs, DVDs, and BDs requires its reflecting surface to be at one of two possible internal depths to return either an in-phase beam or an out-of-phase beam. The path difference between these two depths must be about 1/4 of the wavelength of the laser used as depicted in Figure FB2.7. A dashed-line grid is provided to aid the eye.

As noted, the Michelson interferometer is inefficient. Modern optical disc storage systems employ optics that uses polarization to increase efficiency, virtually eliminating any of the beams returning back to the laser. Diagrams that show the relative position of the optics are presented in the main portion of this chapter. The physics of the geometrical optics used in CD, DVD, or BD players, including polarization are given in Chapter 10. However, we present here the empirical results for one key component, the quarter wave plate, which is made of a birefringent crystal. Calcite is a commonly used birefringent material. When light passes through a birefringent crystal positioned in a specific orientation, it encounters the material having an index of refraction in one plane that differs from that in the orthogonal plane. Consequently, a polarized, monochromatic beam that passes twice through the crystal has its polarization vector shifted by 90°, provided the thickness of the crystal and wavelength are both appropriately chosen. This phenomenon is shown in Figure FB2.8. The physics of this material will be discussed more generally and in greater detail in Chapter 10.

COMPREHENSION VERIFICATION CV2.4

Identify in Figure 2.15 the following: (i) The hub where the disc rests, (ii) the laser and sensor housing, and (iii) the worm screw and motor that drives the housing radially from the central hub to the outer radius and back.

COMPREHENSION VERIFICATION CV2.5

For a DVD, estimate the ratio of the net signal between when the light beam is properly focused on a pit, compared to when it is on a land. What is the relative importance of interference and scattering in the net signal? Assume half of the intensity of the original beam travels to the DVD and the rest directly to the sensor as the reference beam. Also assume the DVD is a double layer.

Answer: First, 50% of the beam going to the DVD does not influence the measured signal since those photons interact with the out-of-focus data surface, returning as many photons in phase as out of phase. Effectively, 0.25 of the original beam intensity participates in the land versus pit areas of the DVD. When the beam is on a land, the net intensity is 75% of the original beam since 0.50 from the reference beam and 0.25 returned from the disc combine constructively.

Referring to Figure 2.18, 0.70 of the in-focus portion of the beam falls on a pit, returning $0.25 \times 0.70 = 0.18$ of the original beam as destructive interference. Another 7% of the in-focus beam is lost (reflected) off the edges of the pit. In other words, 7% of 0.25, or 0.02 is lost. The 23% remainder of the in-focus light falls on a land, returning $0.25 \times 0.23 = 0.06$ as constructive interference. Thus, the intensity contributions from a pit in terms of the original intensity are 0.50 from the reference beam, −0.18 from the destructive interference, +0.06 from constructive interference. The net signal when the beam is over a pit is 0.38 of the original signal or 0.38/0.75 about half of the strength when the beam is over a land. In contrast, the false assumption that DVDs work by scattering the beam into or out of the path leads to a signal ratio of $0.6/0.75 = 0.8$ between pit and land, assuming the pit size compared to beam size is similar to that above. Interference is far more effective than simply absorbing or scattering photons out of the beam. The net number of destructively interfering photons, each negates a photon from the reference beam.

2.5 PHOTOCOPIERS AND FAX MACHINES

Photocopiers use electrostatic attraction to transfer toner to the desired places (imaged portion) on a drum or belt, and then to transfer the toner image to paper. The heart of a photocopier is a photoconductor, a material that can be applied to large areas on a drum or conveyor belt. A photoconductor material has the property that its surface can be charged up and subsequently be neutralized in local areas where light strikes it. (See Fig. 2.20.) Figure 2.21 demonstrates the arrangement of charge on a photoconductor after it has been uniformly charged with electrons and then exposed to an image. The surface charge has been neutralized in those locations where the image was white and still charged elsewhere. *Note*: static image is a mirror of the real image.

There are six basic steps to making a photocopy. The inner workings of a drum photocopier are depicted in Figure 2.22. It starts with uniformly charging a large portion of the photoconductor, using a coronal wire to spray a uniform layer of electrons onto the outer surface of the drum. The physics behind a coronal wire is given in Intro Physics Flashback FB2.4. Next, the photoconductor is exposed to the image on the paper to be copied. Dark areas and dark lines remain negatively charged, while exposure to all of the white areas becomes neutral. This virtual image consisting of electrons is a mirror image of the dark portions of the real image. The exposed portion of the photoconductor is then brought into proximity of a fine powder known as toner. Toner is not ink, but rather a black powder with tiny particulates that intrinsically carry positive charge. A supply of toner is normally kept in a container and delivered mechanically via a belt or other mechanism to the proximity of the latent image on the photoconductor. Here the positively charged toner jumps via an electrostatic force to the negative portions of the

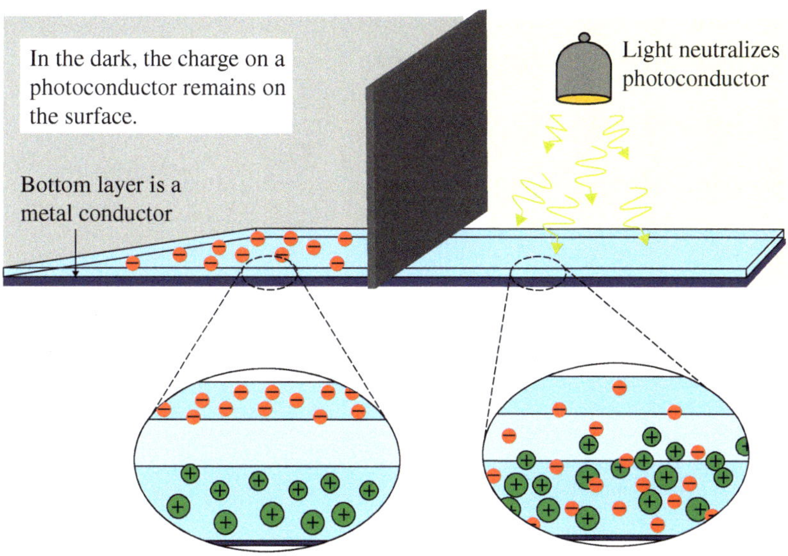

FIGURE 2.20 A photoconductor has the property that it can hold a surface charge until it is exposed to light. Charge is never free to move along the plane of a photoconductor layer.

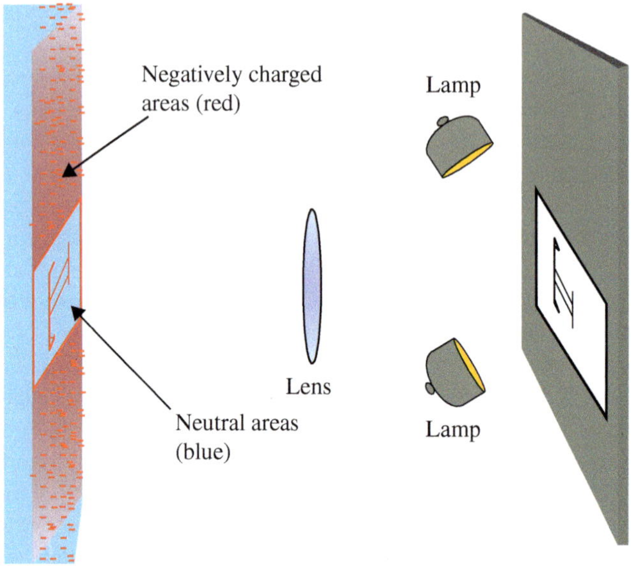

FIGURE 2.21 The locations of charge concentrations are shown on a photoconductor after is has been first uniformly charge and the exposed to the image shown.

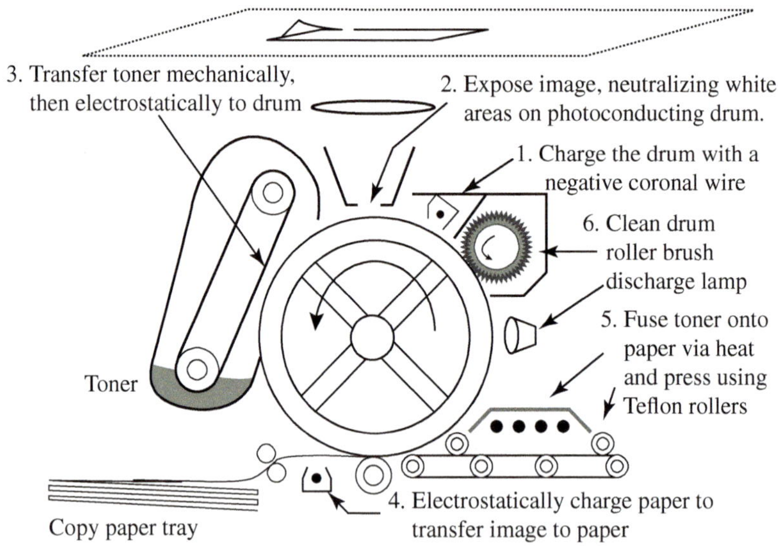

FIGURE 2.22 A schematic representation of a drum photocopier, showing the six stages necessary to make a photocopy.

photoconductor. The toner is weakly attached to the photoconductor at this point since the toner partially neutralizes the remaining charged portions of the photoconductor.

Next, a sheet of paper is fed past a second coronal wire (Step 4 in Fig. 2.22), making it negatively charged. In the process, the paper and the portion of the photoconductor holding the toner image are brought into contact. Once again the toner is electrostatically transferred, this time to the paper. (If the process is halted due to a paper jam or other malfunction, the toner on the paper easily smudges or rubs off.) The particles of toner have wax coatings that are heated and pressed between Teflon rollers to make it adhere to the paper. This final stage is called fusing the toner and is the reason the paper is warm immediately after being ejected from the copier. Fuji and Xerox have created a superior toner, having spherical, smaller-sized particles than those of pervious toners. This high-performance toner enables higher resolution, which is especially necessary for color copying.

Fax machines incorporate many of the same features as a photocopier and many machines now perform both functions equally well. One important requirement of a Fax machine is to sample an image via a scanning process. The image, including any text characters, can be represented by a series of tiny dots spaced to render the original image. Most Fax machines now use camcorder technology, except the image sensor format is a single row of pixels rather than the two-dimensional array found in most digital cameras. Fax machines were originally designed to transmit serially over analog phone lines, sending one row of properly positioned dots after another until the entire image was sent. Today, most telecommunications are digital, enabling simultaneous parallel bit streams as well as data compression techniques that increase the transfer speeds.

INTERESTING TIDBIT TB2.4

The person who invented xerography, Chester Carlson, was rejected by more than 20 separate companies from 1939 to 1944 as he searched for a commercial firm to develop and market the world's first photocopier. Eventually, Carlson found Haloid (later renamed Xerox Corporation), which at the time was a fledgling small photopaper company. While expecting very limited product success, Haloid began sales in 1959 and changed the way people conduct business.

INTRO PHYSICS FLASHBACK FB2.4

Electrostatics

The primary physics that impacts photocopiers is static electricity, the portion of EM known as electrostatics. When electrical charges are stationary (even temporarily stationary), the forces are described by Coulomb's Law. Opposite charges attract while like charges repel. Coulomb's Law states the force on a test charge, q_1, from another charge, q_2, is proportional to the multiplicative product of $q_1 \cdot q_2$ and inversely proportional to the square of the distance between the two. Each of

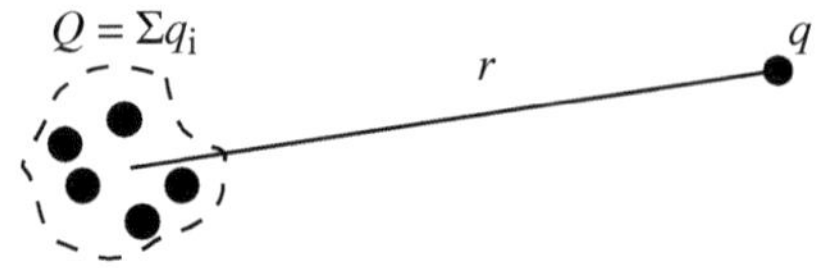

FIGURE FB2.9 The force on a single charge from a group of charges.

the charges, q_1 and q_2, can be positive or negative with force being attractive or repulsive depending on the multiplicative product of the charges. If there are several charges, the net force on the test charge is simply the superposition of the individual forces. In the case where all of the other charges are localized well away from the test particle as shown in Figure FB2.9, the Coulomb force can be approximated as if all of the external charge is concentrated at a single point and the equation becomes thus:

$$\vec{F} = \left(\frac{1}{(4\pi\varepsilon_0)}\right)\cdot\frac{qQ}{r^2} \tag{FB2.1}$$

Here, $\varepsilon_0 = 8.85\times10^{-12}$ $C^2/(N\ m^2)$ and the total charge, Q is the sum of all of the individual charges that are contained in the small volume.

When the number of individual charges becomes large, it becomes useful to generalize the problem. Instead of calculating the collective response from individual point charges, we define a smooth distribution of charge density, ρ, and integrate over a volume. This generalization leads naturally to the concept of the electric field, $\vec{E}$, where the force on a charge, q, is given by $\vec{F} = q\vec{E}(x, y, z)$ and the electrostatic potential, $V(x,y,z)$, a measure of the charge density, is given by $\vec{E} = -\text{Grad}(V)$. *Note*: the gradient (Grad) is simply the partial derivatives of V in three dimensions.

One important class of applications in electrostatics is the spherical conductor. As was noted in Intro Physics Flashback FB2.2, the charge in a conducting material moves to cancel out any internal electric fields so $\vec{E} = 0$ inside a spherical conductor. If the conductor itself has a net charge then all of the excess is found at the surface. Thus, the electrostatic solution exterior to the spherical conductor is independent of whether the conductor is a solid, a hollow shell, or has all of its charge concentrated at its center. Exterior to the sphere, we have

$$V = -\int \vec{E}\cdot dl = \left(\frac{1}{(4\pi\varepsilon_0)}\right)\cdot\frac{Q}{r} \tag{FB2.2}$$

Interior to the sphere,

$$V = \left(\frac{1}{(4\pi\varepsilon_0)}\right)\cdot\frac{Q}{R} = \text{constant}, \tag{FB2.3}$$

where R is the radius of the sphere.

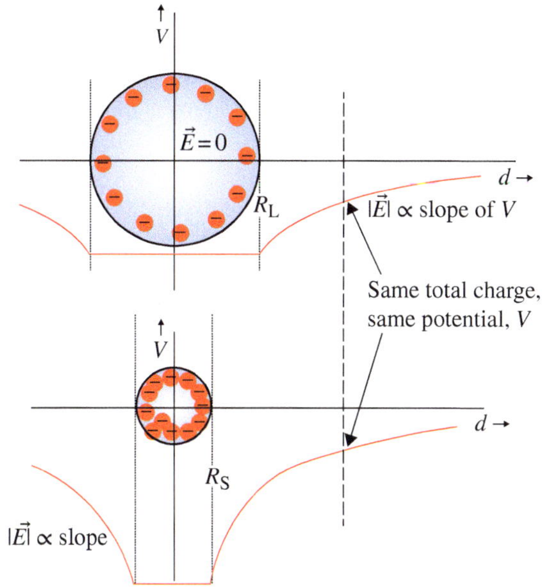

FIGURE FB2.10 Graphs of the voltage potentials for two spheres with same net charge.

Figure FB2.10 graphically depicts two isolated spheres of different radii, but with the same net charge. Plotted as a red curve is the corresponding voltage potential, $V(d)$ as a function of distance for each case. The radius of the sphere shown at the top is $r=R_L$, while the bottom sphere has a much smaller radius of R_S. Both spheres contain the same number of extra electrons, making the $V(d)<0$ in the vicinity of each sphere. Inside each sphere, $V(d)$ is constant so the gradient is zero. However, $V(d)$ in the bottom case is much stronger (more negative) than the upper example since the charge is concentrated into a smaller volume. However, the values of $V(d)$ for the two cases become identical for $d \gg R_L$. The strength of the electric field, $|\vec{E}|$ is proportional to the derivative of this slope since $\vec{E} = -\text{Grad}(V)$. $\vec{E}(d)$ in both cases are equal for $d \gg R_L$ as well.

A couple of interesting insights have emerged. First, the electric field exterior to a charged, smooth, symmetrical conductor can be treated mathematically as a charged point source. (By extrapolation, a long cylinder can be well approximated by a charged thin wire.) Second, the strength of $\vec{E}$ is much higher near a conductor with a convex surface having a small radius of curvature. This is a statement that the force on each of the mutually repulsive charges is greatest for sharp edges or thin wires, making it easy to remove some of the excess charge. This fact is important for photocopiers, which use coronal discharge wires "to spray" charge onto various surfaces.

3

DEVICES ENCOUNTERED IN MODERN LIFE

As noted, there are numerous electronic and photonic devices that are encountered routinely in our everyday activities, each with a foundation of one of a few physics principles. For example, a simple alternating current (AC) electric generator and motor have similar components, converting mechanical-to-electromagnetic energy for the former and visa versa for the later. Both operate on *Faraday's law of induction*, as do transformers, metal detectors, microphones, and electric guitars. In this chapter, we present the physics behind a few key example technologies that are frequently found in everyday life in numerous applications.

3.1 METAL DETECTORS FOR AIRPORTS AND TRAFFIC LIGHTS

Metal detectors make use of the fact that electrons in metals move freely in response to electric and magnetic fields, while electrons in most nonmetals cannot. A current in a loop of wire creates a dipole magnetic field. If the current is steady, Faraday's law does not apply since the magnetic field generated remains constant. If, however, the current is alternating, then the magnetic field will oscillate as well, enabling AC currents to be established inside nearby loops of wire. A metal sheet also behaves essentially as a loop of wire since electrons inside a conducting material move to cancel out all internal $\vec{E}$ fields. Consequently, no net electrical current moves through the bulk material of a conductor, but instead charge is carried only on its surface. To see how a sheet of metal responds as a loop of wire, consider Figure 3.1, an instantaneous snapshot of an AC current in a loop of wire. The AC current pictured in Figure 3.1 is shown at the instance of maximum current, generating a magnetic field with its north pole pointing to the right. Circular eddy currents are set up in conducting materials in the presence of this

Modern Devices: The Simple Physics of Sophisticated Technology, First Edition.
Charles L. Joseph and Santiago Bernal.

Companion website: www.wiley.com/go/joseph/moderndevices

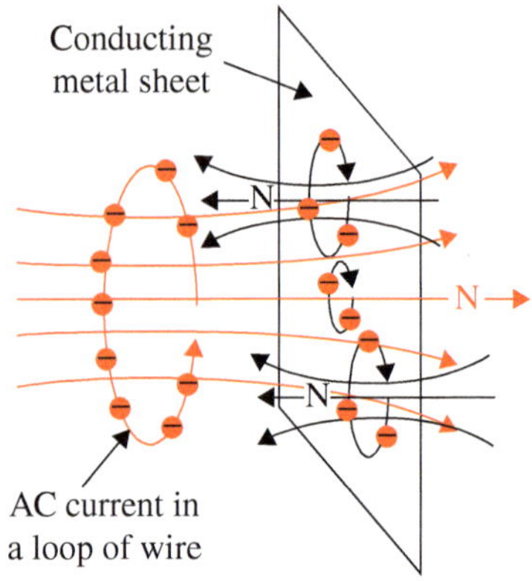

FIGURE 3.1 The resulting counter eddy flows inside a conducting sheet in proximity to an alternating magnetic field.

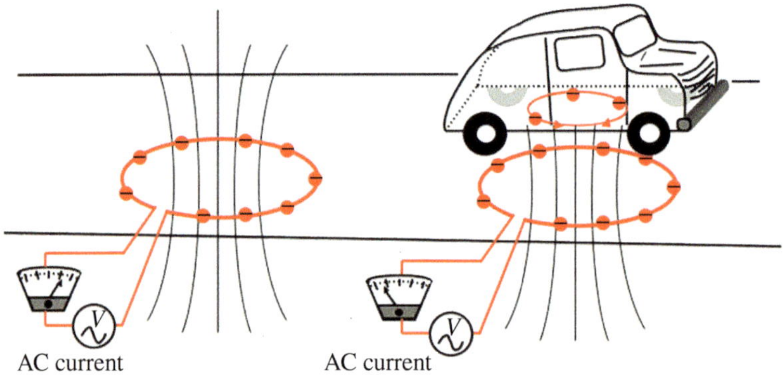

FIGURE 3.2 The basic concept of a metal detector, an automobile sensor in this case.

changing magnetic field. These eddy flows of electrons circulate in the opposite direction to that of the current in the original loop, creating magnetic fields that oppose the original one. At this instance in time, the magnetic north poles from these eddy currents point left. The internal motions of the electrons in the central regions cancel each other out, resulting in a single circular current near the edges that mimics a single loop of wire. Let us reemphasize the induced currents inside nearby conducting sheets always circulate in the reverse direction compared to the original loop. (This is one of many examples that we live in a stable universe. If you push on the universe, it pushes back.)

The design of a metal detector begins with the creation of an alternating magnetic field created by inducing a 10–200 kHz AC current in a coil. For an automobile sensor, often a single loop as depicted in Figure 3.2 is buried in the road surface. The strong AC current within the loop is measured as a baseline. If any significant amount of metal is placed above the loop, negative feedback occurs where the Eddy currents establish an opposing alternating magnetic field, creating an opposing electromagnetic force, *emf*, that partially cancels the original *emf* in the embedded loop. In other words, the impact of the nearby conducting surfaces dramatically reduces the current in the original loop. These two conditions are shown in Figure 3.2 as different values on the AC current meters. Faraday's law is at the heart of most metal detectors used in airport security screening and the sensors in roads to control traffic stoplights.

Other uses of metal detectors include the finding land mines, the detection of guns or knives at security screening locations, the certification of the presence of steel rebar (reinforcing bar) in concrete slabs at construction sites, archaeology, and treasure hunting. Archaeologists, for example, have used metal detectors to map the number of bullets and spent gun shell casings at old battlegrounds to track troop movements and historical accounts of battle strategies. Most of these metal sensors contain a pair of induction-balanced multi-loop coils with integrated circuitry and embedded processors that allow the operator to set sensitivity, discrimination, threshold volume, and track rate speeds along with other parameters. The signals in the two coils become unbalanced whenever small amounts of metal are present. Various metals and alloys have distinctly different phase responses when exposed to AC magnetic fields, enabling metal detectors to discriminate in principle for certain metals while ignoring undesirable ones. However, selective detection of certain alloys remains a technological challenge since some metals (e.g., common tinfoil and gold) have very similar phase responses. A low-frequency (3–20 kHz) non-discriminating mode is capable of canceling the effect of mineralization in the ground, providing greater penetration depth. Many modern detectors have the capacity of continuously switching between the two modes, checking and rebalancing for changing levels of background mineralization.

INTRO PHYSICS FLASHBACK FB3.1

We were taught in our introductory physics class that moving a magnet back and forth through a coil of wire produces an alternating voltage and current in that coil, shown in Figure FB3.1. Moreover, a voltage and corresponding current applied to one or more loops of wire produces a dipole magnetic field identical to a bar magnet. If the current and voltage are constant within the coil, the magnetic field it produces will also be constant and incapable of sustaining a current in a second, nearby coil. This situation is equivalent to stopping the motion of the bar magnet inside the coil. Any initial current induced in a coil will quickly dissipate. Only an alternating voltage applied to a wire coil or the mechanical movement of a permanent magnet with respect to a coil produces oscillatory electric and magnetic fields, which are capable of establishing and maintaining an oscillating current in a nearby wire coil. *Note*: the coil supporting V_{out} on the right of Figure FB3.1 has just as many loops, but does not capture all of the generated magnetic flux, indicating $V_{out} < V_{in}$. These processes, depicted in Figure FB3.1, are simple applications of *Faraday's law of induction* for the electromotive force, ε, expressed mathematically as follows

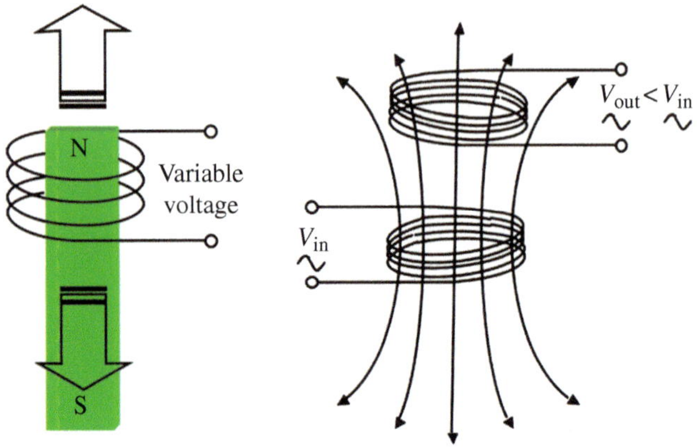

FIGURE FB3.1 Two examples of Faraday's law of induction.

$$\varepsilon = -N\frac{\Phi_{final} - \Phi_{initial}}{t_{final} - t_{initial}} - N\frac{\Delta\Phi}{\Delta t}. \quad \text{(FB3.1)}$$

Here, N is the number of loops in a coil and Φ is the amount of magnetic flux encircled. In fact, any change to the interior magnetic field in one or more loops of wire produces an AC voltage. Figure FB3.2 shows a single loop being mechanically rotated in a uniform magnetic field. The maximum flux occurs when the normal of the loop is parallel to the magnetic field, $\vec{B}$, while the loop encircles zero flux when the angle between the normal and the $\vec{B}$ field is 90°. This is an example of a simple AC generator.

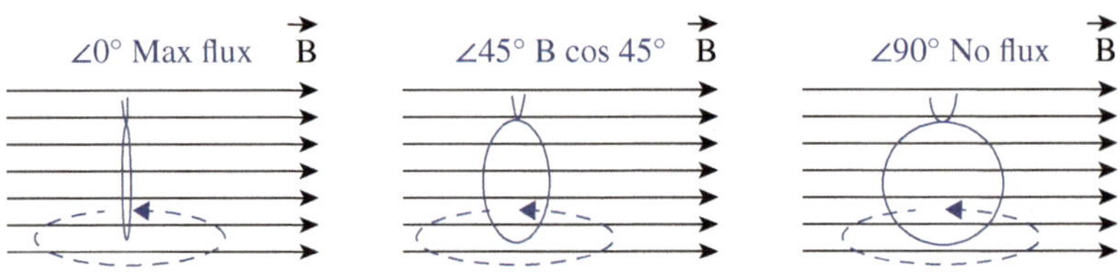

FIGURE FB3.2 The magnetic flux passing through a spinning loop of wire.

One important example of Faraday's law is the electrical transformer, which uses a conducting toroid to transfer virtually all of the magnetic flux from one coil to another. For a transformer, it can be shown that Equation FB3.1 reduces to

$$\varepsilon_P = -N_P\frac{\Delta\Phi_P}{\Delta t} \quad \text{and} \quad \varepsilon_S = -N_S\frac{\Delta\Phi_S}{\Delta t} \quad \text{(FB3.2)}$$

$\Phi_P = \Phi_S$ due to the conducting toroid and assuming both coils have negligible electrical resistance, each *emf* equates to its voltage (i.e., $\varepsilon_P = V_P$ and $\varepsilon_S = V_S$). We obtain the simple transformer equation:

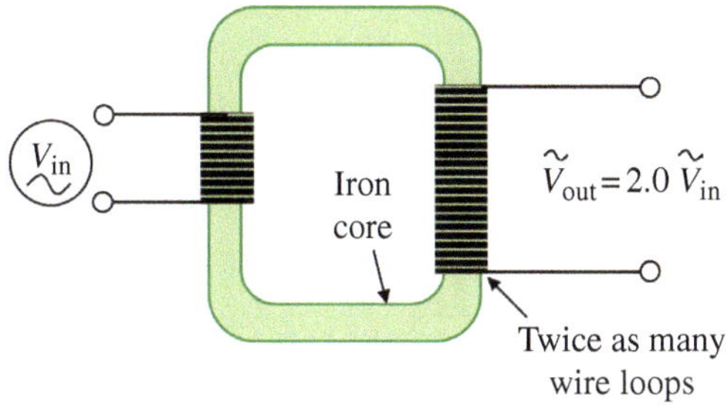

FIGURE FB3.3 A 2 : 1 step-up transformer.

$$\frac{V_S}{V_P} = \frac{N_S}{N_P} \tag{FB3.3}$$

Electrical transformers are used to increase (step up) or decrease (step down) the voltage of an AC. Figure FB3.3 shows the essential components of a simple transformer, a 2 : 1 step up in this case. An iron toroidal ring forms the core of the transformer and is used to capture virtually all of the variable magnetic flux generated from one coil side and deliver it to the other. The ratio of AC voltages is equal to the ratio of wire loops on each side of the core.

Faraday's law of induction is the underlying physics principle for metal detectors used for airport security screening, for detecting automobiles at automatic traffic lights, and for a host of other applications

INTERESTING TIDBIT TB3.1

Most people know what an electric guitar is. However, a "magnetic guitar" might be a more appropriate name since the signal pickup produces magnetic fields that couple to the vibrating steel strings. An electric guitar is yet another example of Faraday's law. In this case, the vibrating steel wires moving through magnetic fields establish small oscillating electrical currents and counter (opposing) magnetic fields that are transferred to the pickups.

3.2 BARCODE SCANNERS, QUICK RESPONSE CODES, AND RADIO-FREQUENCY IDENTIFICATION READERS

Supermarket and other retail scanners have become familiar fixtures since the late twentieth century. Supermarkets in the 1970s were the first to introduce the Universal Product Code (UPC) to increase efficiencies at the cash register, as well as with product inventories, promotional merchandise, and price changes. The effort to create a universal code that would be used initially by a critical mass of retailers and producers was a monumental task. Fortunately, an ad hoc committee of highly respected

FIGURE 3.3 A UPC-A bar code with additional notations showing the locations of digital information associated with producers and products.

corporate leaders in the grocery sector was able to introduce a standardized coding, which enough retailers and wholesalers would implement initially. The incorporation of UCP barcodes expanded rapidly to all supermarket products and across numerous retail industries such as clothing and department stores. Eventually, the UPC system led to a 2009 annual cost saving of \$17B in the U.S. grocery sector alone. The consumer has also benefited from shorter wait times in checkout queues.

A picture of a UPC-A is provided in Figure 3.3, showing its essential features. There are two sets of 10-digit bars, the left half indicating the manufacturer/producer code and the right half identifying the product and its size/quantity. Three pairs of double bars at the start, middle, and end are used as fiducial marks to indicate to the scanner the location of the numerical information. There also has to be clear areas on the left and right of the bar system to prevent false starts or false endings. The product code portion of the label indicates the type of product as well as the quantity inside its container. For example, a UCP bar code attached to a box might identify the producer to be Kellogg's® Cereal, while the bars on the right side indicate it is Corn Flakes®, a 12 ounce box. A table of data inside the retailer's computer is used to assign a price to each item scanned. A second form of UPC label, the UPC-E, was developed for containers that have very limited space such as cans of soda pop. The methods of decoding for both the UPC-A and UPC-E can be found elsewhere.

Next we explore the inner workings of a typical scanner at a supermarket. A UPC scanner must be able to read the product bar code located over the window with a range of orientations and distances. A narrow beam of light, directed to scan several different directions, is used to illuminate the UPC bar code. As in Figure 3.4 (left), a laser beam is sent vertically upward to a pickoff mirror that redirects it to an angled

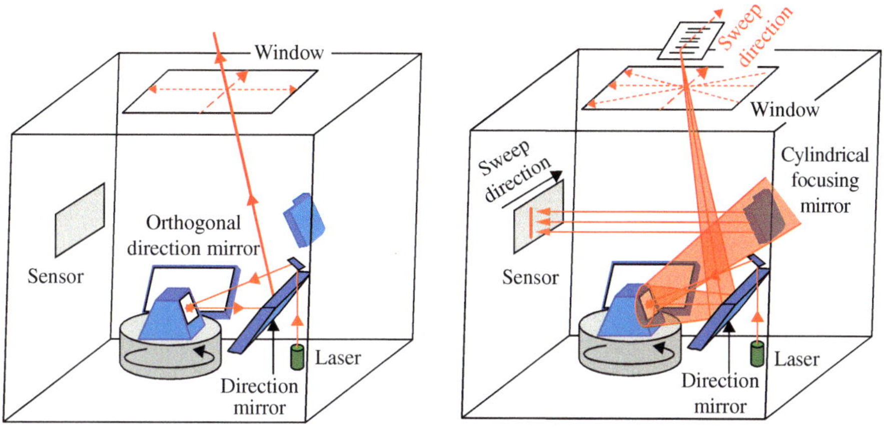

FIGURE 3.4 The internal workings of a supermarket checkout scanner. Left: the illuminating beam as it sweeps across one orthogonal direction mirror. Right: the same, but showing the returned beam including the cylindrical focusing mirror.

mirror on a rotating platform. The rotating mirror sweeps the beam horizontally over a set of five directional mirrors, each orienting the beam across the window in one of four directions. (*Note*: only two of the five direction mirrors are depicted in Fig. 3.4 for simplicity and to avoid confusion.) The left half of the figure shows the path of the beam as it is sweeping from the front of the window to the backside. A dashed arrow shows the sweep path for this particular directional mirror.

Once the scanning beam encounters a barcode as depicted on the right side of Figure 3.4, the barcode scatters a fraction of the light back into the scanner. More light returns if the spot illuminates a white area and less light if the spot is centered on a dark bar. *Note*: the laser beam looses most of its coherency because bar codes are not imprinted on highly polished front-side mirrored surfaces. Thus, the return signal from the originally tight spot is diverging over almost an entire hemisphere, creating a low-intensity background of scattered light that illuminates most of the interior of the scanner. The useful portion of the returned light follows essentially the same path back through the mirror optics, where a cylindrical mirror captures a significant fraction of the returning beam and focuses it onto a sensor. The amount of light captured is set by the areas of the mirrors inside the scanner. In our example, the physical sizes of both the rotating and the cylindrical mirrors determine the fraction of the light that is captured.

There are a couple of important features worth underscoring. First, a cylindrical mirror is used to focus the captured return signal into a column of light (i.e., a bright bar). (Most single-focusing mirrors have simple concaved surfaces that focus the light to a spot. A cylindrical-shaped mirror is used to focus light one-dimensionally into a line of light.) The focal length of the cylindrical mirror and, in turn, its radius of curvature has to be matched to the distance between the mirror and the sensor.

Second, the five directional mirrors surrounding the rotating mirror create four distinct sweep directions as shown on the right side of Figure 3.4. (Two of the directional mirrors create a pair of redundant sweeps: one from left to right and the other right to left.) The complete set of sweeps enables at least one to be somewhat aligned with the barcode over the window. Minor rotational misalignments between the sweep direction and the orientation of the UPC sticker as depicted in Figure 3.4 (right) cause the beam to sample the barcode near its top on one side and the near the bottom on the other side. Any minor rotational misalignment shortens the apparent spacing between the bars. The apparent bar spacing is also shortened by any angular tip of the barcode in the direction of the sweep as well as by added height above the window. This is a primary reason the UPC-A code has three sets of double fiduciary bars.

Other one-dimensional (1D) product identification systems include the Japanese Article Number (JAN) and European Article Number (EAN), both of which are similar to the UPC developed in the United States and have now effectively merged into the single International Article Number, which retains the EAN designation. Similar to the UPC-A and UPC-E, the International Article Number codes have two sizes: the EAN-13 and the less common EAN-8, which is reserved for small packages.

Quick response (QR) code is a matrix system of square dots, a 2D imaging barcode first developed in Japan by Toyota to track parts and subsystem assemblies during the manufacture of automobiles. Any combination of numeric, alphanumeric, or binary/bite modes can be encoded. Various versions of QR codes are shown in Figure 3.5. The QR code uses imaging rather than scanning and is used in a much wider range of applications, including commercial tracking, entertainment and transport ticketing, product/loyalty marketing, and in catalogs. For example, a mobile "app" (application) on a smart phone can be created to decode a company's QR code, also allowing the company to create instant, electronic coupons. Moderate-resolution QR codes such as the 57×57 element (Version 10) have been optimize for cell phone cameras.

QR codes generally have fiducial squares (in this case, a box inside a box) on three corners. This provides the orientation of the QR code to enable the electronic reader to decode the 2D image. Additional, smaller, fiducial squares are also present

(33 × 33) (57 × 57) (177 × 177) Decodable

FIGURE 3.5 The 2D QR code that has become a popular product and service identification code. From left to right: Version 4, Version 10, Version 40, and a damaged QR code from which all information can be recovered. Credit: Compilation from Wikipedia public domain images.

inside the QR matrix of later versions, marking the locations of various types of information. Redundant data is embedded, enabling an error correction algorithm (usually a Reed-Solomon cyclic error correcting code) to recover lost data due to a damaged QR image.

A third form of fast electronic reader is the radio-frequency identification (RFID), wireless identification devices that do not require direct contact or line-of-sight imaging. RFIDs can be injected or implanted in livestock, pets, or even humans. Personnel employed in somewhat hazardous conditions such as workers on an off-shore gas and oil rig, sometimes, wear RFIDs sewn into uniform clothing as a safety measure that allows them to be found quickly in an emergency. Most heart pace-makers are equipped with RFID technology and a defibrillator, allowing a physician to monitor the monthly activity of his patient and to make uninvasive adjustments. Some patients have been unaware that their heart had stopped beating and had been restarted until their doctor told them. Marathon runners are sometimes required to wear ankle or wrist straps with RFID tags that are read at checkpoints throughout the race. Complex RFIDs are now embedded in the passports of many countries, the first being issued by Malaysia in 1998. Other uses of RFIDs include security identification badges replacing cards with magnetic strips, vehicle access control at airports, automatic toll-road collection, inventory tracking of high-value equipment in hospitals that are moved frequently, anti-shoplifting devices, and numerous cases were RFID systems have replaced earlier barcode or QR code technologies. Some automobile service stations are equipped with RFID readers inside their gasoline pumps that link customers to their credit cards. It might become possible one day to bring a fully loaded shopping cart to a grocery checkout counter and have all of the products scanned in situ in less than a second. One primary difficulty at the present time is how to accurately count multiple identical objects. Another possibility might be to check the availability of a product on the shelf without sending someone to look for it.

RFIDs are generally short-distance devices that rely on electromagnetic induction, another application of Faraday's law. There have been other forms of RFID technology such as tags that used capacitive rather than inductive coupling, but the manufacture of capacitive coupled devices ceased in 2001. Some RFID tags contain batteries and are classified as active tags, enabling these tags to transmit data over hundreds of meters independently of any interrogator or reader. However, most RFIDs are passive devices that remain inactive until brought into the vicinity of an interrogating electronic reader. A passive tag is actually powered inductively by the external electronic reader, which is depicted schematically in Figure 3.6. The interrogator reader is effectively a two-way digital radio, continually emitting an electromagnetic wave at a predetermined frequency, and receiving data back from the tag. The RFID tag must contain some arrangement of conducting loop(s) or antenna tuned to capture the changing magnetic field, creating an AC power source. Once powered up, the tag then becomes a transponder, transmitting its data back to the antenna inside the interrogating electronics.

To reiterate, the outer coil on the passive tag serves a dual purpose. First, it captures enough signal from the reader to power the small circuitry and static memory in the center. Second, it serves as an antenna to transmit data back to the interrogator.

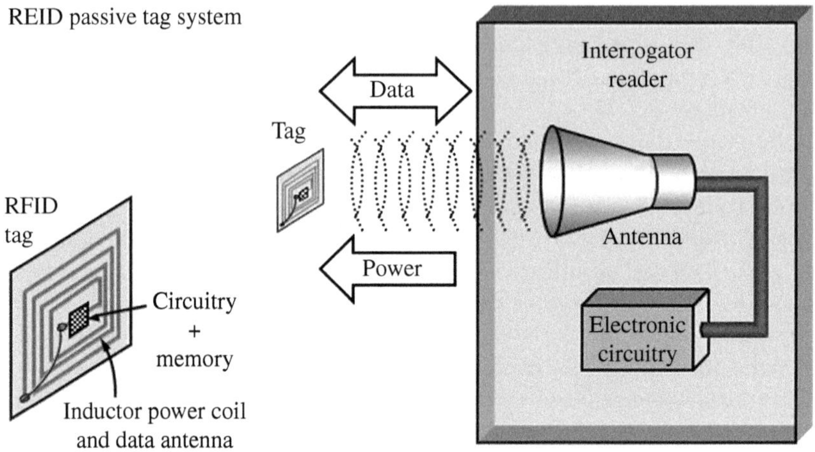

FIGURE 3.6 A schematic representation of an RFID passive tag system.

It should be noted that not all passive tags use a single-loop antenna, although most do. Obviously, the signal from a passive RFID tag is weak and can only be read over short distances of a few meters due to the small amounts of energy available to it. RFID technology has been available almost as long as the UPC barcode has (the 1970s), but the high fabrication cost per tag has limited its use. Various RFID systems operate at different frequencies over low-, high-, or ultra-high-frequency (LF, HF, or UHF) radio bands, causing operational incompatibilities between systems. While there are no universal RFID conventions, an international standard electronic product code (EPC) for RFIDs was established in 2010 around an UHF radio wave. Adoption of the EPC standard was fueled by a significant drop in fabrication cost (tag prices starting at $0.09 USD in 2011) as well as an increase in performance reliability to 99.9%. It is unlikely that the EPC will completely replace UPC and EAN barcodes any time soon due to the complexity of an RFID tag, which will continue to be more expensive than a printed barcode.

One important class of battery-powered active RFIDs are those placed on vehicles to collect toll road charges without the vehicle having to stop. All RFID tags, both active and passive, are especially sensitive to the nearby environment and systems often interfere with other radio devices. The windshields of some automobile models, for example, have partially conducting surfaces, preventing tags being attached to the inside of the window, just behind the rearview mirror. Usually, the toll road collection agency offers an option for an external tag, normally one to be mounted to the front license plate when necessary. These systems are fast and efficient, transferring information over several meters within a few milliseconds of time. The transaction speed enables tagged automobiles to pay tolls while maintaining highway speeds. Nevertheless, many tollbooth collection stations do require all traffic to slow down for safety purposes.

RFIDs are increasingly trending toward miniaturization. In 2009, Bristol University researchers successfully glued micro-RFIDs to live ants to study their behavior in their underground habitat. Hitachi has manufactured a 0.05 mm × 0.05 mm RFID chip, the size of a grain of sand, but the attachment of antennas is a major challenge, limiting the signal range to only a few millimeters.

The ability to read information without direct contact is the principal advantage of an RFID system, but also its major weakness. RFID devices are inherently susceptible to malevolent uses. The packages and luggage being transported, for example, can be scanned efficiently for recently purchased valuables, facilitating theft. Another important concern is identity theft. A third party can electronically record gasoline or diesel fuel purchases via an RFID linked to a credit card during the transaction. The thief can use this information subsequently for his/her own purchases. While the passports issued by many countries now contain RFID tags with numerous biometric information to prevent forgery, unauthorized capture (skimming) of e-passport data has been demonstrated on numerous occasions from as far away as 10 m. Identity thieves with sophisticated electronics can capture an individual's photo, birth date, and other specific personal data. Counter measures to protect the holder of an e-passport include passports with a thin metallic lining on the inside cover to make skimming more difficult and a printed basic access control (BAC) that functions as a personal identification number (PIN). The BAC has to be manually entered into an RFID reader to enable the encryption of any communication between the chip and the interrogator. Similar to computers, these high-tech e-passports will always be vulnerable to malevolent attacks with the advantage seesawing between the skimmers and those seeking to preserve the integrity of the information contained in e-passports.

COMPREHENSION VERIFICATION CV3.1

You observe a man drive up to a gate-controlled parking lot, take out his toolbox, and place it on the metal sensor for an exit gate to open it, and then drive his pickup truck with enough ground clearance to avoid hitting the toolbox into the lot through the exit side. You also observe that the metal sensor for the same exit gate does not detect the presence of most bicycles. Give a plausible explanation why the metal detector senses a toolbox but not a bicycle, assuming both have similar amounts of ferrous metals.

Answer: The metal of the toolbox is closer to the ground sensor and has a larger cross section for an inductive response to the metal detector (Faraday's law) compared to the bike. *Note*: putting the bicycle on its side would dramatically increase its detection efficiency similar to the different coil orientations seen in Figure FB3.2.

3.3 GLOBAL POSITIONING

Any position on a small, flat surface can be read off a grid or can be measured directly with a ruler from a reference point. A precise, 3D point on, above, or below the surface of the Earth, however, requires indirect measurements, most often by sending pulses from several accurately known locations and determining the separate travel times. The transponder-to-receiver distance is simply: $d = v \times t$ and the uncertainty in the distance is $\Delta d = v\Delta t$. (*Note*: the velocity, v, depends on the medium through which the wave propagates, which can be calculated extremely accurately. Relativistic effects are also significant corrections, especially for the on-orbit atomic clocks. A discussion of these corrections is beyond the scope of this text.) Each pulse consists of short bursts of a traveling wave at a specific frequency and wavelength. Electromagnetic (e.g., radio- and visible-light) waves are the only suitable waves that can propagate through the vacuum of space. A Global Position System (GPS) consists of a series of satellites, operating on this principle, which are accurately tracked from the ground and continually send out radio pulses.

The primary technological challenge of this method is the extremely high speed ($\sim 3 \times 10^5$ km/s) that all electromagnetic waves travel, which requires the use of ultra-high-accuracy atomic clocks. The pulse arrival times have to be measured to a ten billionth of a second (10^{-8} seconds) to distinguish between one location and another that is three meters further away. Instantaneous contact with only three satellites would be necessary if each of these as well as the ground GPS receiver, all could measure *absolute* time to the very high accuracy required. While each orbiting satellite does have an atomic clock on-board, it is far too expensive and too bulky to equip the ground receivers with these clocks. Instead, each ground unit GPS has a clock that can measure time sufficiently precise in *relative* rather than absolute arrival times.

A hand-held GPS device also contains a small electronic processor, allowing it to determine its position from four or more GPS satellites. Contact with an extra satellite is essential to provide redundant information that the ground processor uses iteratively to calculate positional errors and any clock drift. To see how this is accomplished, consider the simplified problem of 1D positioning depicted in Figure 3.7. Top: A single satellite sends a burst of signals at t_0, the prescribed time for the next sequence. The burst travels a distance, $d = v\Delta t = v(t_{Ar} - t_0)$, arriving at the position of the "black X" (value ~2) at t_{Ar}. However, the ground clock in this example has experienced a small drift and is now slightly faster than the real time. According to the ground clock, the arrival time is $t_{Ar} + \varepsilon$, where ε is a very small quantity. The Δt of the fast ground clock and the inferred distance both appear larger than these really are, corresponding to the "black X" (value ~1.8). Contact with a second satellite, shown at the bottom, initially produces an ambiguity in the ground position with Satellite A, indicating farther to the right and Satellite B indicating farther to the left. *Note*: the correct position is not simply the midpoint between the two apparent locations since each satellite has a different elevation. Internal calculations indicate the processor clock is slightly fast and an adjustment is made toward the correct time. This process is repeated until the correct time and position (blue X) are achieved and all errors minimized.

Extrapolation to 3D is shown in Figure 3.8. The dotted-gray curved arrow shows the approximate altitude for a GPS orbit taking a US satellite directly over the X position in the plane of the page. In general, the projection of most satellite positions will be lower than the dotted curve, indicating the 3D locations are either closer to—or farther away from—the plane of the page. At least four satellites are required to determine longitude, latitude, and elevation to about 1 m root mean squared (RMS).

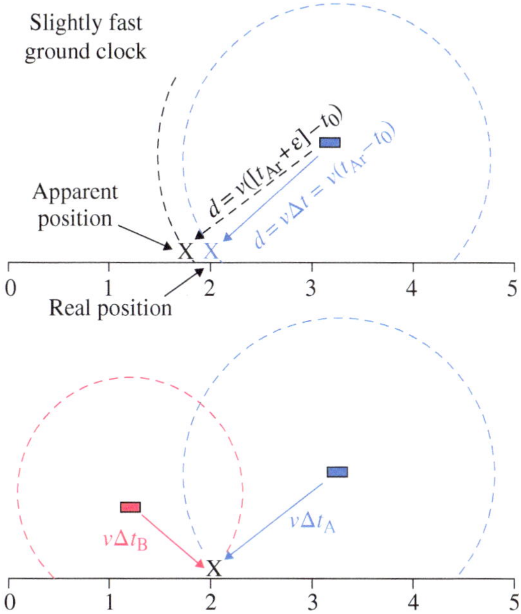

FIGURE 3.7 The problem of determining 1D position using 1 satellite (top) compared to having an extra satellite to cancel errors (bottom).

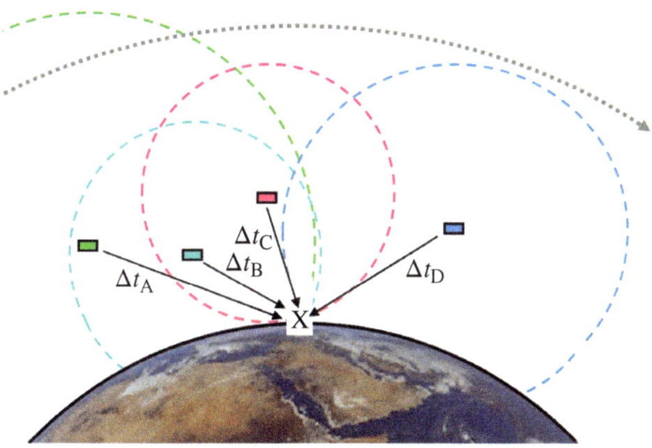

FIGURE 3.8 Contact with four satellites is necessary to calculate longitude, latitude, and altitude to 1 m RMS.

GPS systems can determine if you are above or below sea level, provided the ground unit can still receive the radio signals.

The US military began creating the first GPS system in 1973. It was made available to the public and became fully operational in 1994 with 24 satellites. Each US

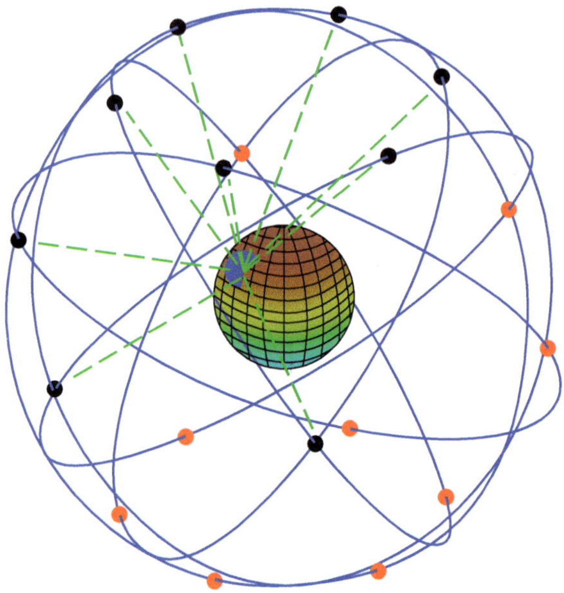

FIGURE 3.9 The distribution of US GPS I satellites. Source: El pak, https://commons.wikimedia.org/wiki/File:ConstellationGPS.gif. CC public domain.

GPS satellite orbits at an altitude that is approximately half as high as geostationary with an orbital period of about 12 hours. The arrangement of US GPS satellites is shown schematically in Figure 3.9. The European Union and a few individual countries have subsequently developed their own GPS systems since the United States has restricted access and has reduced the accuracy of its GPS for non-military uses during national emergencies. The third-generation US system, GPS III, was initiated and funded in 2000. The launches of the first GPS III satellites are scheduled for 2017.

Other GPS systems include GLOSNASS (Russian: GLObal NAvigation Satellite System), which was made public to civilians in 2007 and a three-satellite Chinese navigation system that has been covering China and surrounding regions since 2000. A second-generation Chinese system known as Compass or BeiDou-2 will consist of a 35-satellites, providing global positioning. BeiDou became operational in December 2011 with 10 satellites in operation. In addition, the European Union's Galileo positioning system of four satellites was launched October 2012 for in-orbit verification. The full Galileo system of 30 satellites is expected to be completed by 2019, providing superior position performance at high latitudes and carrying distress transponders to enable the sender to know that help is on its way. Also, there is the Indian Regional Navigational Satellite System, consisting of seven satellites, three in geostationary orbits over India.

3.4 TRANSPORTATION TECHNOLOGIES

Transportation technologies are important to our lives since we live in a global economy where numerous goods are shipped vast distances across oceans and continents. Fossil fuels are the most effective means of energy storage for the majority of cars, buses, trucks, airplanes, and ships. The transportation sector accounted for 27% of the total US energy consumption in 2011, not including the energy used to build those vehicles. (Chapter 21 on power generation contains a more extensive treatment of energy sources including alternatives and consumption.) Worldwide commerce was made possible by large cargo capacity vehicles and by plentiful cheap fossil fuels, which might not remain inexpensive in the future despite the high confidence in hydraulic fracturing by some in finance and energy companies. In this section, we explore the strengths and weaknesses of two modes of vehicle propulsion and then a few alternative fuel technologies. We also investigate the physics behind speed radar guns and high-speed rail.

3.4.1 Internal Combustion Engines versus Electric Motors

Two modes of vehicle propulsion that typically come to mind are electric motors and internal combustion engines using either diesel or gasoline. Obviously, we are ignoring for the time being hybrid cars, steam-driven engines, and a few other modes of transportation. Fossil fueled vehicles are inherently inefficient, utilizing approximately 20% of the available chemical energy from the fuel and collectively being a major source of greenhouse gases. The principal advantages of internal combustion engines have been a readily available supply of fuels, nearly instantaneous turn-key start-and-go operation, and rapid refueling. In contrast, electric vehicles (EVs) are efficient, converting at least 80% of the stored energy into kinetic motion. EVs are less complex, low maintenance (e.g., no need to change the oil), and have higher torque-to-weight ratios providing faster acceleration than the typical fossil-fueled vehicle. In addition, an EV has superior performance than its traditional counterpart for stop-and-go urban traffic, especially if dynamic braking is used to convert much of the vehicle's kinetic energy back into recharging its batteries. However, EVs have important limitations. First, the typical EV can only be driven a limited distance (typically 80–120 km, 50–75 mi) on a single charge before requiring a time-consuming (hours) recharging. While this range is adequate for 90% of all automobile commuters, a combustion engine is required for long trips or to tow trailers. Second, batteries banks require a substantial fraction of the vehicle's volume and add significantly to its weight. Rechargeable batteries (lithium, nickel–cadmium) are expensive and have operational lifetimes of 3–4 years, requiring periodic purchases that typically are a significant fraction of the original purchase price. In addition, batteries contain hazardous materials that must be recycled. Technological advancements of batteries are expected to improve power storage, but only a factor of two at most by 2050, indicating batteries will remain a major limitation for EVs for some time to come.

3.4.2 Alternative Fuels

The most commonly advocated alternative fuels for motor vehicles are ethanol biodiesel and hydrogen. Alga is the most promising biofuel crop since it produces the most energy per biomass and can be grown on marginal land. Some algae plants can increase their mass 10-fold in 24 hours under ideal growing conditions. The technological challenge is to replace fossil fuels in a sustainable manner, while significantly reducing the emission of anthropogenic greenhouse gasses that cause global warming. The 2011 opinion of a science committee of the European Environment Agency indicates biofuels do not necessarily mitigate global warming at least in the short term. Deforestation to grow biofuel crops, for example, increases atmospheric CO_2 initially, even if the forest is allowed to regrow after being harvested. Collectively, alternative fuels might someday be part of the solution, providing fuels and reducing greenhouse gas emissions, but other actions such as serious conservation efforts are essential and offer greater contributions to the overall solution than do alternative fuels. To understand the limitations of alternative fuels, consider biodiesel production, which in the United States is derived from corn. Significant amounts of fossil fuels are used in the fabrication of fertilizers to grow the vegetation, to transform the plant material into fuels, as well as to transport seed, fertilizers, waste, and final fuel products. If the entire US corn crop were converted to biodiesel or ethanol, it would only supply about 7% of the country's energy needs. Moreover, US beef and dairy herds would virtually disappear, as would the US production of corn oils and fructose corn syrup. It is not feasible to change most food crops to fuel crops since worldwide food prices have already increased substantially due to the small-scale conversion that has already taken place. On the other hand, some biofuels can be grown on marginal lands that are not suitable for food crops. Alternative fuels represent at most potentially small steps toward reducing modern society's dependency on fossil fuels, primarily by recycling used materials. For instance, used cooking oils that continue to be waste by-products can be converted into biodiesel.

Another alternative energy source often touted is hydrogen. The conversion from a fossil fuel to a hydrogen economy has been herald by some as "the solution" to our future energy needs. Unfortunately, noteworthy critics such as Joseph Romm, formerly Principal Deputy Assistant Secretary of the US Department of Energy, claims: "A hydrogen car is one of the least efficient, most expensive ways to reduce greenhouse gases." Asked when hydrogen cars will be broadly available, Romm replied: "Not in our lifetime, and very possibly never" (McClatchy Newspapers, May 15, 2007). Similarly, Lux Research, Inc., a consulting firm that provides strategic advice for emerging technologies, issued a report in 2013 that stated: "The dream of a hydrogen economy envisioned for decades by politicians, economists, and environmentalists is no nearer…." Lux Research notes the extraordinarily high capital costs will remain a nearly insurmountable barrier, except for niche applications. Hydrogen-powered forklifts and buses for local mass-transit routes are two relatively successful examples. Stationary power plants fueled by hydrogen may offer the greatest future promise for this technology, a topic we will return to in Chapter 21 on power generation.

Concentrated quantities of liquid or gaseous hydrogen do not occur naturally so these must be manufactured from another source such as from fossil fuels by stripping off the hydrogen atoms from hydrocarbons or from water by breaking hydrogen–oxygen bonds. All methods of hydrogen gas production require more energy creating the fuel than can be retrieved from it later. This statement is simply a consequence of the *Laws of Thermodynamics*. In this sense, hydrogen is an energy carrier similar to batteries, storing energy from recently generated power, rather than from a conventional fuel, containing chemical energy formed millions of years ago. If a supply of pure oxygen is available to combine with the hydrogen, then heat plus water are generated without harmful byproducts. Liquid rocket engines often used cryogenically cold liquid oxygen and hydrogen. However, if hydrogen fuel is burned in air (i.e., an internal combustion engine), small amounts of oxides of nitrogen (NO_x) are also created, which are approximately 310 times more harmful greenhouse gases than is CO_2.

In contrast with hydrogen internal combustion engines, fuel cells are a clean alternative method of consuming hydrogen gas as an energy source, producing DC electrical current. Automobiles and trucks using hydrogen fuel cells are essentially electric vehicles with water vapor being the only emission. The physics of a hydrogen fuel cell, also known as a proton exchange membrane (PEM), is depicted schematically in Figure 3.10. Hydrogen gas is circulated past an anode mesh coated with platinum, which serves as a catalyst to strip electrons from a fraction of the hydrogen atoms. Some of the protons (ionized hydrogen) simply recombine with free electrons in the vicinity, while others diffuse across the electrolyte membrane. Only isolated protons are able to pass through the PEM; H atoms, heavier ions, and molecules are blocked. The following chemical reaction occurs on the other side:

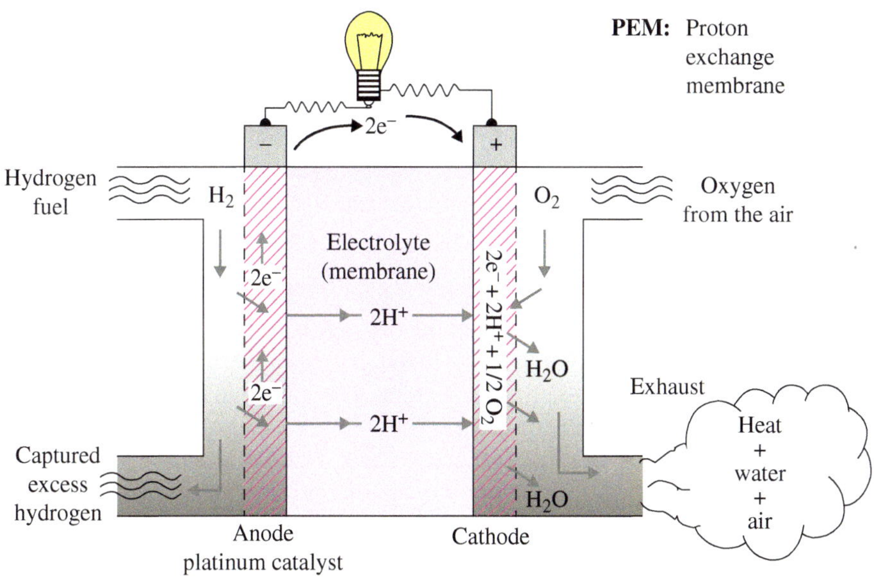

FIGURE 3.10 The schematic operation of a hydrogen fuel cell to produce electricity.

$2e^- + 2H^+ + 0.5O_2 \rightarrow H_2O$ with molecular oxygen being supplied from the air. This reaction only proceeds with any measurable efficiency when the electrons that have passed through the electrical load are available at the cathode. The reaction is also exothermic with heat being a waste product.

There are several shortcomings with hydrogen fuel cells, cost being the greatest challenge. In 2002, the US Department of Energy (DOE) estimated that fuel cells were so expensive that the cost of a hydrogen automobile would be more than $1M USD. One of the cost drivers is the numerous stacks of PEMs required, each with a high-tech membrane and an anode catalyst typically made of rare substances such as platinum. However, new nanometal catalysts made of nickel-tin developed by Honda, for example, have reduced fuel cell costs substantially. Nevertheless, a DOE vehicle cost estimate in 2010 was still $300,000 based on the new nanometal catalysts. Most of the criticism of hydrogen-fueled vehicles is based on economic rather than technological issues. One technological issue with PEMs is freezing. Water vapor is generally present in the air and is created at the porous mesh cathode, which can freeze at temperatures below 0°C (32°F) and can form ice barriers preventing the chemical reaction. Hydrogen fuel cells must be preheated or maintained above freezing prior to operation. Another issue is the lack of infrastructure, the storage, transportation, dispensing, and safety of hydrogen fuel.

3.4.3 Speed Radar Guns

The acronym RADAR is short form of RAdio Detection And Ranging. It has long since joined the general public lexicon, referred to simply as "radar." There are two types of speed radar guns: pulsed and continuous with the latter further divided into two sub-categories of Doppler radar and frequency modulated. Virtually all traffic-control speed radar guns are *continuous* Doppler radar type devices with current guns operating in the microwave K or K_a (US) and K_u (European) bands. Similarly, speed indicating guns used in sporting events or amusement parks are also continuous Doppler instruments. These devices make use of the Doppler effect reviewed in Intro Physics Flashback FB3.2. *Note*: the usage of "Doppler Radar" by television meteorologists is misleading and technically erroneous since most radar equipment employed by air traffic control and the weather services function on multifaceted principles with *pulsed* radar being but one component. Figure 3.11 depicts a dashboard-mounted speed radar gun in use on a moving patrol vehicle. A small portion of the transmitted signal (wavefronts shown in black) goes directly to gun's receiver to serve as a reference signal. As can be seen in Figure 3.11, the spatial angular coverage of the transmission beam must be kept as narrow as possible to minimize the number of return signals, but broad enough to capture both on-coming traffic as well as some stationary objects that are necessary to determine the patrol-car speed. Typically, speed radar guns produce a central beam with 80–85% of its energy in a 12° cone.

Doppler radar guns require signal processing circuitry to sort out and interpret electronically the multiple return signals. A highly idealized plot of the spectrum returned to the dashboard receiver is shown at the bottom-right of Figure 3.11.

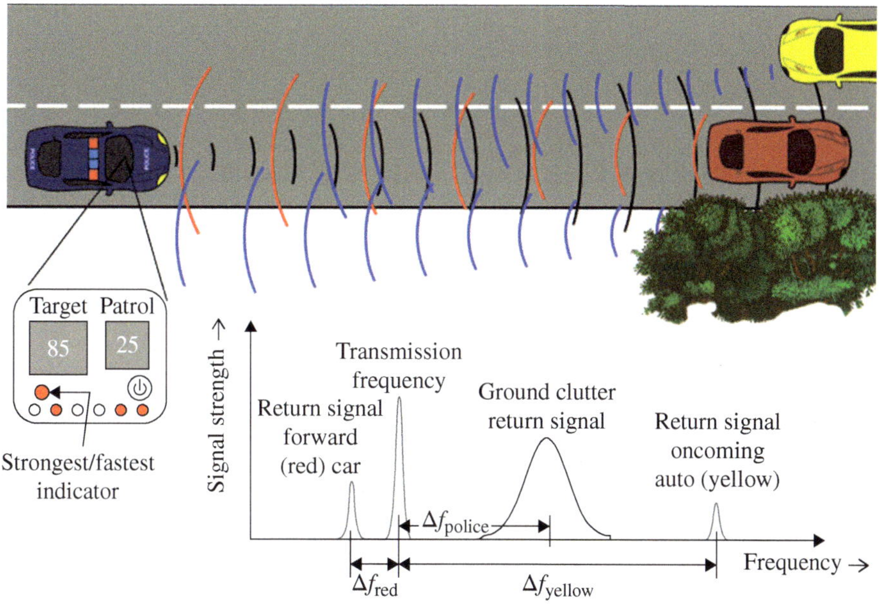

FIGURE 3.11 A microwave beam is emitted from a moving police vehicle (black wavefronts). The return signal has several components, each reflected from various objects on and along the road and each with a corresponding Doppler shift (Δf). Wavefronts spaced closer together (higher frequency) compared the transmission frequency are blue shifted, while those with larger separations are red shifted.

Normally, the transmitter frequency produces the highest peak since part of the transmission signal goes directly to the receiver. Return signal peaks to the right of the reference transmission frequency are referred to as blue shifted, while those to the left are red shifted. (A color shift refers to a change in wavelength, $\lambda = c/f$, where c is the speed of light.) The second most prominent return peak comes from random stationary objects along the side of the road. This spectral feature is usually broad because it represents the return from numerous objects at various distances, some of which have small motions such as leafy branches flexing in the wind. The frequency difference (Δf_{police}) between the reference (transmission) peak and the central peak of stationary objects, which is always blue shifted for front-pointing guns, is used to calculate the speed of the patrol car. On-coming traffic also produces a blue-shifted peak, while the return signals from the same-lane traffic can be either red or blue shifted depending on the relative speeds of the forward vehicle and the police car. As depicted, the red automobile is traveling faster than the patrol vehicle, producing a red-shifted peaked return signal (i.e., wavefronts with larger separations than those of the transmitted signal).

All radar guns have several configuration modes with indicators showing the settings at a glance. We depict the configuration status with light-emitting diodes (LEDs). Most latter models use its digital display to set and show the modes. An important choice selects between identifying velocity of the fastest object or the one with the strongest peak that is usually associated with the closest vehicle. (Most large trucks, however, also return stronger signals than do most automobiles.) The significance of this mode choice is denoted by the fact that its LED indicator is usually larger than the others as well as placed in a prominent location on the back of the gun. The panel display shows the speed of the object, denoted as the target speed, and the speed of the patrol car that might be moving. The actual speed of the target is the speed of the object minus the speed of the patrol car. This value can either be the raw speed of the vehicle with the officer mentally subtracting his own speed, or more often the net (actual) speed that has already been subtracted electronically. The officer also usually has to adjust his velocity estimate according to the *cosine-theta factor* for any vehicle moving at oblique angles. Hills and curves in the road result in cosine-theta factors. (See Intro Physics Flashback FB3.2.)

INTRO PHYSICS FLASHBACK FB3.2

Most people who have waited for a train at a railroad crossing have experienced the change in pitch of the train's horn between its approach and its passing. Constant tones are simple pressure waves, consisting of a series of compression fronts and rarefaction troughs propagating through the air at a constant velocity. If the train horn is approaching, then the distance between each subsequent compression front becomes shortened compared to the spacing when the horn is at rest, since the train is moving closer and closer to you as depicted in Figure FB3.4. After the train has passed, it is moving away from you, causing the distance between subsequent compression fronts to become larger than when the train horn was at rest. The observed difference in pitch or frequency of wavefronts is known as the *Doppler effect* and has numerous applications in modern devices. (The same phenomenon can be observed if a horn is at rest and you, the observer, is the one moving at a constant speed past the source.)

It is important to note that the Doppler effect only applies to the component of velocity approaching or receding from the observer. Consider listening to the horn of a train moving on a large circular track where you are standing at the center of the circle. No Doppler shift will be observed since the train is neither approaching nor moving away from you; it remains at one radius from you at all times. If you subsequently move in one direction, say by a 0.5 radius, some Doppler effect will be observed as the train approaches and recedes from the track location closest to you. The amount of Doppler shifts at the 0.5 radius location, however, will be less than those observed at a place very close to the circular track. Similarly, the automobile closest to a railroad crossing hears the greatest shift in the horn's tone with each subsequent car waiting in the queue hearing a slightly smaller and smaller Doppler effect. The fact that the Doppler shift only applies to the radial component of a velocity is also known as the cosine-theta effect.

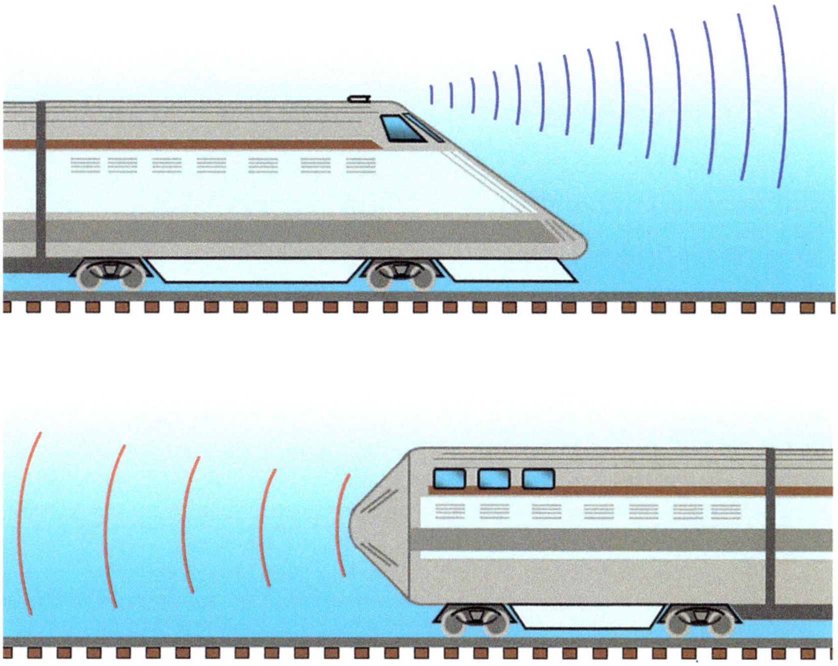

FIGURE FB3.4 The phenomenon of Doppler effect for a horn on a train.

The Doppler effect can be measured for any type of wave. Astronomers, for instance, have been using light (electromagnetic waves) to determine the relative line-of-sight velocities of stars and galaxies for more than a century. Stars emit a continuum of light, punctuated by certain specific colors (wavelengths) that are darker or brighter than adjacent wavelengths. These features, the "UPC Bar Code" impressed in the spectrum correspond to the chemical makeup of the outermost layers of the star. In other words, each type of atom absorbs and emits at a uniquely set of specific wavelengths. Many physical properties of each star such as its temperature, atmospheric pressure, rotation, and radial velocity can be determined spectroscopically by measuring in detail the shape of the absorption profiles. Also, the component along the sightline of the star's relative velocity can be measured from the shift in wavelength of the feature's center. (These concepts will be discussed in detail in Chapter 9 on light sources and in Chapter 12 on spectrographs.) Figure FB3.5 shows schematically the process by which the radial velocities (the cos θ) of three objects are measured.

The velocity of each astronomical object is shown in Figure FB3.5 as a dashed-line arrow with a length indicating its relative speed. In this example, Star 1 is moving away from the observer (red shifted), while Star 2 has a larger overall velocity but only a small portion of its velocity is in the direction toward the observer (blue shifted). The Moon is shown in Figure FB3.5 as moving tangential to the observer's line of sight. It reflects light from the Sun so the Moon reflects

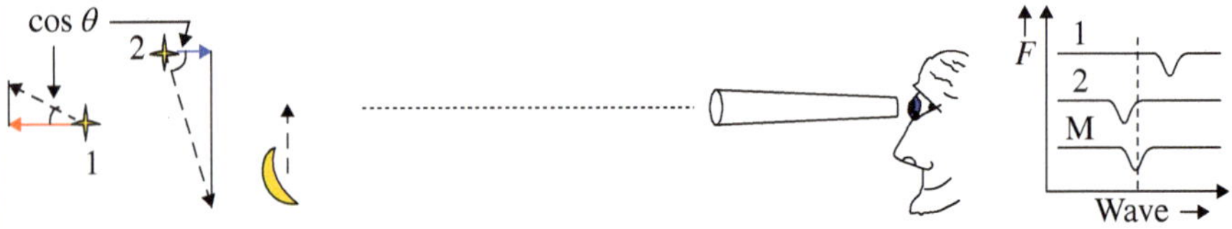

FIGURE FB3.5 The basic method that astronomers use to measure the radial velocities (i.e., the component of velocity that is towards or away from the observer) of stars and other astronomical objects. A dashed-line arrow denotes each velocity.

the same spectral features as the Sun. *Note*: all of the astronomical objects depicted in Figure FB3.5 are so far away from Earth that their relative locations cannot be placed to scale on any display medium. Thus, all would be in the same field of view and all could be observed simultaneously without tipping the telescope. The same portion of spectra of all three objects with offsets for clarity is graphically represented at the right. A vertical dashed line marks the rest wavelength (i.e., no radial velocity component). As can be seen, the spectral feature for the Moon is centered on this wavelength; Star 1 has the largest Doppler effect, red shifted in this case; Star 2 is blue shifted, but not as much as Star 1 is red shifted.

The framework of a hand-held radar gun is shown in Figure 3.12 with cross-sectional diagrams viewed from the side and back. Some of the internal components such as the batteries in the handle, the antenna horn, and the electronics card are shown as dashed or dotted lines. The operation of this gun is very similar to the dashboard-mounted version, except hand-held guns are normally employed while stopped. A trigger on the handle activates a continuous wave beam, which locks the last measured speed when released. The back end of the gun is simpler, only displaying the speed of the object. Portable speed guns are often equipped with an audible alarm, alerting the operator that a vehicle with a speed exceeding his threshold has been detected. The officer can adjust this threshold speed from one speed-trap location to another. This feature enables the officer to focus on the traffic rather than to be preoccupied by staring at the digital display. If there are multiple vehicles in his range, the officer has to determine visually which one is offending. The range control is separate from the audible maximum speed threshold, which will be discussed shortly.

Up to this point, we have been discussing a highly idealized operation of a radar gun. In reality, the raw return signal is more complex and messy, consisting of a series of weak peaks against a continuum containing noise, especially for dashboard models. For traffic control, a radar gun operator has to have specialized training to produce meaningful readings. He or she must know how to interpret the results based on the radar gun's configuration and the traffic conditions. Moreover, the officer must be able to select the correct operating modes that are appropriate for each traffic situation and terrain. Poor choices will lead to bogus results and false returns. The anatomy of a speed radar gun along with a schematic block diagram of its circuitry is given in Figure 3.13. A typical raw return spectrum is also plotted in the lower left,

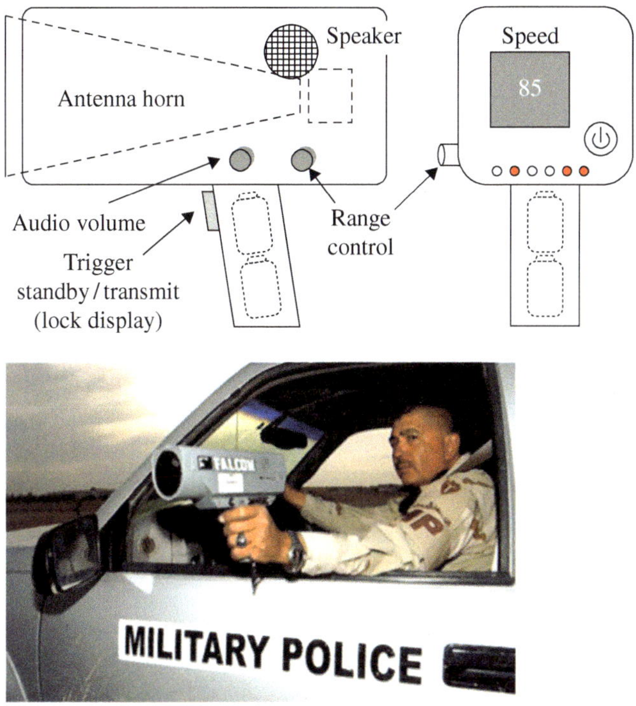

FIGURE 3.12 Basic components of a common portable speed gun. Source: US Army Master Sgt. Lek Mateo. Picture of Sgt. Jessie DeLarosa at Tallil Air Base, Iraq.

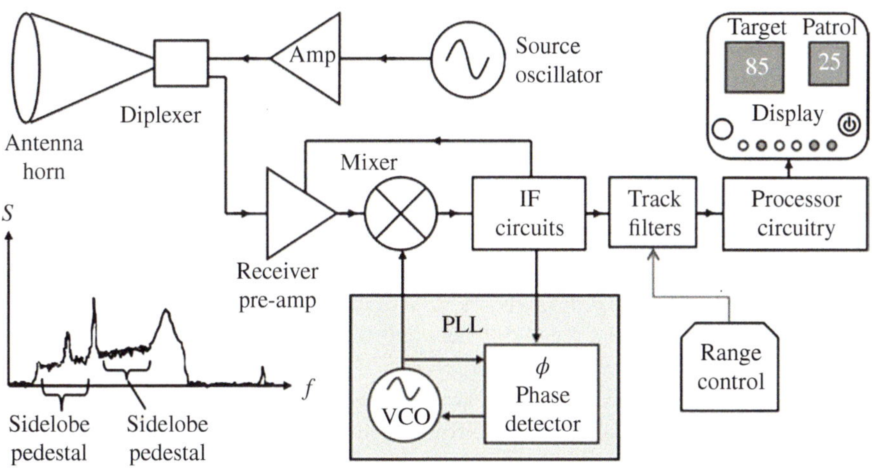

FIGURE 3.13 Block diagram comprising key electronics components of a speed radar gun. Plotted at the lower left is the return signal, S, as a function of frequency, f, including noise and clutter scattering. The signal forms a spectrum consisting of discrete peaks and a continuum.

revealing the challenges for the gun's electronics processor and the operator. System noise comprises not only random fluctuations but also a pedestal signal from side-lobe clutter caused by signal reflections from objects well outside the central beam. Recall the primary central beam encompasses 80–85% of the energy, leaving 15–20% of the energy to scatter from angles much larger than the nominal 12° cone.

The key components of a dashboard-mounted speed radar gun start with the source oscillator, shown at the top center of Figure 3.13. Its oscillatory wave is sent to a power amplifier and subsequently to the diplexer attached to the transmitting/receiving antenna horn. The diplexer is a passive device, which sends virtually the entire signal out the antenna horn and directs the returning reflected microwaves into receiving/processor circuitry, which first goes through the receiver preamplifier, a low-noise amplifier. The signal direction of the diplexer is not perfect, however, and a small fraction of the outgoing oscillation couples into the receiver circuitry, where it is used as a reference. Next the high-frequency microwave signals must be converted into a corresponding set of intermediate-frequency (IF) signals. The IF circuitry reproduces the receiver signal spectrum but at a lower range of frequencies. This conversion effectively spreads the signal spectrum along the horizontal axis (the f axis). The IF circuits also sample the single frequency of the source oscillation, sending a portion of it to the phase-lock loop (PLL). A PLL, a common circuit found in numerous applications, is used to determine small drifts in a single-frequency oscillation, track it, and to return a DC voltage that can be used to compensate for any drifts. A PLL consists of two primary components, a phase detector that generates a DC voltage proportional to the small difference in the frequencies, Δf, between two oscillations, and a voltage controlled oscillator (VCO), which creates an oscillatory wave with a frequency proportional to its input DC voltage. The output of the VCO is fed back into the phase detector, causing the VCO to lock its frequency to that of the source oscillator, following any frequency drifts. The DC voltage of the phase detector can be used to calibrate the IF spectrum continuously. A mixer, depicted as an "X" inside a circle, combines the RF (radio frequency) of the source oscillator with the corresponding single-frequency from the VCO. The IF circuitry, now locked to the source oscillator, down converts the RF spectrum to an intermediate-frequency spectrum. The IF circuit also sends a feedback signal to the preamplifier to adjust its gain, continually regulating the amplitude so that the features are neither too weak nor saturated.

Finally, the range control of a radar gun is the input to the track filters, which eliminate (filter out) all portions of the spectrum except for one or two small intervals chosen by the operator. In other words, the radar gun has range controls, selecting one or two ranges of velocities for processing while reducing the amplitude of the rest of the spectrum to zero. For example, consider the very weak blue-shifted return signal from on-coming traffic, shown at the right of the S–f plot in Figure 3.13. This small feature corresponds to the return from yellow automobile in Figure 3.11 with a velocity proportional to Δf_{yellow} minus Δf_{police}. As the yellow car approaches, its return signal first increases due to an increasing $1/r^2$ cross-sectional projection and then decreases as the vehicle moves out of the cone of the central beam. A range of distances and velocities can be measured for each lane of traffic. The track filters isolates a range of frequencies of interest and sends this portion of the spectrum to the digital processor. If two return signals are present in this range, the processor determines which one of the two to display.

Police radar guns are calibrated using a tuning fork. The gun often has a switch to put it into a test mode and the tuning fork must be appropriately matched to the radar's frequency. An invalid tuning fork can lead to errors of as much as 14% (i.e., registering 57 km/s for a vehicle actually traveling 50 km/s).

A Lidar speed gun, which is an alternative to radar gun, relies on the principle of time of flight between several pulses of a laser beam. Its functionality resembles that of a GPS system, except only the relative positions are needed to calculate a velocity. When the trigger is pulled, a 30 ns pulse is sent followed by additional pulses every millisecond. Each subsequent pulse arrives earlier than the previous one for a vehicle approaching. A principle advantage of a Lidar speed gun is the tight pencil beam (a 4-milliradian cone), avoiding the confusion multiple returns from several objects. Lidar's principle disadvantage is the same tight pencil beam, requiring the officer to use a telescopic sighting scope. For this reason, radar speed guns continue to be widely used.

3.4.4 High-Speed Rail

The transportation of people via high-speed rail (HSR) is advantageous compared to air transport, especially for distances of 200–1000 km (~130–625 mi.) where airport security and check-in processes offset the higher maximum speeds of air travel. (The fastest way to travel between London and Paris is via HSR through the Channel Tunnel.) HRS travel is less often impacted by adverse weather conditions, compared to air travel. In addition, a typical European HSR has a maximum passenger capacity that is 13% greater than a six-lane highway, even though it requires only 40% as much land. More importantly, the energy efficiency of HSR is at least four times better than, for example, a Boeing 737 and twice that of a typical midsize automobile with two passengers, assuming a train ridership that is 60% of capacity. The maximum speed that a freight or passenger train can safely traverse is set by a number of physical conditions including track quality such as smoothness and track curvature as well as external factors such as crossing signal controls. In the United States, passenger trains are limited to 59 mph and freight trains to 49 mph on tracks without block signal systems.

A number of improvements to the railroad itself are necessary for HSR. For example, HSR requires smooth tracks and curves with a large turning radius (i.e., not a tight curve). Continuous welded rail, which is perhaps the most significant modern roadbed technology, is used to form a continuous several-kilometers-long rail, providing a smooth ride, increased rail strength, and needs less maintenance than traditional rail. Continuous welded rail, also referred to as ribbon rails, reduces friction between the train wheel and the rail as well as avoids misalignments between sections and decrease vibrations. Conventional rail from the nineteenth and twentieth centuries used sections of rail (12–24 m in length) that were connected together with short pieces called fishplates that bolted onto the side of the two rail segments. Small gaps are left between each rail section to allow for thermal expansion, which give trains passing over jointed tracks their "clickety-clack" sound. In addition, HSR use advanced switches with very low entry and frog angles. (Frog angles are the number of degrees between two sets of track that are crossing each other.)

Moreover, high-speed trains employ a number of key technologies to improve efficiencies and prevent the train from coming off the rails. These include improved aerodynamic design to reduce drag from air friction, but avoiding excess (wing-effect) lift. Noise is also reduced at high speeds with improved aerodynamic designs. Other technologies to improve efficiency and safety are regenerative braking and dynamic weight shifting. As discussed previously, regenerative braking converts the kinetic energy of motion into electrical energy, a highly effective way of slowing down. A weight-shifting controller senses when one set of wheels such as the outside wheels during a turn are carrying a larger percentage of the weight than normal and moves a central massive block to rebalance. This action stabilizes the train and mitigates any adverse impact of centripetal or centrifugal forces generated when the train is traveling somewhat too slowly or fast on a banked curve.

4

VACUUM SYSTEMS: ENABLING HIGH-TECH INDUSTRIES

Vacuum technology is needed for a wide variety of manufacturing methods, many of which only require rough (i.e., low) vacuums that are easily achieved. These include, among others, the production of light bulbs, the molding of composite plastic components, some flight instruments for aircraft, the hard coatings of engines used in Formula One racecars, and removing contaminants from air conditioning systems before charging with refrigerants.

Other fabrication applications require high vacuum ($<10^{-6}$ of an atmosphere) or ultrahigh vacuum (UHV; $<10^{-10}$ atm) systems with particular attention being paid to contaminants. Instruments and electronics operated in outer space all need to be tested in HV conditions prior to launch. The very best vacuums are required for semiconductor processing (especially for ion implantation), ultraviolet photolithography, and molecular beam epitaxy (MBE) growth of thin films. Moreover medical devices used in radiation treatments (e.g., high-power Klystrons, LINAC) all require UHVs. Other UHV environments are needed for the operations of X-ray photoelectron spectrographs, mass spectrometers, electron microscopes, Auger electron spectroscopes, and particle accelerators, among others. Thus establishing an excellent vacuum, while having all of the necessary equipment inside it is a critical, enabling technology for many high-tech instruments and industries.

The SI unit for pressure is the Pascal (Pa) where $1\,\text{Pa}=1\,\text{N/m}^2$ ($\text{kg/(m s}^2)$) and one standard atmosphere of pressure, P_{atm} is approximately 100 kPa. A pressure measurement can be referenced either as an absolute pressure, where a volume containing no atoms or molecules has a $P=0$, or compared to atmospheric pressure, where the over pressure is the total value minus 1 atm. Pressures greater than 1 atm use the latter reference and tend

Modern Devices: The Simple Physics of Sophisticated Technology, First Edition.
Charles L. Joseph and Santiago Bernal.

Companion website: www.wiley.com/go/joseph/moderndevices

to be measured in the United States and Canada in pounds per square inch (psi) or in Europe in kilogram-force per square centimeter. Tire pressures, for example, are measured as the net overpressure above atmosphere. Consequently, the measured tire pressure will be higher on the top of a mountain than it was on the valley floor. That same tire pressure will be higher on a hot summer afternoon than on a cool, early morning.

Vacuum pressures are most often given in absolute pressures and measured in units of Torr (mm of mercury) or in millibar, where 1 mb ~ 1 Torr. For laboratory vacuum systems, lower pressures indicate higher quality or harder vacuums. There are no universal designations for various levels of vacuum pressures or the exact range of pressures associated with each, but generally UHVs have pressures of less than 10^{-8} Torr, high vacuums have $10^{-8}<P<10^{-4}$ Torr, and rough or low vacuums are $10^{-4}<P<760$ Torr (1 standard atm).

INTERESTING TIDBIT TB4.1

The number of molecules in a very small 1 L chamber at one atmosphere of pressure is 6×10^{23}, Avogadro's number. The best vacuums that can be achieved in the laboratory is approximately 10^{11} Torr (10^{-17} atm), which means there are still six million molecules roaming around in this small chamber. The vacuum of outer space is a 1000 times better than anything that can be achieved on Earth, having 1 atom/cm^3.

INTERESTING TIDBIT TB4.2

Otto Von Guericke invented a vacuum pump in 1650. To demonstrate the force of air pressure, he joined two hemispheres having a diameter of 0.51 m, pumped the air out, and then harnessed a team of eight horses to each hemisphere. Von Guericke showed the horses could not pull the enclosure apart, but the two hemispheres were easily separated once the air was replaced. He repeated the challenge 3 years later using 24 horses with the same result.

4.1 VACUUM CHAMBER TECHNOLOGY

Creating a vacuum is simply a matter of pumping the gasses out of a sealed container, known as a chamber or tank. One of the most familiar devices for creating a vacuum is the glass bell jar, which is still used. Figure 4.1 depicts a typical large, floor-standing vacuum chamber. *Note*: a common alternative configuration for a large, cylindrical vacuum tank has a length greater than its diameter and often has a horizontal axis of rotation. These very large chambers are usually made of stainless steel rather than glass, giving the chamber designer the freedom to position access ports and windows of various sizes as well electrical, mechanical, or other feedthough flanges to accommodate the specific operational needs for that chamber. Some of

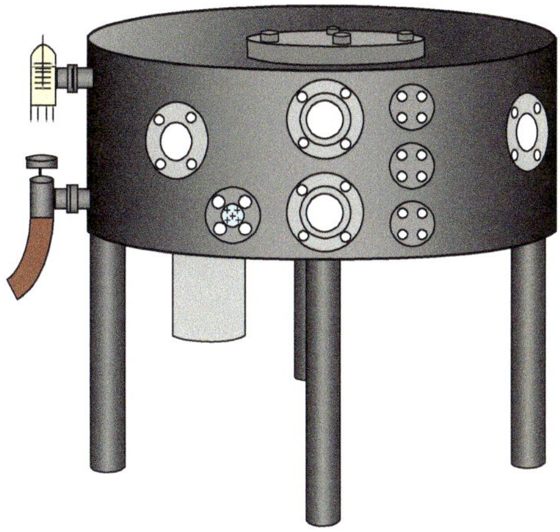

FIGURE 4.1 A large floor-standing vacuum chamber. At the left is an ion vacuum gauge (top) and valve with rubber hose to roughing pump (bottom). Numerous access ports (electrical feedthroughs, window ports, and blanks) are shown on the circumference. A cryopump is shown attached underneath chamber.

these flange ports may have solid blanks, space holders that can be used at a future time for introducing new attachments with expanded or different functionalities.

In Figure 4.1, a valve connected to a rubber hose that goes to the roughing pump is shown at the left (bottom). Above this valve is an ion gauge used to measure the residual pressure once HV or UHV is established. At the bottom and toward the back of the tank (light gray) is the chamber's main vacuum pump, which are frequently located beneath the chamber. There normally is an inside baffle just above the main pump to protect it from tools or other foreign objects falling into it.

HV and UHV chambers are pumped to working vacuums in two stages with a crossover pressure (CP) being determined by the chamber size and the maximum rate of the main pump. First, most of the air and gases are removed with a roughing pump, which establishes a low vacuum by removing more than 99.9% of the gas. Next, the main pump takes over, bringing the chamber down to its operational vacuum level and then maintaining it. The CP can be calculated by $CP = CV/V$, where CV is the crossover pumping speed in Torr-liters and V is the volume of the chamber. For virtually all HV or UHV applications, the main pump will be quickly overwhelmed, stop functioning, and maybe damaged, if the inrush of residual gas is too large. As a precaution against premature exposure of the main pump to the chamber, tanks generally have a minimum of two (and usually three) types of pressure gauges and an electronic controller capable of closing various valves. The operating ranges for some pumps and gauges are given in Figure 4.2, none of which are capable of spanning the 15 orders of magnitude in vacuum pressures of laboratory chambers. These ranges are set by the physics of the pump and gauge designs, a few of which are discussed in latter sections.

The level of vacuum that can be established and maintained by a chamber is set by the balance between the pumping speed of the main pump and the collective leak

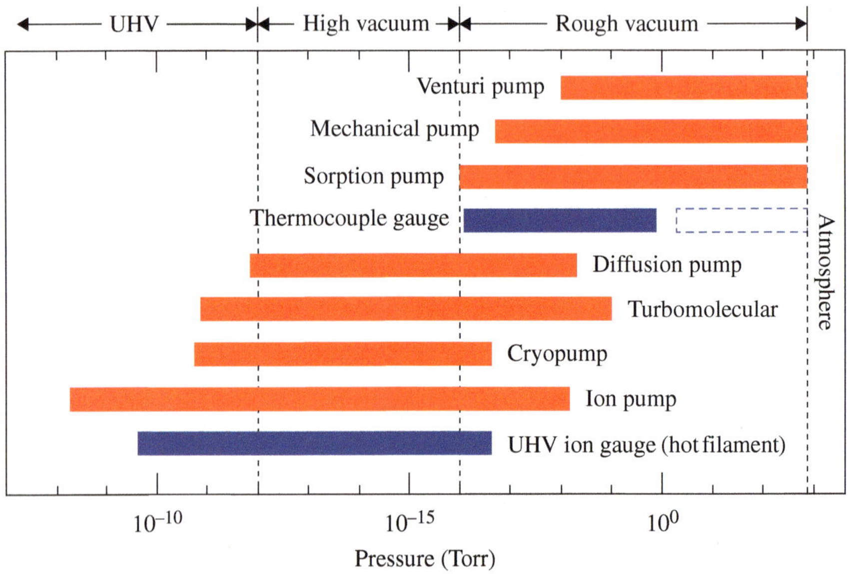

FIGURE 4.2 The normal operating ranges of various type of pumps (solid bars) and gauges (patterned bars). The three classifications of vacuum are shown at the top.

rates from all of the seals associated with ports, flanges, and feedthroughs. All vacuum chambers, even the most tightly sealed ones have microleaks that allow minute amounts of gas to seep back slowly into the cavity. Apart from tank defects, the primary sources of microleaks originate in the seals associated with the attachment of blanks, flanges, and ports. Obviously, a given pump cannot achieve the lowest pressure displayed in Figure 4.2 if the chamber has an unusually large number of ports or if the some of the port seals are of poor quality. Conversely, somewhat harder vacuums than those listed can be realized for particularly well-sealed chambers, especially if additional measures are employed.

There are two basic types of seals used for connecting two vacuum tank pieces together: (i) rubber or viton O-rings pinched between two metal surfaces and (ii) copper or silver-plated copper gaskets sandwiched between two surfaces with hard knife edges. Figure 4.3 shows a cross-sectional diagram of these two types of sealing surfaces. Vacuum tanks often employ a combination of O-ring and gasket seals. In the best UHV chambers, any fixtures having an O-ring are isolated from the main chamber by a UHV valve, relegating O-rings exclusively to the initial roughing stage.

While there are several standard configurations for O-rings and gaskets, as well as a number of vacuum quick-connection flange systems, we consider here only two examples for simplicity: the ASA O-ring and the CF Conflat® gasket systems. A sample of these is pictured in Figure 4.4. For ASA-style seals, one component part has a rectangular-shaped profile cut into its surface to hold the O-ring. This groove has to have a smooth surface and be shallower than the cross-sectional diameter of the O-ring, but sufficiently wide that the deformed O-ring does not contact both sides of the channel.

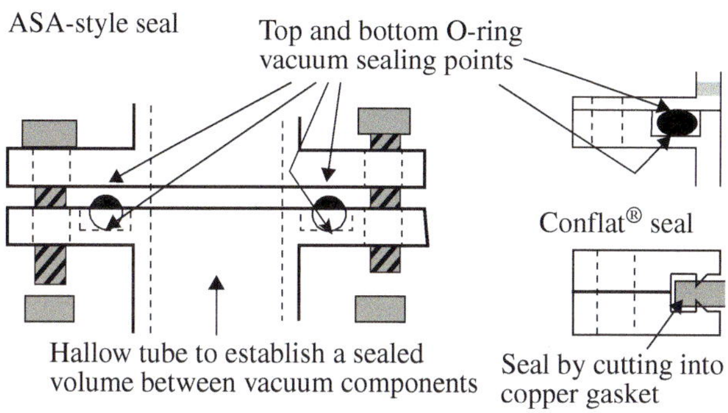

FIGURE 4.3 Cross-sectional diagrams of the two type of vacuum sealing mechanisms. The sealing surfaces as depicted are on the top and bottom surfaces. The groove must be wide enough to allow the O-ring to deform, making a seal. For a gasket, the harder knife-edge flange cuts a sealing grove into the softer gasket material.

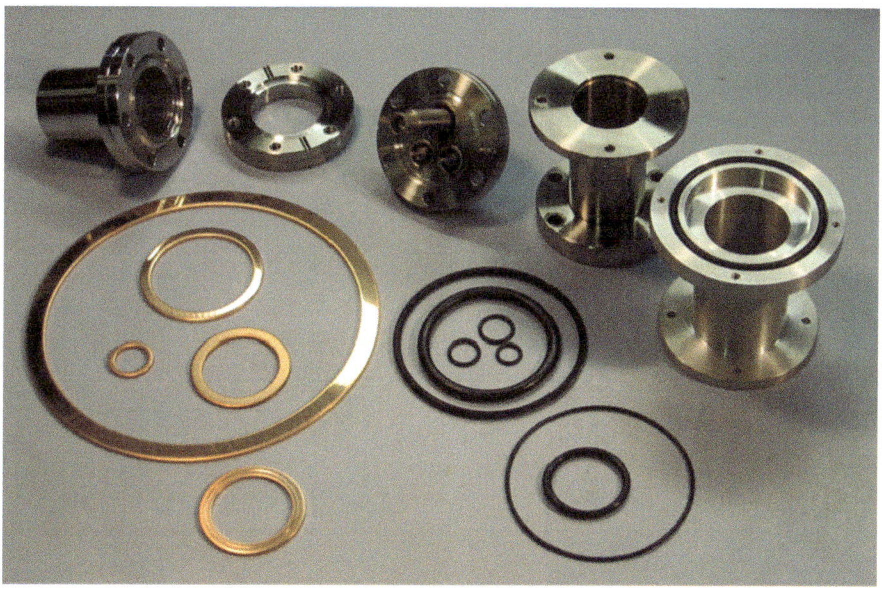

FIGURE 4.4 An assortment of O-rings and copper gaskets along with a flanges. One feedthrough flange with three electrical connectors is shown at top center.

(ASA grooves come with length, depth, and width specifications for each O-ring.) The other component to be vacuum mated has a polished-smoothed flat surface. When the two parts are bolted together, the O-ring becomes compressed into an ovular shape with two vacuum sealing points being on the top and bottom of the O-ring as shown in Figure 4.3. ASA rubber O-rings may be reused multiple times, but eventually wear out once the rubber begins to lose its elasticity. A vacuum tank using O-rings can achieve HV but rarely UHV operation provided it has a sufficiently large pump.

Copper gaskets must be used to achieve the best UHV operation, although one or two O-ring seals might be employed initially during the preliminary rough-out pumping. Copper, being a relatively soft metal, is used with stainless steel (SS) Conflat flanges in high or ultrahigh vacuum systems. When two CF flanges are bolted together, the SS knife-edges deform both surfaces of the copper under a plastic flow. The copper in the gasket fills in any imperfections in the knife-edge surfaces while becoming work hardened. The spring-like quality of the work-hardened copper maintains an excellent seal over a wide range of temperatures from −196 to 450°C. However, if the gasket is to be exposed to elevated bake temperatures repeatedly, silver-coated gaskets should be used to avoid oxidation of the copper, which may compromise its effectiveness. Metal gaskets are normally only used once and replaced each time the seal needs to be broken.

Contamination control is a critical issue, especially for UHV applications. Once a HV is established, any contaminants slowly boil off, spreading and attaching these foreign materials to all surfaces inside the chamber. In some cases, the introduction of a "dirty" component into a UHV chamber can permanently degrade the chamber's ability to maintain UHV. No combination of cleaning with solvents and vacuum baking will restore the chamber to its original condition.

There are two primary sources of undesirable foreign materials: (i) residual hydrocarbons (oils and greases) on the surfaces of equipment used in the chamber that have not been adequately cleaned and vacuum baked, and (ii) the introduction of equipment inside the chamber that are not rated for HV applications. Many everyday materials (especially most plastics, tapes, wire insulation, glues, etc.) are not rated for vacuum systems. These materials continually outgas harmful vapors for months or years under vacuum conditions, eventually becoming brittle. Hydrocarbons, including the oil from humans, are among the most difficult to remove from a vacuum chamber once introduced. Objects and equipment are normally cleaned in a multistep process, which differs from one vacuum system to the next and which varies according to the object needing to be cleaned. A metallic part, for example, may start simply with soap and water. Once the piece is sufficiently free of dirt and oils, repeated cleaning of the part with acetone followed with isopropanol will remove virtually all remaining hydrocarbons. (One must use caution when using solvents since various solvents may chemically attack some vacuum-qualified parts.) The most effective cleaning method uses an ultrasonic cleaner, which bathes the part in the solvent (acetone or isopropanol) while vibrating the part with ultrasound waves. To prevent recontamination, nitrile gloves as well as protective gowns, masks, or other garments are often used in handling vacuum equipment.

If a vacuum chamber sits idle for any length of time, considerable amounts of contaminants (especially H_2O) can be adsorbed onto the interior surface of the chamber. As the tank is pumped, these contaminants boil off from the interior surfaces, initially retarding the speed of achieving the desired vacuum levels. For this reason, the chamber and equipment to be placed inside the chamber are often heated under very low pressure, a process known as vacuum baking.

Pumps are classified into two types: gas transfer and gas capture. A foreline pump must first be used on all HV or UHV chambers to achieve a rough vacuum. In a gas transfer type, a foreline pump is used to remove the exhaust from the main pump. An advantage of the gas transfer type is its ability to operate for prolonged periods of time (months). Gas transfer pumps can be subdivided into two categories: displacement and momentum transfer. The simplest pumping system to consider is a mechanical gas displacement type such as a piston in cylinder. During each cycle, the piston collects some residual gas from the chamber and then compresses it to a pressure greater than 1 atm where it vents to the atmosphere. Mechanical pumps, however, are only used to establish rough vacuums and used as foreline pumps. A second type of gas transfer pump imparts momentum to gas molecules, forcing these to move in one direction and concentrating the gas in a small volume. These pumps can only be used as HV or UHV pumps. An example of a momentum transfer pump is the turbomolecular pump, which uses high-speed turbine blades to force molecules to one side. The small volume that is slightly overpressurized by the turbine blades is normally connected to a mechanical gas transfer pump, which then moves the gas to the external environment. The other class of vacuum pump, a gas capture type, simply traps or freezes residual gas that comes in contact with the pump. Gas capture pumps are very effective, but eventually become full and cannot adsorb any additional gas. At that point, the pump has to be taken offline and rejuvenated, usually baked out before it can be placed back in service. One exception, the ion pump, a highly efficient capture type, achieves UHVs under ideal conditions down to 1×10^{-12} Torr ($\sim10^{-18}$ atm) and can operate for extend periods. However, when the ion pump requires rejuvenation, it is a major overhaul rather than a simple in-house procedure.

COMPREHENSION VERIFICATION CV4.1

Calculate the net force of the air pressure on a vacuum chamber lid that is 1.5 m in diameter when 90% of the air is removed. Then calculate the same net force when the vacuum pressure is 1% of atmosphere and for 0.001 of atmosphere (~1 Torr). Do these values represent a lot of force? Relate this force to some benchmark value.

Solution: the surface area is approximately $2\pi r^2$ where $r = 0.75$ m. Atmospheric pressure is 101 kPa or (14.9 psi). The outward pressure from the partial vacuum inside the tank is 1/10th that of the external, resulting in an effective pressure of 0.9. Thus, the net force is

$$F_{\text{Net}} = 0.9 \times A \times p = 0.9 \times 3.53 \text{ m}^2 \times 101 \text{ kPa} = 321 \text{ kN} = 321{,}000 \text{ N}. \qquad \text{(CV4.1)}$$

To calculate the net force for a vacuum pressure that is 1% of atmosphere, substitute 0.99 for 0.9 in the equation. Thus, $F_{\text{Net}} = 352{,}000$ N and $F_{\text{Net}} = 356{,}000$ N, for 0.01 and 0.001 of atmosphere, respectively. Observation: no matter how much harder of a vacuum that is pulled, the total force on the surface will not rise above 356,000 N.

How much is 350,000 N? A typical young adult male in the United States weighs about 170 lb or equivalently approximately 77 N. Thus, the force on our vacuum tank lid is equivalent to about 4500 young men being supported by it. A large, fully loaded semi tractor/trailer on US highways weighs 20 tons (~18,000 N) so our lid is supporting close to the equivalent of 19 large trucks. Is it any wonder that teams of horses could not pull apart the two components of a vacuum chamber? (See the tidbit on Otto Von Guericke.)

4.2 PHYSICS OF SOME VACUUM GAUGES

A **mechanical gauge** is often used in the initial pump-down stages. Sometimes, its sole function is to indicate whether or not the roughing pump has been started and the necessary valves have been closed. These gauges usually are narrow tubes sealed at the dial end by a thin metal plate, which flexes against the restorative force of this thin piece. The flexure is in one direction if the tank pressure is greater than the room pressure and in the opposite direction if the chamber is below atmosphere. The amount of flexure is proportional to the difference between the chamber and outside pressures. A circular dial is most commonly coupled to the metal surface. Mechanical gauges are capable of measuring vacuum pressures down to approximately 99.7% of an atmosphere (~2 Torr). The range of a typical mechanical gauge, denoted by the dashed-line bar in Figure 4.2, is shown along side the graphical bar for the thermocouple gauge. Note the gap in pressure measurement coverage from 0.8 to 2 Torr. Few vacuum fabrication processes require precise knowledge of pressures in this range. Those that do need to use more sophisticated and costly sensing technology.

A **thermocouple (TC) gauge**, as depicted schematically in Figure 4.5, is perhaps the most widely used, since these cover the range of crossover pressures in most HV and UHV systems and since these gauges are adequate for many HV applications. Thermocouple gauges, capable of $10^{-4} < P < 0.8$ Torr measurements, are often used with control electronics to switch automatically between the roughing and main pumps. These pressure sensors and control electronics protect against premature exposure of the tank to the main pump and quickly isolate the pump from the chamber should a problem develop.

The physics behind a TC can be understood in terms of the responses of various metal alloys to temperature. When two ends of a wire are held at two different temperatures, a small voltage potential is observed between the two ends. If two separate alloy wires are subjected to the same disparity of temperatures, one will generate a

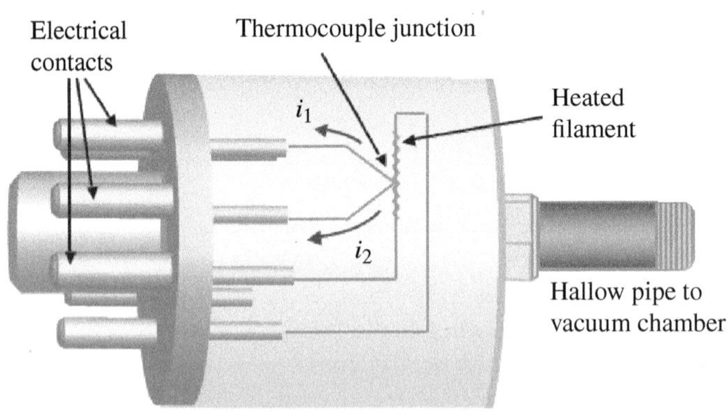

FIGURE 4.5 The anatomy of a thermocouple (TC) gauge. The interior volume of the gauge has the same vacuum as the chamber, usually being connected through a hallow pipe (right) with a threaded end. The resistance of the TC is set by rate of cooling, which is proportional to the amount of residual gas.

slightly higher voltage than the other, a difference measured on millivolt (mV) or microvolt (μV) scales. A thermocouple junction is created if the two ends sharing a common temperature are also connected electrically. As shown in Figure 4.5, two wires of separate alloys are brought together to form a thermocouple junction and connected to a heated filament. Each wire generates a distinctly separate voltage potential, but in a manner that produces opposing currents, $i_1 < i_2$ in the figure. The net current $(i_2 - i_1)$ accurately measures the temperature of the thermocouple junction. This device is transformed into a vacuum pressure measurement by the following method. Energy is added continually at a fixed rate to the heating filament. At the same time, convective cooling from gaseous molecules coming in contact with the filament sets the amount of cooling of thermocouple junction. The amount of cooling goes down as the density of gas molecules drops. That is, the temperature of the filament increases as the pressure drops, since the energy being supplied to the filament is constant while the efficiency of energy removal is being reduced.

There are several combinations of alloys commonly used to make thermocouple junctions. The most frequently used in HV chambers is the K type, consisting of Chromel versus Alumel wires. (Chromel is an alloy of 90% nickel and 10% chromium, while Alumel is 95% Ni, 2% Mn, 2% Al, and 1% Si.) *Note*: most vendors of thermocouple vacuum gauges quote 10^{-3} Torr for the lowest pressure that can accurately be measured. Indeed, this is the lowest pressure where the thermocouple and associated electronics produce linear and precise measurements. Nevertheless, a K-type thermocouple remains reasonably accurate, but with less precision and less linearity, down to 10^{-4} Torr.

The UHV sensor of choice is the **hot cathode ion gauge** illustrated in Figure 4.6. It consists of three electrodes all functioning as a triode. The voltage across the resistive hot filament is typically 30 Vdc and generates a 10 mA (0.01 Amps) current of thermionic free electrons. These free electrons are attracted toward the grid, which biased at approximately +150 to +200 Vdc. The grid collects some electrons, but most pass between the individual coils with more than enough energy to ionize any

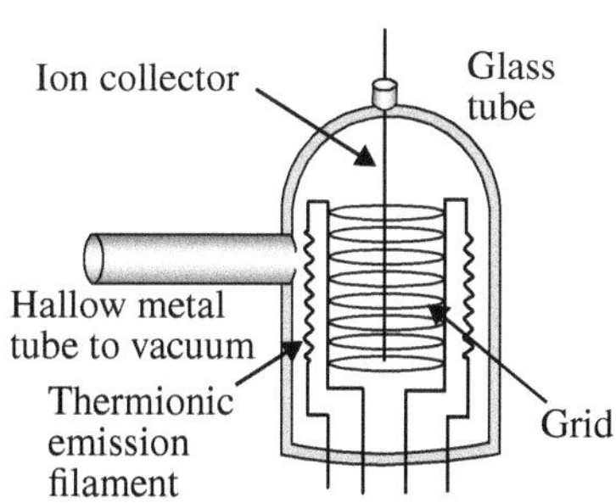

FIGURE 4.6 A hot cathode ion gage functions by passing a current and resulting voltage drop through a resistive material that heats up, emitting electrons into the vacuum. A series of rings connected to the positive volt side of the DC voltage, accelerating the free electrons toward the center. While these rings collect some electrons, many pass through, ionizing the residual gas. The current between the ion collector and the grid is proportional to the residual pressure.

residual gas molecules encountered in the interior. Once ionized, the positively-charged molecules drift towards the ion collector, the central anode wire biased at 0 Vdc. The ions pick up an electron from the central wire, returning the molecule to its neutral state. The current supplied to the ion collector to replenish its lost electrons is measured in picoamps ($\times 10^{-12}$ A) and is proportional to density (pressure) of the molecules in the gauge.

The hot cathode ion gauge, depicted in Figure 4.6, is a Bayard–Alpert type. It is enclosed in a glass tube, mounted externally to the tank usually via a Conflat flange. This configuration is inherently vulnerable to catastrophic damage should any external equipment accidentally strike it. A shield, fabricated from a hard metal mesh to allow convective cooling, is sometimes built to protect the tube gauge. An alternative is to use a "nude gauge," consisting the three electrodes without the glass enclosure, inserted directly into the vacuum chamber. Nude gauges may be used if there is space available and the tank operation is not disturbed by stray light coming from the filament, which functions as an incandescent lamp.

While hot cathode ionization gauges have linear response over 10^{-4} to 10^{-9} Torr, all ion gauge measurements are seriously affected by gas composition. For example, He gas only produces 0.18 of the signal that N_2 gas does, according to a report in the *Journal of Vacuum Science Technology*. The low-pressure limit of a hot cathode ion gauge is set by soft X-ray emission originating from the grid by electrons striking it with energies of approximately 180 eV. In turn, a small portion of these X-ray photons will hit the ion collector wire, knocking off a photoelectron. The induced current of photoelectrons from the collector wire is indistinguishable from the current created by ionized molecules striking the wire. The range of pressures that can be measured by a Bayard–Alpert-type gauge can be extended down to 10^{-10} Torr simply by using an ultrathin ion collector wire, which has a smaller cross section to the soft X-rays.

4.3 LOW VACUUM VIA VENTURI, MECHANICAL, OR SORPTION PUMPS

Venturi pumps are clean, simple devices to remove 95–98% of the gases from relatively small vacuum chambers. The venturi phenomenon occurs when air or possibly any gas is forced to move more rapidly through a restriction, causing a significant drop in pressure of the gas. The same observation is made when a person blows across the top of a strip of paper, lifting it up from its limp position. Figure 4.7 is a schematic representation of a venturi pump. Typically, compressed gas (e.g., air and nitrogen) at pressures of 60–100 psi is supplied at the left. The pressure drops to a small fraction of an atmosphere as it passes through the restriction, pulling gas up from the bottom. All of the gas, both from the pressurized source and from the vacuum chamber, is then vented to the right through a muffler. The flow rates from the compressed gas range from 15 to 280 L/min (0.5–10 cfm, cubic feet per minute). Venturi pumps can often achieve vacuum levels of approximately 2500 Torr, removing 95–98% of the tank gas.

Mechanical vacuum pumps come in several designs (e.g., piston, rotary vane, and diaphragm), all being a displacement pump. Some use special vacuum-rated oils

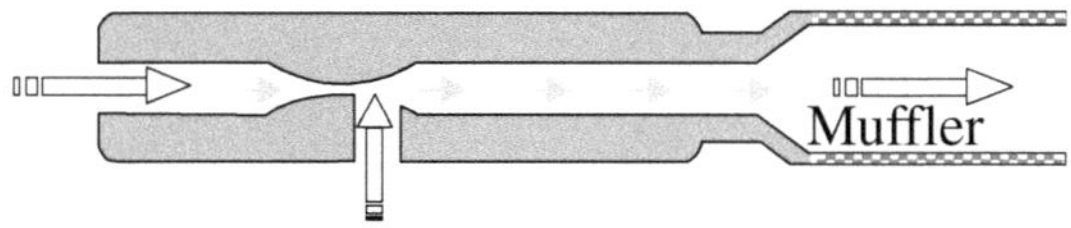

FIGURE 4.7 In a venturi pump, a gas flows through a restriction, causing the pressure to drop. An opening (bottom) pulls air from the volume to be evacuated. Only low-quality rough vacuums can be established with this device.

having low vapor pressures as an internal vacuum seal and lubricant. These are known as wet pumps, while others (e.g., diaphragm and screw type) are often operated dry. Generally, wet designs pull harder vacuums than dry ones, but trace amounts of the pump oil coat the interior walls of the connecting hoses over time, a process known as back streaming. Attaching a wet roughing pump through an extended hose and minimizing the length of time the pump operates near its lowest pressures reduces the rate of contamination. A foreline trap can also be used as a second barrier inhibiting these oils from reaching vacuum chamber. Moreover, hoses and traps should be replaced on a scheduled basis when used as roughing conduits to UHV chambers. Despite its contamination drawback, wet mechanical pumps are widely used because these devices can sustain prolonged operation that is necessary to support HV transfer pumps and do not require a steady supply of consumables.

Sorption pumps are clean, powerful tools for achieving rough vacuums right down to the edge of HV levels. Sorption (sometimes known as cryosorption) pumps are gas capture systems that have limited operating periods before becoming full. Multiple sorption pumps can be used in parallel to increase the pumping speed as well as extend operating times. Pictured in Figure 4.8 is a portable two-sorption-pump system, equipped also with a venturi pump, TC and gauge, and three isolation valves. A sorption pump is simply a metal container filled with synthetic zeolite pellets having pores to maximize surface areas. A Styrofoam or stainless steel Dewar sleeve surrounds the sorption pump and is filled with liquid nitrogen, cooling the entire zeolite-filled canister down to 77 K (−196°C). The pump often has internal metal fins to facilitate the transfer of thermal energy between the zeolite pellets and the liquid nitrogen in the Dewar, since zeolite is a poor conductor. Gases entering a cold sorption pump condense or adsorb onto the zeolite and interior wall surfaces, removing these from the gas phase and leaving a vacuum. *Note*: O_2 and N_2 are liquid at these temperatures.

Typically, the bulk zeolite material has surface area of 500 m^2/g, giving each sorption pump a 60,000 Torr-liters capacity (or equivalent of 80 L of atmosphere). Sorption pumps capture most gases well with the exception of hydrogen, helium, and neon, which do not condense at liquid nitrogen temperatures. Nevertheless, these hard-to-pump gases do respond to the cold temperatures, causing these molecules and atoms to linger inside the pump, essentially being pumped. The noble gases (e.g., He and Ne) are less problematic for sorption pumps if another pump first roughs these. The ultimate vacuum pressures that can be achieved with sorption pumps are 10^{-2} to 10^{-3} Torr. The sorption pump system pictured in Figure 4.8 is used as the main pump for a small chamber. The attached venturi pump serves as a roughing pump. When one or both of the sorption pumps become full, the Styrofoam cryogenic Dewar is

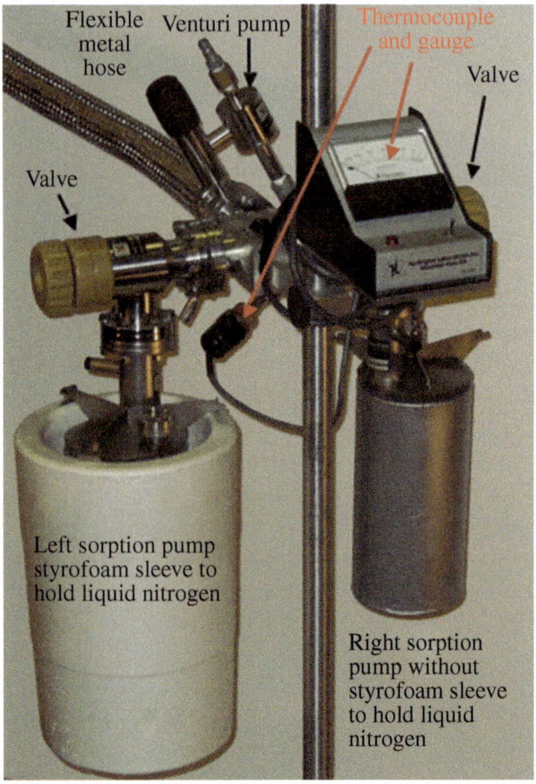

FIGURE 4.8 A pair of sorption pumps along with supporting equipment is shown. These pumps function by cooling the residual gas from the chamber to the point where it condenses to liquid form. The pictured pump station has valves so one or both sorption pumps can be used and gauges to measure two stages of vacuum.

removed and a heater sleeve is attached, which bakes the canister and drives the accumulated gas out the relief valve. In this case, the relief valve is simply a rubber stopper inserted into a small horizontal tube extending from the neck of the sorption pump. The pump becomes operational again, once the stopper is inserted and the pump is allowed to cool. Multiple sorption pumps are sometimes used to rough down very large UHV systems, since several groups of sorption pumps distributed along extended beamlines can quickly and cleanly evacuate a large volume.

4.4 HV VIA DIFFUSION, TURBOMOLECULAR, OR CRYOGENIC PUMPS

While the popularity of the **diffusion pump** has declined somewhat in recent years, it still remains the most prevalent high vacuum pump since it offers high pumping speeds, high throughput, high forepressure tolerances, low ultimate pressures,

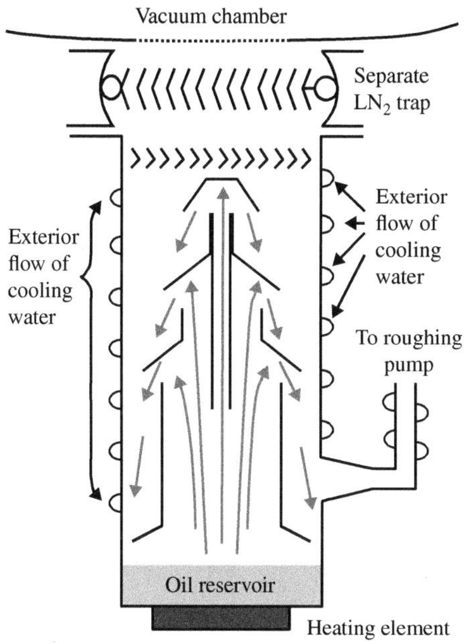

FIGURE 4.9 A schematic representation of a molecular diffusion pump is shown. A heating element causes a special oil of large, complex molecules to boil, sending small amounts of oil upward as depicted by the gray arrows. The oil strikes deflectors and is gravitationally pulled back toward the oil reservoir, dragging residual gas molecules down to the lower portion of the pump. A roughing pump continuously removes the slightly over-pressurized gas caused by the oil flow.

excellent back-streaming characteristics, and long-term reliability. The ribbed structure on its exterior gives the diffusion pump its iconic distinctive appearance. A diffusion pump, depicted schematically in Figure 4.9, is a momentum transfer device of the gas transfer type. This pump uses very expensive, high-performance oil, which is heated from the bottom causing it to boil sending oil vapor and small droplets upward. The interior structure and walls are cooler, condensing the vapor into additional droplets and redirecting the oil droplets of highly complex hydrocarbons downward in the outer regions of the pump volume. The downward flow of small oil droplets concentrates the residual gas at the bottom of the pump. A roughing pump, normally a wet mechanical one, continually drains off the slight overpressure caused by the flow of oil. The pump requires continual cooling, usually a flow of water around the exterior, to maintain the ongoing vapor-to-liquid phase transition of the oil and to prevent any scorching of the oil. Diffusion pumps have mechanical traps in the form of chevron fins as a precaution against back streaming of oil into the main chamber. In addition, a separate trap, cooled by liquid nitrogen (LN_2), is sometimes inserted between the main pump and the chamber, providing an extra cold-surface trapping. A midsized diffusion pump with an LN_2 trap can reach 10^{-8} Torr vacuum levels and pumping speeds of 5000 L/s.

One principal advantage of a diffusion pump is its ability to pump hydrogen and helium as well as reactive gases that would destroy other high vacuum pumps. Diffusion pumps are also relatively vibration-free and noise-free systems, important for many vacuum depositions of optical coatings. It normally takes about an hour of preparation before the pump is up to operating temperatures, thermally stabilized, and can be exposed to the vacuum tank. A diffusion pump must be itself roughed prior to starting the heating of its oil. Once HV is obtained, the pump can be operated continuously for an indefinitely long period of time. Shutdown procedures are also necessary to insure the coolant is not turned off prematurely.

Turbomolecular pumps (colloquially referred to as Turbo pumps) are simply a set of turbine blades spinning a very high (20,000–90,000 rpm) revolutions per minute. The device is a momentum-transfer type of pump. An atom or molecule that randomly collides with a blade is given an impulse, sending it towards the pump exhaust. The compression ratio of the chamber side to the exhaust side of a stack of stators can be 10^8 for N_2, signifying a chamber vacuum of 10^{-12} Torr can be achieved with a foreline pressure of 10^{-4} Torr. For H_2 and He, however, the compression ratio is only approximately 10^3, indicating the pump has a much harder time pumping these gases. In practice, pressures of 10^{-7} to 10^{-10} Torr and pumping speeds up to approximately 3500 L/s can be achieved with turbo pumps. UHV levels can be reached by using a small turbo pump to remove the exhaust from the large turbo, and then a mechanical pump to eliminate the final exhaust.

Most turbomolecular pumps require special grease or oil lubricants for the drive motor, but these are located on the exhaust side of the stack of blades. Back streaming is not important in a turbo pump since any oil molecules encountering the stators are forced back toward the exhaust and removed. Some turbo pumps have dry motors with magnetically levitated bearings. These are used in truly dry vacuum chambers and are exhausted by other dry pumps.

Turbo pumps are used extensively in leak detection systems. A common leak detector consists of a He sensor connected to the exhaust of a turbo pump. Working either by itself or in parallel with the chamber's main pump, a high vacuum is pulled. (The optimal vacuum pressure for operating a leak detector is near the highest pressure that the turbo pump can withstand without damage, since the detection speed is proportional to flow rate through the chamber.) Leaks are detected by injecting small amounts of helium at exterior points around the chamber where leaks are most likely to form and then waiting to see if the He finds its way into the chamber and subsequently arrives at the pump's exhaust. There is approximately a 2–3 second delay from the time the He is dispensed through a small wand and some of it arrives at the pump. The search for HV leaks can only be performed for a limited amount of time before small excesses of He concentration form throughout the laboratory and the detector senses He gas continuously. In most laboratories, the search can resume after 30 minutes, assuming a normal air exchange. The process is slow and tedious, especially if the search for leaks has to cover a large surface.

A Cryopump, depicted schematically in Figure 4.10, is an oil-free HV pump of the gas capture type. Cryopumps, properly known as cryogenic pumps, are similar to sorption (cryosorption) pumps, except portions of the pump are substantially colder. The

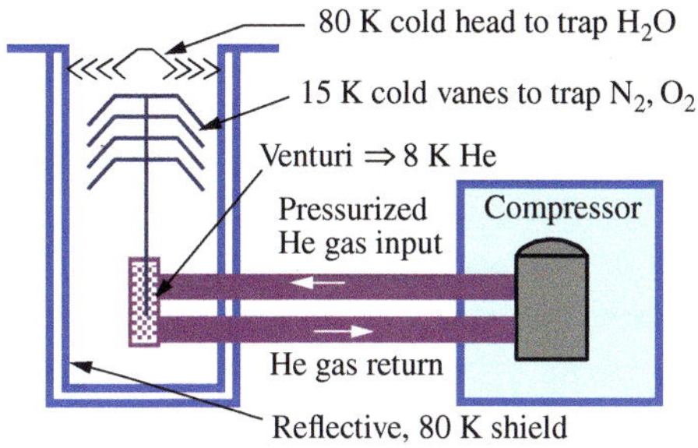

FIGURE 4.10 A cryogenic pump operates by dramatically changing the pressure of He gas at two points in the cycle. The sudden drop in the He pressure causes it to go from approximately room temperature to about 10° above absolute zero. The helium is connected to a series of vanes, which become sufficiently cold to freeze the residual gas from the vacuum chamber.

basic physics behind the cryopump is to create an ultimate refrigerator and attach a cold finger to a series of progressively larger cold surfaces. The primary requisite is to get various surfaces sufficiently cold that various gas constituents are frozen or adsorbed onto one of several surfaces and held there for extended periods. All cryopump designs have several surfaces held at various temperatures so that the pumping load is distributed. These pumps employ all three forms of thermal isolation found in consumer thermos bottles. To be able to maintain an 8 K central temperature, the coldest surfaces need to be shielded from warmer ones. This temperature isolation extends to the gases being pumped from the vacuum chamber. The first surface encountered by molecules entering the pump is an 80 K cold head, which removes water and hydrocarbons, as well as partially condenses gaseous N_2 and O_2 to liquid. Next, the molecules run into 15 K vanes that freeze most gases, especially N_2, O_2, and Ar. The undersides of these inverted cup-shaped vanes are often coated with specially processed charcoal or zeolite, which at 15 K is sufficiently cold to hold helium, hydrogen, and neon.

The refrigerant is 99.999% pure helium gas, pressurized to about 160 psi by a compressor with a return pressure of 60–80 psi. Similar to the consumer air conditioners and refrigerators described in Chapter 2, a compressor circulates the refrigerant around a closed path that includes a throttle region in the form of a Venturi mesh, causing an abrupt drop in pressure and temperature. The He drops from room temperature to approximately 8 K and back over a distance of approximately 20 cm. (Cryopump manufacturers have begun advertising 4 K systems, providing even better gas capture properties.) In contrast to ordinary refrigerators, the refrigerant remains a gas throughout the entire cycle.

It normally takes about 2 hours to start a large cryopump. These pumps require extensive roughing to vacuum pressures of approximately 50 μm (~5×10^{-2} Torr) prior to starting the compressor and refrigerator. (The turn-on pressure should not be confused with the CP, the crossover pressure.) Too much gas initially inside the pump at turn-on will establish a convective-energy transport path, which will prevent the refrigerator from achieving cryogenic temperatures. In other words, gas molecules will transport thermal energy from the warm outer regions of the pump to the Venturi mesh, warming it up and releasing any gas molecules starting to be adsorbed on cold surfaces. In contrast, if the vacuum pressure at turn-on is sufficiently low, there are not enough molecules to transport the heat from the warmer surfaces and the cooling from the refrigerator is the dominant process. Any residual gas inside the pump becomes frozen to various cold surfaces as the temperature drops. Once completely cold, the pump has a much greater capacity to adsorb new gas. It normally takes 1–1.5 hours for the refrigerator to come down to its cryogenic operating temperature. The external compressor is water cooled in larger systems.

The crossover point for cyropumps is between 100 and 500 Torr-liters, depending on pump size. If the total amount of gas in the chamber (volume × pressure) is too large, various cold surfaces will be warmed by the inrush of gas, shutting down the pumping process. Cyropumps can be operated continuously for a few weeks under normal conditions between regenerations. The reconditioning is simply a matter of letting the pump warm up and then using the foreline pump to evacuate the cryopump. First, purging with dry nitrogen and externally warming the pump above room temperature assist this process.

4.5 UHV VIA ION PUMPS

Ion pumps are the best choice for UHV chambers, since these pull the hardest vacuums, as well as are clean, vibration free, and can be baked. Ion pumps also have low power consumption and long operating lifetimes despite being a gas capture type pump. The structure of an ion pump, shown in Figure 4.11, consists of a cluster of short hollow tubes in the center that collectively form the anode, two plates made of titanium or tantalum on either side that form the cathode, and two exterior magnets forming a strong dipole parallel to the axis of the tubes. The stainless steel cathode tubes are biased between +3000 Vdc and +7000 Vdc while the cathode plates are held at ground. Atoms and molecules that enter the pump become ionized primarily through collisions with free electrons. Ions are accelerated towards either cathode plate, striking with sufficient velocity to sputter Ti or Ta atoms as well as ejecting additional free electrons. The sputtered Ti or Ta atoms coat all surfaces inside the pump including the walls of the anode tubes and pump casing. At the same time, the free electrons accelerate toward, moving in tight circular spirals around the magnetic field lines. There are three physical mechanisms to capture gas: chemical reaction, ion burial, and neutral burial. Free Ti atoms that have been sputtered are chemically reactive with most atmospheric constituents (N_2, O_2, H_2O, CO_2, CO, and H_2). A gas molecule with a Ti atom attached is readily adsorbed onto stainless steel surfaces or absorbed into the titanium plates, a process known as gettering. Noble gases such as He and Ar are somewhat more difficult to pump, but become trapped via burial processes beneath subsequent layers of sputtered Ti. Some ion pumps are operated as "triode pumps" in which one of the Ti plates is replaced with a slotted Ta plate. The ability to pump H_2 is reduced to some extent with a triode configuration, but the pumping efficiencies of the noble gases are enhanced.

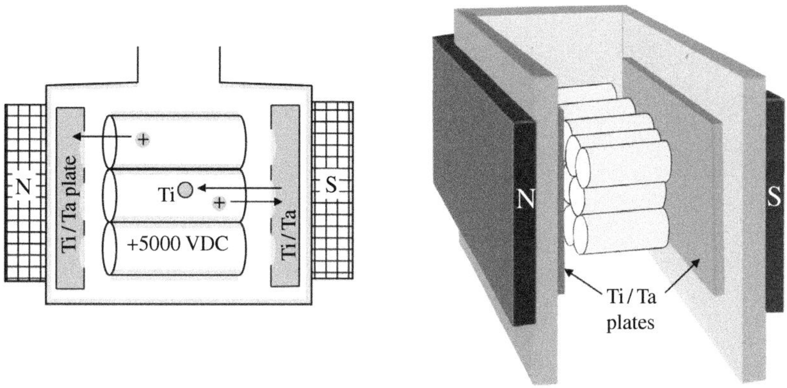

FIGURE 4.11 Ion pumps produce strong internal electrical fields, which accelerate the electrons and positively charged molecules. Many of these charges strike titanium or titanium and tantalum plates releasing a few Ti or Ta atoms, which chemically bond with gas molecules and then become adsorbed onto the interior walls of the pump in a process known as gettering. The sequence of events also produces more ions, which continue the pumping process.

The strong dipole magnetic field ($800<\vec{B}<2000\,\text{G}$) and the tubular shapes of the stainless steel anode tubes causes the free electrons to loiter inside the tubes, making tight circular trajectories. This structural combination enhances the likelihood of each electron encountering a gas molecule and ionizing it to start its capture process.

The largest ion pumps have pumping speeds of 300–500 L/s, while typical UHV chambers in industry and in research laboratories use approximately 100 L/s systems. Miniature ion pumps are sometimes attached to portable instruments to maintain UHV. Ion pumps typically are rated to have a 40,000–80,000 hours operating life at 10^{-6} mbar (~10^{-6} Torr). Normally, this equates to several years of operation before a pump has to be replaced or refurbished. Only larger pumps can be effectively refurbished since the interior Ti/Ta plates need to be replaced.

It is frequently necessary to bake the entire UHV chamber, including the main pump. For many applications, a 200–250°C bake is adequate to achieve the necessary cleanliness. Ion pumps can generally be baked up to 350°C with magnets attached and up to 450°C with the magnets removed. Cables, however, even bakeable cables should never be exposed to temperatures above 220°C.

5

CLEANROOMS, AN ENABLING TECHNOLOGY

Cleanrooms are controlled environments where airborne pollutants, microbes, and particles are regulated to specific limits. For example, a single particle inadvertently coming in contact with an integrated circuit (IC) chip during fabrication can render that chip nonfunctional. For this reason, airborne contaminates must be kept to a minimum to keep production yields high. Historically, cleanrooms were classified according to the number of particulates per cubic foot. Specifically, the most common Class 100, 1,000, and 10,000 cleanrooms indicated upper limit on the number of particles in a cubic foot. This obsolete characterization, which is still widely used unofficially, was replaced in 2001 by an international standard that reflects the modern requirements for cleanrooms. These ISO 14644-1 standards, listed in Table 5.1, place limits on the numerical counts per cubic meter for each range in particle size. Cleanrooms and the procedures for maintaining these environments are customized according to each application. For example, the garments required for an ISO 7 (Class 10,000) cleanroom differ substantially from those in an ISO 5 (Class 100). *Note*: ISO 14698 cleanroom standards refer to bio-contamination and are not considered further in this section.

There are three principle processes used by cleanrooms to maintain a pristine environment: (i) prevention of external contaminants being introduced via personnel and equipment entering the room, (ii) exchanging external air through a laminar flow with extensive filtration, and (iii) removal of existing contaminants through special cleaning procedures. A typical floor plan (Fig. 5.1) reveals the multistage steps used to minimize external contaminants being brought into the cleanroom. An important first step is to remove and trap shoe particulates using treated fiber mats and an electric vacuuming shoe brush. Next, personnel enter the gowning room where they cover themselves with garments, boots, gloves, and hoods. A completely covered individual including overalls, as shown in Figure 5.2, is colloquially referred to as

Modern Devices: The Simple Physics of Sophisticated Technology, First Edition.
Charles L. Joseph and Santiago Bernal.

Companion website: www.wiley.com/go/joseph/moderndevices

TABLE 5.1 ISO 14644-1 Cleanroom Standards

Class	Maximum Number of Particles/m³						FED STD 209E Equivalent
	≥0.1 μm	≥0.2 μm	≥0.3 μm	≥0.5 μm	≥1 μm	≥5 μm	
ISO 1	10	2					
ISO 2	100	24	10	4			
ISO 3	1,000	237	102	35	8		Class 1
ISO 4	10,000	2,370	1,020	352	83		Class 10
ISO 5	100,000	23,700	10,200	3,520	832	29	Class 100
ISO 6	1,000,000	237,000	102,000	35,200	8,320	293	Class 1,000
ISO 7				352,000	83,200	2930	Class 10,000

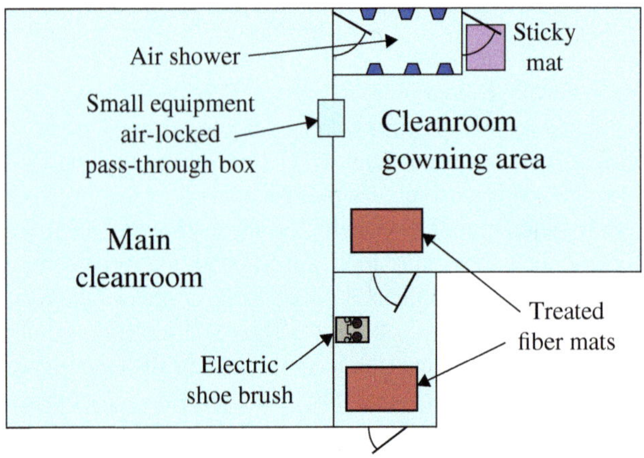

FIGURE 5.1 A typical floor plan for a cleanroom, showing the multiple decontamination stages used prior to entering the cleanroom.

wearing a bunny suit. These outfits consist of tightly woven fabrics that retain the numerous particles per minute that humans continuously shed. The fabrics also do not hold static charge, which would attract particles that could be deposited elsewhere. *Note*: troublesome areas (e.g., footwear and headwear) sometimes have two layers of covering and personnel are required to dress in a particular order. Once adequately covered, the individual walks across a sticky mat with an adhesive top layer and into an airlock room called the air shower. The mat, consisting of multiple layers of disposable thin films, removes residual particulates still remaining on the bottom of the special footwear. The air shower, consisting of a number of jet nozzles, removes contaminants that might have gotten onto the exterior of the garments. Figure 5.2 shows the inside of an air shower. Some cleanrooms also have an airlock box, where small equipment that has been thoroughly cleaned can be passed without having to be hand carried by a gowned person. The main cleanroom is overpressurized so that any contaminants that have escaped these mitigation techniques and become dislodged will flow away from the cleanroom.

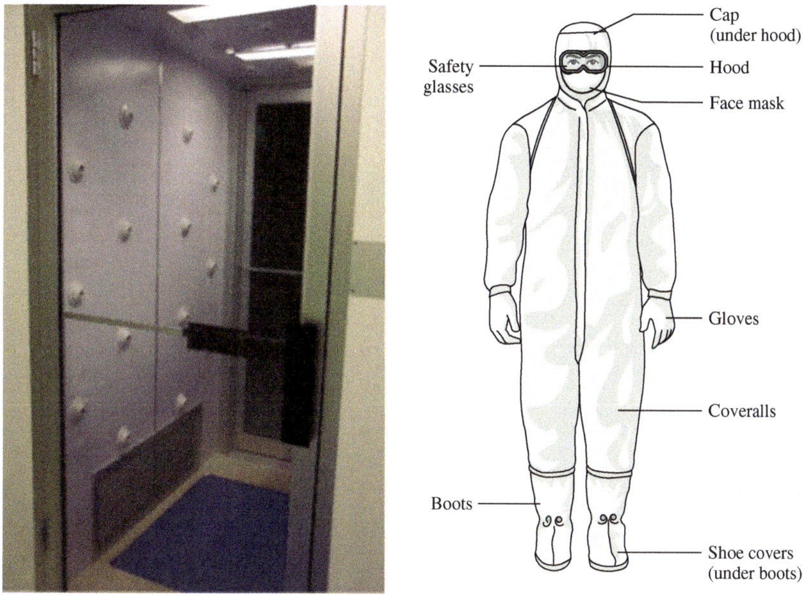

FIGURE 5.2 An individual completely garmented and an air shower that is used prior to entering a cleanroom.

In addition, all items to be brought into a cleanroom are screened for suitability. Pen and paper, for example, must meet specific standards. Writing tablets as well as cleaning tissues must be made of paper with anti-static and low-lint characteristics. Printed documents such as operational procedures are often covered in clear plastic, encapsulating each sheet and attached in loose-leaf binders. All such coverings must be static free and not emit volatile aerosols. In some circumstances, personnel are not allowed to wear cologne, perfumes, mascara, or a variety of other beauty products. Finally, the paint or wall surfaces as well as particulates generated via friction (e.g., from wall sockets or from cooling fans inside computers and other electronic equipment) must be scrutinized and controlled.

The second essential process to maintain a sufficiently pristine environment is extensive air filtration. Two strategies for air filtration are shown schematically in Figure 5.3. The vast majority of cleanrooms incorporate laminar airflow, designed to drive any residual particulates toward the floor and hold those there until the surface is cleaned. A key component is the use of high-efficiency particulate air (HEPA) filters, which were first developed for the Manhattan Project of World War II (WWII) and commercialized in the 1950s. Air turbulence tends to knock off or kick up particulates on surfaces, causing these to become air borne again and potentially contaminating devices under fabrication. A single 0.5 μm particle, 1/200th of the diameter of a human hair, can destroy an integrated circuit. In recent years, a new ultra-low particulate air (ULPA) filter has been introduced. ULPA were developed specifically for cleanrooms where the generation of large amounts contaminants were unavoidable. In these few cases, the airflow is deliberately made turbulent, stirring up surface particulates and

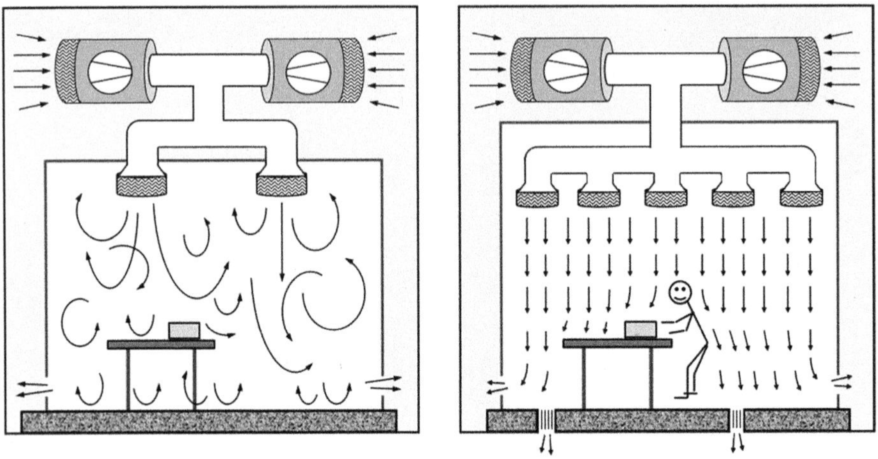

FIGURE 5.3 Two approaches to cleanroom design. Left: high turbulent flow that kicks up loose particles and forces these out the exit vents. Right: laminar flow, the more common type of cleanroom, moves residual loose particles to the floor and traps these there until removed by routine cleaning procedures.

using a large volume flow of air to sweep these to ULPA filters. Laminar flows, however, generally provide superior air quality and ULPA filters have begun to be used in conjunction with HEPA filters to provide the best performance for commonly configured cleanrooms.

The final key process to maintaining a pristine environment is frequent and regularly scheduled cleanings. Filters are replaced according to fixed schedules, while gowns, floor and table surfaces, and floor mats are routinely cleaned or replaced. Some garment apparels are disposable, being used once and discarded. Once again, all chemicals used in the various cleaning procedures are scrutinized and restricted as are the materials (e.g., mops and wipes) used to clean surfaces. Some cleanrooms go so far as to require the use of deionized water. There are even procedures and schedules for trash removal, including material requirements for the receptacles.

6

SOLID-STATE ELECTRONICS

Commercially available electronic equipment such as radios began the transition from vacuum tube technology to solid-state components in the 1960s. Initially, these devices consisted of individual components, including transistors and diodes along with resistors, capacitors, and inductors. During the 1970s, integrated circuits (ICs) came into their own with various IC chips, containing hundreds of transistor, capacitor, and resistor components, performing a single or at most a few functions. An assortment of IC chips along with some external components could be mounted on to a circuit board and several boards wired together to perform various tasks. By the 1980s, large-scale integration of IC chips were possible, containing hundreds of thousands of components. The application-specific IC (ASIC) and the field-programmable gate array (FPGA) became widely available during this decade. This enabled engineers to create custom chips for specialized tasks. For instance, the central processors that enabled personal computers were among the first ASIC chips, and the cell phones today contain an ASIC to perform its specific functions. The trend continued during the 1990s with very large-scale integration (VLSI), containing millions of components. By 2008, billion-transistor processors were commercially available.

The semiconductor industry continues to increase circuit complexity and speed, while minimizing the power consumption. Size reduction of individual components is a critical enabling factor since all gains in speed and improved functionality require substantially more electrical power (and waste heat) unless offset by smaller parts. A small capacitor, for example, can be charged up to a particular voltage with less energy than can a large capacitor. The impact from any defects in miniaturized components, however, is more severe and alters overall performance to a greater degree than these do in large ones. Tight specifications are required to achieve these

Modern Devices: The Simple Physics of Sophisticated Technology, First Edition.
Charles L. Joseph and Santiago Bernal.

Companion website: www.wiley.com/go/joseph/moderndevices

reductions in scale without a dramatic increasing failure rates and loss of reliability. High-quality materials and better processing methods have to be employed.

Despite the enormous complexity of multimillion-component modern circuits, each retains a number of very simple but important physical properties that can be approximated (only for a specified range of frequencies) by a simple combination of three passive components: resistors, capacitors, and inductors. The electrical response of complex circuitry, known as its impedance, can be described quite effectively by a few parameters on a specification sheet, enabling anyone to predict how the circuitry will respond to various inputs and how well it will interface with various other electronic circuits. In general, the most complicated electronics equipment can be treated as a "black box" where an input is transformed to an output without having to know its inner workings. One particularly important parameter is its frequency response, since digital or analog information must be transmitted via a series of pulses or via amplitude modulation or frequency modulation of an oscillating wave. The frequency response is a rough indicator of the rate that information can be transferred.

Other important parameters of any circuitry are: noise, stability, and reliability. While most difficulties associated with any of these parameters requires the assistance of an electrical engineer, basic physics can still provide important clues to the various sources of any problem. External noise sources can arise from a connection to another piece of equipment or through the air from an electromagnetic force (*emf*) emission from inferior equipment in proximity. In addition, tiny amounts of noise are introduced internally from a variety of sources (e.g., Johnson noise from a current flowing through a resistor) and can be magnified by active components such as a transistor. Such noise can often be damped (filtered) simply by a judicious choice of a capacitor at a proper location. Conversely, the sudden appearance of high-frequency noise might indicate a capacitor has failed (or is failing if the noise is intermittent). As we shall see, certain combinations of components inherently oscillate in both voltage and current, a necessary quality for many circuits, but one that can lead to instability. Well-designed circuits usually have additional circuitry to provide negative feedback, built-in inefficiencies working in opposition to the main circuit to prevent instability or a runaway condition. Reliability is primarily an empirically measured quantity, although electrical engineers can estimate the overall dependability of new circuitry from an analysis of known failure rates of various components and fabrication technologies.

IC or VLSI circuits contain the vast majority of the electronic components, ranging from active elements such as transistors, to passive components such as capacitors and resistors. A schematic drawing of a capacitor design for an integrated circuit, depicted in Figure 6.1, shows the thin film layers necessary in its fabrication. The silicon oxide (specifically silicon dioxide, SiO_2) layer is the insulating material, separating the two conductor planes (solid gray). The SiO_2 layer is only 10–20 nm thick, corresponding to a thickness of 100–200 atoms, in the region of the capacitor and nearly four times thicker between the two sets of leads that connect the capacitor to the rest of the circuitry. The conducting "wires" in an integrated circuit are actually foil strips and the large separation between conducting layers minimizes the amount of unintended stray capacitance.

IC and VLSI circuits are the limiting factor in overall circuitry lifetime since these chips contain the vast majority of the circuit components and since the physical sizes of

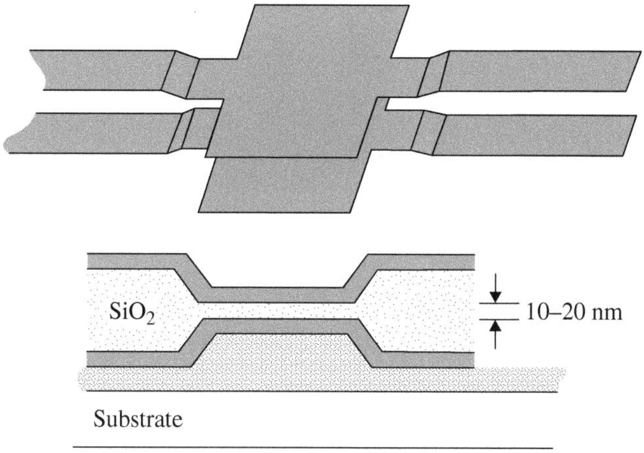

FIGURE 6.1 A schematic layout of a capacitor on an integrated circuit.

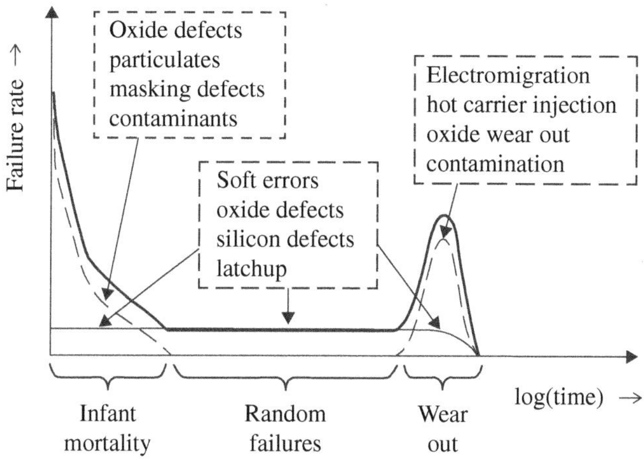

FIGURE 6.2 The "bathtub" failure rates for integrated circuits, including VLSI circuits. The physical processes responsible for various individual failures are listed in dashed-line boxes.

these individual components are much smaller than any external discrete parts. A plot of the overall failure rate has a characteristic known as the "bathtub" curve. The failure rate, plotted in Figure 6.2, is broken into three time domains: (i) infant mortality, (ii) random failure, and (iii) wear out. Listed in dashed-line boxes are the physical mechanisms responsible for the three types of failure. Each box has an arrow pointing to its individual failure curve, plotted either as a dashed or a thin solid line. The overall failure rate (bold line) is the combined rates at any given time, resulting in a bathtub shape curve. *Note*: the horizontal axis is the logarithm of time, indicating the wear-out period is actually longer than is the infant mortality. Also, if the circuitry survives its infant mortality period, there is a high probability it will function properly for a long time. Some of the physical processes that contribute to the failures will be discussed in Chapter 7.

INTERESTING TIDBIT TB6.1

A "black box" in science and engineering refers to a system or device (usually a complex one) where nothing is known about the contents inside it, but can be viewed simply as an input and an output with the black box providing the transfer characteristics.

COMPREHENSION VERIFICATION CV6.1

Problem: Use the "bathtub" graph of failure rates (Fig. 6.2) to estimate the implied lifetimes, assuming the infant mortality timescale is 1 month of operation.

Answer: The important things to note are the horizontal axis is logarithmic and the relative horizontal lengths associated with the infant mortality, random failures, and wear out. For a logarithmic scale, each constant segment of length is 10 times larger than the previous one. The distance along the horizontal axis attributed to random failures is about 2× that of the infant morality period, implying the random failures period is 10^2 times longer (i.e., $10 \times 10 \times 1$ month) or 8.3 years of use. In a similar fashion, the wear-out period has a length on the horizontal axis that is 0.5× that of the infant mortality period, but starts at the end of the random failure interval corresponding to the beginning of the 10^3-month scale. (If the length on the horizontal axis were equivalent to the infant mortality length, the wear-out period would be 1000 months or 83 years. However, we observe it to be only about half that length.) If you have examined log graph paper or have experience dealing with logarithmic numbers, then you know that the minor tick marks inside of each 10× segment are not evenly spaced. The ticks for numerals 2 and 3 along any 10× interval occur at 0.3 and 0.5 of the distance from its start with four through nine bunched into the right half of the segment. Alternatively, you could perform a series of trial-and-error guesses, using your handheld calculator to determine empirically that $\log(2) = 0.3$ and $\log(3) = 0.48$, which is close to 0.5. Thus, we estimate that the wear out interval is 500 or (0.5×1000) times as long as the infant mortality interval. (One should never be too precise when one is inferring information from a graph without numerical scales on the axes.)

In summary: Infant mortality: 1 month

Random failures: 8 years

Wear out: 40 years

6.1 CONDUCTING, SEMICONDUCTING, AND INSULATING MATERIALS

Solid-state conductors, semiconductors, and insulators are formed from either single crystal or polycrystalline materials. (Amorphous materials are occasionally used in semiconductor applications, but these represent a small minority.) A crystal is a solid with its atoms having a regular geometrical arrangement that repeats in a periodic way, its lattice. A polycrystalline material consists of a large number of small, single crystals, each with a lattice randomly oriented with respect to one another. Many elements (e.g., Si, C, Ge, and most metals) and some chemical compounds (e.g., NaCl, GaAs, and SiO_2) readily assimilate into polycrystalline materials when solidifying from the liquid form. Solid-state devices made of single crystals have the highest quality performance. To create a single crystal of any reasonable size, it has to be grown very slowly using extremely pure source materials.

The valence electrons in an element or compound determine the lattice structure of these solid-state materials. These atoms are connected to adjacent atoms via covalent bonds, resulting in a very complicated interlocking framework. As a system of atoms, the discrete quantum mechanical (QM) energy states of a crystalline material differ from those of its constituent elements. Permitted QM states that are closely spaced in energy are called bands since collectively these approximate a continuum of states. Two particularly important energy bands in all crystals are the valence and the conduction bands. Electrons in the valence band remain localized within the lattice, while those in the conduction band are very weakly bound to any small group of atoms and are free to move when an electric potential (voltage) is applied. In all crystals, there is always a bandgap, a forbidden energy range, between the valence and conductions bands. This bandgap is small for conducting metals, midsized for semiconducting crystals, and large for insulating materials.

Calculating the electron quantum mechanical states of a crystalline material to determine in detail its electrical properties can be daunting. Fortunately, we can use information contained in the *Periodic Table of the Elements* as well as the results from *Fermi–Dirac (FD) statistics* to understand the general properties of any solid-state material. FD statistics, which applies to specific types of QM systems such as the electrons in a crystalline lattice, can be used to infer the probability distribution of electrons populating the covalence and conduction bands.

The Fermi probability function, $F(E)$, plotted in Figure 6.3, is the probability of finding an electron populating the various quantum mechanical states as a function of energy. $F(E)$ equals 1 for all tightly bound inner subshells of individual atoms, which are effectively isolated from the lattice structure. $F(E)$ equals 0 for electrons in states with energies well above the conduction band. For the intermediate states of the crystal, containing portions of both the valance and conduction bands, $0 < F(E) < 1$. In other words, the innermost atomic subshells are full and closed, so the probability of finding an electron in each of those quantum mechanical states is simply one. Those tightly bound electrons do not interact with other atoms or the lattice electrons. Only the electrons that populate the valence and conduction bands are able to

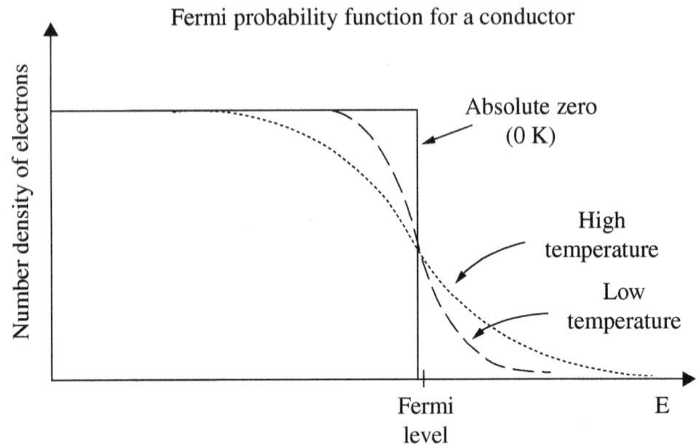

FIGURE 6.3 The probability of finding an electron occupying a quantum mechanical (QM) state as a function of energy in a crystalline conductor. Three probability curves are plotted, representing three separate temperatures. Similar curves exist for insulators and for semiconductors, although the Fermi levels differ.

transition between states as a function of the material temperature and its forbidden energy bandgap. *Note*: the absorption of a photon can also excite temporarily an electron from any level to a higher one. However, photon excitation from an inner shell requires X-ray energies, which normally are a negligible phenomenon.

The QM rule that electrons fill the quantum mechanical states, starting with the most tightly bound ones, proceeding upward, and if necessary partially filling the next state, is only strictly correct if the material is at absolute zero (0 K, −273.15°C) where all motions cease. The Fermi level is the value where the potential energy from the $\vec{E}$ fields of neighboring nuclei is equal to the average diffusion energy per electron. In practical terms, when the crystal is at absolute zero, electrons occupy all of the QM states up to Fermi level and none above it. For temperatures above absolute zero, the Fermi level marks the energy where $F(E)=0.5$ and it occurs midway between the valence and conduction bands. As seen in Figure 6.3, the sharp edge near the Fermi level increasingly softens as the temperature is elevated. Electrons in the innermost subshells continue to populate these subshells fully, regardless of temperature. Keep in mind this type of probability curve is only valid if the material remains crystalline; if the temperature is too hot, the lattice structure beaks down as the material becomes liquid.

Armed with an understanding of the statistical probability of finding an electron in various energy states, we now return our attention to the QM energy levels of various crystals. Remember, these crystalline states differ from those of its constituent atoms. Figure 6.4 shows the electron populations in various energy bands for three classes of crystals: semiconductor, metal conductor, and insulator. In each type of crystal, there is a bandgap between the conduction and valence bands, ranging from small gaps for metal conductors, to midsize for semiconductors, to large for insulators. An electron cannot absorb any energy that would place it inside the bandgap since it is quantum mechanically forbidden. Moreover, there are discrete energy

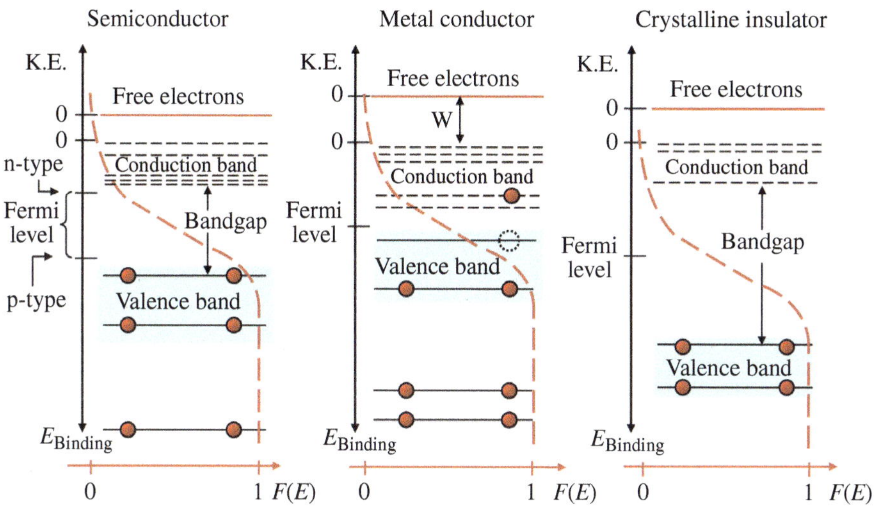

FIGURE 6.4 Energy diagrams of the quantum mechanical states, including the conduction and valance bands, for three types of crystals. For each diagram, the Fermi probability function (dashed red line) is plotted for the same warm temperature. One electron in the metal conductor diagram is shown thermally excited from the valence band (denoted by dotted circle) to conduction band.

levels within the various bands, each being able to hold a maximum of two electrons, one with spin up and the other with spin down.

While it is busy, Figure 6.4 encapsulates into a single picture all of the basic physics required to understand how solid-state circuits work. Plotted are the energy states, including the very important valance and conduction bands. The vertical axis is the total energy (E_{total} = P.E. + K.E.) where the potential energy is also referred to as the binding energy. There are two "0" tick marks. The bottom one is where the electron has gained just enough energy to escape the solid material. However, once it departs the surface, it leaves behind a positive charge, attracting it back to the solid. Thus, an extra amount of energy known as the work function, W, is required if the electron is to escape permanently. An ejected free electron has a kinetic energy that is equivalent to the total energy imparted to the electron minus the energy to escape its valance band and minus the amount W (i.e., K.E. $= E_{Imparted} - E_{Binding} - W$). (Semiconductors that have been converted to n-type or p-type (to be defined later) have different Fermi Levels than pure semiconductor crystals.) Overlaid is the Dirac probability function, plotted as a red, dashed line. *Note*: the $F(E)$ axis is horizontal rather than vertical as it was in Figure 6.3. The three identical $F(E)$ curves represent the same warm temperature (25°C < T < 40°C). A value of $F(E) = 1$ occurs at the right for each of the three diagrams, while zero occurs at the left. Also *note*: the Fermi level occurs half way between the valance and conduction bands with the $F(E)$ curve having the same shape but being offset vertically. (A 100°C change or more in temperature is required to alter the shape of the Fermi probability curve perceptibly.)

Combined with the QM states shown in the figure, the electrical behavior for each of these crystalline materials become evident.

Consider the metal conductor (middle diagram), having an odd number of electrons. The Fermi curve indicates the probability of finding an electron in the highest states within the valance band is somewhat less than unity (say $F(E) \approx 0.85$), and the probability of finding it within the conduction band is $F(E) \approx 0.15$ (15%). (*Note*: these probability values are heuristic, for illustrative purposes only.) In this particular example, one electron is shown missing from the valence band (dotted circle) and is depicted thermally excited into the conduction band. In the absence of an external electric field, most of the electrons in the conduction band will simply return back to the valance band, statistically maintaining the relative fraction of electrons in the two bands. When a voltage is applied, electrons in the conduction band flow through the crystal lattice. In contrast, insulators (right diagram) have a large forbidden bandgap, one that only overlaps with the extreme tails the Fermi probability function. As a result, virtually none of the electrons in an insulator have been thermally promoted to the conduction band, strongly inhibiting any electrical current. It should be realized, however, that a sufficiently high voltage across an insulator could cause a breakdown of the material and loss of its insulating properties. When this occurs, the insulator is usually permanently damaged.

As seen in Figure 6.4 (left), the bandgap of a semiconductor represents a midsized gap between the two energy bands. The size of this gap is 1.1 eV for silicon, 0.72 eV for germanium crystals, and is 3.4 eV for wide-bandgap GaN crystals. (*Note*: a wide band gap in solid-state semiconductors is still midsized compared to the forbidden gap of crystalline insulators (e.g., 9.0 eV gap for SiO_2).) In the semiconductor depicted, the probability curve indicates a small percentage of electrons are thermally excited to the conduction band at any one time. At room temperature (RT), the number of electrons in the conduction band is 8.6×10^9 and 1.5×10^{13} electrons/cm^3 for pure silicon and pure germanium, respectively. These numbers correspond to tiny fractions of the atoms with an electron in the conduction band of 1 in 10^{13} for Si and 2 in 10^{10} for Ge. While a Ge crystal has an inherently higher percentage of electrons in its conduction band compared to a Si crystal, it only produces a 1.6 mA current in the presence of a 10 V/cm electric field, too small for practical applications.

Doping, the process of adding specific impurity atoms during crystal formation, is the only practical method to obtain the desired electrical performance from a semiconductor. A discussion of the electrical behavior of doped semiconductors is given in Section 6.3. In the meantime, consider Figure 6.5, showing two portions of the *Periodic Table of the Elements*. As is well known, elements with similar chemical properties exist along a column. For example, copper, silver, and gold in column 1b (highlighted in lime green) are the three elements with the highest conductivity (least electrical resistance) and each has a single valance electron beyond a closed d subshell. C, Si, and Ge all occur in column 4a (green), each having four valence electrons and forming semiconductor crystals with a "cubic" lattice. Elements on either side of the 4a column make excellent impurity dopants with N, P, As, Sb, and Bi providing donor electrons and B, Al, Ga, In, and Tl being good acceptor sites for electrons in crystal lattices made of column-4a atoms. The fraction of dopant atoms is small ($<10^{-3}$).

I	II				III	IV	V	VI	VII	VIII
1a	2a		1b	2b	3a	4a	5a	6a	7a	8
^{1}H Hydrogen		Transition elements								^{2}He Helium
^{3}Li Lithium	^{4}Be Beryllium	Metals			^{5}B Boron	^{6}C Carbon	^{7}N Nitrogen	^{8}O Oxygen	^{9}F Fluorine	^{10}Ne Neon
^{11}Na Sodium	^{12}Mg Magnesium	(3b, 4b, 5b, 6b, 7b, 8)			^{13}Al Aluminum	^{14}Si Silicon	^{15}P Phosphor	^{16}S Sulfur	^{17}Cl Chlorine	^{18}Ar Argon
^{19}K Potassium	^{20}Ca Calcium	•••	^{29}Cu Copper	^{30}Zn Zinc	^{31}Ga Gallium	^{32}Ge Germanium	^{33}As Arsenic	^{34}Se Selenium	^{35}Br Bromine	^{36}Kr Krypton
^{37}Rb Rubidium	^{38}Sr Strontium	•••	^{47}Ag Silver	^{48}Cd Cadmium	^{49}In Indium	^{50}Sn Tin	^{51}Sb Antimony	^{52}Te Tellurium	^{53}I Ioadine	^{54}Xe Xenon
^{55}Cs Cesium	^{56}Ba Barium	•••	^{79}Au Gold	^{80}Hg Mercury	^{81}Tl Thallium	^{82}Pb Lead	^{83}Bi Bismuth	^{84}Po Polonium	^{85}At Astatine	^{86}Rn Radon

FIGURE 6.5 Portions of the Periodic Table. Elements highlighted in various colors critically important to semiconductors. Copper, silver and gold, the materials with the best conductivity, are highlighted as well.

Moreover, chemical compounds formed by combining 3a elements (highlighted in light brown color) along with 5a elements (blue) form an important class of semiconductors. These compounds often referred to as III–V semiconductors (after the old-style Roman numeral labels of the Periodic Table). III-nitrides in particular have become a multi-billion dollar, high-tech industry. This class of crystals is often doped with beryllium, magnesium, and carbon, among others.

The specific details of doping and electron populations for each and every type of crystal are not as important as an understanding of the basic physics involved in the choice of semiconductor materials. It is also useful to have a general idea of the region of the Periodic Table where key elements used in semiconductor exist and why. (Everyone knows from popular press accounts that semiconductor and silicon industries are one and the same. This is a useful method to identify the column in the Periodic Table containing single-element semiconductor crystals. Semiconductor crystals of compounds are formed by various combinations of elements from columns on either side such as GaAs or InN.)

INTERESTING TIDBIT TB6.2

A useful benchmark number to remember is $1/40\,\text{eV} = 0.025\,\text{eV}$, which is approximately the average kinetic energy of atoms or free electrons in equilibrium at room temperature. The bandgap of silicon is 1.1 eV, another useful benchmark number since the majority of solid-state devices are made of it. (1 eV is the energy an electron obtains as it drops through a 1-V potential. To remember the conversion: $1\,\text{eV} = 1.6 \times 10^{-19}\,\text{J}$, it is the charge of an electron ($1.6 \times 10^{-19}\,\text{C}$) multiplied by 1 V.)

INTRO PHYSICS FLASHBACK FB6.1

Valence Electrons and the Periodic Table

From introductory classes in chemistry and physics, we learned that the electrons surrounding the nucleus of an atom are separated into quantum mechanical shells and subshells, each with a discrete binding energy. Electrons populate the quantum states of each shell and subshell surrounding the nuclei, starting with the most-tightly bound inner shell and proceeding outwardly subshell-by-subshell. As an example, Figure FB6.1 shows the distribution of electrons as a function of energy for a silicon atom. (*Note*: the binding energy of some subshells actually falls within the range of energies of another major shell. The reader need not worry about the ordering nor the numerical calculations of individual subshells. Visualize instead a simple series of discrete energy bands associated with each subshell.)

In all atoms, the most inner shell can hold a maximum of two electrons. Additional, subshells are subsequently filled in the order from greatest-to-least binding energy with each being completely populated prior to an electron occupying another subshell. No two electrons can simultaneously occupy the exact same location and energy, a rule known as the *Pauli exclusion principle*. If an atom has exactly the right number of electrons to fill its outermost s^2 p^6 subshells, it is inert and does not easily react chemically with other atoms. The hydrogen atom with only one electron is unique in that its electron subshells and associated energy levels can be computed from a simple formula. For all atoms with multiple electrons, the energy levels are modified due to the mutual electrostatic repulsion between electrons, making the situation complex computationally. Fortunately, the

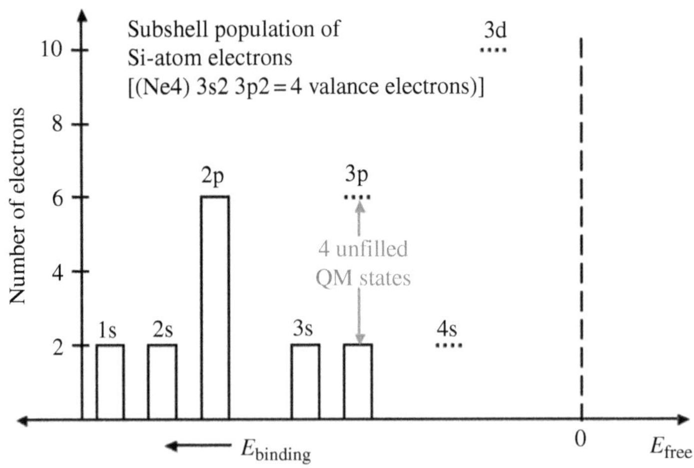

FIGURE FB6.1 The population of quantum states in a silicon atom, which has a total of 14 electrons. The subshells 1s, 2s, 2p, and 3s are completely filled containing 2, 2, 6, and 2 electrons, respectively. Two valence electrons occupy the 3p subshell. *Note*: the positions of the subshells along the binding energy axis are not to scale.

structures and energies for all of the electron shells and subshells have been computed and organized into the *Periodic Table of the Elements*.

Once a subshell has its maximum number of electrons, then that subshell becomes closed, partially shielding any outer electrons from the electrostatic forces of the nucleus. Any electrons that are insufficient in number to fill the next outer shell are called *valence electrons* and these govern the chemical properties of the atom such as the formation of isolated molecules or the various lattice structures in a crystal. For example, the neutral oxygen atom, needing two additional electrons to complete its outer most subshell, can bond with another oxygen atom to form gaseous O_2, each completing its subshell by sharing two electrons (covalence bonds). Alternatively, an oxygen atom can combine with two hydrogen atoms to form H_2O, effectively stealing the sole electron from each H atom (ionic bonds). Both carbon and silicon are in the same column of the Periodic Table and have four valence electrons, making either element a −4 oxidizer (needing four electrons) or a +4 reducer (giving up four electrons). C and Si have virtually identical properties such as indistinguishable crystal lattices, although the Si—Si bonds are not as strong as the C—C bonds.

6.2 RESISTORS, CAPACITORS, AND INDUCTORS

Resistors, capacitors, and inductors are passive components of an electronic circuit, but each type is vital to overall functionality, stability, and interconnectivity. As noted, all circuits inherently contain some amount of resistance, capacitance, and inductance, which collectively is referred to as its impedance. Resistance, measured in Ohms, is a property of a material that impedes or opposes the flow of electrical current. Empirical measurements indicate most materials have some resistance to electrical currents that remain constant over a wide range of voltages and currents if the temperature remains fixed. Materials with an extraordinary high resistance are called insulators, those with a small resistance are called conductors, and those in between can be used to form resistors. Most resistors range from a few Ohms to several mega-Ohms and the resistance of any material is proportional to its length. Resistors inside IC chips are usually fabricated from simple polysilicon (polycrystalline silicon), which is resistive if not doped. Most metals are good conductors with the best being silver, followed by gold, copper, and aluminum. (*Note*: any measurable voltage applied directly across a conductor (i.e., a short) results in a very large current that causes it to heat quickly, making it more resistive and violating the assumption of a fixed temperature. Normally, an actual resistor or other circuitry with resistance is placed in series with a conductor such that the conductor contributes a negligible portion of the total resistance and a correspondingly inconsequential voltage drop.)

Capacitors normally are constructed of two conducting surfaces in close proximity to each other. The capacitance, measured in Farads, is enhanced dramatically if a very thin dielectric material is sandwiched between the two surfaces. With the exception of power capacitors, most have a capacitance between 10^{-12} and 10^{-6} F (pF to μF).

(Savvy experts tend to refer to a pico-Farad (pF) as a "puff.") In addition to the intentional placement of capacitors, all circuits have various amounts of stray capacitance. Two conductors (wires) always interact capacitively with each other. If the two wires cross without contacting each other, the capacitive coupling is very small or negligible. Wires that run parallel to each other over long distances have significant capacitance. In fact, the cross talk between the signals in a multiple-wire cable caused by capacitive coupling is a major limitation in telecommunications. This statement is valid even if the individual wires are coaxially shielded, which dramatically reduces direct capacitive cross talk.

Many portions of various electronic circuits have a performance response dominated only by resistors and capacitors. These portions, known as resistor–capacitor (RC) circuits, have important common properties, making it straightforward to compare one to another and easy to understand their basic performance. Intricate RC circuits, however, must first be reduced from its complex groupings of resistors and capacitors to the equivalent resistance and capacitance of a minimal number of components. For example, Figure 6.6, can be simplified into one or two resistors plus one or two capacitors. The methods used to simplify an arbitrary circuit with an impedance, represented by a set of complex numbers, are beyond the scope of this text, but a review of the formulas to combine resistors only or capacitors only is given in Intro Physics Flashback FB6.2 and FB6.3. One important practical use of resistors is to

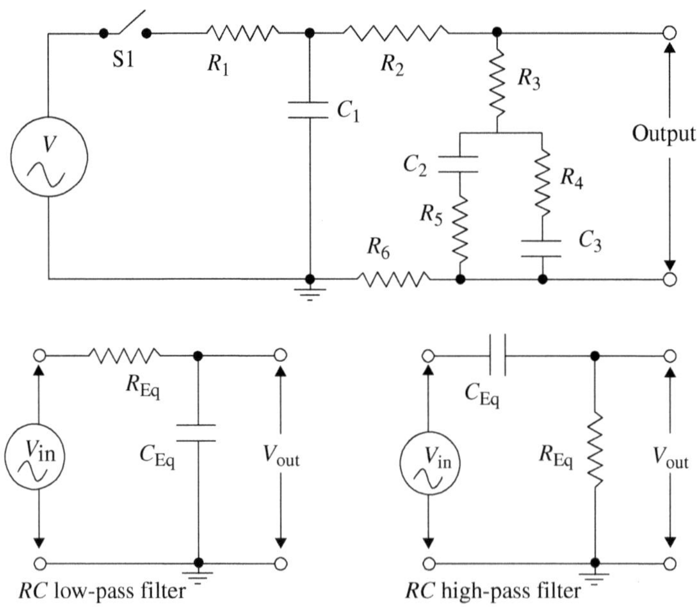

FIGURE 6.6 A representation of an *RC* circuit with a complex set of resistors and capacitors (top). Groups of resistors and capacitors can be simplified to an equivalent circuit with far fewer components, perhaps to a single resistor and one capacitor. When an *RC* circuit can be approximated by one of the two cases at the bottom, it performs as a pass filter.

use two or more in series to provide several discrete DC voltages from a single power supply. This application of resistors is known as a voltage divider and its importance will be obvious in the next section dealing with transistors in circuits.

If the arrangement of electronic components depicted at the top of Figure 6.6 can be approximated by one of the two shown at the bottom, then the circuitry responds as a high- or a low-pass filter. The gain (V_{out}/V_{in}) as a function of frequency, f, for the two cases takes the forms plotted in Figure 6.7. For the low-pass filter, the output voltage tracks exactly the input voltage from $f=0$ (DC voltage) through low-frequency variations ($f \ll f_0$). At sufficiently high frequencies, the gain begins to roll off, approaching zero for $f \gg f_0$. The opposite behavior occurs for a high-pass filter, which passes very high frequencies, but attenuates low ones. In both cases, the defined break point where the gain falls to 0.707 $(\sqrt{2})$ occurs at its natural frequency, defined as $2\pi f_0 = 1/\sqrt{RC}$. Notice that at sufficiently high frequencies, all capacitors respond as a "short," passing those components of the electrical voltage and current identically as a wire. This fact is very useful for the analysis of circuits, enabling the individual to identify those capacitors likely to filter unwanted ripples from those that pass time-variable voltages. You can also use this basic fact to recall easily the RC arrangement for a low and a high pass filter. To obtain the circuit's response in the high frequency limit, mentally replace the capacitor in the circuit diagram with a wire. For the high-pass filter, this substitution allows V_{in} to be transmitted to the V_{out} location. For the low-pass filter, V_{out} is shorted so $V_{out}=0$ at high frequencies.

Up to this point, we have ignored induction in electronic circuitry. The most generalized impedance contains inductance, L, capacitance, C, and resistance, R, terms. Any electrical charge moving over any path (i.e., a current) establishes a magnetic field, which in turn, interacts with those same charges to resist any changes in current. This is the definition of inductance. A current moving through a wire that has been coiled has a significantly larger inductance than does the same current moving through a straight wire of the same length. Nevertheless, the motion of a single electron and its corresponding magnetic field does cause a tiny amount of inductance, known as *self-inductance*.

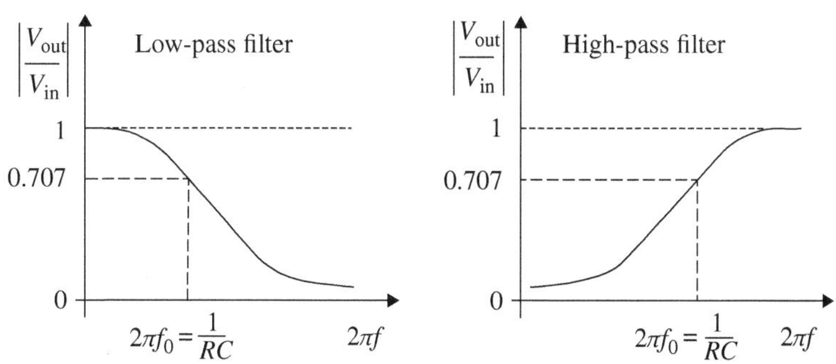

FIGURE 6.7 The gain as a function of frequency for a low-pass filter (left) and a high-pass filter (right).

Circuitry that contains all three types of impedance are known as an *RLC* (or sometimes a *LRC*) circuit. Whenever, the resistance, inductance, and capacitance are in series within a loop of circuitry, that portion forms a simple harmonic oscillator (SHO) with a resonance frequency, $f_0 = 1/2\pi\sqrt{LC}$. *Note*: f_0 does not depend on resistance. The existence of resistance, however, makes it a damped SHO. Oscillator circuits are particularly important since these are necessary for clocks, including the clocking of data streams, for radios and TVs to tune to various stations, and since virtually all circuitry must operate in electronic environments that have rapidly transitioning voltages. The characteristic *R*, *L*, and *C* of any "black box" of complex circuitry is an indicator of how fast the system can operate and whether it will interface well with other black boxes.

Consider Figure 6.8 to understand the physical processes at work in an *RLC* oscillator. The process is started when switch S1 is closed and S2 remains open. The capacitor, *C*, is charged up to a voltage, +*V*, equivalent to that of the battery at the left. At the same time, current starts to flows through *R* and *L*, but quickly ceases since the loop is not complete. Next S1 is opened followed by S2 being closed. Now the circuit starts to oscillate. The charge on *C* causes a current to flow through *R* and *L*, but *L* acts like a flywheel and the current is small initially and builds slowly. A maximum current through *L* occurs just as *C* becomes completely discharged ($V=0$). Simultaneously, a magnetic field established by the inductor (similar to an electromagnet) is at its maximum and as the current starts to diminish, the magnetic field works to sustain the flow of current. The effect of a continued flow of current after the capacitor has been discharged is to draw the top of *C* to negative voltages until it reaches a value of $-V$, exactly opposite of the battery. At that point, the current through *L* has dropped to zero and *C* is at $-V$. The process then reverses with current starting to flow through *L* due to the $-V$ charge on *C*. While being 90° out of phase with each other, the electric current and voltage oscillate sinusoidally continuously with an amplitude that diminishes with time in proportion to *R*, which dissipates energy in each cycle.

It is important to keep in mind that *not* all *RLC* circuits are harmonic oscillators, but inadequately designed *RLC* circuits tend to be somewhat unstable with a tendency to oscillate spontaneously. Figure 6.9 shows two *RLC* examples, both driven by an alternating voltage from the left side and a time-variable output voltage labeled V_{out}

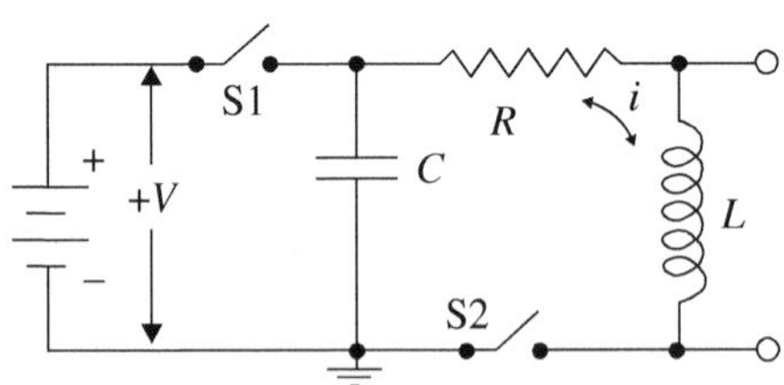

FIGURE 6.8 The right-most loop forms a damped SHO. This simple circuit starts with the charging of C with switch S1 closed and S2 open. Subsequently, when S1 is open and S2 closed, the loop oscillates in current, *i*, and voltage, *V*, with an amplitude that slowly diminishes.

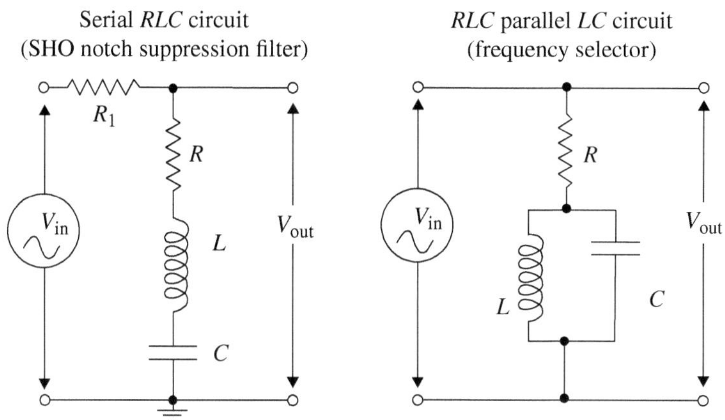

FIGURE 6.9 Two examples of *RLC* circuits.

at the right. The serial *RLC* circuit (left) is the familiar driven, damped SHO. The resistor labeled R_1 is necessary to isolate the V_{in} generator from V_{out} since the *RLC* portion of the circuit amplifies any oscillatory signal flowing through it. The presence of R_1, however, causes the *RLC* circuit to become a notch filter; it suppresses the gain around its natural frequency reaching a minimum of $V_{out}/V_{in} = R/(R + R_1)$ at f_0. A parallel *RLC* is shown on the right side of Figure 6.9. This type of circuit is used to select one frequency (that of f_0) when V_{in} contains overlapping signals with various frequencies. A version of this type of circuit, equipped with an adjustable capacitor, is used to tune to a radio or TV station.

Most electronic equipment (and for that matter, almost all systems found in nature) are too complex to analyze all of the underlying physics in detail. Nevertheless, the most complicated apparatus (even those dominated by active components such as transistors) can usually be treated either as a black box or separated into a small number of subsystems, each treated as a black box with a simple input impedance and another output impedance. In this manner, each black box is well approximated over a range of frequencies by a combination of the passive components, *R*, *L*, and *C*, enabling one to predict its operational properties. Once it is related to a particular application of physics, it is easy to construct conceptual analogs (e.g., a mechanical vs. an electronic SHO) to gain additional insights. This common form of simplification also makes it possible to calculate how well two separate electronics packages will interface with each other.

To envision the approach, consider Figure 6.10, representing a complex package of circuitry. The package is broken down into three subsystems, *Z*1, *Z*2, and *Z*3 (dotted lines), each performing a separate function and each with its own impedance. More often than not, we need not concern ourselves with the internal workings of any of the three separate groupings of circuits or the *Z*1–*Z*2, *Z*1–*Z*3, or *Z*3–*Z*2 interactions. Instead, the entire package (outlined with dashed line) can be treated simply as a single black box with output impedance, Z_{out}, that appears to the outside world as a generalized *RLC* circuit. *Note*: the output might not be

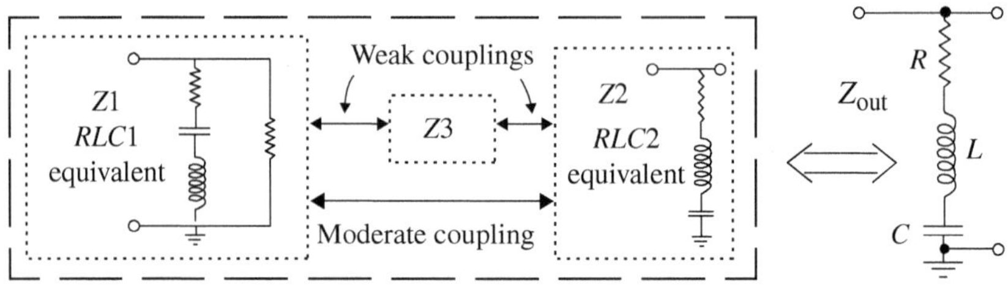

FIGURE 6.10 A conceptualization that complex circuitry can be reduced to a single, simple *RLC* circuit.

designed as an SHO and the inductance, L, might be quite small, the minimal amount caused by stray inductance from the board layout. Regardless, Z_{out} contains all of the crucial information to answer a wide variety of questions. For example, how fast can information be passed from this package? How many devices can this circuitry drive in parallel without overloading it, assuming we know the input impedances of the loads?

A data stream consists a series of 0's and 1's, corresponding say to a series of $V=0$ and $V=+1$ V states. All electronic circuits including data processors have a natural response time or frequency. If the arrival speed of impulses of 0's and 1's is well matched to a band of frequencies surrounding its natural frequency, f_0, of the data processor, then the data will flow readily across the interface. If the two circuits do not have comparable f_0's, much of the energy in the data bits will be reflected back toward the input source, interfering with the arrival of subsequent bits and potentially causing data losses. We acknowledge that digital data streams are clocked and electronic interfaces can be used to reduce the data transfer speed to accommodate the slowest device, but these transfers are significantly less efficient than the natural speed of the slowest of the data storage or data manipulation circuitries. This problem is a major reason that computers at universities, government laboratories, and commercial firms need to be upgraded every few years, especially if connected to a local-area wide network (LAWN) which are continually becoming faster and faster.

INTRO PHYSICS FLASHBACK FB6.2

Equivalent Capacitance and Resistance

We learned from introductory physics classes that the effective resistance from several resistors can be represented by a single resistor. The equivalent resistance is simply the sum of the individual resistances if the resistors are in series and when in parallel, the reciprocal of the equivalent resistance is the sum of the individual reciprocals ($1/R_{eq}=1/R_1+1/R_2+\cdots$.). Similar equations can be used to simplify a complex arrangement of capacitors to an equivalent capacitance. These simplifications, along with formulas the equivalent resistance, R_{eq}, and equivalent capacitance, C_{eq}, are summarized in Figure FB6.2.

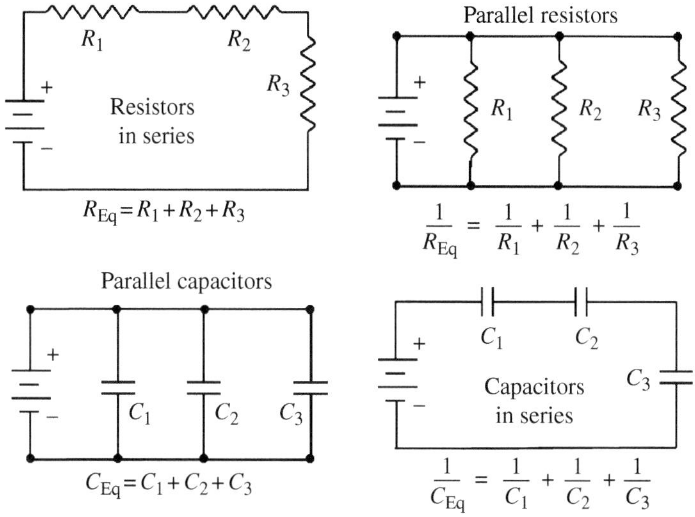

FIGURE FB6.2 Summary formulas for adding in series or in parallel the resistances or capacitances of simple *RC* circuits.

INTRO PHYSICS FLASHBACK FB6.3

The Simple Harmonic Oscillator

The simple harmonic oscillator (SHO) along with its mathematical solution is an extraordinarily powerful tool, since the periodic motion of objects, electrical voltages of many circuits, and numerous examples in nature, undergo simple harmonic oscillation or very close approximations of it. The SHO solution can be used to interpret, to analyze, and to understand a very wide range of applications. Once you are familiar with the response of an SHO in one form (say a mass on a spring), you can easily conceptualize cause and effect reactions of any other SHO such as those in electronic circuits.

Simple harmonic oscillations exist in virtually all cases where an equilibrium state exists, the system has inertia (including electrical), and a restorative (toward equilibrium) force exists. Three examples of SHO are given in Figure FB6.3, ranging from mass on a pendulum, a mass attached to a spring, and a simple circuit consisting of a capacitor and an inductor. (Strictly speaking, all of these cases are SHO only for small perturbations. That is, the amplitude, A, is small, requiring either a minor kick or a small initial displacement to get things going.) The SHO solution is so powerful that it can also be used to estimate various properties such as the natural frequency even for oscillatory responses that depart from SHO. In many cases, the solutions are simply the SHO solution plus a small perturbation to account for the nonlinear component of the restorative force.

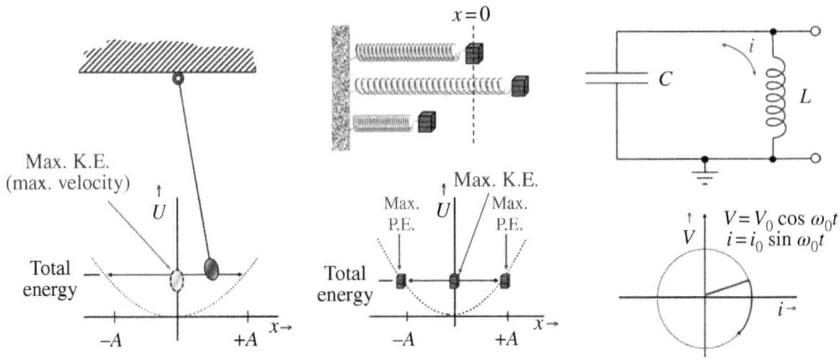

FIGURE FB6.3 Three examples of the simple harmonic oscillator, two mechanical ones and an electronic circuit one.

Inertia is an important requirement of any SHO system. If very little or no mass is attached to the spring or at the end of the pendulum, the restorative force simply snaps back to its equilibrium, perhaps with some overshoot. If the SHO has inertia, then the restorative force converts potential energy to kinetic energy as it approaches equilibrium, while kinetic energy is converted back to potential energy as it moves away from equilibrium. In a simple electric circuit as shown in Figure FB6.3, the inductor, L, resists any change in the current, i, and the capacitor, C, supplies the restorative force in the form of the electric potential (i.e., the voltage). The capacitor sinusoidally charges back and forth between $+V_0$ and $-V_0$ while the current, which is 90° out of phase with respect to the voltage, does the same through the inductor. *Note*: an SHO is an idealized case that assumes there is no dissipative force (friction in a mechanical SHO, resistance in a SHO circuit). This is an excellent approximation in many circumstances.

The steady-state solution for all simple harmonic oscillators describes the trigonometric sinusoidal function, which is equivalent to an object moving on a circle as observed from an edge-on view. If we position ourselves directly underneath a pendulum looking up, we would only see left-right motions. Mathematically, the solution is $x = A\ \sin(2\pi\ t/P + \phi)$ where t is time, P is the period, the time necessary to repeat the cycle, and A is the amplitude, the largest distance either left or right. The portion of the cycle when we first started our observation is the phase, ϕ. We note that the cosine of an angle is simply the sine of that angle plus 90° (i.e., cos[angle] = sin[angle +90°] or sin[angle] = cos[angle −90°], indicating either the cosine or sine function with appropriate initial phase angle can be used as the solution to a SHO). If we start plotting (watching) the pendulum as it passes directly over us going from left to right, then the phase is zero ($\phi = 0°$). The plot of the displacement of a SHO against time is simply the dashed line curve in Figure FB6.4. The period, P, the time interval necessary before one segment of the plot starts to repeat can be measured from one peak to the next, from one trough to the next, from one zero crossing going positive to the next, or any arbitrary starting point. Fundamental properties of an SHO such as P are easy to calculate. For the

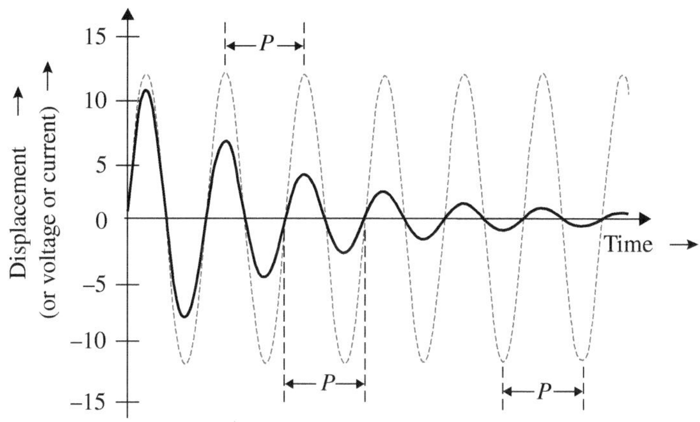

FIGURE FB6.4 Displacement versus time plots for a simple harmonic oscillator (red dotted line) and for a damped SHO (bold line). The period of oscillation, *P*, referenced from three starting points is shown.

LC circuit, $P = 2\pi\sqrt{LC}$. $P = 2\pi\sqrt{m/k}$ for a mass on a spring and $P = 2\pi\sqrt{L/g}$ for a pendulum, where m is the mass, k is the spring constant (of force), L is the length of the pendulum wire or string, and the constant $g = 9.81\ \text{m/s}^2$ is the acceleration of gravity at the earth's surface.

The resonant period or equivalently the natural frequency ($f_0 = 1/P$) are indicative of the how the SHO will respond to repeated external stimuli. For example, very little effort needs to be applied to keep a child on a swing set going high into the air, if each push occurs when the swing is at the correct phase (i.e., at the end of its backward motion). A few missed pushes here and there will have little impact on the motion of the swing. The individual giving the pushes is said to have excited the natural frequency, f_0, of SHO pendulum-like motion. If the pushes were to occur with a repetitive rate that differs from the f_0 of the SHO or at random points in the cycle, then some pushes would slow the swing down and the average amplitude of the swing will be much smaller and variable.

The SHO is an idealized case that does not account for dissipative losses. Real harmonic oscillators always have at least some small amount of dissipative force, whether it be in the form of electrical resistance or friction in mechanical systems. Even a simple wire, which is a good conductor, has some resistance. These systems, known as damped simple harmonic oscillators, have an oscillating amplitude that diminishes over time until all motions cease. The bold line curve in Figure FB6.4 plots the oscillatory movement of damped SHO, having a fairly strong amount of dissipation.

To maintain the oscillation, energy from an external source must be supplied to compensate for any dissipative losses. Such systems are referred to as driven simple harmonic oscillators. An example of a driven mechanical SHO is given in Figure FB6.5, where a wheel is rotated at a constant rate, piston-style driving the

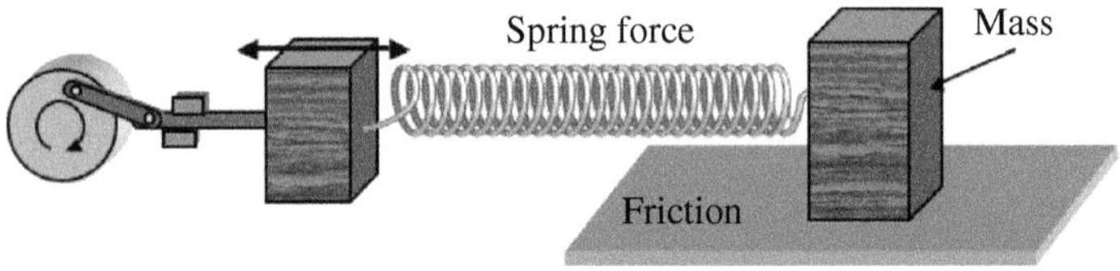

FIGURE FB6.5 An example of a mechanical driven SHO.

left block back and forth. The coupling between the drive wheel and the left block has tight tolerances, insuring left block's amplitude of motion is fixed by the travel of the piston rod, and its frequency of oscillation strictly corresponds to the rotational speed of the wheel. The right block attached to the spring forms the SHO, which is driven by the left block and is also damped by friction between its mass and the surface over which it slides.

There are some important steady-state properties of a driven SHO that are relevant to a general understanding how complex systems will interact with each other. First, only the mass of the right block and the spring force determine the natural frequency that optimizes the amplitude (maximum back and forth motion) of this SHO system. If the drive wheel rotates much slower or much faster than f_0, the maximum range over which the right block slides will be significantly smaller than it would be at its natural frequency. Moreover, the motion of the right block will be out of phase with the left block, indicating the two blocks will be moving in opposite directions during a portion of each cycle. This phase difference imparts additional forces on the drive mechanism, causing it to work hard and perhaps wear out prematurely. When the drive wheel operates close to the natural frequency, both blocks are move in the same direction at the same time (in phase) with the total range of the right block being greater than that of the left one (an amplification). If the amount of dampening (friction in this case) is relatively small, the amplification will be very large. Conversely, a large dampening leads to a small (but still >1) amplification. *Note*: the discussion above is only for steady-state conditions, once all transient phenomena have ceased and the system has had time to settle down to continuous operation.

6.3 DIODES AND TRANSISTORS

Diodes and transistors are active components in solid-state circuitry, providing the nonlinear performance necessary for amplification or switching. For example, a diode creates nonlinear resistance/conductance in contrast to a passive resistor that retains these properties over a wide range of voltages and currents. The most widespread use of a diode is to pass current in one direction but essentially not in the other. Transistors are ubiquitous in ICs and VLSI chips, forming the fundamental blocks of modern circuits. Transistors are used in solid-state circuitry for a wide variety of purposes, but primarily as either an amplifier or a switch. Amplifier applications, which typically use bipolar transistors with three electrodes, include among others, TVs, mobile phones, radio transmitters, and stereo sound systems. Field effect transistors (FETs) with four electrodes are used in digital logic gates and in switched-mode high-power supplies. (An individual logic gate may contain up to about 20 transistors.) A key requirement of both diodes and transistors is the sharp boundary separating crystal layers doped with charge acceptor atoms and with charge provider atoms, known as n-type and p-type, respectively.

Prior to a treatment of diode and transistor structures, it is instructive to examine the electrical properties of a crystal that has been doped with only one type of impurity. If the dopant consists solely of donor–electron atoms, then it is n-type with the dominant charge carrier being electrons. (The "minority" charge carriers for an n-type material are holes—electrons missing from the lattice. The ratio of majority-to-minority charge carriers is typically ~1000.) The schematic representation of Figure 6.11, an n-type semiconductor, illustrates the lattice structure for a Si crystal. Each Si atom shares one of its four outermost electrons with each adjacent neighbor, so that every atom in the lattice has eight shared valance electrons surrounding it. These valance electrons are represented by red spots and the valance bonds by the solid black lines between the atoms in the figure. The material has been doped with a small percentage of phosphor atoms, which have *five* valance electrons, one more than required by the lattice. These fifth electrons are also represented by red circles, but without a bonding line to an atom. In a Si lattice, each P atom is an electron donor, requiring about 0.01 eV of additional energy to promote this fifth electron to the conduction band. In other words, the Fermi level is 0.01 eV below the conduction band as depicted in the energy diagram of Figure 6.4. Virtually all of these fifth electrons are in the conduction band at RT, since the average kinetic energy of atoms and free electrons is 0.025 eV at RT.

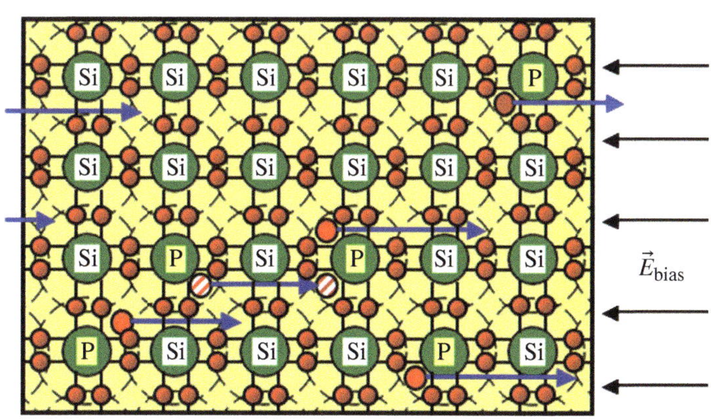

FIGURE 6.11 A schematic of the lattice structure of a silicon n-type semiconductor. The impurity dopant element in this case is phosphor, having a fifth valance electron and being donor atoms.

If the n-type material represented in Figure 6.11 has a voltage differential applied from side to side, then those electrons in the conduction band move relatively freely in response. The electric field points from high to low voltage, right to left as pictured. Recall positive charges move with the electric field, $\vec{E}$, while electrons, which have a negative charge, move opposite to it (i.e., up the potential). Electrons arrive from the left in the figure, move through the portion of the lattice shown, and exit to the right. The process is not 100% efficient, however. Some of the electrons passing close to one of the positively charged P ions with its electron missing may be captured by that ion and temporarily removed from the current flow. This inefficiency is represented in Figure 6.11 by the white-red hashed circles near the bottom center of the lattice, denoting the electron exchange between the two P atoms. The locations of the positively charged P atoms (the holes), which move in the opposite direction to the electrons, are the minority charge carrier. Thus, an n-type semiconductor behaves similar to an ordinary resistor when a voltage is applied, albeit one with minimal resistance.

Alternatively, if the impurity atoms have only three valance electrons (e.g., boron, aluminum, gallium, or indium), then there will be too few valance electrons in the silicon lattice to complete all of the crystal bonds. Figure 6.12 depicts the lattice structure for this p-type semiconductor. Referring once again to the energy diagram of Figure 6.4, the Fermi level is 0.01 eV just above the valance band for these boron acceptor atoms. A free electron combining with a boron atom is equivalent to an electron jumping up from the valance band to this Fermi level, which is very favorable at RT. Moreover, the thermal energy required to promote this same electron back to the conduction band *is not* favorable, locking it to the B atom. Effectively, the boron atoms pull over an electron from a nearby silicon–silicon bond, leaving electron holes in the lattice as shown in Figure 6.12 (white spots). All of the B atoms are negatively charged and are surrounded by eight lattice electrons, while some of the

FIGURE 6.12 A schematic of the lattice structure of a silicon p-type semiconductor. The impurity dopant element in this case is boron, having only three valance electrons and being acceptor atoms.

Si atoms only have seven lattice bonds, one missing—a hole. These boron ions remain negatively charged and locked into lattice. This p-type semiconductor also acts as a low resistance material, but the majority charge carriers are holes rather than electrons. Semiconductor scientists and engineers treat holes as if these are positively charged particles moving through the crystal, when in fact all positively charged Si ions remain fixed within the lattice. The movement of a hole actually occurs when an electron jumps from a nearby lattice position to fill an existing hole. While it is the electrons making the actual jumps, the locations where there are a net positive charge (a hole) are moving oppositely. Hence, it makes sense to treat holes as the equivalent to positively charged particles and to classify holes as the dominant charge carrier in a p-type semiconductor. In Figure 6.12, the black arrows indicate the movements of holes, indicating the locations of the electrons undergoing a swap in lattice locations. New holes are created at the right edge of our crystal when the electrical contact removes electrons, sending these back to the rest of the circuit. (*Hint*: To keep the nomenclature straight, relate the majority charge carriers to the dopant type: n-type for negative (electrons) and p-type for positive (holes).)

The p–n boundary of a conventional diode demonstrates the physics behind the nonlinear electrical behavior of all active components. A schematic representation of a silicon p–n diode is given in Figure 6.13. The left half of the diagram has been doped with electron-acceptor atoms, making it p-type, while the right half has been doped with electron-donor atoms, making it n-type. Majority charge carriers are represented inside circles and minority carriers simply by the sign (+ or −) of the charge. (Recall the ratio of minority-to-majority carriers is ~10^{-3}.) The boundary between p- and n-type materials, known as the junction, must be sharp and relatively free of the crystal imperfections that reduce transfer of majority charge carriers across the junction. It is impossible to make good p–n junctions by mechanically compressing or gluing the two sides of a diode together. Instead, there needs to be an abrupt change in the doping materials during crystal growth.

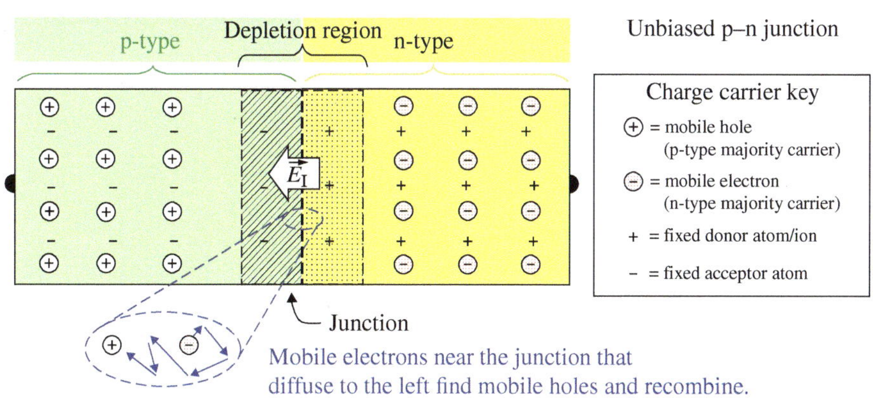

FIGURE 6.13 An unbiased p–n junction diode. Random thermal drifts of the majority charge carriers result in a depletion region (~1 μm thick) with an internal electric field.

As we have seen, most of the majority carriers in any doped semiconductor are in the conduction band and free to move throughout the material. These majority carriers randomly drift around the lattice even without the presence of an external electric field. Consider the conduction band electrons in the vicinity of junction on the n-type side. If these electrons move away from the junction, then the n-type material becomes locally more negative and mutual electrostatic repulsion forces some of the excess electrons back toward the junction. Alternatively, if some of these electrons drift toward the junction and cross over it to the p-type side, these electrons eventually find mobile holes (the majority carriers) and recombine with those. Identical processes occur on the opposite side of the junction for the mobile holes. Thus, the random drifting of both major charge carriers leave a depletion region on both sides of the junction, containing only the fixed donor and acceptor ions and resulting in an internal electric field, $\vec{E}_I$, that points from the n-type to p-type crystal. Mobile charge carriers that subsequently drift into the depletion region are swept back away from the junction by $\vec{E}_I$. The size of this depletion region is typically 1 μm (about 10,000 atoms thick).

Inside a circuit, a p–n diode responds in one of two ways, depending on whether it is forward or reversed biased. These two conditions are depicted in Figure 6.14. A battery as oriented plus a resistor in the top diagram are connected to the diode to produce a forward biased electric field, $\vec{E}_B$, one that is opposite to that of $\vec{E}_I$. If $\vec{E}_B$ is

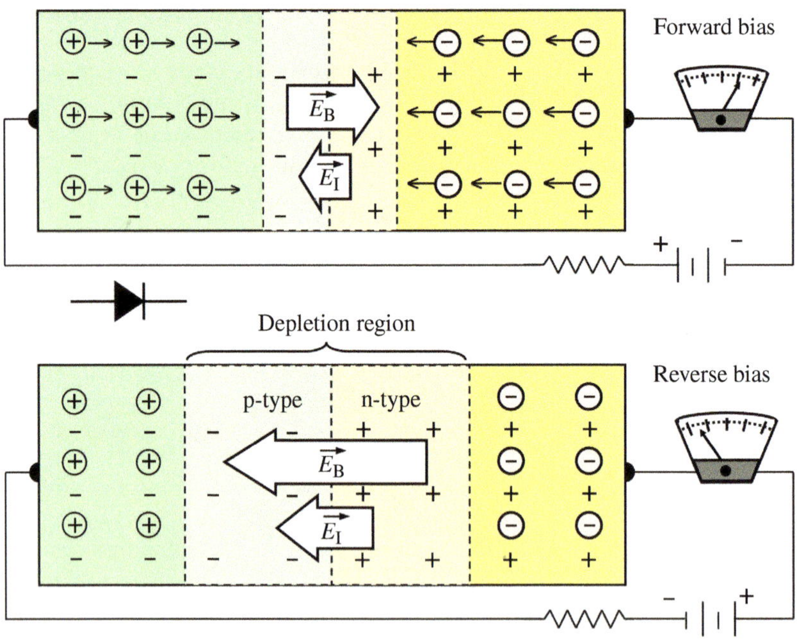

FIGURE 6.14 Diagrams for the two operating conditions of a p–n diode. Current flows freely in the forward direction but not in the reverse. The circuit symbol for a p–n diode is presented at the left between the two diagrams.

larger than $\vec{E}_I$, then the depletion region disappears and both majority carriers (holes and electrons) flow across the junction where these combine with their opposite counterpart. New replacement holes are created at the left by the removal of electrons via the electrical contact, while additional electrons are supplied at the right. *Note*: there is a voltage threshold (~0.5 V for silicon) above which current flows freely and below which it does not. This voltage threshold corresponds to the condition of $\vec{E}_B \geq \vec{E}_I$. The current flow above threshold increases rapidly as a function of voltage.

If the polarity of the battery is switched (bottom diagram), then the diode is reversed biased and the current meter on the right reads a low value regardless of the external bias voltage. In this case, $\vec{E}_B$ and $\vec{E}_I$ point in the same direction, creating a large depletion region that acts as an insulator. The small amount of current that does flow while reversed bias is nearly constant as a function of voltage up to the point where a catastrophic breakdown occurs. The flow of current through a p–n diode as a function of bias voltage, plotted in Figure 6.15, is known as a characteristic *i–V* curve. The electronic symbol for a p–n diode is provided as a reminder that positive charge flows from p- to n-type material (and electrons oppositely). As can be seen, the current for a reverse bias ($V<0$) is nearly constant, while it increases rapidly for a forward bias ($V>0.5$ V). For sufficiently large voltages on either extreme, the electrical properties and the material itself begin to breakdown, leading to a runaway condition. The two breakdown voltages, denoted V_{break}, *are not* symmetrical about $V=0$.

We are now ready to examine the physics and operation of a bipolar transistor. Two schematic representations along with the corresponding circuit symbols are given in Figure 6.16 for a pnp and an npn bipolar transistor. A bipolar transistor has three distinct parts, the emitter, base, and collector, each of which is biased individually. The bias voltages or "biases" of the crystal are taken to mean separate DC voltages applied at all times to the base, emitter, and collector to make the transistor function properly. Normally, the base–emitter junction is forward biased, eliminating any depletion region, while the base–collector junction is reversed biased with a depletion region and an internal electric field pointing towards the collector. The current flows indicated in Figure 6.16 represent the directions positive charges move.

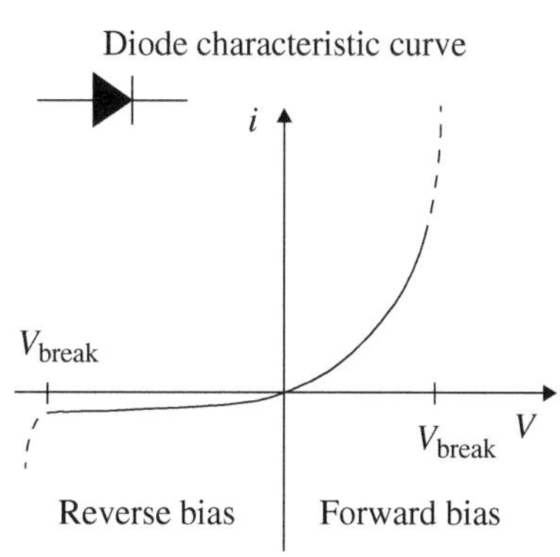

FIGURE 6.15 The characteristic *i–V* curve for a p–n diode.

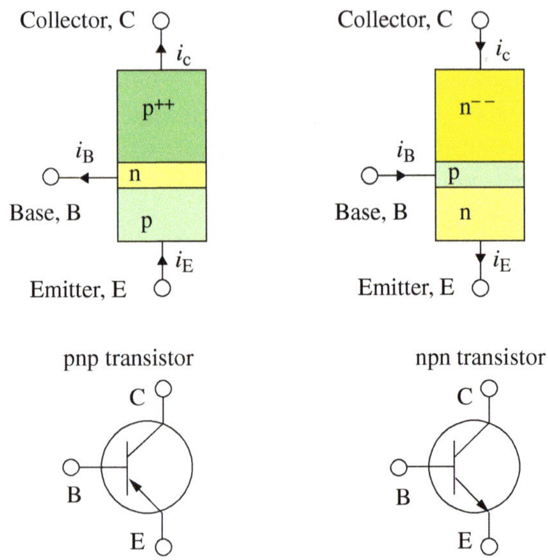

FIGURE 6.16 Schematics and circuit symbols for two types of bipolar transistors.

Observe that in both cases, the inward or outward flow of charge is the same for both the base and the collector, which is opposite from the emitter. Conservation of current requires $i_E = i_B + i_C$. In other words, the current flowing into the emitter equals the current leaving from the base plus the current departing via the collector.

A bipolar transistor is analogous to butting back-to-back two p–n diodes in opposite directions and we might expect a large current between the base and the emitter based on our previous discussions of forward versus reverse biased diodes. This expectation, however, is absolutely incorrect because there are some critically important differences between a transistor and the two-diode analogy. Explicitly, the crystal layer comprising the base must be thin compared to the diffusion length (i.e., the mean free path) of the majority charge carrier from both the emitter and collector. Another important distinction is the doping levels in the collector, which are higher than those in the emitter. To remind us of this feature, we used a "p^{++}" and "n^{--}" designation on the collector side. The collector also has to be physically larger than the emitter because it has to carry more current and dissipate more heat. (The transistor can be destroyed by swapping the emitter and collector leads since the biases will be inappropriate for the respective levels of doping as well as the relative internal resistances incorrect.) While there are some distinct performance differences between pnp and npn transistors, we shall restrict our attention from now on to the pnp type for simplicity.

In a pnp transistor, mobile holes are the majority carrier in the emitter and collector, which are both p-type. Holes injected from the emitter into the base may participate in one of three processes: (i) combine with the majority carriers in the base—mobile electrons, (ii) diffuse through the base to recombine with mobile electrons being injected via the electrical contact to the base, or (iii) diffuse across the base region into the depletion layer and be swept by $\vec{E}_I$ into the collector. The very thin layer of the base in bipolar transistors is the essential feature, making the transistor a useful device and insuring process (iii) is far more probable than either (i) or (ii). In actual transistors, this thin base also guarantees that $i_C \gg i_B$. In fact, 98 or 99% of the holes injected into the base from the emitter are swept into the collector, implying only 1–2% must leave via the base since $i_E = i_B + i_C$.

To serve as an amplifier, the output of a transistor must be greater than its input. The gain or amplification, A, which is usually defined as output divided by input, is given by $A_V = V_{out}/V_{in}$ for voltage gain or as $A_i = i_{out}/i_{in}$ for current gain. A design of a single-transistor amplifier is given in Figure 6.17. The device structure of this transistor (lower left) shows an asymmetrical depletion region between the base and collector caused by the increased levels of dopant in the collector. (If the impurity level in the collector were the same as those of the base and the emitter, then the depletion region would extend equally into the base and the collector, requiring an unacceptably thick base crystal that would then function as a pair of back-to-back diodes.) A depletion region does not exist between the base and emitter because it is forward biased.

A straightforward circuit design for a single-transistor amplifier with a common emitter configuration is given in Figure 6.17 (upper right). The necessary bias

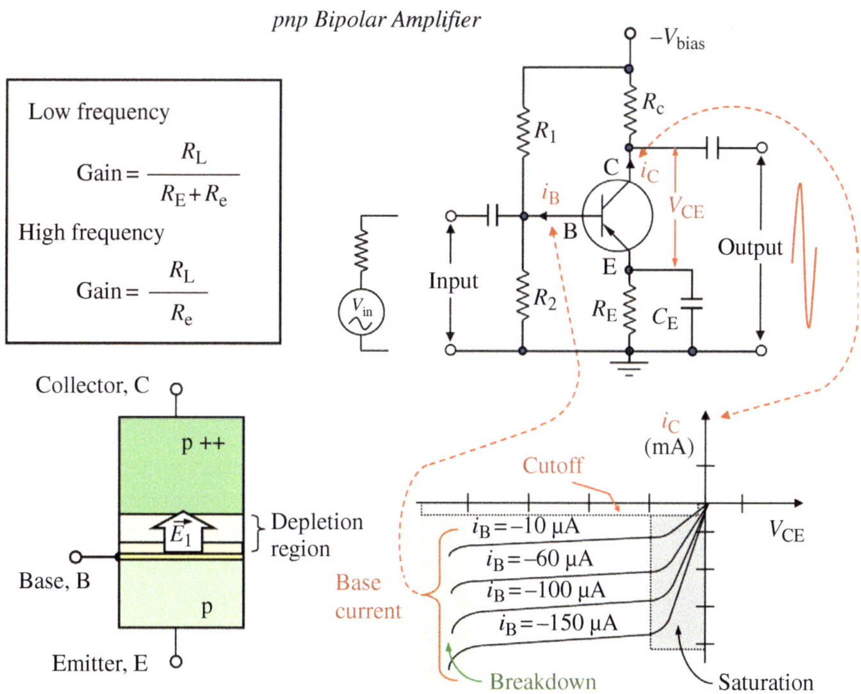

FIGURE 6.17 The design of a single-transistor amplifier.

voltages are supplied from one negative source ($-V_{bias}$) through a set of voltage dividers. (See Intro Physics Flashback FB6.4 for a treatment on voltage dividers.) For example, resistors R_1 and R_2 establish the DC voltage bias at the transistor base. R_C and R_E bias the collector and the emitter, respectively. A judicious choice of resistor values (bias voltages) determines transistor performance, including the current at the base, i_B, prior to any input signal. A family of characteristic curves for several possible values of i_B, are plotted in Figure 6.17—bottom right. Each curve represents a reverse-biased p–n junction diode for with a particular DC bias and corresponding base current, i_B, between 10 and 150 μA. The collector current, i_C, is measured in milliamps (i.e., in thousands of microamperes), consistent with $i_C \gg i_B$. We are now ready to discuss the physical processes that lead to amplification in a bipolar transistor.

An input signal sent to the base will produce amplification at the collector, provided all of the holes from the emitter pass into the collector. The base–emitter junction is forward biased so the crossing current is a very sensitive function of the base–emitter voltage. (See Fig. 6.15.) There is already a significant current flowing from the emitter into the base where it is immediately swept into the collector along with a small current from the base. If a pulse of voltage is sent into the base, temporarily more electrons flow in or out of the base, dramatically changing the amount of current between the emitter and base. The current flowing through collector is comparably increased since virtually all of the charges crossing the emitter–base junction are swept via the electric field of the depletion region. The already large current between the emitter and the collector follows any change to the base, but as an intensified signal.

In Figure 6.17, a sinusoidal input voltage is sent through a capacitor to the base. (The capacitor will transmit any varying signal, but block any DC current.) The input signal is small and only produces very minor changes to the base current, which is equivalent to moving along the nearly horizontal portions of one of the characteristic curves (plotted bottom right). As can be observed, the voltage between the collector and emitter, V_{CE} (labeled in red), changes dramatically for a small change in i_C, the collector current. The gain or amplification is simply V_{CE}/V_{in}, which is again coupled through a capacitor so that only the variable component of the signal is transmitted as an output. There are two ranges of gain, one low and one high frequency, set by the RC circuit component containing R_E and C_E. (See Section 6.2.) This results in two equations for the gain given in Figure 6.17 (upper left). At high frequencies, C_E becomes a conducting shunt to R_E, essentially making $R_E = 0$ for any rapid voltage changes. Typical gains for bipolar transistor amplifier are approximately 5 for low frequencies and approximately 200 for high frequencies.

There are of course, many solid-state amplifier designs, including those with two or more transistors, npn instead of the pnp discussed, and others with the base or with the collector tied to the common. We have not discussed a number of other important considerations (e.g., feedback and noise control) as well. Also, we have only discussed simple p–n junction diodes, leaving out tunnel diodes or zener diodes (also called avalanche for reference). Our goal was to provide a basic physical understanding of the p–n junction as it relates to typical diodes and bipolar transistors. Many introductory textbooks on electronic circuitry devote whole chapters to the various transistor designs.

INTERESTING TIDBIT TB6.3

In 2002, approximately 60 million transistors were fabricated for each man, woman, and child on Earth. Source: J. Turley (2002).

6.4 FET, JFET, MOSFET, CMOS, AND TTL

The reader should be aware that there are many device architectures used in solid-state semiconductors and most can be identified readily by universal acronyms. The most common subgroups are field effect transistor (FET) and metal–oxide–semiconductor (MOS), leading to hybrid acronyms such as nMOS, pMOS, CMOS, JFET, and MOSFET. FET and MOS always form root names inside any larger acronym, allowing anyone to identify the basic operational physics of that class of devices. Other letters are used only to modify these two roots. For example, nMOS has a MOS architecture based on n-type semiconductor and the "p" in pMOS stands for p-type material. CMOS, which we will describe later, now dominates the digital semiconductor industry. There is also transistor–transistor logic (TTL), which are logic circuits based on the bipolar transistors discussed in Section 6.3. TTL dominated logic circuits until about 1975 and now are only used in niche applications requiring less than a few hundred transistors. We begin with a discussion of FETs, followed by MOS, and end this section addressing some of the advantages and disadvantages of the various device types.

An FET is a special class of transistor with important advantages over conventional pnp or npn transistor designs, including a higher input impedance, lower noise, and higher resistance to nuclear radiation damage. The principal disadvantage of an FET compared to bipolar transistors is its limited gain, which is less than ideal for some amplification applications. Basic FET designs and electronic circuit symbols, given in Figure 6.18, have three components: the source, gate, and drain. In contrast with bipolar transistors, FETs have four electrical contacts, two of which are connected conductively to each other to form a single gate on both sides of a semiconductor channel. The source current, i_S, is always greater than- or equal to- the drain current, i_D, and the direction of positive charges is shown by arrows.

A good analogy of an FET is the flow of water through a pipe having a valve. For simplification, we once again restrict our discussion to one type of FET, the p-channel FET as pictured in Figure 6.19. If the gate contacts are unbiased (i.e., left floating) and a voltage is applied between the source and the drain as depicted in the left diagram, then the holes (the majority charge carriers) move through the channel easily since it is essentially a crystal with a single dopant. (See Fig. 6.12 and Section 6.3.) Similar to a fully open valve, which allows water to flow with minimal impediment, the FET responds in this situation as if it is a very-low-ohm resistor. If, however, we add a second battery (right diagram) to bias separately the gate, then the p–n junctions are activated and depletion regions are formed. The FET now acts as a high-ohmic resistor since many of the mobile holes originally in the regions near the junctions have recombined with mobile electrons to form the depletion regions, leaving far

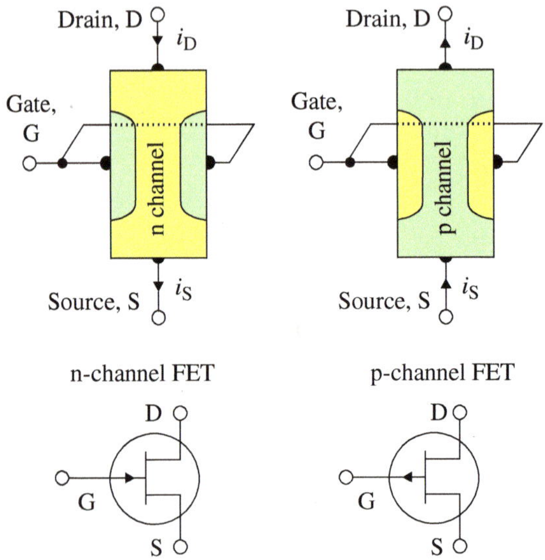

FIGURE 6.18 Schematic drawings and circuit symbols for an n-channel and a p-channel FET.

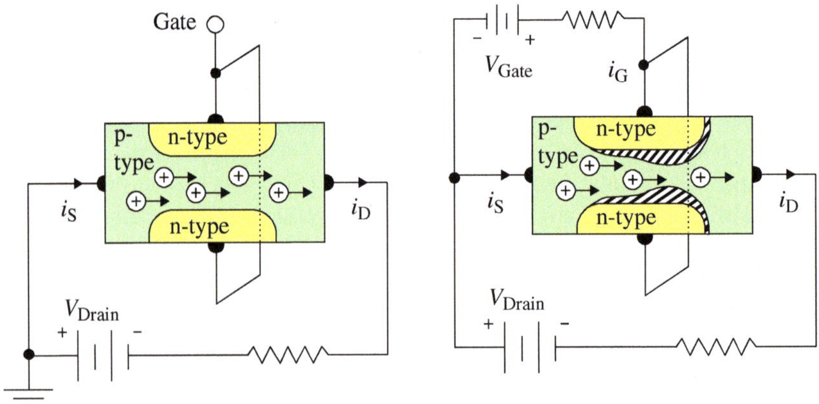

FIGURE 6.19 A p-channel FET that has been biased, allowing a current to flow. If the gate electrode is "open," then current flows freely (left). If a voltage is applied to the gate (right), then a depletion region forms (black-white dashed area), restricting the current.

fewer charge carriers in the p-channel for the flow along the length of the channel. *Note*: the thickness of the depletion regions is not uniform through the channel since mobile charge carriers flowing through the channel neutralize some of the fixed atoms in the first part of the depletion region and there are fewer carriers towards the end of it. The size of the depletion regions increase as a function of the gate voltage and $i_D = 0$ if V_{Gate} is sufficiently large. In practical applications, the separate

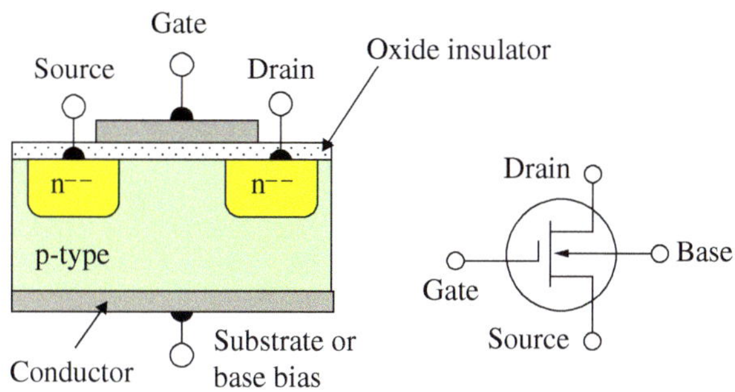

FIGURE 6.20 Device architecture of an n-type MOSFET (or nMOS) and corresponding circuit symbol.

bias voltages for the source, gate, and drain are not provided by batteries, but instead created using a single DC power supply and appropriate resistors in series to form voltage dividers. Thus, an FET is an efficient design to turn on and off the flow of current rapidly, making it an excellent digital switch for computers, logic circuits, and solid-state data storage devices among others.

A MOS is an alternative and more common FET architecture, also known as MOSFET. Figure 6.20 shows the basic device structure along with its circuit symbol. Following the previous color scheme and notation, the n-type material under the source and drain contacts are heavily doped. (Heavily doped in this context still means much <1% of the atoms.) Also *note*: the similarities of—and the few differences between—the MOSFET circuit symbol compared to that of the previous FET. (Circuit symbols reflect the operational physics and can be a helpful recall tool.) The MOSFET has a gate electrode separated by an oxide insulator, extending horizontally over the channel volume and partially covering the source and drain semiconductor volumes. As we shall see shortly, a MOSFET capacitively couples to the various types of semiconductor underneath, modifying the relative density of majority charge carriers in response to its applied voltage. The circuit symbol similarly reflects the capacitive coupling.

The two operational states of an n-type MOSFET are depicted in Figure 6.21. If the gate voltage is negative (right), then a large concentration of mobile positive charges are drawn close to the oxide barrier in the central region below the gate and depletion regions are formed on both p–n junctions. A series of "–" symbols on the gate electrode signify its voltage potential. Combined with the positive mobile charges beneath the silicon oxide insulator, this portion of the device is effectively a capacitor. An "OFF" state of the logic switch occurs in this case and any current flow between the source and the drain is strongly inhibited. The FET in the "OFF" state functions as two oppositely oriented back-to-back diodes since the width of the central region is too large to behave as an npn transistor. Any voltage difference between the source and drain only increases one or the other depletion regions.

In contrast, the "ON" state of the FET in Figure 6.21 occurs when the gate voltage become sufficiently positive, allowing current to flow easily between the source and drain. (An intermediate positive voltage will still produce a relatively low resistance

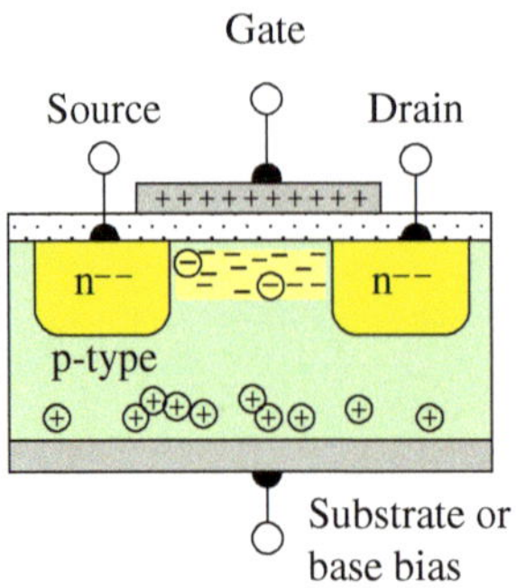

FIGURE 6.21 The operational states of a MOSFET logic switch. A positive gate voltage establishes the "ON" state (left), allowing current to flow easily between the source and drain. A negative gate voltage produces the "OFF" state (right).

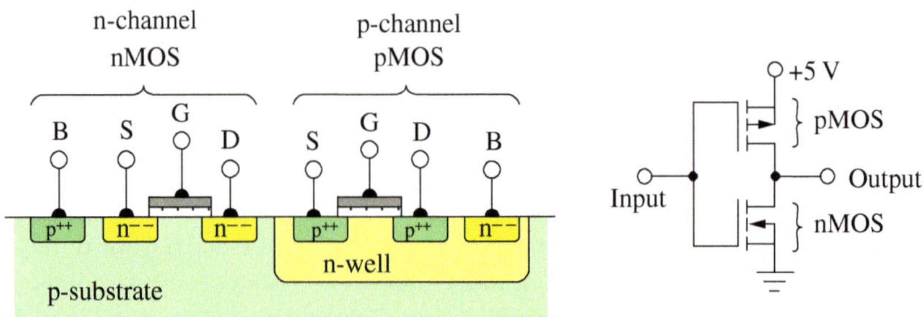

FIGURE 6.22 CMOS device architecture of an FET and its circuit symbol.

path, but one that is not nearly as efficient as a large $+V$.) The strong $+V$ on the gate electrode, once again acting as a capacitor, pushes the major charge carriers in the p-type material away from the gate and toward the bottom of the device. This "ON" state leaves the negative minority carriers and draws in some of the mobile negative carriers from the drain and source volumes. The impact effectively creates an n-type channel in the central region just below the gate (left figure), which enables the free flow of current between the source and drain. Similar to the single-dopant behavior observed in Section 6.2, the charge carriers do not encounter a voltage barrier and can flow freely in response to a voltage difference between the source and drain. The channel in a MOSFET only exists during the "ON" state in contrast with a conventional FET where it is permanent.

CMOS stands for complementary-symmetry MOS. This type of FET consists a symmetrical pair of p-type and n-type MOSFETs working together. More generally, all CMOS electronic components (e.g., an FET, a capacitor, or a resistor) are each comprised of a pair of nMOS and pMOS sections. The CMOS FET is pictured in Figure 6.22 along with its circuit symbol. (*Note*: the pMOS portion of the FET has a volume labeled "n-well." This small volume is doped n-type and its "well" designation indicates it serves as a pool of negative mobile charge carriers.) As with previous FETs, the gate electrode capacitively couples to the substrate material. Pairs of FETs are combined in parallel operation to form a single FET. The "complementary-symmetry" notation indicates each pair has identical components made from p- and n-type material, giving the device high noise immunity and low static power consumption. In a CMOS FET, a significant amount of power is only drawn when it changes state (i.e., from "ON" to "OFF" or visa versa).

CMOS has become the dominant architecture and industry standard for digital circuitry due to its high noise immunity and low power consumption especially when the circuitry is idle. Low power consumption implies not much waste heat is generated and is the main reason CMOS is the architecture of choice for VLSI chips. TTL, based on the older bipolar transistor, is still used in niche digital applications that require only several hundred transistors. Historically, a TTL device consumed substantially more power than a comparable CMOS device when idling, but its power consumption increased more slowly than it did for the CMOS device as a function of increased clock speed. This TTL advantage has all but disappeared as scale sizes of circuit components have decreased. Nevertheless, CMOS circuitry has always been sensitive electrostatic discharge (ESD) and care must be taken not to destroy a CMOS chip when handling it. TTL logic is less sensitive to ESD damage.

Finally, there are other FET architectures that we have not discussed. For example, the junction FET (JFET) is used as a voltage-controlled variable resistor (or a switch if operated only at the extremes of its range). Many of these devices can be easily be found on various webpages and the associated physics interpreted from a single device diagram, using the knowledge gained here. Also, the word, "metal" contained in the MOS acronym can be somewhat of a misnomer since polycrystalline silicon layers have often been used to create the necessary conducting paths and electrodes. The advantage of using polysilicon as a conductor is its ability to self-align precisely with the edges of other portions of an electronic component. Sometimes, the term insulated-gate FET (IGFET) is used instead of MOSFET. However, MOS with actual metal gates have regained popularity since the on chip conductors must have the lowest resistance possible for high-speed applications.

COMPREHENSION VERIFICATION CV6.2

Problem: Consider Figure 6.22. As is typically done, the circuit symbol at the right identifies input and output connections without relating these specifically to the gate, drain, or source. Relate the eight leads (B, S, G, D) from both the nMOS and pMOS sides (left side of the figure) to the output and input of the circuit symbol on the right. Explain.

Answer: The input is capacitively coupled to the pMOS and nMOS sides in the circuit symbol, indicating the input must be the gates, so both leads labeled "G" connected to each other must be the input. Since current in a FET always flows from the source to the drain, both "D" leads combine to form the output.

INTERESTING TIDBIT TB6.4

For several decades, the semiconductor of choice is silicon. Wide-bandgap materials in particular gallium arsenide potentially offers significant advantages over silicon and has been touted for years as the "wave of the future." However, as of yet GaAs does not form good semiconductor-to-insulator interfaces, making this material currently unsuitable for MOSFETs. Engineers and scientists invested in silicon research and production enjoy the following retort: "Gallium arsenide, the wave of the future. It always has been and always will be."

Other wide-bandgap materials (e.g., AlGaN and SiC) have had success. Traffic lights are rapidly being replaced with new ones that use light-emitting diodes made of GaN, eventually saving the US $5B/year in energy. AlGaN has become a multi-billion dollar industry and continues to grow rapidly.

6.5 SUMMARY

The semiconductor industry continues to increase circuit complexity and speed. Very large-scale integration (VLSI) chips now boast transistor counts measured in the billions. The trend toward miniaturization continues to be a critical enabling factor necessary to achieve increased functionality and speed, while minimizing power consumption. Tight device tolerances and quality control in turn are essential to obtain reductions in scale without a dramatic increasing failure rates and loss of reliability.

Semiconductors are made of polycrystalline materials with single-crystal chips offering the highest quality and device performance. Native semiconductor materials must be doped with impurities to obtain the required electronic properties. The physics of conductor, semiconductor, and insulating crystals has been discussed along with the requirements for impurity dopants. VLSI chips contain active and passive components. Transistors and diodes are active, while resistors and capacitors are passive. (If needed, very small inductors can be fabricated on chip, but larger ones are too bulky.) All circuitry, no matter how complex, can be treated as a black box with an input response of a simple combination of an inductor, resistor, and capacitor (*LRC*) and an output reduced to another *LRC* impedance. This fact enables individuals to estimate the range of speeds as well as the loads that can be attached to the black box.

Finally, there are two types of transistor: bipolar junction and FET. The physics leading to the performance capabilities have been revealed. Initially, the bipolar transistor was the most common, but the FET has dominated both digital and analog circuits since the late 1970s. MOSFET, in particular the CMOS FET is used virtually universally. Although there is, perhaps a dizzying array of device architectures, knowledge of the basic operating principles (i.e., the physics) of the two architectural categories presented in this text enables the reader to organize unfamiliar variants and understand how these function.

INTRO PHYSICS FLASHBACK FB6.4

Voltage Dividers

During an initial presentation to students, batteries are often depicted in circuits to conceptualize the application of various bias voltages at specific points in the circuit. However, the use of multiple batteries to provide several biases in general is not feasible. Batteries are bulky, heavy, and wear out at different rates altering the relative bias levels. Instead, bias voltages are supplied via voltage dividers from a single DC power supply. A voltage divider is simply two or more resistors in series with electrical contact points between the resistors. The voltage at any one of these contact points remains fixed, provided any load attached to it has a resistance that is much larger than those in the voltage divider.

The current, i, passing though a resistor equals the voltage across it divided by its resistance ($i = V/R$). Equivalently, we may write $V = iR$. If there are two or more resistors in series, all carry the same current so that $V_{total} = \Delta V_1 + \Delta V_2 + \Delta V_3 + \cdots = iR_1 + iR_2 + iR_3 + \cdots$. The fractional voltage drop across each must be proportional its resistance divided by the total resistance. Examples of a two-resistor and a three-resistor voltage divider are provided in Figure FB6.6, along with the appropriate equations to determine the DC bias voltage at points (A) and (B).

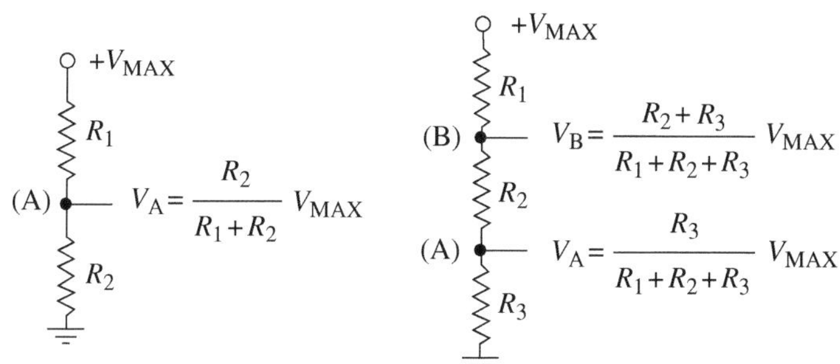

FIGURE FB6.6 Two examples of voltage dividers.

7

HIGH-TECH SEMICONDUCTOR FABRICATION

Semiconductor fabrication requires a number of important technologies, including vacuum systems, cleanrooms, crystal and amorphous thin-film growth methods, photolithography, and etching techniques. The technologies necessary to fabricate the increasingly complex, highly integrated circuits, require progressively more precision, cleanliness during fabrication, and miniaturization. The basic problem of fabricating integrated circuits (ICs) is that millions of components (e.g., transistors and resistors) must be laid down as uniform layers of conducting, semiconducting, or insulating films across the entire device. Any crystalline semiconducting layer must be doped with appropriate atoms to make the material accepting of electrons (p-type) or other atoms that are electron donors (n-type), all while minimizing dislocations and impurities. Then, the most recently deposited film in those physical regions that do not require this layer must be removed while leaving the remaining layers intact and undamaged. Moreover, most films are deposited at elevated temperatures and each film has its own coefficient of thermal expansion, introducing stresses and strains into the film layers.

7.1 THIN FILMS

Thin films are classified as any amorphous or crystalline layers that are less than 10,000 Å (1,000 nm) thick, which have been fabricated on a substrate or on an existing film. Typically, layers of 10 nm or more can reliably be put down without any significant gaps in the coating. The basic problem of depositing a thin film is to vaporize the appropriate atoms in approximately the correct percentages and to transport the molecules, atoms, or ions to the substrate. Once there, the atoms must be condensed onto the surface with enough residual kinetic energy that the atoms can migrate if necessary

Modern Devices: The Simple Physics of Sophisticated Technology, First Edition.
Charles L. Joseph and Santiago Bernal.

Companion website: www.wiley.com/go/joseph/moderndevices

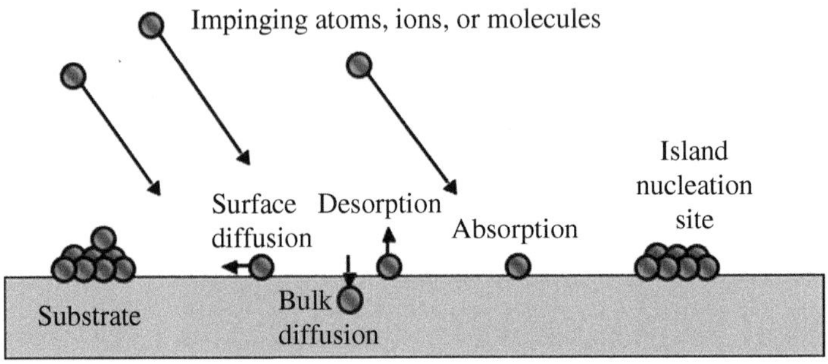

FIGURE 7.1 The vapor deposition process for the formation of an amorphous film.

to the appropriate locations. (If the process is a chemical rather than physical deposition, then a chemical reaction must also take place on the surface, one that releases the carrier atoms or molecules back to the gaseous state and retains the necessary atoms with sufficient kinetic energy.) Figure 7.1 shows the deposition process for an amorphous material. An impinging atom strikes the surface of the substrate, whereupon it can stick immediately (adsorption) or leave the surface (desorption). Alternatively, an impinging atom may migrate into the substrate (bulk diffusion) or move along the surface until it encounters and binds to other atoms (surface diffusion).

For most film depositions, the initial growth begins in isolated island nucleation sites. Additional gaseous atoms either clump to form new islands or contribute to—and enlarge—existing ones. The process occurs in nonequilibrium. The deposition is usually conducted under high or ultrahigh vacuums (UHVs) to insure the impinging atoms have a long mean free path. The elevated tenuous gaseous atoms are naturally hot, and the substrate is often maintained at somewhat elevated temperatures as well to facilitate surface diffusion. However, the substrate cannot be too hot or else virtually none of the impinging atoms will remain on the surface.

Single crystal or polycrystalline films can be grown for many materials. Silicon, carbon (organic), and most metals, for example, inherently tend to form polycrystalline structures. To create single-crystal devices, most depositions are fabricated using a crystalline wafer as a lattice template, a process known as *epitaxial* growth. If the intrinsic lattice spacing of the film atoms is well matched to that of the substrate and the film deposition occurs under favorable conditions (e.g., slow deposition rates and UHV), then a single crystal forms with minimal internal stresses. The substrate or previous epilayers form a location guide for newly arriving material. (See Fig. 7.2.) If there is lattice mismatch between the two layers, a polycrystalline film is usually formed with grain boundary defects between the localized crystals. The range of acceptable temperatures of the epilayers is far more critical than it is for the growth of amorphous films. Also, crystalline growth usually requires a harder vacuum environment than does an amorphous growth.

Substrate and vapor temperatures are important for the density and overall quality of the film being deposited. Thin film formation and growth, depicted in Figure 7.3, can follow one of several paths depending on the deposition rate and substrate temperature during the growth. If the temperature of the substrate is too hot, new islands

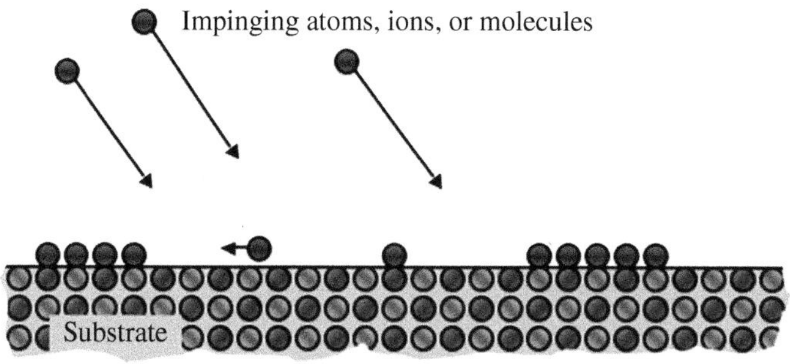

FIGURE 7.2 A vapor deposition process for the formation of a polycrystalline or a single-crystal film.

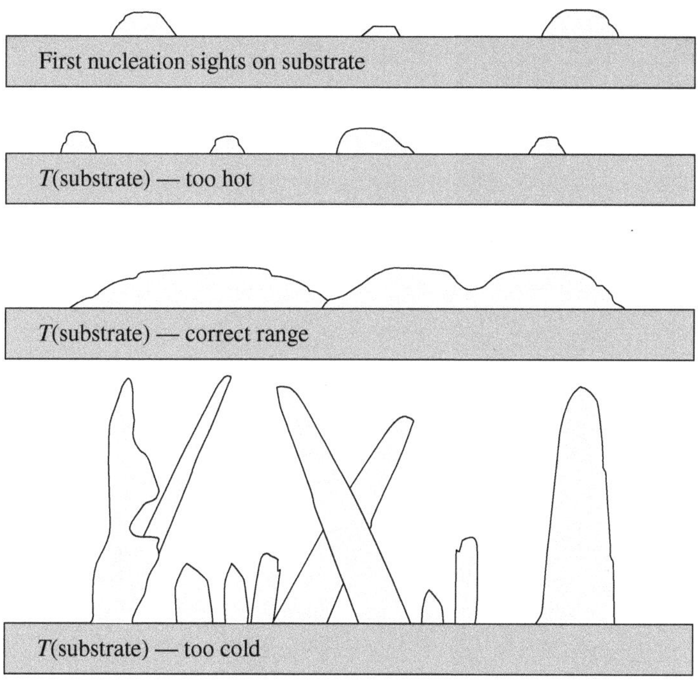

FIGURE 7.3 The quality of a thin film is influenced by the temperature of the substrate during deposition. The objective is to have a substrate temperature that forms a dense, uniform film.

will form continuously, but will eventually sublime back to the vacuum and the film will never cover the surface. If the temperature is too low (cold), then atomic migration (diffusion) will be severely inhibited, preventing the growth of a uniform dense film in favor of a fluffy low-density one, which will have poor electrical and optical properties. In this case, features resembling stalagmites form and grow at all angles, shielding localized regions from additional material deposition. If *T*(substrate) is

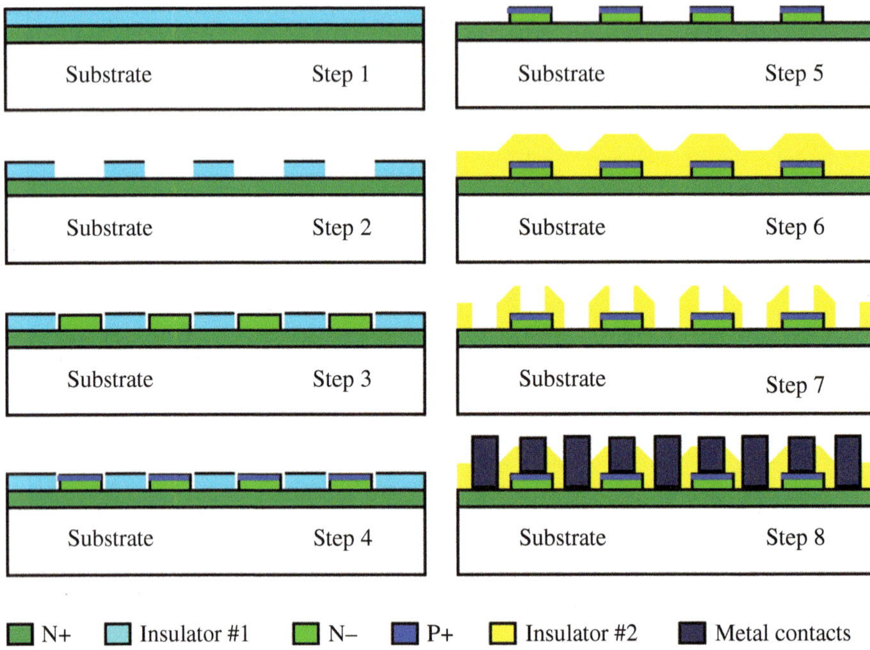

FIGURE 7.4 An example of the device fabrication steps used by the Naval Research Laboratory to create 2D image sensors from wide-bandgap materials of AlGaN. Source: Dr. Charles Eddy, Naval Research Laboratory, Washington, DC. Reproduced with permission of Dr. Charles Eddy.

close to the ideal temperature, then nucleation sites grow and merge with others to form a dense, relatively uniform film with a limited number of grain boundaries. These high-quality films produce the best optical and electrical properties.

To make solid-state devices via thin films, it is necessary to deposit multiple layers, including conducting, insulating, and semiconducting, and then to remove the unwanted portions of the latest film by etching. As an example, Figure 7.4 shows the basic steps to make a particular wide-bandgap image sensor. Steps 2, 5, and 7 require the selective removal (etching) of materials, which in turn involves several intermediate steps associated with photolithography to deposit a mask that is resistant to the etching.

Many methods of device formation introduce defects and damage to the overall structures and the films, requiring high levels of quality control. Moreover, most films have significantly different coefficients of thermal expansion and are deposited at various temperatures. These processing steps result in three types of film flaws: point, line, and planar defects. (See Fig. 7.5.) A point defect can be a normal crystalline atom trapped at an incorrect lattice point (a self-interstitial atom) or one that is missing, a lattice vacancy. A substitutional impurity atom at the lattice point or an interstitial one between lattice points can also cause a point defect. Grain boundaries form planar defects when two or more crystalline nucleation islands grow to merge with each other. Dislocations (line defects) can be threading or straight lines and are usually caused by the stresses and strains within the films when cool after

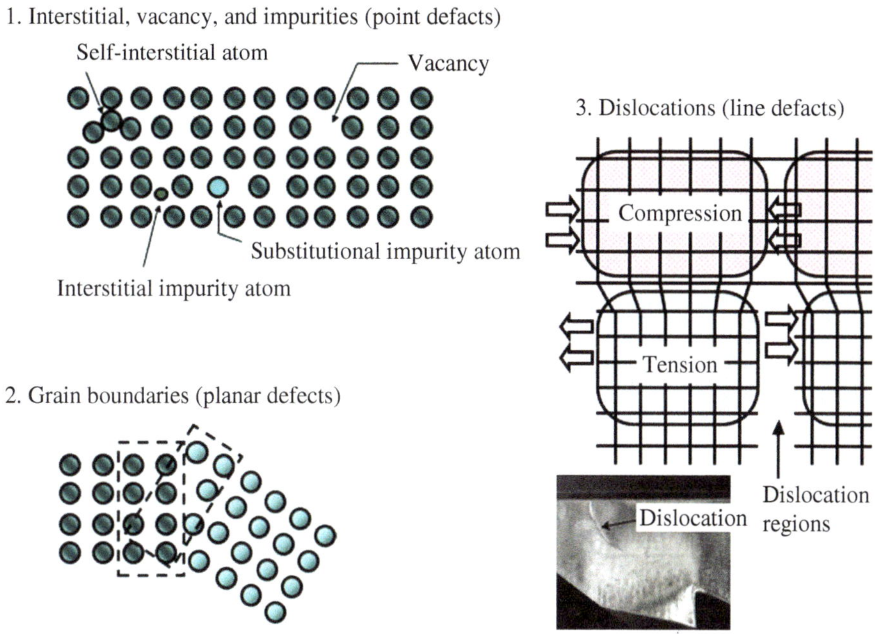

FIGURE 7.5 The three types of materials defects. The image of a dislocation in an AlGaN film. Source: Dr. Charles Eddy, Naval Research Laboratory, Washington, DC. Reproduced with permission of Dr. Charles Eddy.

processing. Some crystalline materials inherently have larger separations between separate lattice locations than do other film structures. As shown in Figure 7.5, the atoms near the boundary between two crystalline films that have mismatched lattice spacings introduce tension and compression strains between the lattices with small and large separations, respectively.

Each class of defects plays a major role in altering electronic performance of the materials, introducing systematic changes as well as reducing the operational lifetime and reliability of the circuitry. This statement is increasingly important as the number of components multiply and the physical size of each constituent is reduced. The circuitry is less tolerant to defects, each of which represents a larger fraction of each component.

INTRO PHYSICS FLASHBACK FB7.1

Thermal Expansion

The stresses and strains experienced by thin films can best be conceptualized by examining the response of bimetal strips to variations in temperature. A bimetal strip, consisting of two distinct thin metal layers fused together, is depicted in Figure FB7.1. These strips are used in thermostats to switch on and off a furnace or central air conditioner. The strip bends in one direction as the temperature

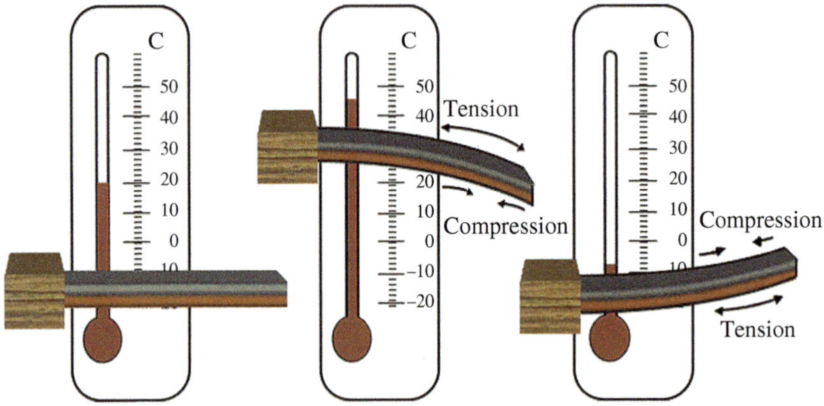

FIGURE FB7.1 Shows the response of a bimetal strip to different temperatures.

drops and in the other for elevated temperatures due to the differences in the coefficients of thermal expansion. As the temperature fluctuates, each metal contracts or expands at a different rate, introducing tension in one and simultaneously compression in the other since the two metals cannot slip at the common interior surface. If the two layers are sufficiently thin, the combined stress plus compression forces bend the strip. The stresses and strains intensify as the thickness of either or both increase. The two metals can and do separate if the thermally induced stresses are sufficient to break the bonding between the two materials.

Thin films are deposited at various temperatures (some elevated) onto a substrate, which is sufficiently thick that it severely restricts the amount bending flexure within multilayer films. Each film layer as well as the substrate has a distinct coefficient of thermal expansion. Moreover, most film layers are very thin compared to the bimetal layers and many materials are less deformable than are typical metals, causing severe internal stresses and strains often resulting in defects in the films.

7.2 THIN-FILM DEPOSITION METHODS

Thin-film depositions can be classified into two types: chemical and physical depositions. For the former, a chemical reaction takes place in the fluid (liquid or gas) precursor on the deposition surface. For instance, electroplating is a process where an electrical current is applied between two electrodes in a salt water solution. The metal of the anode is sacrificed, forming a charged metallic salt that is transported to the cathode piece being plated and the metal atoms from the salt are subsequently deposited. Most thin films are created via vapor deposition processes that take place under high or UHV. These too are classified as chemical vapor depositions (CVDs) or physical vapor deposition (PVD), each having a number of subgroup classifications. (See Table 7.1.) For example, metal organic chemical vapor deposition (MOCVD) is a subset of CVD.

TABLE 7.1 Types of Vapor Deposition

Physical Vapor Deposition (PVD)	Chemical Vapor Deposition (CVD)
Evaporation:	Plasma-enhanced CVD (PE-CVD)
Thermal	Atomic layer deposition (ALD)
Electron-beam (e-beam)	Metal-organic CVD (MOCVD)
Sputtering:	
RF	
DC magnetron	
Pulsed laser deposition (PLD)	
Molecular beam epitaxy (MBE)	

Perhaps, the simplest of these conceptually is a thermal evaporator, where the material to be deposited is placed in a crucible and an electric heating element surrounding the crucible heats the material until it liquefies and then vaporizes, coating everything inside the vacuum chamber. Some PVD evaporators use beams of energetic electrons or pulsed laser ablation "to boil off" (sublimate) material from a target, which then coats all surfaces in the chamber. Sputtering is another PVD process, which accelerates a beam of ions (normally of an inert gas) toward a target. There, the plasma vaporizes material at a rate of a few atoms per microsecond, coating the various surfaces. One advantage of sputtering is that the temperature of the target material can be kept relatively low. There are too many PVD and CVD techniques along with all of the hybrid variants to be discussed in this book. Instead, we provide four common processes as representative examples.

One of the oldest and simplest vapor deposition methods, depicted in Figure 7.6, is a bell jar vacuum chamber. It is used for non-refractory materials. A pump to provide high (10^{-4} to 10^{-6} Torr) vacuum is attached to a vacuum collar, a flat polished plate with feedthroughs and a vacuum pump port. A glass or metal dome with a ring of soft, non-outgassing material attached forms a vacuum seal between the dome and collar. One important advantage is the easy and open access to the contents inside the vacuum chamber that is obtained by simply lifting the dome off the plate. The evaporation apparatus consists of crucible that will not react chemically with the material to be deposited. The crucible often rests in a coil of tungsten wire, which serves as the heating element. This tungsten coil is attached to wires, running through ceramic insulating standoffs and connecting to an external low-voltage, high-current, DC power supply. (The coil performs much the same way that an ordinary incandescent light bulb does. When electricity passes through tungsten, a relatively high-resistant metal, it gets hot and emits visible- and IR-wavelength light.)

The insulating standoffs help prevent the formation of a conducting film, creating an electrical short to ground, if the deposition material is conducting. The surface or surfaces to be coated are usually positioned sufficiently above the crucible to facilitate a uniform thickness and faced downward to avoid spilling the molten material in the crucible.

Another thermal PVD method, depicted schematically in Figure 7.7, is the electron beam evaporation. Similar to a bell jar thermal deposition, the substrate(s)

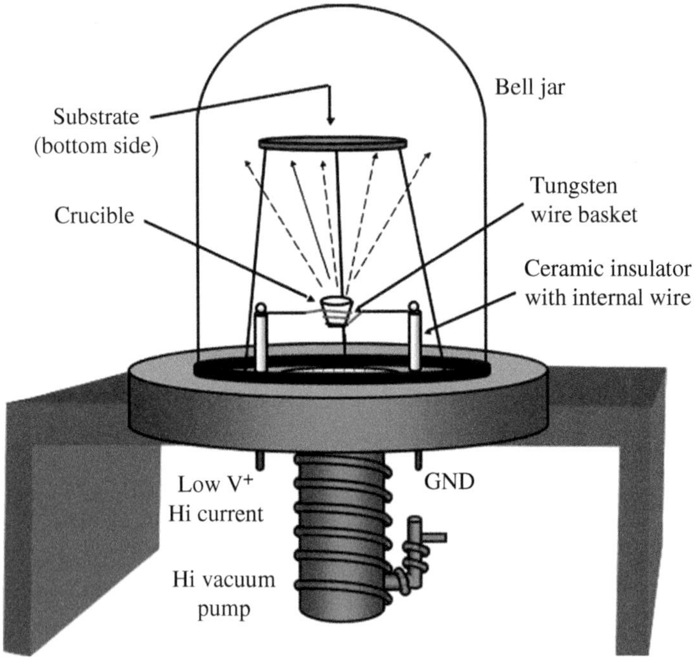

FIGURE 7.6 A simple bell jar evaporator system used for the deposition of non-refractory materials.

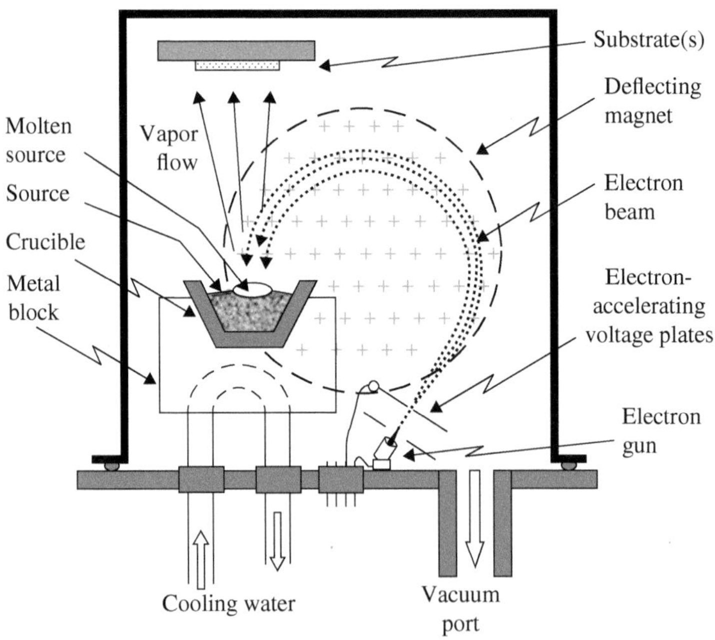

FIGURE 7.7 A vacuum chamber designed for thin-film deposition using e-beam evaporation. This method is used for refractory materials.

are positioned above the crucible and face downward. In contrast, e-beam evaporation is used to deposit refractory materials. An electron gun creates a beam of free electrons, which are accelerated and focused into a relatively collimated beam via a set of voltage plates. These electrons then enter a portion of the vacuum chamber with a uniform magnetic field (into the page in Fig. 7.7) and begin to gyrate around the field lines. The accelerating voltage and magnetic field are chosen such that the material to be evaporated inside the crucible intercepts the circular path of the electrons. The energy from the impinging electrons liquefies and vaporizes the central top portion of the material, which then evaporatively flows toward the substrate(s) where it condenses onto the surface(s). Overall temperature of the crucible is maintained via heat exchange between the crucible and a metal block surrounding the crucible. A regulated flow of cool water is used to transport heat away from the metal block and out of the chamber.

One particularly important thin film method is MOCVD, metal organic chemical vapor deposition, extensively used commercially in the mass production of devices requiring high-quality films. MOCVD is also a preferred method for wide-bandgap devices made of III–V materials since alternative techniques often result in unacceptable levels of device damage. (III–V refers to the columns in the *Periodic Table of the Elements*. Some important III elements are Al, Ga, and In, while N, P, and As are examples from column V. In particular, III-nitrides (e.g., AlN, GaN, AlGaN, and InGaN) have become in recent years a multibillion dollar industry, being driven especially by new requirements by the military and high-tech companies.)

A schematic layout of an MOCVD chamber and supporting equipment is provided in Figure 7.8. As with all CVD systems, a chemical reaction takes place on the surfaces of the substrates and a portion of the carrier molecules along with any unused gases must be removed via a vacuum pump. *Note*: most MOCVD chambers have a gas recovery capability attached to the vacuum pump exhaust so that the excess gases can be captured, separated and recycled. A key feature of any MOCVD method is the requirement to combine several gases into a mixture, including the pickup of molecules by bubbling one or more gases through a liquid known as the alkyl source. There are usually multiple 5-foot-tall high-pressure (2400 psi when full) gas cylinders linked together to provide continuous flows of the gases. These cylinders are often stored outside or in another room. Some of the gases present a health hazard to personnel, and there should be hazmat sensor/alarms located in multiple locations.

MOCVD requires carefully controlling the relative flows of each constituent gas as well as the overall rate of the mixture into the chamber. A controller is used to monitor and adjust the individual flow rates entering the chamber. The gas pressure exiting the chamber is measured using a baratron, a vacuum gauge based on changes in electrical capacitance resulting from flexure of a thin metal plate when exposed to unequal pressures. The controller adjusts a throttle valve to the vacuum pump based on the pressure readings from the baratron. The mixture of gasses is often dispensed through a set of showerheads and the platform holding the substrates is rotated to insure a uniform deposition of material. In conclusion, the temperature of the substrates can be managed several ways. In the example shown in Figure 7.8, the platform holding the substrates contains carbon-based dipole molecules that can be

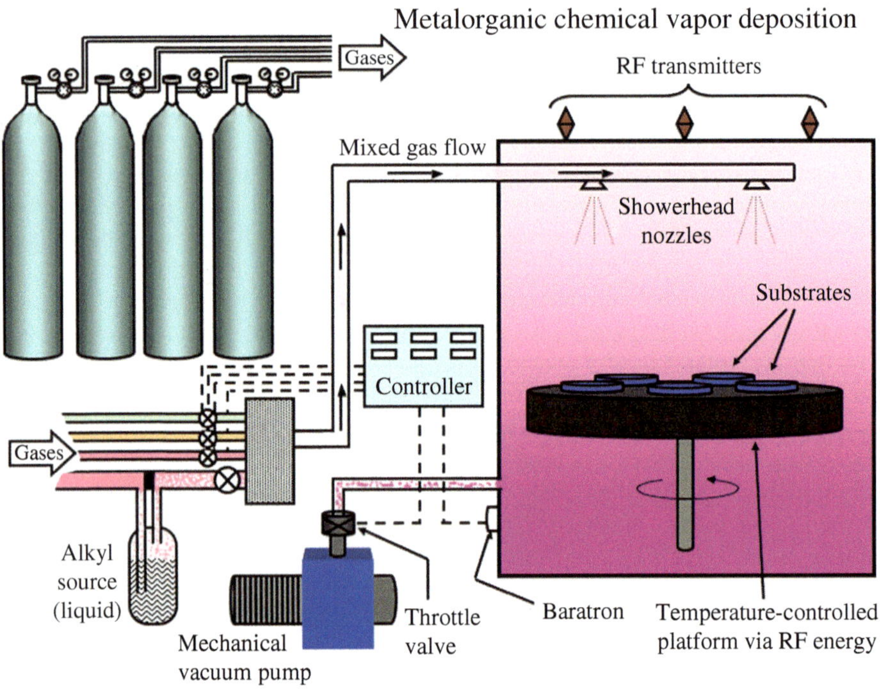

FIGURE 7.8 A schematic representation of a MOCVD deposition system. The showerheads and substrates are in vacuum, while most of the supporting equipment is not.

excited by radio frequency (RF) waves similar to the operation of a microwave oven. RF waves are added as needed to maintain the platform temperature.

Finally, molecular beam epitaxy (MBE) is used to deposit high-quality single-crystal films. MBE is unique in that the operator can simultaneously monitor and control film growth at the atomic level. In other words, it is capable of depositing layers that are a single atom thick at a time. To accomplish this level of sophistication, the chamber operates at elevated temperatures ($600 < T < 1200°C$), high vacuum or UHV ($10^{-10} < P < 10^{-7}$ Torr), and with slow deposition rates (typically < 1000 nm/h).

The most critical tool in the MBE process is the reflection high energy electron diffraction (RHEED) probe. RHEED is a technique where a grazing-incident electron beam scatters off the surface material, creating an interference pattern indicative of how uniformly (or irregularly) the lattice sites are being filled. A basic layout of an MBE chamber, given in Figure 7.9, has a few features in common with other systems. For example, the substrate is rotated to assist the formation of a uniform film. An MBE chamber, however, has several distinctive components. The first is the RHEED probe as already noted. The dotted line in Figure 7.9 between the RHEED gun and sensor demonstrates the grazing incident nature of the electron beam. Another attribute of MBE chambers are the multiple effusion cells with doors that allow various gases to seep into the vacuum chamber at the appropriate times. The system has a series of plates surrounding the chamber

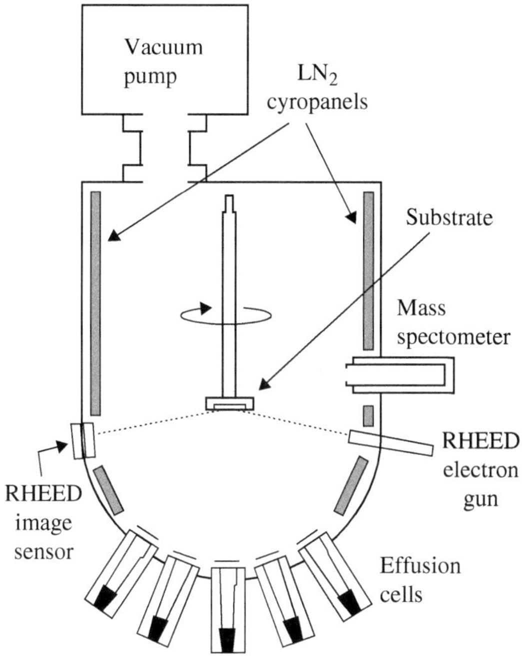

FIGURE 7.9 A schematic of a basic MBE chamber.

that are maintained at liquid nitrogen (LN_2) temperatures. These serve as extra vacuum pumps since most molecules will freeze to any cryogenic surface and remain there throughout the deposition process. These plates assist the primary pump maintain UHV even when the effusion cells are injecting gas. Auxiliary equipment often includes a mass spectrometer to evaluate the relative abundances of residual gas constituents from the effusion cells and to verify the absence of contaminants.

In summary, we have described thermal and e-beam evaporation methods, along with the MOCVD process and a cursory treatment of a basic MBE system. We will discuss in greater detail MBE in Section 7.3. Nevertheless, there has not been a treatment of ion sputtering or laser ablation techniques, which involves creating a plasma (electrically charged gas) and then transporting the plasma to the deposition site (substrate) using electrically charged plates.

INTERESTING TIDBIT TB7.1

Bell Labs, which was the first to create a working solid-state transistor, was the first to introduce molecular beam epitaxy (MBE) in 1970. Bell Labs introduced modulation doping for wide-bandgap materials specifically, GaAs/Al(Ga)As in 1978.

7.3 HIGH-PURITY CRYSTALS VIA MBE

It is useful to examine an actual operating MBE chamber to gain greater insights beyond those obtained in the previous section. Our intent is to demonstrate to the reader that with a little practice, he or she can start with the basic physics implied by the diagrams, combine a small amount of logic, and be able to understand unfamiliar technologies as well as be able to identify the essential components without getting mired in all of the details. Figure 7.10 contains a picture and a schematic of Professor Sean Oh's MBE chamber at Rutgers University. Do not be daunted by the large number of ports and equipment attached to the main chamber, many more than were labeled on the simplified, basic schematic of Section 7.2. The number of effusion cells is 12 rather than 5, according to the diagram on the right. Many of the component equipment are not identified in either figure, suggesting these elements may play supporting roles that can be ignored for the time being. Also, what is the function (purpose) of the two sets of hollow cathode lamp plus PMT on the opposite sides of the chamber?

To begin, identify any component that appears multiple times. In this case, there are 12 effusion cells, which you might expect all (or most) will look similar regardless of the type of gas being used. Taking note that the diagram is two-dimensional (2D) while the actual chamber is 3D, the effusion cells are likely the long tubes extending furthest from the main chamber. Some of the effusion cells appear to have a black cylinder at the end of the tube, while others have a similar cylinder with a silver color—possibly an aluminum exterior. The chamber also has several viewports (windows to see into the chamber). These can be identified as glass-like ports with relatively short tubes into the chamber. Any port used for electrical feedthroughs will not likely be a viewport. One special viewport probably appears to have a phosphor coating to see the RHEED interference pattern from the electron gun, judging from the diagram. The RHEED pattern might be the two bright spots on a black disk at eye level to the gentleman in the white sweatshirt.

There appears to be a motor at the top of the picture, which might be used to rotate the substrate. This assumption might be reinforced by the fact that all of the effusion cells appear to be either radial to the main chamber or slanted upward, which would be the case if the substrate is held upside down in the chamber similar to other thin-film deposition systems. From knowledge obtained in Chapter 4, one can confirm that the chamber is suitable for MBE processing since the attachment ports have Conflat seals that are necessary for UHV operation. In addition, the voltages and the amounts of currents being delivered to the chamber through each electrical port can be estimated, once you have gained some experience.

As the previous two paragraphs describe, some simple logic and basic understanding of physics can be used to obtain significant familiarization. The reader should be mindful, however, that some of the conclusions were drawn from assumptions. If the technology is important to you, it is always best to confirm such information with an expert or an independent source. That leaves the following question: What purpose does the two pairs of lamp plus PMT serve? (A PMT, short for photomultiplier tube, is a non-imaging or at best a very-limited-imaging light

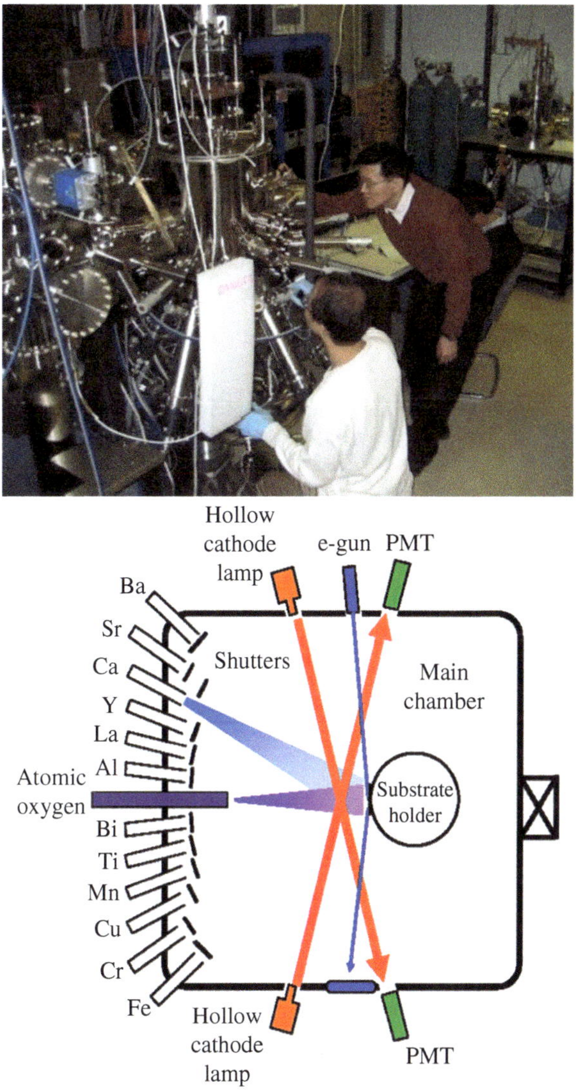

FIGURE 7.10 A picture of an operational MBE deposition chamber at Rutgers University and a simple schematic of its main chamber. Source: Prof. Sean Oh, Department of Physics, Rutgers University. Reproduced with permission of Prof. Sean Oh.

detector.) The key clue appears on the other side of the chamber, the hollow cathode lamps (to be discussed in Chapter 9 on light sources). A hollow cathode lamp is a special class of discharge lamp that produces a single wavelength (one color) of light, used for the spectral analysis of one particular gas. (Some hollow cathode lamps are capable of being tuned to a few distinct wavelengths and can be used for to evaluate a few discrete elements.) From this information, we can infer that each pair of lamp

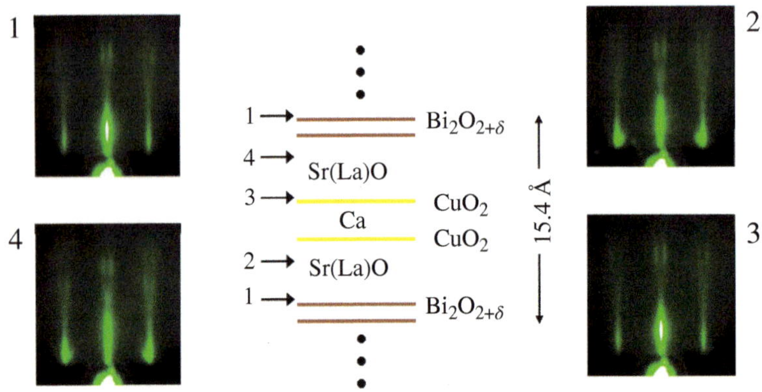

FIGURE 7.11 RHEED interference patterns from four distinctly different films. Source: Prof. Sean Oh, Department of Physics, Rutgers University. Reproduced with permission of Prof. Sean Oh.

plus PMT is used to measure the residual gas fraction inside the chamber from one or more of the effusion cells. In other words, a greater amount of gas between the lamp and sensor results in less light arriving at the PMT since more of it is absorbed. This inference is consistent with the absence of a mass spectrometer in the diagram, since the lamp+PMT pairs serve that function. Keep in mind, the lamp+PMT can only measure the partial pressure of only one gas at a time. The fact that two lamp+PMT pairs are depicted in the diagram might suggest these represent a few lamp+PMT pairs. Alternatively, similar chemicals are used in research laboratories to grow many of the thin films. It is possible that two lamps could provide all of the necessary abundance diagnostics, two gases at a time and retuning the lamps between each deposition layer. However, the limitations imposed by nature are rarely so kind to the experimenter.

Let us now move onto an examination of a few specific example of MBE depositions and the associated RHEED patterns. Observe the subtle differences in the interference pattern, shown in Figure 7.11, as layers of Bi_2O_2, Sr(La)O, and CuO_2 are sequentially laid down. Layers 1 and 3, Bi_2O_2 and CuO_2, respectively, have a narrowly peaked, centrally extended, vertically oval bright spot, while layers 2 and 4 do not. The subtle differences between any two of these patterns suggest any RHEED analysis of film quality should be left to an experience experimenter or a quality control engineer. Also, the five layers (layer 1 is repeated) combined are only 15.4Å thick. How many atoms thick does this imply? The vast majority of all atoms are approximately 1Å (0.1 nm) in diameter, making this unit a good benchmark number to learn. The average layer thickness is thus three atoms and the combined five layers are 15 atoms thick.

Figure 7.12 shows the MBE growth of a multilayer film along with some basic RHEED interference patterns for various classes of film. Layers 1 and 3 in the figure are deposited at elevated temperatures, 850 and 800°C, respectively, while the amorphous growth in step 2 and the polycrystalline growth of Al in step 4 are done at room temperatures. A bright single spot with two small lobes is the RHEED pattern for the

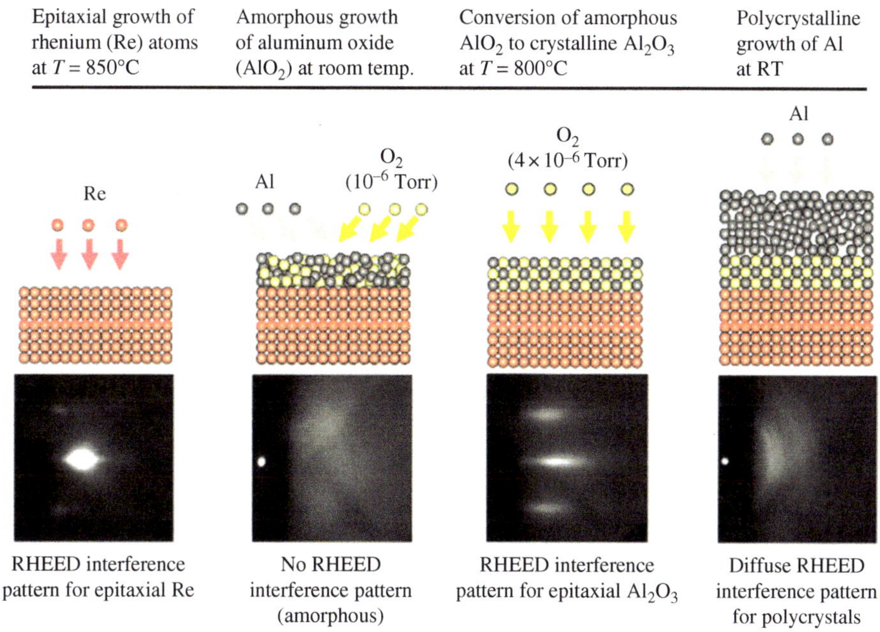

FIGURE 7.12 MBE thin-film depositions and corresponding RHEED interference pattern for significantly different internal film structures. Source: Prof. Sean Oh, Department of Physics, Rutgers University. Reproduced with permission of Prof. Sean Oh.

epitaxial growth of Rhenium (Re) atoms in this particular MBE chamber. Next, an amorphous aluminum oxide layer is grown on top of the Re crystal. This layer requires aluminum atoms and a 10^{-6} Torr partial pressure of O_2, deposited at room temperature. There is no interference pattern (only random scatter) during this stage since all amorphous materials by definition have no crystalline internal structure. The amorphous aluminum oxide (AlO_2) is then converted to crystalline Al_2O_3 at 800°C with a partial pressure of 4×10^{-6} Torr of molecular oxygen in step 3. Finally, this multilayer adds polycrystalline aluminum at room temperature, resulting in a RHEED interference pattern that has structure but is diffuse. While each microcrystal of the polycrystal produces constructive interference, the pattern from each is out of phase with the others leading to partial destructive interference.

7.4 PHOTOLITHOGRAPHY AND ETCH TECHNIQUES

Each time portions of a film have to be removed during the fabrication of a solid-state circuit, a series of intermediate stages associated with photolithography and an etch must be taken. For example, three of the processing steps that were shown in Figure 7.13 (Steps 2, 5, and 7) actually represented a multiprocess procedure. Figure 7.13 depicts a common wet etch and photolithography method used for silicon-based devices.

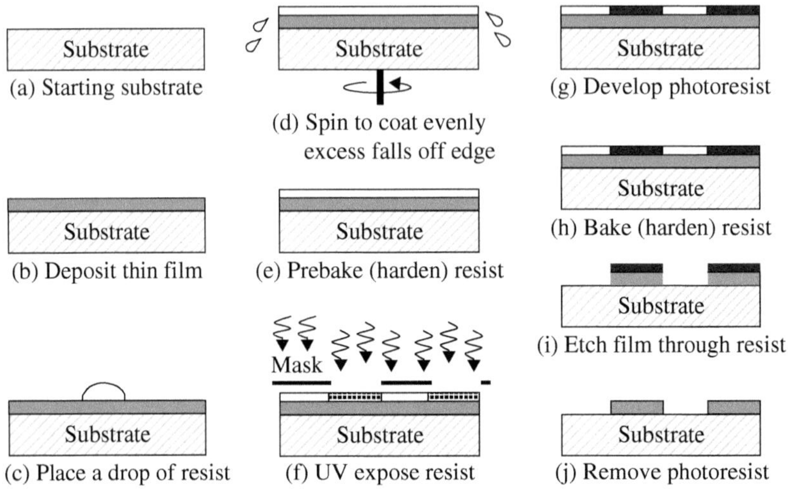

FIGURE 7.13 The multistages of depositing a film, adding a photoresist material, exposing the resist material through a mask to activate select portions, and wet etching away those unwanted portions of the film.

To wet etch away unwanted portions of a film, one first grows a film and places a drop of photoresist liquid. (The liquid material, once it is hardened and exposed to ultraviolet (UV) light, will be resistant to the etching process. Hence it is called photoresist.) After the drop is placed on the surface, the substrate is rotated (typically ~3000 rpm) to spread the resist material relatively evenly with any excess falling off the edges. Unfortunately, a perfectly uniform thickness is rarely achieved; the thickest portions usually occur at the edges of the wafer. The substrate is then prebaked to harden the photoresist. Next UV light is exposed selectively through a mask, the entire surface is developed, and once again baked. Some photoresists are positive, becoming insoluble to the developer during UV exposure, while others are negative, becoming soluble to the developer during UV exposure. Which is depicted in Figure 7.13?

The film is now ready to be wet etched, which occurs through the photoresist. In our example, those unexposed regions etch easily, while the portions exposed to UV light resist etching. Some photoresist materials (over the exposed regions) still remain after this next-to-last step. A separate chemical reaction is necessary to remove it, prior to the next film deposition.

It is important to realize there are several photoresist materials, none of which are suitable for all wet etchings since various photoresist materials can chemically react with particular thin films and render the film useless. (*Note*: the vast majority of solid state circuitry are silicon based, but there is a large and rapidly growing market for devices made of various wide-bandgap materials.) In addition, some procedures are dry rather than wet etches. All etching applications have natural limitations and methods to mitigate any shortcomings. One common difficulty, shown in Figure 7.14, is the tendency of wet etches to undercut the boundaries protected by the photoresist. Care must be taken to let the etching process go long enough to insure layer 2 is

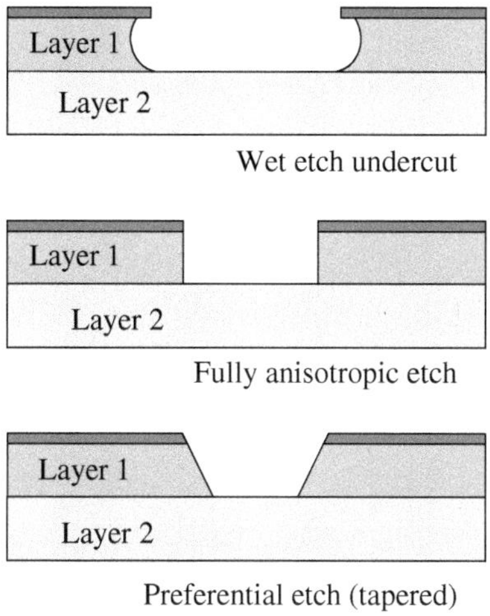

FIGURE 7.14 The profiles of a device edge from a wet etch (top) and two dry etch processes (middle and bottom). All etching methods pose limitations.

exposed, but short enough to limit the amount of undercut. Flow rates, the thickness and chemical makeup of layer 1, and etch time are all-important parameters, which are different for each type of film.

Dry etching, the bombardment of the surface with ions, normally results in less damage and greater etching control than do most wet etches. However, not all materials are suitable to be dry etched and unacceptable levels of damage at the component edges may occur, depending on the film. In some circumstances, a tapered profile called a preferential etch is used to mitigate edge effects. A preferential dry etch, depicted at the bottom of Figure 7.14, comes at the expense of needing more of the wafer's real estate, which may already be at a premium due to the ever increasing complexity and miniaturization of modern circuitry.

Further, metal interconnects have microscopic imperfections, some of which can be introduced during subsequent processing. Examples of the imperfections found in metal layers are shown in Figure 7.15. Obviously, there are microscopic irregularities in any long metallic strip, including uneven edges, pinholes, and pinspots. Expansions and contractions of the underlying films can cause a break in a conducting line. If the interconnects are made of a metal alloy that is not sufficiently stable at elevated temperatures, dendritic growth can occur, occasionally forming a bridge to another conductor that shorts the two conductors together. The device metallurgy is especially important if the device is to operate at high temperatures. In some circumstances, shorts and breaks can be repaired after the device has been fabricated. A bridge can be eliminated by discharging an appropriately sized capacitor through the two wires forming the short, causing the bridge between these to burn out (similar to blowing a fuse). Fixing a break buried beneath other layers is far more difficult and a plan of action can only be formulated after the problem area is inspected with a microscope.

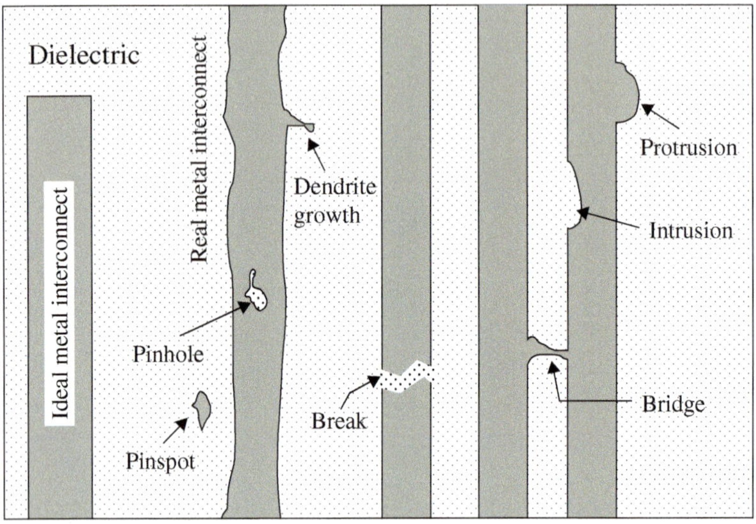

FIGURE 7.15 Some of the microscopic imperfections associated with metal interconnects.

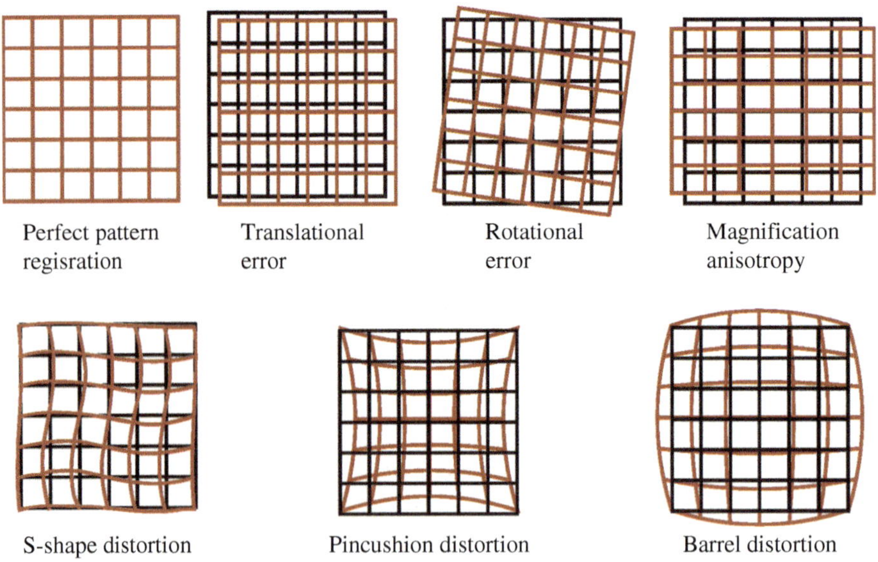

FIGURE 7.16 Types of registration and distortion errors associated with photolithography.

In addition to the difficulties and errors introduced by other processing steps, photolithographic registrations and image distortions are a major limitation in device fabrication. Figure 7.16 shows highly exaggerated forms of these errors, ranging from translational and rotational errors of one mask to another, to differences in image distortions that may vary between separate masks. The black grids in the figure represent a perfect one-to-one match between two photolithographic images, while the overlaid red grids show an accentuated mismatch. The critical point is: each device component may be only 1 μm wide and its integrity from one mask to the next must be maintained over a 4 inch (102 mm) wafer. In other words, a placement of any single 10^{-6} m-wide portion of one film must be retained everywhere across 0.1 m diameter area, indicating a localized positioning error of one part in 10^5 could result in a complete mismatch between two layers needed to make say a transistor.

To insure the integrity of any 1 μm-wide component, we must have a repeatability tolerance of one part in 10^6. Comparing this specification to an LCD display of a laptop computer, which typically has 1440×900-picture elements, the photolithography tolerances need to be 1/1000 of the smallest dot that can be displayed on the laptop. A typical laptop display of approximately 1000×1000 pixels is a good benchmark number for comparison in a wide variety of situations.

7.5 IN SITU AND INTERMEDIATE-STAGE TESTS

The device fabrication based on thin films is expensive and complex, requiring repeatable positional precision of each photolithographic image and meticulous attention to its depositions. It is no wonder that in situ measurements are made during thin-film depositions and quality control tests are done prior to proceeding to the next fabrication stage. In this section, we describe some of these test and evaluation techniques. We have already discussed the RHEED measurements in MBE depositions.

One of the most widely used analytical tools employed to evaluate thin films is the X-ray photon spectrometer (XPS), depicted in Figure 7.17. The XPS produces both photoelectrons and Auger electrons. The energy of the former depends on the energy of the incident photons, while the Auger electrons are independent of the energy of the input photons. A monoenergy or monochromatic X-ray source illuminates the sample, releasing both types of electrons. These electrons are electrostatically focused and subsequently retarded (i.e., slowed down) before these enter a large, dome-shaped capacitor. The electrostatic field, created by the voltage difference between the two curved plates, bends the path of the free electrons toward a microchannel plate (MCP) intensifier and image sensor. The curved capacitor is attached to a power supply that sweeps through a range of voltages, measuring the relative number of electrons emitted from the surface of the sample as a function of the emitted kinetic energy. For a given voltage level, only those electrons within a small range of speeds (kinetic energy) will pass through the hemispherical capacitor and be sensed at the other end. Electrons with higher speeds will strike and be absorbed by the outer plate, while slower ones strike the inner shell. The chemical make up along with other quantum physical effects such as spin-orbit coupling can be determined from XPS. Any atomic differences as a function of depth into the material are determined by the tip angle (θ in Fig. 7.17) of the X-ray source. Grazing incident X-rays sample the top-most layers of the film and angles closer to the surface normal penetrate deeper.

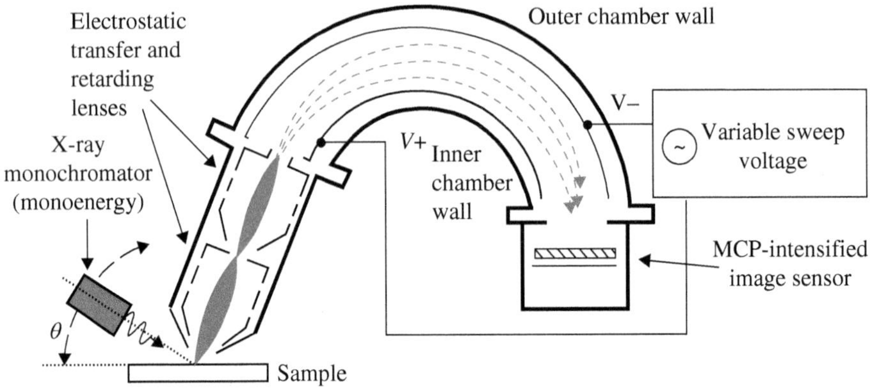

FIGURE 7.17 A schematic representation of an X-ray photon spectrometer (XPS) used extensively to evaluate thin films.

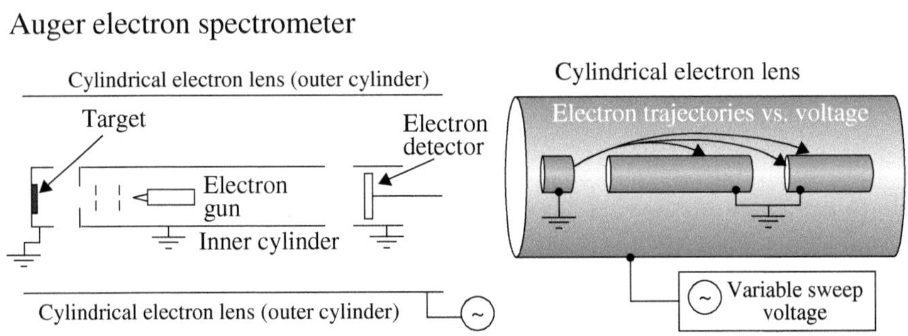

FIGURE 7.18 A cylindrical electron mirror and a cross-section drawing of an Auger electron spectrometer (AES) The cylindrical mirror is used to isolate electrons within a small range in kinetic energies.

Another surface analyzer is the Auger electron spectrometer (AES), which is less expensive than an XPS but produces data that are far more difficult to interpret. An AES uses a cylindrical electron lens as shown in Figure 7.18, consisting of two long, hollow metal tubes on a common axis to form a capacitor. The inner one has two cylindrical slits separated by a distance larger than its diameter. Electrons from an on-axis e-gun are accelerated through a pair of plates with holes onto the target (sample), again producing photoelectrons and Auger electrons that are emitted within a cone. Some of the electrons pass through the inner cylinder's slit into the gap space between the two cylinders where the paths bend according the voltage on the capacitor. Similar to the curved capacitor of an XPS system, only those free electrons within a small range of kinetic energies can pass through both slits at any given time. Once again the outer cylinder is attached to a variable voltage sweep power supply to measure the distribution of electrons as a function of kinetic energy. The sample is probed with electrons rather than X-rays, so the sample is less

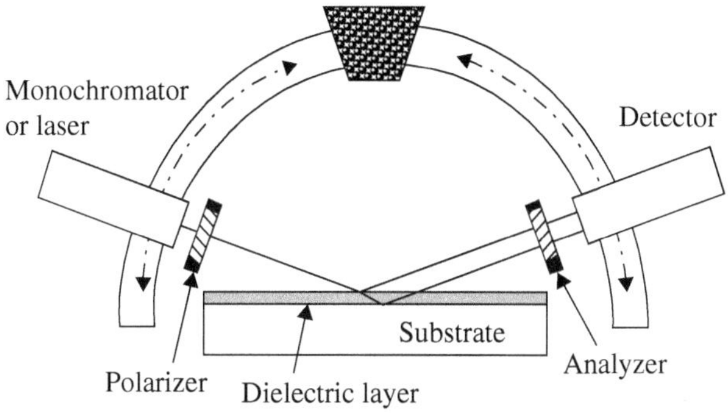

FIGURE 7.19 A schematic setup of an ellipsometer.

susceptible to electron depletion than an XPS provided the streaming rate of electrons from the e-gun is comparable to rate of photoelectron and Auger electrons created. An optional low-intensity ion source may be placed to one side near the target to prevent the target from becoming negatively charged.

Both the XPS and AES techniques are conducted in UHVs to mitigate scattering from gas molecules. Both can measure chemical concentrations down to the 0.1% level, one part per thousand. Other techniques (e.g., secondary ion mass spectrometry (SIMS)) are necessary to obtain trace concentrations down to the part per million (ppm) levels. The energy resolution of an AES can be improved if the inner cylinder has an appropriately located, third slit, allowing the free electrons to pass one more time through the central region before being detected. Both XPS and AES systems damage the surface being studied, depending on the total exposure. Another analyzer based on the photoelectric effect is the photoemission spectroscopy (PES), which is usually uses ultraviolet photons that do not produce Auger electrons. The apparatus is very similar to the XPS.

An ellipsometer, depicted schematically in Figure 7.19, is an instrument of choice for real-time in situ measurements of dielectric thin films as these are being deposited. The instrument makes use of the fact that reflection at a dielectric interface depends on the polarization of the light, while the transmission of light through the layer changes the phase of the incident wave according to the index of refraction of the layer. This information enables both the refractive index and the thickness of the film to be determined simultaneously. The refractive index is a measure of film quality. Dense dielectric films with the correct chemical ratio, will produce the appropriate index of refraction. All other film parameters will not. A visible wavelength ellipsometer can measure thicknesses between 1 nm (10 atoms) and several microns and can be used to monitor growth rates as well as to indicate when the growth has reached the desired depth. Many ellipsometers use a common helium/neon laser, operating at 632.8 nm, as well as a polarizer, an analyzer, and a detector. The analyzer is simply a second polarizer, which is used to determine the rotation that best nulls the signal arriving at the detector. Some systems contain optional compensators, quarter wave plates that enable the light to be varied from linearly, to elliptically, to circularly polarized wavefronts.

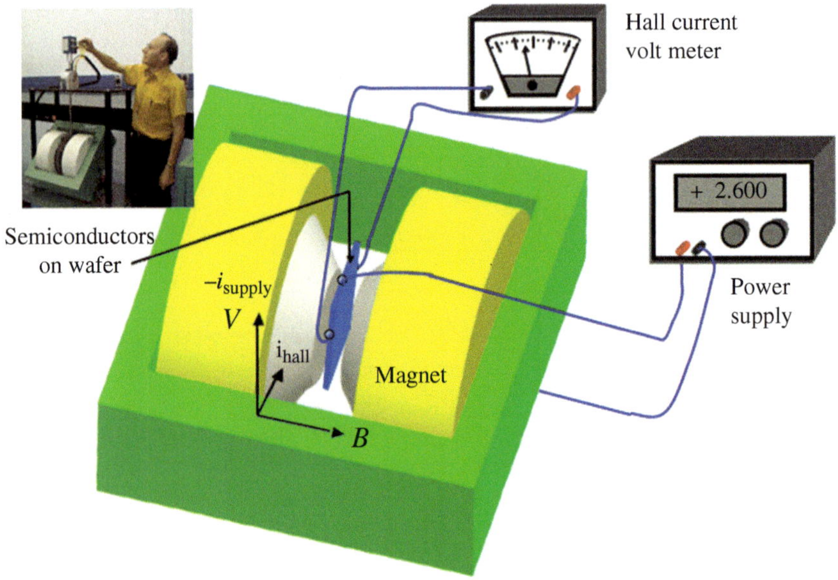

FIGURE 7.20 A schematic representation of Hall effect testing. Photo inset: Bob Thurber of National Institute of Standards and Technology in preparation to make a Hall effect measurement. Source: W. Robert Thurber, National Institute of Standards and Technology. Reproduced with permission of Dr. W. Robert Thurber.

The angle of the light source plus polarizer with respect to the normal of the substrate must always equal that of the detector plus analyzer, since the angle of incident equals the angle of reflection. Any change to one must correspond to the same change on the other side. Ellipsometry is an indirect method, requiring amplitude and phase changes to be converted into an index of refraction in the form of complex numbers. There are numerous variations of ellipsometry including single wavelength versus spectroscopic, standard versus anisotropic, and others.

Finally, a *Hall effect* measurement is an important evaluation tool, especially well suited to evaluate the semiconductor films. The Hall effect makes use of the fact that electrical charges moving inside a magnetic field experience a force at right angles, the magnetic component $q\left(\vec{v}\times\vec{B}\right)$ of the Lorentz force. Figure 7.20 is a schematic representation of a semiconductor on a wafer being tested inside a magnet. When a power supply produces a current flowing through a rectangle, the magnet forces some of the charge to the edges at right angles where it accumulates, generating a voltage potential in the direction orthogonal to the current flow. The Hall effect is different for the various charge carriers and is used to study the details of conduction in semiconductors. It can also be used to determine n-type versus p-type semiconductors, along with the free carrier density and the carrier mobility of the material, making it a powerful probe of material quality.

INTRO PHYSICS FLASHBACK FB7.2

The Photoelectric Effect and Auger Process

Observations, dating back to the late ninetieth and early twentieth centuries, have demonstrated the photoelectric effect and the Auger process, respectively. The photoelectric effect, first observed in metals and later in nonmetals, is a process where free electrons are released from the surface of a material due to an incident beam of light. These free electrons are known as photoelectrons and can only be ejected if the wavelength of the light is sufficiently short. Red, orange, and yellow light, for example, do not release electrons no matter how intense the beam becomes. Photoelectrons from sufficiently short-wavelength light appear virtually instantaneously once the surface is illuminated, even for very low intensity beams. The photoelectrons have kinetic energies that increase proportional to $1/\lambda - 1/\lambda_0$, where λ_0 is the threshold wavelength necessary to produce free electrons. Specifically, the maximum kinetic energy of the photoelectrons is

$$(\text{K.E.})_{\text{max}} = h(f - f_0) = hc(1/\lambda - 1/\lambda_0) = hf - W, \qquad \text{(FB7.1)}$$

where f and f_0 are the frequency of the light and threshold frequency, respectively, h is the Planck constant, and c is the speed of light. Each term in the equation has the units of energy and $W = hf_0$, known as the work function, is the minimum amount of energy required to remove an electron from the surface.

The photoelectric effect supplied a key piece of information, demonstrating the quantum nature of light, the photon, and supporting the theory of the wave–particle duality of matter and energy. Photoelectric measurements can only be understood in terms of particle-like photons, each having a unique energy associated with its wavelength. There are two means to increase the incident energy in a light beam: (i) increase its intensity and (ii) shift to a shorter wavelength (bluer) beam. The first method does not produce photoelectrons for long-wavelength light, invalidating the wave nature for this class of experiments. The second method creates free electrons virtually instantaneously, indicative of particle-like collisions with the surface atoms. Those photons with energies greater than the threshold (work function) create photoelectrons while those that are less energetic do not. For the former case, an increase in beam intensity results in more photoelectrons per second, all with an average velocity (kinetic energy) corresponding to the excess energy above the work function. Both observations are indicative of blue photons being more energetic than red ones.

Electron kinetic energies arising from the photoelectric effect continue to increase into the ultraviolet and X-ray band passes. X-ray photons have sufficient energies to eject a tightly bound electron from an inner shell of the atom. Whenever an inner shell electron is expelled, a secondary phenomenon occurs called the Auger effect. Auger electrons are more energetic than ordinary photoelectrons and come off the surface with discrete energies associated with the internal electron structure of the atoms, making Auger electrons a particularly useful probe of the surface material. Both processes are depicted in Figure FB7.2.

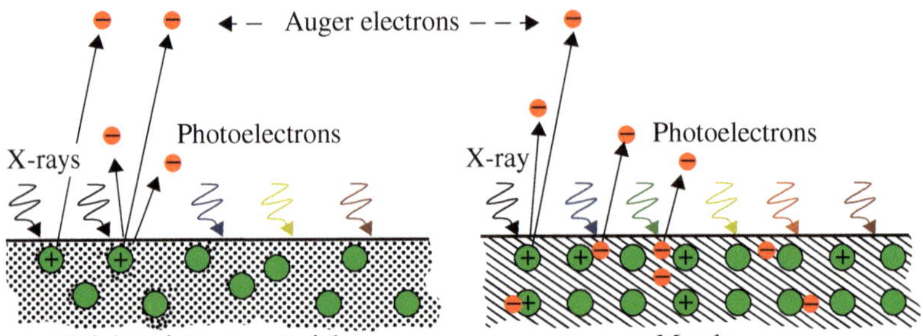

FIGURE FB7.2 The photoelectric and Auger processes are observed in many materials. The kinetic energy of the ejected photoelectrons is proportional to the energy of the incident photons above some threshold. For X-ray photons, photoelectrons and Auger electrons are emitted. Auger electrons have discrete kinetic energies, indicative of the atoms from which these came.

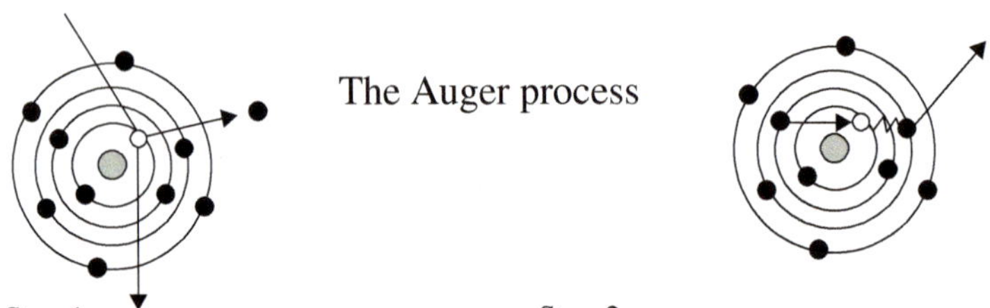

FIGURE FB7.3 The Auger effect where an X-ray or energetic free electron impinges upon an atom to eject a photoelectron with a range of kinetic energies and then a second Auger electron with a discrete energy.

Photoelectrons are most easily ejected from materials with weakly bound electrons. Such materials include common metals where electrons in the conduction band are very loosely attached to any one atom. The phenomenon is observed for photon energies of approximately 2.2 eV or greater ($\lambda \sim 560$ nm or shorter, green light and bluer) in those materials with the smallest work function of $W \sim 2.0$ eV. Other materials have large work functions ($2.0\,\text{eV} < W < 7.0\,\text{eV}$) with thresholds at blue and ultraviolet photons. Photoelectrons come off the surface with a range of angles and a range of velocities. In contrast, X-ray photons also induce Auger electrons with discrete kinetic energies that are higher than those of photoelectrons as depicted in Figure FB7.2. Auger electrons can be used to probe the chemical makeup of various materials. The Auger process as shown in Figure FB7.3 depends on the shell structure of the most tightly bound electrons. An X-ray or a collision with an energetic free electron first ejects a photoelectron. The kinetic energy of this photoelectron is much smaller than the binding energy of the inner shell since the conservation of energy dictates the kinetic energy will be the total energy of the colliding X-ray photon minus the potential energy (i.e., binding energy). Once the atom has a vacancy in an inner shell, another electron from a less-tightly bound shell drops to fill the vacancy, again producing a photon. The internally generated photon then interacts with the remaining bound electrons to eject a second one with a discrete energy corresponding to the difference in shell energies.

INTERESTING TIDBIT TB7.2

While Albert Einstein's biggest contributions to physics were his theories of special and general relativity, he won the Nobel Prize for Physics in 1921 for his contributions to the photoelectric effect and Brownian motion. These latter efforts helped to establish the particle nature of light, the photon.

INTERESTING TIDBIT TB7.3

The Hall effect is an important diagnostic tool for evaluating semiconductors. In recent years, several solid-state devices based on the Hall effect have supplanted earlier equipment, significantly enhancing performance. One example is the electronic ignitions that have replaced mechanical breaker points in automobiles.

COMPREHENSION VERIFICATION CV7.1

Question: MBE is claimed to be done at elevated temperatures, but there are cryogenically cold plates at several internal locations inside the chamber. Can these two conditions coexist? If yes, explain. If no, why not? Is MBE processing very wasteful of energy? (Make use of benchmark numbers in you analysis.)

Answer: Useful Benchmark Numbers

T(LN_2) (air liquefies)	77 K, −196°C, (−321°F)
Water freezes	0°C
Water boils	100°C
Max T of household oven	220°C (~430°F)
T of blast furnace (molten iron)	900+ °C

The difference in temperature between the cryogenic plates and the residual gases is about 1000°C ($\Delta T \sim 850-[-196]$). At atmospheric pressure, there would be a huge condensation of hot gas onto the plates. Convection is normally the most efficient heat transportation mechanism. It is the primary cooling mechanism to prevent computers from overheating. However, both hot and cold regions can exist simultaneously in the high or ultrahigh vacuums used by MBE. The number of molecules per volume in high vacuum is 10^{-7} of the density at one atmosphere, reducing the convective cooling or heating efficiencies by a similar factor.

7.6 DEVICE STRUCTURES AND IC PACKAGING

After the circuitry has been fabricated on a wafer, there are several more fabrication stages that must occur before a circuit is ready for distribution or use. First, individual integrated circuits must be separated (cut) from the others that may have been fabricated simultaneously on the wafer. The standard wafer size at the beginning of the twenty-first century has a 4-inch diameter, although several foundries have added 6 inch wafer capability. The left side of Figure 7.21 represents an example wafer layout, containing six devices of four separate types. There are two large devices and four smaller ones, three of which are labeled T2-devices.

An individual circuitry is called a die, once it has been cut from the wafer. There are normally electrical contact pads and input/output (I/O) buffers to connect the internal circuitry to external circuitry. The die often contains on-chip diagnostic circuitry as a quality control measure to evaluate a limited set of the device functionality. In the development stage, the die might be placed on a probe station where several (<20) individual electrical contacts can be made to conduct a few rudimentary tests by hand. Alternatively, the die might be installed in an automated probe station where all of the electrical contact pads are engaged and a thorough evaluation is made.

Device yields are never perfect, with some fraction having to be discarded. This statement is particularly valid for customized, one-off special circuits where the yields are often less than 50%. The greater the physical size and the greater the complexity of the circuitry, the lower the yields are. For large science-grade custom

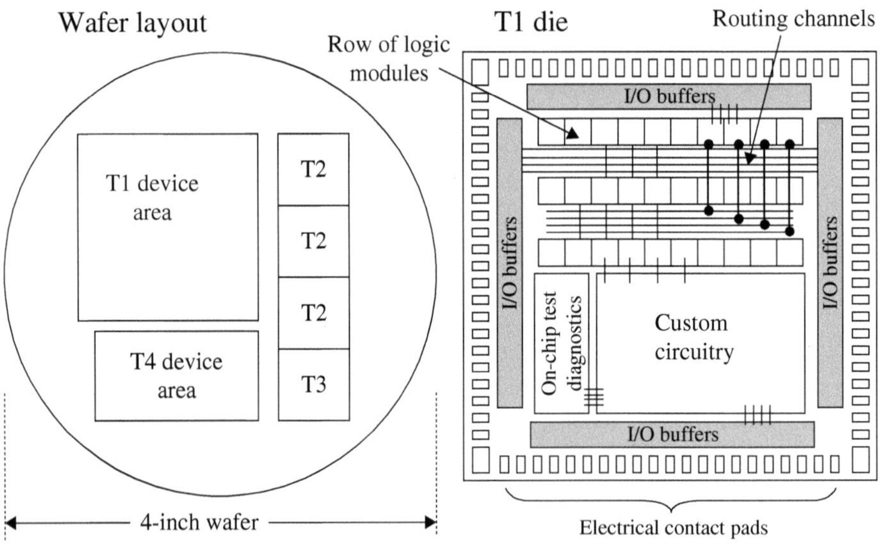

FIGURE 7.21 A wafer layout, having six devices of four types (left) and the T1 device once it has been cut out of the wafer (right). The device is referred to as die after it is removed from the wafer and before being packaged for use.

image sensors, a foundry run of 20 wafers with one die per wafer is made, of which one or two chips will be suitable for the application and another two to three are engineering grade—functional but out of specification. In this case, only a few of the dies are ever packaged into a functional device.

Packaging entails two things: encasing the die in either a plastic or ceramic skin and making permanent electrical contacts to interface the die to external circuitry. In the past, the electrical contacts were to a set of pins, one of a handful of standardized sizes. These chips were mechanically inserted into socket holders on a circuit board. Today, the process is usually more integrated with a circuit card being populated with chips automatically attached and connected by a machine.

8

MATERIALS SCIENCE—INVALUABLE HIGH-TECH CONTRIBUTIONS

Materials science is really a cross-disciplinary applied physics field, essential to numerous technologies and the ongoing improvement of many modern devices. The impact of material scientists to the physics of modern devices has largely been subsumed in various topics throughout this book. However, it is worthwhile to highlight briefly the important contributions made by this field of study as well as underscore some not found elsewhere in this text.

Macroscopically, materials with new properties, ranging from super hardness to time-dependent or environmental-dependent deformation, are being developed. New polymers and composite materials as well as functionally and graded materials are other examples having novel bulk properties. On the microscopic scale, material scientists are at the forefront of growing new nanomaterials, wide- bandgap semiconductors, and thin films, among others. Some materials are being created that are capable of molecular self-assembly. One of the most promising developments is the ability to synthesize materials and structures that mimic traits found in living creatures, a field of study known as biomimetry. For example, the silk produced by spiders has much greater tensile strength and resiliance per unit mass than steel. Each strand of spider silk has a complex protein structure, consisting of a combination of three components: strong, springy, and flexible sections. Synthetic materials based on this type of three-protein architecture may one day be used in parachute lines, suspension bridge cables, medical sutures, artificial ligaments, and body armor. In 2011, three researchers at the Swiss Federal Institute of Technology described a concept for a bridge that can carry out self-diagnostics and self-repair based on biomimetic characteristics that they described as tensegrity. Still other examples of biomimetry embody the same atomic physics as DNA, viruses, or bacteria to develop nanowires, nanotubes, quantum dots, and other nanotechnological devices. More

Modern Devices: The Simple Physics of Sophisticated Technology, First Edition.
Charles L. Joseph and Santiago Bernal.

Companion website: www.wiley.com/go/joseph/moderndevices

generally, the material classifications are biomaterials, carbon, ceramics, composite materials, glass, metals, nanomaterials, polymers, refractory, semiconductors, thin films, and functionally and graded materials.

Furthermore, surface science and engineering play critical roles in the functioning of modern devices, whereas instrument performance for most of the twentieth century depended solely on the bulk characteristics of various materials. Miniturization of many photonic and electronic instruments have compelled many components to rely on surface rather than bulk properties of the constituent materials. For instance, a high-precision electronic resister uses a very thin film to control its ohmic value to tight tolerances. These thin-film resistors, however, are not able to survive large unintended currents compared to the less accurate carbon resistors. Another example of recent surface engineering is the development of a new type of super-sticky adhesive material based on the biomimicry of the gecko's foot. Each toe on its foot contains numerous tiny ridges called lamellae, which have very small setae (hairs) that in turn, split into extraorinarily minute sections called spatulae. The design of the gecko foot dramatically enhances the van der Waal's attraction, the residual electrical charge between two surfaces, allowing the gecko to stick to almost anything. The resulting adhesive material has already found applications in attaching Global Positioning Systems (GPSs) and smart phones to the windshields of automobiles. *Note*: the adhesive action can easily be disingaged, but in only one directional motion.

8.1 THE USE OF COMPOSITE MATERIALS

Composite materials, also known as composites, are materials made of two or more constituent materials, resulting in an end product that has significantly different physical or chemical properties from any of its component substances. For example, carbon fibers reinforced with an epoxy polymer are used in aircraft, multicrew racing sailboats, statospheric balloon gondolas, and spacecraft. These carbon fiber–reinforced polymers are much stronger while being significantly lighter than any metal alloy, allowing strong but lightweight structures to be built. Fiberglass (glass fibers reinforced with plastic—either polyester or epoxy) is a composite, which was developed during World War II as a radome—a weather-protecting dome for radar antennas. It is not as strong as carbon-fiber composites, but it is lightweight, strong, robust, and can be fabricated with a variety of surface finishes. Fiberglass has been used in a wide variety of applications, ranging from telecommunications, boats, aircraft, hot tubes, surfboards, storage tanks and piping, to home construction. Plywood, made of thin sheets of wood with the grains oriented 90° from one layer to the next, and concrete, formed from stone agregate and cement, are also two familiar composites that have been in use for quite some time. Add to the latter, rebar (rods of steel having ribs) and the composite becomes reinforced concrete. Concrete itself has excellent resistance to compression forces, but cannot withstand large tensile forces that pull it apart. In contrast, steel has superior tensile strength compared to its compressive strength. Concrete expands as it dries, causing the steel to be under some tensile stress and the concrete in turn under some compressive stress. In effect, the reinforced concrete has moved its net zero stress point to take advantage of each component's contributive strength.

Most composite materials are generally anisotropic, having tensile or compression strengths that depend on the directional orientation of the forces. These materials have superb performance in applications where the stress–strain forces occur primarily in one direction. In contrast, many simple materials such as aluminum or steel are examples of isotropic materials. Failure of composite materials can occur through shocks, impacts, or repeated cyclic stresses. Recurring temperature variations and differences in the contraction/expansion rates between constituent components, for example, can cause delamination, the separation between two layers, or can cause fiber pull out, where individual fibers separate.

8.2 THIN-FILM MULTILAYERS

Many of the techniques used by the semiconductor industry to grow or deposit thin films were discussed in Chapter 7, and device architectures based on these thin films were examined in Chapter 6. Refer to those chapters for material definitions, physical properties, and deposition methods. Material scientists continue to make important improvements to thin-film development and to epitaxial techiques, including homoeptiaxy, heteroepitaxy, heterotopotaxy, and pendeo-epitaxy, for amorphus, crystalline, and polycrystaline materials. Wide- bandgap materials, in particular, have received extensive research investments in recent decades.

One topic, however, not adequately covered in this text is the advancement of thin films for optical components. For example, multilayer films have been deposited on optics for many decades as interference filters, polarizatizers, long and short bandpass filters, antireflection coatings, and diachroic filters. An antireflection layer is applied to camera optics to increase the transmission efficiency while reducing glare and scattered light from these surfaces. The uses of bandpass filters will be discussed in Chapter 12 and polarization in Chapter 10. All of these coatings have distinctly different indexes of refraction and thicknesses to pass or to reflect the desired light of specific wavelengths while rejecting out-of-band wavelengths. The sharpness of the wavelength transition between in-band and out-of-band reflectivity/transmission has improved dramatically in the twenty-first century due to more precise control of the deposition processes. Perhaps Raman spectroscopic studies have benefitted the most from the very sharp cut-ON and cut-OFF edges provided by these new optical films. Moreover, the use of color, such as in image display systems, have traditionally relied on the natural pigmentations of various materials. Twienty-first-century researchers also use biometry, colorization based on microstructure surfaces similar to the Morpho butterfly wings or colorization based on the reflective cells of the cephalopod skin of the octipus, cuttlefish, and squid. These biometry solutions may create a new type of low-energy TV screen. Chapters 10, 12, and 15 discuss relevant concepts and techniques for optics, diffraction, and color in display systems, respectively.

8.3 NANOTECHNOLOGY

Solid-state devices such as computers and other electronic devices over the past three decades of twentieth century consisted of circuit components (transistors, resistors, capacitors, etc.) with sizes that were measured in microns ($\mu m = 10^{-6}$ m). The emerging field of nanotechnology is the manipulation of materials and the fabrication of devices on the nanometer ($nm = 10^{-9}$ m) scale. In other words, nanotechnology pertains to device structures with sizes of 1–100 nm in at least one dimension. (Atoms have diameters of ~1 Å (i.e., 0.1 nm), which implies devices can be as small as 10 atoms across in nanotechnology—a natural barrier preventing further miniaturization.) The scale of nanotechnology also indicates quantum mechanical effects are very significant, forcing nanotechnology into a research category rather than a technology that is driven by particular technological goals. This new research category encompasses numerous research and technologies that occur below the size threshold of the quantum realm.

There is a wide range of potential applications for nanotechnology, both militarily and commercially. Important future uses include among others nanomedicine, nanotoxicology, green nanotechnology, and regulation. Significant future nanomaterials comprise fullerenes, carbon nanotubes, nanoparticles, nanowires, and quantum dots. A single bit (a "1" or a "0") in a computer, for example, might ultimately be reduced to whether or not a single atom is caged inside a nanotube. At the start of 2014, the United States has created the National Nanotechnology Initiative (NNI) and has

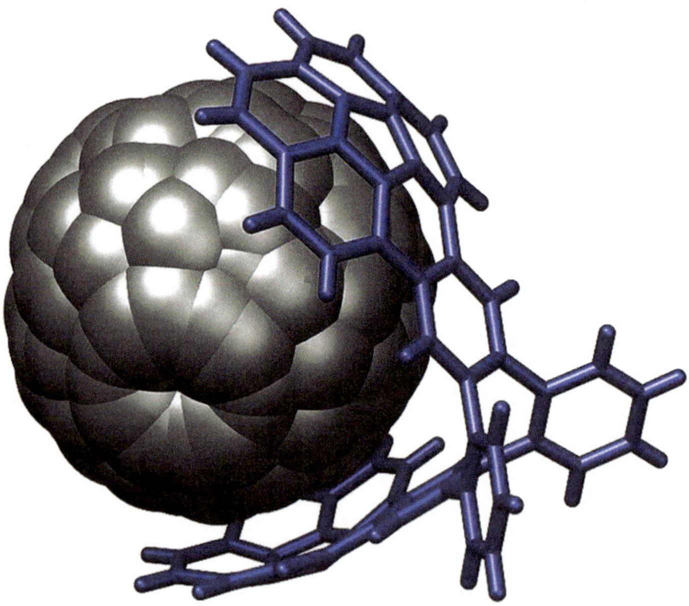

FIGURE 8.1 Molecular tweezers clasping a C60 fullerene. Source: Reprinted with permission from Sygula et al. (2007). © 2007 American Chemical Society.

invested \$3.7B USD in nanotechnology, while the European Union has invested 0.9B Euros and Japan the equivalent of \$750M USD.

We conclude this chapter with an exciting example of nanotechnology, molecular tweezers, having two arms that are capable of latching onto a single molecule. The term was first introduced by Howard Whitlock and popularized by Steven C. Zimmerman in the mid-1980s to early 1990s. Figure 8.1 shows a crystal structure of molecular tweezers, consisting of two corannulene pincers clasping a C60 fullerene (Buckyball molecule). The authors, A. Sygula, F.R. Fonczek, R. Sygula, P.W. Babideau, and M.M. Olmstead, refer to these tweezers in the *Journal of the American Chemical Society* as "A Double Concave Hydrocarbon Buckycatcher."

9

LIGHT SOURCES

Light sources come in two forms: continuum and emission line. The former produces light over a wide range of wavelengths (colors) without any large gaps in color. Continuum light sources can be further subdivided into thermal and nonthermal sources, depending on whether the radiation field is coupled to the temperature (internal atomic velocities) of the material or not. Emission-line lamps produce light only at a set of discrete wavelengths. The specific set of emission lines are unique to each particular chemical, consisting of atoms with electrons populating identical orbital shells. In contrast, thermal light sources are independent of chemical composition.

Light is a form of electromagnetic (EM) energy that can be released or absorbed by an atom. It exhibits the wave–particle duality, exhibiting both particle characteristics and wave behavior. Light is made up of many small particle-like packets called photons that have energy and momentum but no mass. The equations for energy and momentum of a single photon are as follows:

$$E = hf = \frac{hc}{\lambda} \tag{9.1}$$

$$p = \frac{h}{\lambda} \tag{9.2}$$

Here, h is Planck's constant (6.63×10^{-34} J s), c is the speed of light (2.9979×10^{8} m/s), λ is the wavelength, and f is the frequency of the light wave.

Light waves also come in a wide range of wavelengths known as the electromagnetic spectrum. The wavelength of visible light refers to its color with red being the longest and violet being the shortest. The electromagnetic spectrum spans far beyond visible light, however, ranging from radio waves at the longest wavelengths through Gamma

Modern Devices: The Simple Physics of Sophisticated Technology, First Edition.
Charles L. Joseph and Santiago Bernal.

Companion website: www.wiley.com/go/joseph/moderndevices

rays being the shortest wavelengths (highest frequencies). The most energetic Gamma rays have energies and momentums that are more than twice that of the rest energy of an electron. Consequently, some Gamma-ray photons have sufficient energy to decay, producing a electron–positron pair of particles. Alternatively, the annihilation of an electron and a positron in a collision can produce one or more Gamma rays. X-rays have energies comparable to the binding energy of the innermost electron shells in heavier atoms. Thus, the removal of one of the most tightly bound electrons in an atom requires the absorption of an X-ray. Electron transitions between the weakest bound orbitals generally correspond to infrared photons, although in a few cases, transitions between very weakly bound electron shells correspond to microwaves. Most transitions between the bound states of an electron in an atom occur at ultraviolet (UV) wavelengths with visible wavelengths containing the second largest number.

All EM waves have common properties and it is common to refer to "the color of photons" that are not visible, although this use is not correct strictly speaking. For the most part, "color" of non-visible EM radiation is referred in terms of relative quantities. For example, a physical process that may boost the energies of X-ray photons might be referred to as having made the photons bluer or as becoming harder (as in increased energy). Figure 9.1 schematically shows the EM spectrum. Also shown are the logarithmic scales of wavelength, energy, and temperature associated with each type of EM radiation.

Each specific wavelength of a photon has a unique energy, momentum, wavelength, and frequency as related in Equations 9.1 and 9.2. *Note*: the huge range (some 12 orders of magnitude) of these parameters, spanning from radio waves to gamma rays, requires the scales be logarithmic to be placed on a single graphic. A temperature scale is also provided in Figure 9.1, indicative of the condition when matter and the EM fields are coupled and in thermal equilibrium.

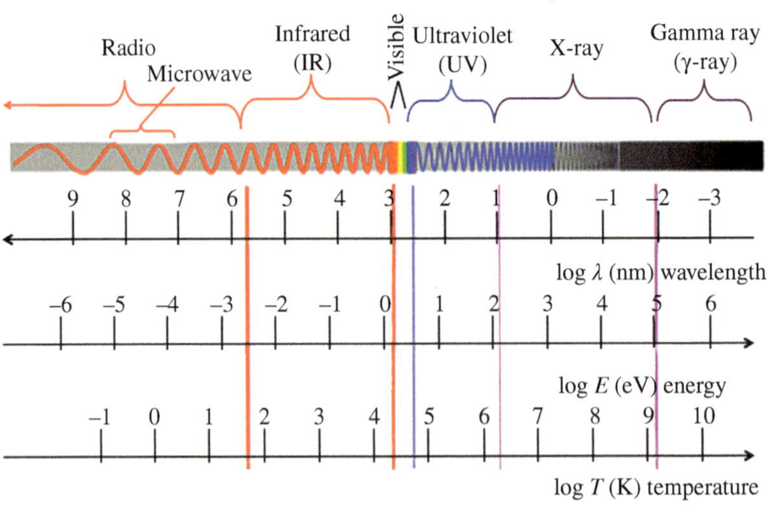

FIGURE 9.1 The electromagnetic spectrum with corresponding logarithmic scales of wavelength, energy, and temperature.

Figure 9.2, plotted on a linear scale, shows an expanded view of the visible portion of the EM spectrum. Each color has a different range of wavelengths. Yellow, for example, only spans about 20 nm, while red covers 130 nm out to 750 nm. (A range of wavelengths is known as a bandpass or a band.) The sensitivity of the human eye (as well as most visible light sensors) falls off at both extremes, making deep reds or deep violets appear darker than these really are.

All matter continually absorbs and re-emits EM energy and for energies $1.65\,\text{eV} < E < 3.1\,\text{eV}$, the EM waves are in the form of visible light. The minimum and maximum energies for visible photons correspond to wavelengths of 750 and 400 nm, respectively. The atoms and molecules of all materials also have internal motions (vibrations or rotations), and often these atoms have varying amounts of translations. Here we use translational motion to denote an individual velocity of an atom or molecule relative to its neighbors, rather than the bulk flows that gasses and liquids might have. For solids, the scale of any atomic translation is very small, less than the distance to the next atom, and the atom always experiences a restorative force. Thus, translational motions in solids are vibrational in nature. Individual atomic or molecular speeds in a bulk material are never the same from one atom to the next. At any moment in time, some atoms will have larger than average speeds, while others less. The relative number of atoms as a function of speed forms a statistical distribution, which remains constant despite continual collisions between atoms that slow down some while speed up others. The amount of atomic motion is directly related to the temperature of the material with the atoms and molecules in warmer materials having higher average velocities both internal and translational than those contained in cooler materials.

Since all materials (even neutral ones) contain electrical charges, atoms emit or absorb EM waves resulting from accelerations of these charges. A charge particle (e.g., electrons and protons) that experiences any acceleration must absorb or emit light to conserve both momentum and energy. This absorption and emission is normally the consequence of an electron within an atom transitioning from one orbital shell to another, while also giving a small kick to the atom. However, each atom continually exchanges energy between itself and other atoms around it through collisions, sharing energies of motion. For gasses of sufficient density and for liquids and solids, the EM radiation field is coupled to the internal motions of the atoms, resulting in thermal radiation once equilibrium is achieved.

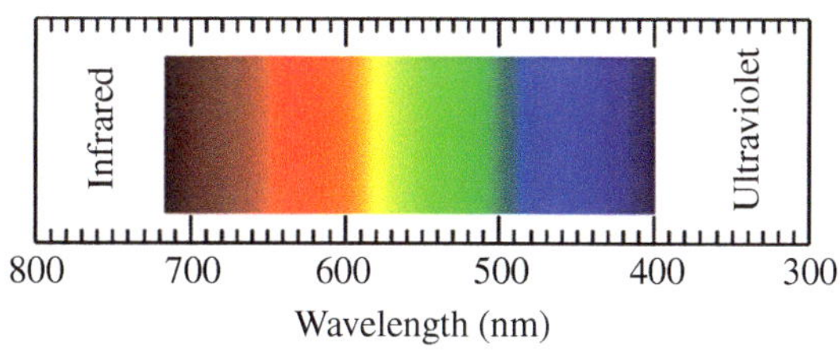

FIGURE 9.2 The visible portion of the electromagnetic (EM) spectrum, showing the color and corresponding wavelengths.

Thermal radiation is characterized by three physics laws: Planck's law of radiation, Wien's displacement law, and Stefan–Boltzmann law, corresponding to Equations 9.3, 9.4, and 9.5, respectively.

$$u(\lambda,T)=\frac{8\pi hc}{\lambda^5}\left(\frac{1}{e^{hc/\lambda kT}-1}\right)\left(\text{J s}^{-1}\text{nm}^{-1}\right) \quad (9.3)$$

$$\lambda_{\max}T = 2.8979\times10^{-3}\ (\text{mK}) \quad (9.4)$$

$$\frac{P}{A}=\varepsilon\sigma T^4\left(\text{Wm}^{-2}\right) \quad (9.5)$$

Here, h is Planck's constant, c is the speed of light, λ is the wavelength, ε is the emissivity (often $\varepsilon \approx 1$), and $\sigma = 5.67\times10^{-8}$ (W/(m^2 K^4)) is the Stefan–Boltzmann constant. T is the absolute temperature measured in Kelvin [T(°C) = T(K) + 273.15]. The Planck equation (Eq. 9.3) describes the distribution of the energy per second per unit wavelength (i.e., the power per unit wavelength) originating from a material at a single temperature. This distribution of energy is also known as blackbody radiation. Note that Equation 9.3 is independent of the atomic mass, meaning all materials at a given temperature, regardless of chemical makeup, will absorb and reradiate the same distribution of photons as a function of wavelength. This is an idealization, albeit a very powerful one, of the real spectrum observed in detail.

Figure 9.3 shows the Planck distributions (blackbody curves) for a number of temperatures. The number of photons at all wavelengths increases as the temperature

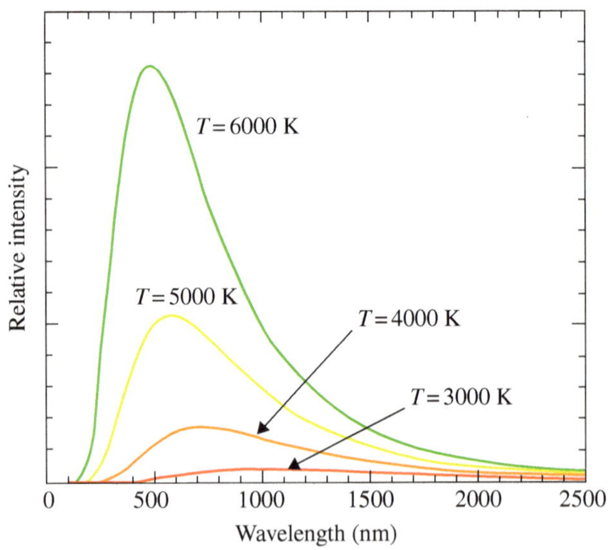

FIGURE 9.3 Blackbody curves for various temperatures, showing the intensity of infrared, visible, and ultraviolet wavelengths.

increases with the number at short wavelengths increasing the fastest. This trend is known as a color index by astrophysicists and an increase in temperature is said to shift an object's blackbody radiation towards the blue. Wien's displacement law (Eq. 9.4) can be used to calculate λ_{max}, the wavelength where the blackbody curve reaches its peak. The curves in Figure 9.3 are color coded according to the apparent color that the bulk material would have. The surface temperature of the Sun is 5800°C and has its peak luminosity is at green wavelengths. (The Sun appears to our eyes to be yellow only because humans have greater sensitivity to orange and red than they do to blue and violet.) Also, an object emitting blackbody radiation with a temperature of 2000–3000°C (3600–5400°F) appears red even though it radiates far more light in the infrared than in the visible. Figure 9.4 shows the luminosity curves of various blackbody temperatures with respect to the range of visible light wavelengths. *Note*: the vertical scale has been expanded in the middle and bottom plots for better visualization. The relative intensities on a common scale are given in Figure 9.3. Three UV subgroups from medical terminology are shown. For progressively shorter wavelengths, these are UVA ($320<\lambda<400$ nm), UVB ($280<\lambda<320$ nm), and UVC ($\lambda<280$ nm).

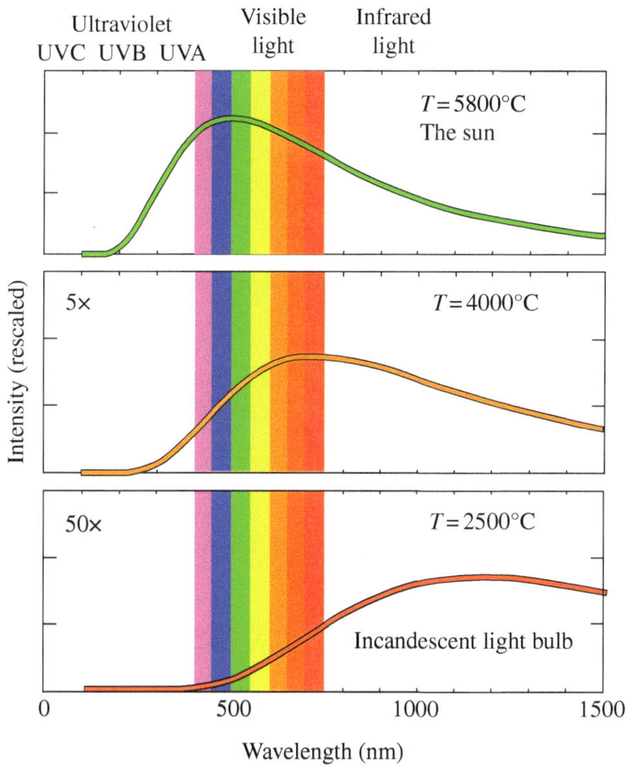

FIGURE 9.4 The distribution of emitted photons as a function of wavelength for blackbody radiation. Individual curves to the temperature of the radiating material.

Finally, radiometry is the measure of the total EM radiation, while photometry is only that portion of the EM radiation that is detectable by the eye. The Stefan–Boltzmann law (Eq. 9.5) is radiometric, covering the total power radiated including wavelengths outside the visible band. P/A, the total power radiated per unit surface area of a blackbody is proportional to the body's absolute temperature, T, raised to the fourth power. P/A is called irradiance and has units of Watts per meter squared (W/m^2). Radiance, an absolute measure of brightness or intensity, is the irradiance per solid angle (i.e., per steradian). Using the Stefan-Boltzmann law and the irradiance from the Sun, it is possible to estimate the mean temperature on the Earth or other planets in our solar system. For Earth, the temperature estimate is about 60°F, assuming approximately 30% of the light is simply reflected back into space. The actual average temperature is closer to 68°F as the result of naturally occurring greenhouse gases.

When comparing light sources for the home or workplace use, one should compare the luminous flux (lumens) rating rather than the power rating (wattage) of the bulb. The amount of lumens is the energy being emitted at visible wavelengths that can be seen by the human eye. More precisely, the luminous flux is the radiant flux weighted by the sensitivity of the typical human eye as a function of wavelength. Any photons created with wavelengths longer than 750 nm or shorter than 400 nm do not contribute to the luminous flux (i.e., have zero weight) since the sensitivity of the eye is zero in those bands. The wattage in contrast is the amount of electrical power required for operation of the light source and does not account for the inefficiencies encountered in converting electrical energy into visible photons. An archaic definition of luminous intensity, rarely used in the twenty-first century, is the candela (cd). The candela was a standard unit established when candles were the primary source of artificial indoor lighting. One candela (or candle) is 1/683 W/sr of the radiant intensity, measured at a wavelength of 555 nm.

9.1 INCANDESCENT LAMPS

An incandescent lamp bulb is shown in Figure 9.5. Its basic operation is to pass an electrical current through a resistive filament, usually tungsten. The free electrons in a metal accelerate when a voltage is applied, causing these electrons to bump into

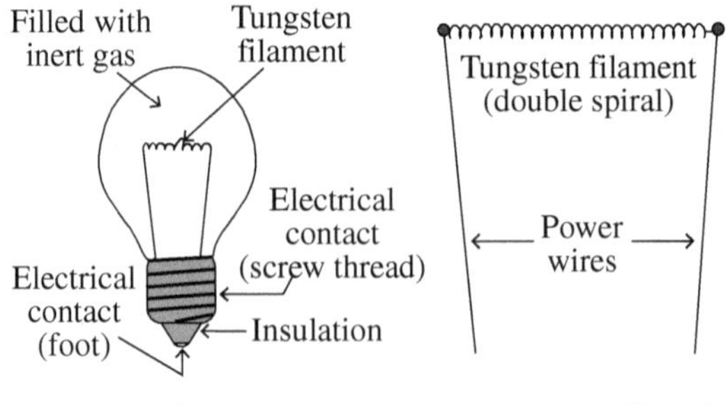

FIGURE 9.5 The structure of a standard incandescent bulb (left) with major components labeled. An enlarged view of the filament structure is shown at the right.

atoms in the filament and agitating the atoms to an increased temperature. The filament goes from room temperature to 2500°C (4500°F) in less than a second, increasing its resistance and reducing the flow of current, which retards any further increase in temperature. This hot filament produces light that is very nearly blackbody. The interior side of the glass bulb is usually frosted to diffuse (scatter in all directions) the light from the filament.

A filament is under considerable stress when in use and its requirements are very challenging since most materials at the operating temperature readily burn in the presence of oxygen. Historically, there have been two approaches to prevent the self destruction of the filament. The first bulbs invented by Thomas Edison used a carbide filament and the volume of the bulb evacuated. The vacuum inside the bulb extended the life of the carbon filament by depriving the combustion process of oxygen and preventing a rapid burning of the carbon.

The second approach, which is widely used today, incorporates a tungsten filament and a glass bulb filled with an inert gas, usually argon. Tungsten (Ti), which is the most durable conducting material, remains solid up to 3410°C (6170°F), but slowly sublimes at the 2500°C (4500°F) operating temperature. This filament is made of a long incredibly thin length of tungsten metal. In a typical 60-W bulb, the tungsten filament is about 0.5 m (20 inches) long but only 25-μm (0.001 inches) thick. (The thickness is approximately that of a human hair.) The tungsten is arranged in a double coil to fit it in a small space. That is, the filament is wound up to make one coil, and then this coil is wound to make a larger coil. In a 60-W bulb, the coil is about 2 cm (0.8 inches) long.

In a modern light bulb, the inert argon gas greatly reduces the loss of tungsten. When a tungsten atom sublimes from the filament, it collides with an argon atom and is often bounced back toward the filament, where it rejoins the solid structure. The Ti atoms do not react or combine with the Ar gas since it is inert. Trace amounts of the tungsten in an operating bulb, however, make it to the inner surface of the glass bulb where it can be absorbed and permanently lost to the filament. When the bulb eventually burns out, the remnants of the tungsten can be seen by a gray discoloration, usually a spot on the bottom of the bulb as its oriented in its socket. This spot marks the coolest part of the glass bulb and is the portion of the inside surface where it is most difficult for an Ar atom to dislodge a Ti atom, potentially returning to the filament.

Unfortunately, the efficiency of converting electrical energy into photon energy in conventional incandescent lights is poor, only a few percent. Only 10% of the total radiant energy produced occurs at visible wavelengths. (This can be seen as the ratio of the area under the curve marked by visible light to the total area under the curve in the bottom plot of Figure 9.4, which extends to the right far beyond that plotted.) There are other loss factors, including the efficiency of converting electrical energy into a hot filament as well as the losses of heat by conduction and convection. All of these mechanisms reduce the overall efficiency multiplicatively.

In most circumstances, the convection process is so efficient that it is the dominate mechanism for transporting thermal energy. However, the hot filament emits most of its energy in the infrared, which is invisible to the human eye and is able to escape directly to the outside world unimpeded. Nevertheless, the Ar gas does convectively

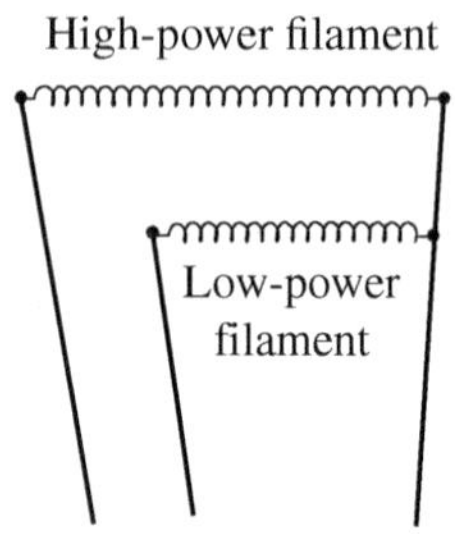

FIGURE 9.6 Filaments in a three-way bulb.

transport a tremendous amount of heat from the filament to the walls of the glass bulb in a standard lamp, making it too hot to touch.

To increase the luminous flux and required electrical power, a longer Ti coil is used. For example, a 100 W bulb is brighter than a 60 W because it has a longer filament. Three-way bulbs have two separate filaments, one small low-powered and one large high-powered, as shown in Figure 9.6. The three light levels are achieved from dimmest to brightest by powering (i) only the low-power filament, (ii) only the high-power filament, and (iii) both filaments simultaneously.

Halogen lights are also incandescent bulbs that recycle the tungsten back to the filament, but more effectively than do the conventional ones. The bulb is smaller, operates at a somewhat higher temperature, and is filled with bromine/iodine/oxygen gasses, compared to a standard bulb. Tungsten atoms sublime from the hot filament, but then readily combine chemically with the gasses and are soon re-deposited back on the filament to prolong its life. The filament of a halogen bulb typically operates at 2800°C, 300°C above ordinary lamps. As a result, the bulbs are brighter and bluer than conventional bulbs. The efficiency of converting electrical energy to light can be as much as 9%, compared to the approximately 3% of its standard equivalent. Clean gloves are often used in handling a halogen lamp to avoid getting oils on its surface, which can produce cooler spots on the bulb surface and significantly shortening its operating life. Most skin oils, grease, or synthetic oils are very efficient at absorbing and reradiating infrared and visible photons to the point where these can create relatively cooler spots on a halogen bulb, allowing Ti atoms to adhere and remain on the glass surface.

9.2 GAS DISCHARGE LAMPS

Gas discharge lamps make use of the orbital structure of the atom. A gas is sealed into a bulb with electrodes as shown in Figure 9.7 and an arching high voltage is applied temporarily across the electrodes to ionize the gas. Once the gas is partially ionized, the voltage is reduced to sustain a current through the plasma gas. The ions drift in one direction, while the electrons move in the opposite direction, creating collisions between the ionized and neutral atoms as well as free electrons with atoms as shown schematically in Figure 9.8. These collisions cause the electrons that are

FIGURE 9.7 A 15 kW xenon short-arc gas discharge bulb used in popular IMAX theater projectors. Source: Atlant, https://upload.wikimedia.org/wikipedia/commons/9/9e/Xenon_short_arc_1.jpg. Used under CC-BY 2.5 https://creativecommons.org/licenses/by/2.5/deed.en.

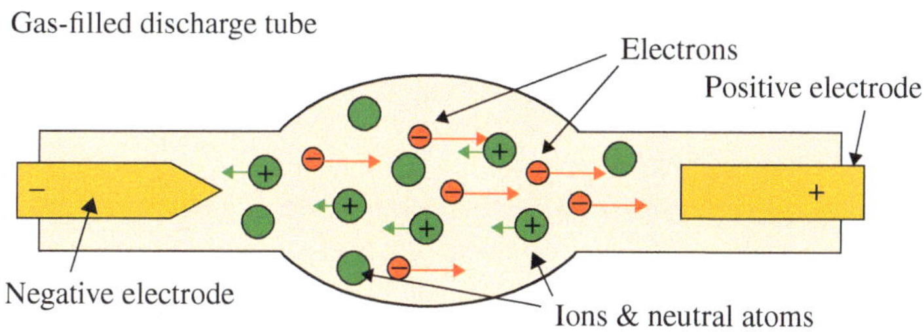

FIGURE 9.8 Schematic operation of a gas discharge bulb. *Note*: the drift velocity of the electrons is larger than those of the ions. The random thermal component of the velocities is not shown.

still bound to a nucleus to excite to higher orbitals. When these electrons decay back to the original orbitals, photons at specific wavelengths and energies are released. Most gasses emit several emission lines according to the internal orbital shell structure of its atoms. The energy of the photons is the difference between the orbital energies; the brightest emission lines are called resonance lines. Small energy differences lead to infrared photons, while large changes result in UV ones. If the gas is sufficiently dense, weaker transitions are enhanced and the emission lines become pressure broadened. For some types of gas (e.g., Xe and deuterium) at sufficiently high densities (i.e., high pressures, $p \sim 1$ atm.), the broadened emission lines merge into one another, forming a continuum light source.

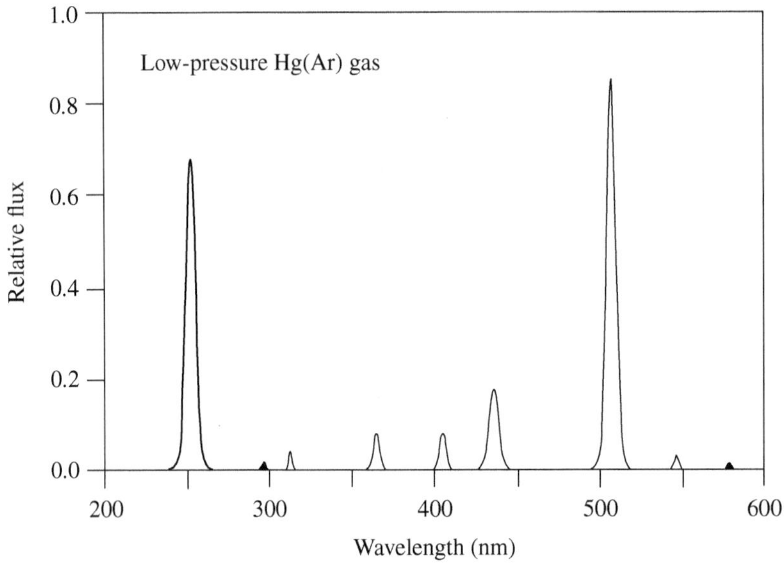

FIGURE 9.9 A plot of the light output from a wavelength calibration lamp, containing low-pressure gas of argon and mercury. Source: Adapted from the Newport-Oriel Instruments catalog.

While gas discharge lamps can be operated with either alternating or direct currents, most research laboratory sources use a DC power supply with a separate igniter circuit. The initial starting spark is triggered manually by pushing a momentary switch. These high-voltage igniters often emit an EM pulse, capable of destroying nearby devices that are sensitive to electrostatic discharge (ESD). Discharge bulbs are inherently unstable, requiring highly regulated laboratory DC power supplies. These power supplies are regulated to 0.1% or better to achieve good radiometric performance, the property of maintaining a precise intensity. A discharge lamp takes a few moments to start operating, and it normally flickers before completely failing.

Discharge lamps emit white light as well as a variety of specific colors. These last longer and are often whiter than incandescent bulbs. Three familiar discharge lights widely used in private applications are fluorescent bulbs used for indoor commercial lighting, sodium lamps used to illuminate roads and parking lots, and neon signs used in bars and restaurants. Low-pressure argon plus mercury gas lamps, which have a series resonance lines throughout the near-UV and visible bandpass, are frequently used in research to calibrate the wavelength scale on a spectrograph. Inexpensive, small versions of these Hg(Ar) lamps are often formed into very narrow U-shaped tubes and are called Pen-Ray Lamps. A graphical plot of the flux as a function of wavelength from a Hg(Ar) calibration lamp is given in Figure 9.9.

One important variant of the gas discharge lamp is the hollow-cathode lamp (HCL), producing discrete spectral lines and often being used as a frequency standard for tunable lasers. An HCL, depicted schematically in Figure 9.10, is similar to other gas discharge lamps. It consists of a sealed volume of gas, (usually a noble gas

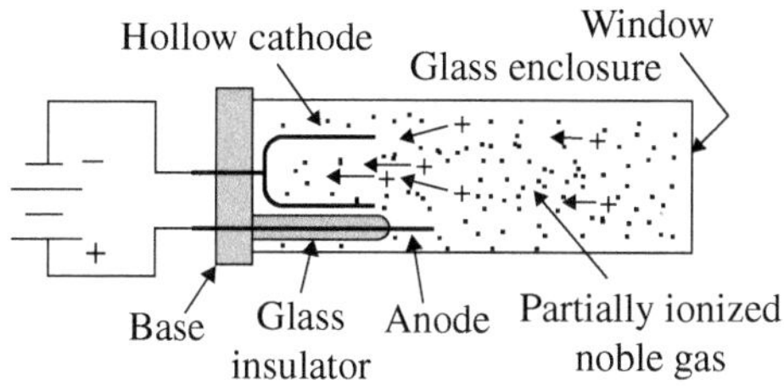

FIGURE 9.10 A schematic drawing of a hollow cathode lamp.

in this case) with two electrodes, a cathode and an anode, to ionize part of the gas. Initially, ions only flow directly between the cathode and anode, collisionally exciting and heating the noble gas until the entire volume becomes a partially ionized plasma. A hollow cathode lamp gets its name from the cup-shaped cathode, which is made from the element or elements of interest. Ions throughout the lamp are accelerated toward—and collide with—the cathode, sputtering some the metal atoms into the gas phase. Both the sputtered atoms and noble gas atoms undergo collisions that cause excitation followed by photon emission. The cup-shaped cathode helps to focus the light out the window end. The window material can be borosilicate glass, although quartz, magnesium fluoride (MgF_2), or lithium fluoride (LiF) are required for UV, vacuum UV, and far UV transmission, respectively. (Vacuum UV (VUV) is taken to mean $120\,nm < \lambda < 200\,nm$, and far UV, which is also a part of the VUV, we define as $90\,nm < \lambda < 120\,nm$. These defining VUV ranges are set by the approximate short-wavelength cutoffs of MgF_2 and LiF materials commonly used for VUV windows.)

The hollow-cathode lamp produces well-defined and very-narrow (0.005 nm full width half maximum, FWHM) emission lines, much narrower than can be produced by most spectrographs or monochromators. Moreover, the strength of a single emission line can be enhanced relative to all others produced by the HCL. Similar to other gas discharge lamps, an HCL uses a highly regulated DC power supply after the start up phase. Its DC supply, however, operates at a low voltage (typically 5–20 V) and low currents ($i < 80\,mA$). If the voltage is set to a level that causes the ions to gain a kinetic energy over one mean free path that exactly matches the energy of an atomic transition, then the emission line corresponding to that transition will be strongly enhanced. Some HCL lamps also use external light sources to assist in the single line enhancement.

Hollow-cathode lamps are used in atomic absorption spectroscopic studies and fluorescence spectroscopic investigations. HCLs are also employed extensively in molecular beam epitaxial growth of semiconductor thin films as seen in Section 7.3.

9.3 FLUORESCENT LAMPS

Fluorescent lights are one class of gas discharge lamps used historically in commercial buildings. Various research groups began pioneering fluorescent as well as neon types of gas discharge lamps in the middle of the nineteenth century, especially in

Germany and subsequently in France. The Europeans were the first to commercialize neon lights. Thomas Edison and others who eventually joined forces to form General Electric Corporation also experimented with fluorescent lighting some years later than the Europeans, but GE abandoned its efforts for several decades due to the success of their incandescent bulbs. It was recognized early on that fluorescent bulbs produced light 3–4 times more efficiently than do incandescent sources as well as have operating lives that are approximately 2.5 times longer. The principal disadvantage of a fluorescent light is the complexity of its design and the corresponding higher cost to manufacture than its incandescent counterpart.

Fluorescent lamps contain low-density gas and operate on alternating current (AC) as turnkey systems. Most fluorescent lights have gas pressures of 0.3 atm primarily of argon and smaller amounts of mercury. Mercury (Hg) is the key excitation element in a fluorescent bulb, emitting more than half of its total luminosity in 254 nm wavelength UV emission followed by weaker emission at 185 nm. Argon contributes several additional emission lines in the visible band. The UV photons are converted to visible light by a powder coating on the inside of the glass bulb. Additional circuitry, especially external components in older fluorescent fixtures, is necessary for easy ignition and to stabilize lamp output inexpensively.

Long-tube fluorescent bulbs differ from most gas-discharge lamps in two principal ways. First, there are two electrical contacts on each end of the tube. Second, the inside of the tube is coated with a fluorescent powder (often a mixture containing phosphor), which converts UV photons to visible light. The efficiency of converting UV to visible photons is not perfect and a small amount of UV radiation is emitted, despite further attenuation by the glass itself. Over years of operation, fluorescent lights are capable of lowering the transmission of UV windows (e.g., quartz and fused silica) of laboratory detectors and filters. Care should be taken to shield ultraviolet equipment from exposure to fluorescent lights when there is a long-term hiatus in its use.

Fluorescent tubes without this phosphor are known as black lights because most of the energy is emitted in the near ultraviolet, which is invisible as well as harmful to the human eye. Black lights are sometimes used in disco-dance facilities, for sun tanning, and used in biological applications (e.g., germicidal lamps, curing dental fillings, and the detection of blood at crime scenes).

The components of the traditional fluorescent lamps are shown in Figure 9.11. The starter circuit contains a small volume of gas, a capacitor and a bimetal switch. When AC power is applied, the small gap in the starter begins to arch allowing significant current to flow through it and through the filaments, heating both the starter and the ends of the fluorescent tube. The initial heating of the filaments causes electrons to boil off the surface to begin ionizing the gas in the tube. The heating of the starter also causes the bimetal switch to expand, closing the contact and shutting down the arch inside the starter. As the starter cools without the internal arc, the bimetal switch reopens. By this time, however, the gas inside the fluorescent bulb is ionized, creating a lower path of resistance between the two filaments and virtually no electrical current then flows through the path containing the starter. A fluorescent bulb is a *negative differential resistance* device, meaning the electrical resistance drops as current flows, enabling even more current to flow. A fluorescent lamp would

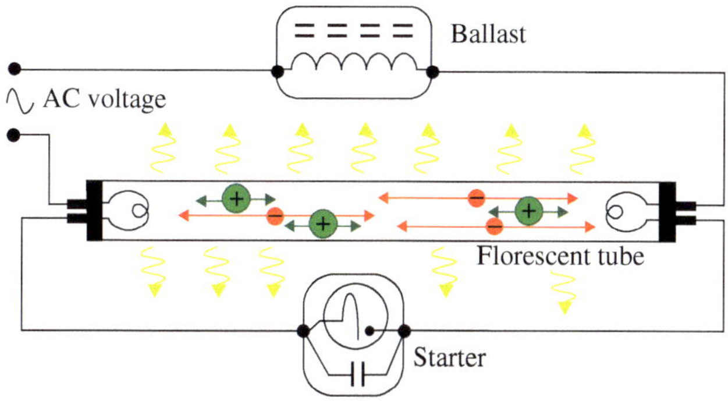

FIGURE 9.11 Schematic of a traditional fluorescent lamp. The starter, pictured at the bottom, contains the bimetal switch and a capacitor. The two coils at either end of the tube are the filaments used to apply AC voltage through the gas. Light is produced through collisional excitations followed by returns to ground states that liberate a photon. *Note*: the gas molecules "slosh" back and forth over shorter distances and in the opposite directions to the free electrons. Source: Inkwina, https://upload.wikimedia.org/wikipedia/commons/e/e1/Fluorescent_Light.svg. Public domain.

quickly destroy itself if a constant AC voltage were supplied. A ballast circuit, which is basically an inductor, is used to prevent this runaway condition and to smooth out the inherent fluctuations that would cause the lamp to flicker during operation. This ballast electronically performs the same function that a mechanical flywheel serves in an automobile engine.

A ballast for AC current applications (also referred to as a choke) consists of a winding of wire on a laminated magnetic core to form an inductor. (It may also contain a capacitor.) An inductor sets up magnetic fields in its core and in the process, an inductor resists rapid changes in the current of a circuit. The ballast has to be properly sized for the type of AC current (e.g., 220 V vs. 110 V), its frequency, and the size of the fluorescent bulb. Historically, many ballasts hum loudly as do some transformers. In most commercial applications, the lights were sufficiently far from personnel and clients that it had rarely been problematic. Improved fabrication techniques and other changes in tube design have effectively silenced these once loud devices. In particular, many bulbs employ electronic ballasts where solid-state transistors alter the AC voltage into a high frequency (10 kHz) form that is faster than the relaxation time for the mercury ions to recombine. This type of ballast limits the current and causes the tube to operate close to its optimal conditions, making the system far more efficient in terms of photons per input electrical energy.

Fluorescent lamps are sensitive to temperature, often failing to ignite outdoors in the middle of winter at northern latitudes. The older style fluorescent bulbs of the 1960s used to take several seconds to start, during which these consumed more than 10 times the energy per second compared to constant operation. Thus, it was inefficient to turn off the lights when exiting a room for brief periods. Instant turn-on devices do not suffer these inefficiencies and multiple cycling of the power is acceptable.

FIGURE 9.12 Two types of compact fluorescent light bulbs.

Figure 9.12 shows two configurations of compact fluorescent light bulbs (CFLs). Multiple tubes or a single tube in a helical coil provides enough surface area to provide light from a small volume. Compact fluorescent bulbs also contain a small drop of liquid mercury, which is located near the hot electrode. This Hg drop is partially vaporized during operation, dramatically increasing the light output of the bulb. As the CFL cools after operation, the mercury is returned to its original location via an amalgamating spot near the electrode. The operating life and the total luminosity of a compact fluorescent are reduced if it is used while oriented nonvertically, making it difficult for the excess gaseous Hg atoms to return to the amalgamating spot. CFLs generally last 2.5 times longer and consume 1/3 as much energy per second as an incandescent bulb of equal luminescence.

9.4 LIGHT EMITTING DIODES

A light-emitting diode (LED) is simply the same semiconductor diode that was discussed in Chapter 6, except that it has been doped with a direct bandgap material, allowing it to emit a visible wavelength photon when the electron drops into a hole at the junction. The schematic drawing of an LED (Fig. 9.13) is almost the same as seen previously. Identical to ordinary diodes, a p–n junction is created, which allows current to flow easily when biased in the forward direction but not in the opposite. Most semiconductor materials such as silicon or germanium have indirect band gaps where the electrons, recombining with holes at the junction, lead to a series of infrared transitions of the electron down to its ground state.

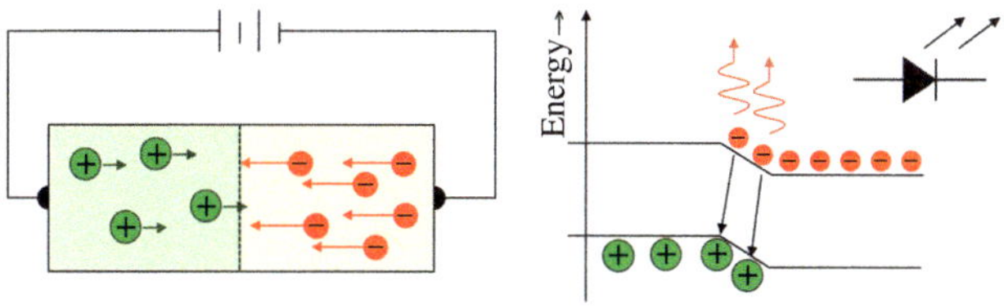

FIGURE 9.13 On the left, a schematic representation of a light emitting diode (LED) and a corresponding energy diagram (right), showing the electrons and holes.

LEDs have been used for decades as indicators on the front of various electronics boxes. More recently, LEDs have been used extensively for taillights on automobiles and trucks. Another specialty lighting need particularly suitable for LEDs are plant growth lights in greenhouses. Panels consisting of an array of LEDs can be configured for particular requirements of the plants. Red LEDs preferentially promote root growth, while blue LEDs are ideal for enhancing chlorophyll and plant leaves. The exact mixture of blue, red, and white LEDs on a plant growth panel can be optimized for particular types of plants. For example, red and yellow leafed plants absorb blue more efficiently than most plants, but reflect most yellow and red photons. Moreover, plants need a different balance of colors at various stages of growth. A particular advantage of an LED for plant growth is the efficiency of creating useful growing light without tremendous amounts of heat, which would exacerbate cooling and water requirements of the greenhouse. Also, plant growth LEDs do not produce UV or infrared light that is harmful to most plants. LEDs are starting to find applications in general lighting as well. Applications requiring strip lighting and spot or floodlights are well suited for LED substitutions. An LED lamp produces approximately twice as many lumens as a CFL and six times many as an incandescent. An LED has a long operating life of at least 30 years and can be operated over a very wide range of temperatures and in harsh environments.

While far more efficient than either incandescent or compact fluorescent lights, one major drawback has been that the narrow beam of each LED. LED lighting has specialized uses as replacements for incandescent bulbs. These include pendant light fixtures, floodlights, and flashlights. Another use is as grow lamps in greenhouses, where arrays of LEDs are placed in shallow box structures and hung over growing plants. One important advantage for plant growth is the narrow wavelength ranges of most LED bulbs. Photosynthesis is stimulated by blue and by red LEDs without energy being wasted to create unwanted heat or to generate other unnecessary wavelengths. Plants in fact have evolved to reflect much of the solar spectrum to prevent the plant from over heating.

9.5 X-RAY SOURCES

X-rays are used in medicine, airport and freight security, inspection for microcracks in oil and natural gas pipelines, among others applications. While there are several methods to generate X-ray radiation, we will restrict our focus here to only one, deferring the other methods to later sections. The two principal methods are (i) hit a piece of metal with a several keV electron and (ii) use a synchrotron to generate a spectrum of visible, UV, and X-ray electromagnetic beams. (X-rays were first discovered in cathode-ray tubes (CRTs) when photographic plates where accidentally left near a CRT.) CRT tubes, which have been used for early TV sets and for oscilloscopes, heat an electrode called the cathode inside an evacuated tube. When hot, the cathode boils off electrons that are accelerated and deflected to strike a phosphor-coated screen in a continuous scanning mode.

Medical X-ray machines use the same basic technique as shown in Figure 9.14. An electronic gun (red) is heated inside a vacuum cavity by passing electrical current through the cathode-heating element. This gun or cathode is held at a few hundred volts below ground. Excess electrons from the hot cathode then boil off and are accelerated towards a tungsten anode in the shape of sloped disc, which is held at a few thousand volts above ground. The beam of accelerated electrons (outlined in red) strike the disk that is constantly rotating to prolong its life. (The electrons would quickly sputter a cavity in a stationary

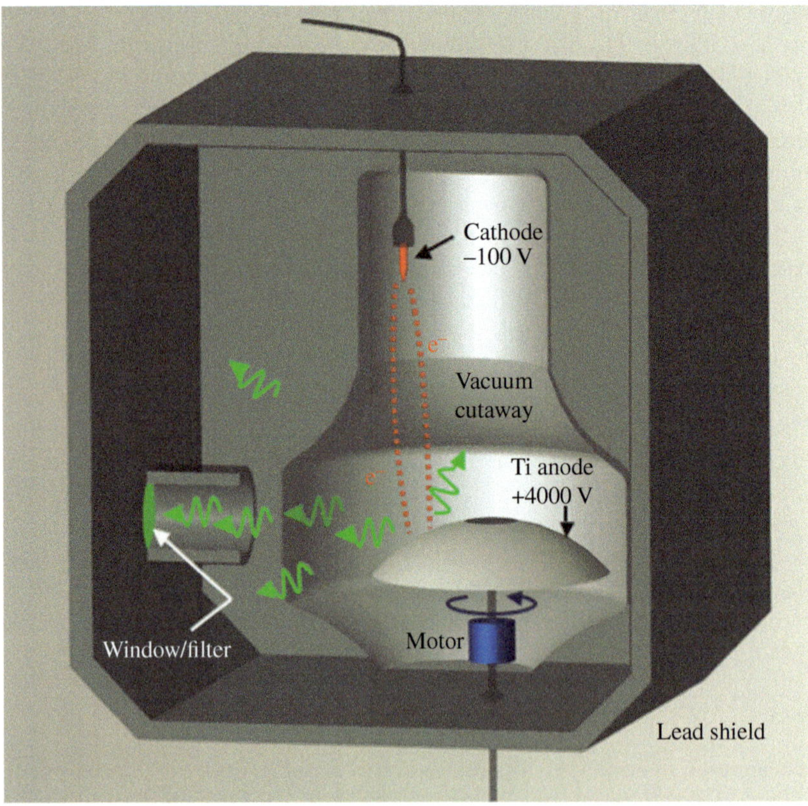

FIGURE 9.14 The essential components of a medical X-ray machine.

anode, rendering it useless.) Most of the energy from the electron beam are lost to heat, with only 1% being converted into X-ray photons.

An X-ray is produced when an electron impinges an atom in the anode, knocking off a tightly bound electron in an inner orbital. One of the remaining electrons then drops from an outer to the inner shell, releasing an X-ray photon. The angle that the electrons strike the anode is critical as most of the X-rays emerge between 60° and 90° with respect to the line between the cathode and the impact spot. The X-rays exit out the window and through a hole in the lead shield. (The paths of the X-rays are shown in Fig. 9.14 as green lines.) The window, which also serves as a bandpass filter, is necessary to maintain a vacuum.

Traditional medical X-ray machines used shadow imaging. That is, the beam is sent through a portion of the body, sandwiched between the X-ray source and either film or a solid-state sensor immediately on the other side of the body part. The dense bone material absorbs the X-rays more effectively than does the soft tissue, casting a shadow. Soft tissue can also be imaged, provided an absorbing die is ingested or inject into the tissue. Various organs can be imaged with similar dies that chemically attach to other compounds. For example, to examine the upper gastrointestinal track, the patient swallows barium sulfide and then X-rays are taken. Traditional X-ray machines create two-dimensional (2D) images. Computer tomography (CT) scans or computed axial tomography (CAT) scans produce 3D images of the tissue. Normally, the patient lies on a table that moves the individual into position where multiple X-ray beams scan through the tissue of interest simultaneously. Image reconstruction is done in a computer to create a 3D image. (See Chapter 19 for more details.)

9.6 LASERS

The origin of the name, laser, came from an acronym: LASER—Light Amplification by Stimulated Emission of Radiation. A maser, which was invented before the laser, also shares the same acronym except "Microwave" is substituted for "Light." Both use the same basic physics, but at different wavelengths. When most atoms absorb a photon, an electron "jumps" to a higher energy shell and then quickly returns to its ground state by emitting one or more photons. There are two non-laser processes involved in triggering the return to its ground state. The atom may undergo spontaneous emission, which usually occurs on timescales of approximately 10^{-8} s. Alternatively, the atom may experience a collision with another atom, triggering the release of photons. Certain atoms, however, have a few electron orbital shells that are metastable, remaining in those excited states significantly longer than both the collisional and spontaneous emission timescales. Electron transitions from metastable states are referred to as forbidden transitions (i.e., these are still possible but do not proceed as quickly as do allowed transitions).

If a photon of the correct energy encounters one of the atoms in an excited metastable state, it stimulates the atom to transition toward its ground state while emitting another photon of the exact same wavelength and phase as the impinging one. This interaction

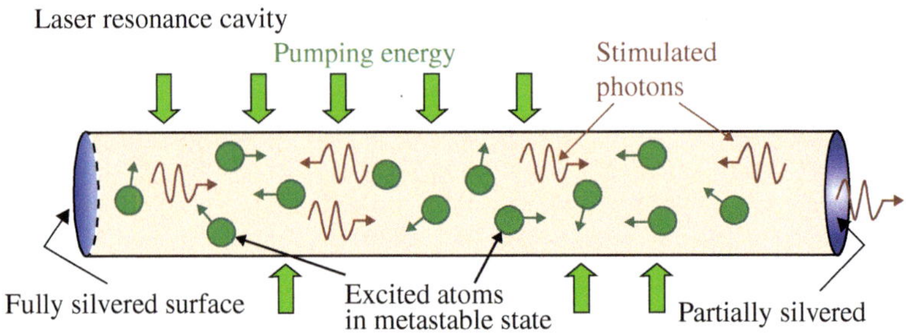

FIGURE 9.15 Laser resonance cavity using a gas medium.

is said to create coherent light, photons with the exact same energy and exact same phase. Thus, there are two requirements for a laser: (i) an energy source to excite most of the atoms into metastable states,[1] creating a *population inversion*, and (ii) establish a resonance cavity with a preferential path for the photons.

These two requirements for a laser are depicted in Figure 9.15. One end of the cylinder is fully silvered to create a good mirror, while the other end is partially silvered so that most photons reflect back into the laser tube, while others escape. The atoms inside the cavity are excited to their metastable states by an external energy source that often completely surrounds the length of the cylinder. This process of creating a population inversion in a laser is called *pumping*. Photons traveling horizontally as pictured have a high chance of encountering an atom, causing it to release a photon exactly identical to the original photons. In turn, both of these photons have long path lengths that enhance the chance to stimulate additional identical photons as the originals. When the photons reach the partially silvered end of the resonance cavity, some of the photons escape and most reflect back into the cavity. The reflected photons continue the process of amplifying the light beam by stimulating additional atoms to create more photons that are coherent. Occasionally, an atom undergoes spontaneous or collisional emission, producing some photons in the laser cavity that are not traveling along the horizontal axis direction. These photons can also stimulate a few photons, but all are quickly lost as these escape out of the cylindrical sides of the tube. The overwhelming majority of the photons stimulated in a laser cavity are coherent (in phase and direction) with the horizontal direction.

The lasing process exponentially multiplies the number of coherent photons, which also depopulates the number of atoms in the metastable state, until equilibrium is established. To replenish the pool of atoms in the excited, metastable states, energy is continually pumped (injected) into the cavity. The amount of pumping energy usually exceeds the laser output by roughly a factor of 10 to maintain the population inversion.

The resonance cavity has to be exactly an integral number of half wavelengths to prevent destructive interference. Lasers require active feedback to maintain the precise

[1] In recent years, it is possible to create lasers from materials that make use of atomic transitions that are slower than most but, strictly speaking, are not metastable.

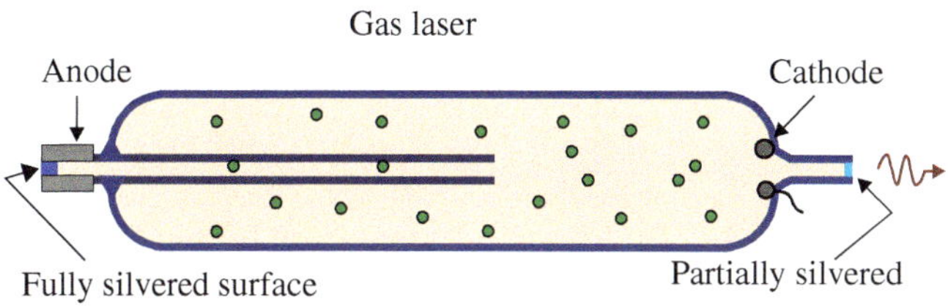

FIGURE 9.16 A schematic drawing of a gas laser. The pumping energy source uses the same gas in the gas discharge lamp configuration.

alignment of the reflecting surfaces as environmental conditions change. In fact, many lasers now use alternative optical surfaces rather than the plane parallel ones shown in Figure 9.15, since the later configuration have tolerances that are extremely difficult to maintain. Phase coherence does drift somewhat, but at any instance in time the light beam emerging from the right-end surface in Figure 9.15 will be coherent. Lasers often have a coherence length specification, which is a measure of the timescale for the frequency of the light to drift multiplied by the speed of light, c. This coherence length is applicable only in the direction of the laser beam. Spatial coherence lengths are defined separately as well, characterizing the phase differences side-to-side across the beam.

A schematic diagram of a gas laser is shown in Figure 9.16, consisting of a glass-enclosed tube filled with a low-pressure (1 Torr = 1/760 atm) of an appropriate gas (e.g., HeNe and Ar^+). The volume is larger than the resonance cavity to provide a reservoir of gas. The pumping mechanism incorporates the gas in a discharge lamp configuration.

Laser light is the purest single-color (monochromatic) source available, making it particularly suitable for a wide range of high-precision applications, including optical data storage devices, high-speed fiber optics data transmission, LASIK eye surgery, holography, photolithography, interferometry, LIDAR, distance ranging, among others. For example, lasers can measure the distance to the moon to an accuracy of a few inches and have demonstrated the orbit of the Moon around the Earth is slowly increasing in size. Lasers are also ideal tools for moderate-precision requirements such as optical pointers, annealing and cutting tools, hair removal and bloodless surgery, barcode scanners, and guided munitions.

There are numerous types of lasers. The media used for population inversion include solid-state, semiconductor, gas, excimer, dye, and free-electron, among others. Strictly speaking, a semiconductor laser is a solid-state laser, but it is customary to separate semiconductors into their own class in laser terminology. A ruby crystal (chromium crystal) doped with corundum was the first solid-state laser, while a yttrium aluminum garnet (YAG) laser is more common. YAG is a crystal made of yttrium aluminum garnet ($Y_3Al_5O_{12}$) most often doped at the 1% level with Nd and denoted as Nd:YAG. The primary emission from an Nd:YAG laser is at a wavelength of 1064 mn in the near infrared. An excimer laser is powered by a chemical reaction involving an excited dimer, which is a short-lived dimeric or hetrodimeric molecule. An excimer molecule is formed from two atoms with one in an excited electronic state. Commonly used compounds used in an excimer laser are ArF, KrCl, KrF, XeCl,

TABLE 9.1 Laser Classifications

Class	Risk Level	Effects
I	Safe	Totally enclosed inside a larger device
II	Safe	Insufficient intensity to damage before the blink reflex protects eye
IIIa	Moderate	Up to 5 mW will cause eye damage in approximately few seconds
IIIb	High	Immediate, severe eye damage
IV	Very high	Can burn skin, fire hazard, possible to shatter glass

and XeF, all emitting at ultraviolet wavelengths. HeNe, ionized Ar, CO_2, and nitrogen are widely used in gas lasers as the gain medium. Regardless of the lasing medium, all lasers must also meet the resonance cavity requirement.

Lasers are operated in two modes: continuous wave and pulsed. In continuous wave operation, energy is pumped continuously and the intensity of the emerging laser beam is relatively constant. For pulsed mode operation, the intensity of the beam switches cyclically between on (full intensity) and off (zero intensity). The peak energy output of a laser beam can be maximized by pulsing, making it particularly suitable for laser ablation and other applications requiring extremely intense beams. Low-power lasers are typically air cooled, while high power normally require water cooling. Lasers are classified from I to IV, according to their intensity level or alternatively the risk hazard to a human eye. These are given in Table 9.1.

9.7 SYNCHROTRON LIGHT SOURCES

We will discuss the architecture and performance of cyclotrons and synchrotrons in Chapter 22 under the general topic of particle accelerators. For the current purpose, it suffices to say that these particle accelerators force charged ions or electrons to move in circular motions. Synchrotrons have become important light sources for state-of-the-art corporate, government, and university research. The principle advantage of synchrotron accelerators as a light source is their ability to produce intense X-ray and UV radiation beams with highly precise intensities. This radiation is nonthermal. While these light sources are used primarily for research, their importance can be seen on the large number of facilities worldwide. The United States has 14 public-accessible synchrotron facilities, each providing two or more photon beams for a wide range of research activities. Europe and Asia each have 25 comparable major facilities and there are a few others elsewhere. The largest and most used radiation source in the United States is at the Brookhaven National Laboratory on Long Island, NY.

9.8 SUMMARY OF LIGHT SOURCES

The photon has become as important to advanced technologies as the electron. The manipulation of the photon requires numerous electromagnetic radiation sources with a wide variety of properties and wavelength bands. The photon differs from an

TABLE 9.2 Blackbody Temperatures and Colors

Object	Temperature	Peak Color
Heat lamp	500°C (930°F)	Infrared
Stove top burner		Appears: dull red
Candle flame	1,700°C (3,100°F)	Orange/yellow
	Temperature gradient	Not in equilibrium
Bulb filament	2,400°C (4,500°F)	Infrared
		Appears: white/yellow
Sun's surface	5,800°C (10,500°F)	Green/blue
		Appears: yellow
Surface of most massive stars	20,000°C (36,000°F)	Ultraviolet
		Appears: white/blue

electron in that a photon transmits energy and momentum, but no mass. Moreover, two streams of photons can pass through each other, both emerging undisturbed. In contrast, two electrons cannot exits at the exact same location at exactly the same time, a property of the electron known as the Pauli exclusion principle. Radiance is a radiometric measure of the total luminosity across all wavelengths from radio to gamma ray. Luminance is the portion of the radiance occurring only at visible light wavelengths and weighted by the typical sensitivity of a human eye.

As demonstrated, EM radiation can be broken into thermal and nonthermal sources, depending whether or not the atoms are completely coupled to the radiation field and are in equilibrium. Thermal sources are used at visible and infrared bands, but technical difficulties severely limit their use in all other bandpasses. Blackbody (thermal) radiation is characterized by the Planck law, which is independent of the type of atoms and which the apparent color depends on its temperature. It is important to keep in mind that blackbody radiation is an idealization, a simplified approximation of the physical processes. If one looks in detail at a spectrum of a thermal radiation source, there will be absorption or emission features indicative of its chemical makeup. Nevertheless, a blackbody curve is a very power tool to estimate the total power, temperature, and other physical conditions. Some examples of blackbody temperature along with the apparent color and the wavelength band where the peak radiance occurs are given in Table 9.2.

The most important distinction for a radiation source is whether it produces a continuum or set of discrete wavelength emissions. X-ray machines, synchrotrons, LEDs, gas discharge tubes, and lasers are all examples of non-thermal radiation sources in that the spectral intensity as a function of wavelength is not well fit by a blackbody curve. As can be seen from Table 9.2, a thermal source would have to exceed the surface temperature of the sun to have any chance producing significant amounts of ultraviolet or X-ray radiation. Any light source that hot would rapidly destroy itself and any surrounding devices, vaporizing most of the materials. Similarly, a thermal radio-band source would have to be quite cold, as would the other apparatus involved in the experiment. (Everything at room temperature is effectively an excellent infrared and radio wave source, producing copious amounts of contaminating radiation.) Fortunately, there are several nonthermal physical processes suitable for producing X-rays and ultraviolet radiation as well as radio waves. These light sources require the acceleration of ions and especially electrons.

Synchrotrons and high-pressure gas discharge tubes produce continuum flux over a range of wavelengths. In contrast, low-pressure gas discharge lamps and lasers produce photon emission at discrete wavelengths, making these valuable tools for wavelength calibration and the wide variety of interferometric applications seen in the section on lasers. X-ray machines and LEDs also produce emission over a limited range of discrete wavelengths, but each emission bandpass is not nearly as narrowly restricted in wavelength as are those from lasers or low-pressure gas discharge tubes. For example, the bandwidth of a single emission line, known as its monochromatic width, of a low-pressure gas discharge lamp is $\Delta\lambda < 0.3\text{nm}$, while the bandwidth of a typical LED is $\Delta\lambda \sim 50\text{nm}$.

Each class and subclass of electromagnetic light source is indispensable for a particular set of technological needs and research continues on most types to enhance the performance of each. There is an ongoing need for better, more efficient light sources across the entire EM spectrum. Improved sources, offering greater precision and other unique properties are enabling technologies, permitting the advancement of biological, physical, and engineering disciplines.

COMPREHENSION VERIFICATION CV9.1

For a halogen lamp, is the radiation thermal or nonthermal? The peak flux occurs at what wavelength? What color does this the peak flux occur? Does the color at the peak flux correspond to its apparent color? Explain. Is the operating temperature of the filament hotter or colder than the Sun's surface? Is it hotter or colder than the filament of normal incandescent bulb? Explain.

Answer: A halogen lamp is a thermal light source, one operating around 2800°C, about 300 above that of an ordinary incandescent bulb. We can use Equation 9.4 to determine the wavelength of the peak flux, but first we must convert the temperature to the Kelvin scale: $T = 2800 + 273.15 = 3070\,\text{K}$. *Note*: the initial significant figure is only two decimal places so our answer is given in no more than three decimal place accuracy. Thus $\lambda_{max} = 9.43 \times 10^{-7}\,\text{m} = 943\,\text{nm}$. Using Figure 9.2 for the best accuracy, the peak output is in the infrared, which is invisible to the human eye. From experience, we know these bulbs appear bluish-white so the overall color does not agree with the color at its peak flux.

The Sun has a surface temperature around 6000 K. Thus, the operating temperature of the filament in a halogen lamp is significantly cooler than the Sun, but hotter than that of an ordinary incandescent bulb. The Sun appears yellow, while the halogen appears bluish-white because the portion of the EM spectrum visible to our eyes. For the halogen, we are only sensing the relatively flat tail of the flux distribution at wavelengths well away from the peak flux. For the Sun most of the light is from the portion of the flux distribution near its peak.

10

SOME BASIC PHYSICS OF OPTICAL SYSTEMS

Propagation of visible light, that is, electromagnetic (EM) radiation in the range 400–700 nm and its interaction with matter can be understood in many circumstances with a few concepts and numerical parameters. Naturally, the same rules apply for radiation in the infrared and ultraviolet (UV) regions of the spectrum, but in this chapter we concentrate on EM radiation that can be detected by the human eye. Using the wave model of light, a broad range of phenomena can be easily explained: refraction, reflection, and diffraction. Refraction and reflection occur when light passes from one medium to another, for example, from air to water or from air to glass, while diffraction can be described as the modification of wave fronts by obstacles, particularly when the sizes of the obstacles are comparable to the wavelength of light. While refraction and reflection are best described with rays or trajectories perpendicular to wave fronts (geometrical optics), diffraction in its general form requires a description in terms of waves (wave optics). In addition, when rotational symmetry can be assumed, as in the passage of light through most lenses or its reflection off telescope mirrors, the description is further simplified (Gaussian optics).

Another property of light that we will consider in some detail is polarization. Polarization is a property of transverse waves and finds many applications in optical systems. We will also discuss briefly the phenomenon of *birefringence* and its relation to polarization.

After presenting the general law of refraction at the beginning of this chapter, we will restrict our description of geometrical optics to the linear case whereby all angles involved are small (paraxial approximation). Only at the end, we will consider aberrations, that is, deviations from linear or from Gaussian optics, including spherical aberration, astigmatism, etc.

Modern Devices: The Simple Physics of Sophisticated Technology, First Edition.
Charles L. Joseph and Santiago Bernal.

Companion website: www.wiley.com/go/joseph/moderndevices

10.1 REFRACTIVE AND REFLECTIVE OPTICS AND THEIR USES

The simplest device for producing images is a pinhole or small aperture cut into a box, a "camera obscura," which only works with very bright scenes on the outside. For this reason, lenses and mirrors are needed for collecting and deflecting light efficiently in cameras and other optical devices.

Light is deflected and reflected when passing from one transparent medium to another, for example, from air to glass. If the angle of incidence is defined as the angle between the incident ray and the normal to the surface, and the angle of reflection is similarly defined, then the law of reflection simply states that

$$\text{Angle of incidence } (\theta_i) = \text{Angle of reflection } (\theta_r) \tag{10.1}$$

Refraction, on the other hand, obeys Snell's law:

$$n_1 \sin \theta_1 = n_2 \sin \theta_2. \tag{10.2}$$

Here, θ_2 is the angle of refraction and n_1 and n_2 *are the indices of refraction* of the transparent media. Figure 10.1 illustrates Snell's law.

The index of refraction depends on the electrical and magnetic properties of the medium and ranges from 1.00 for vacuum to 2.42 for materials like diamond. Since

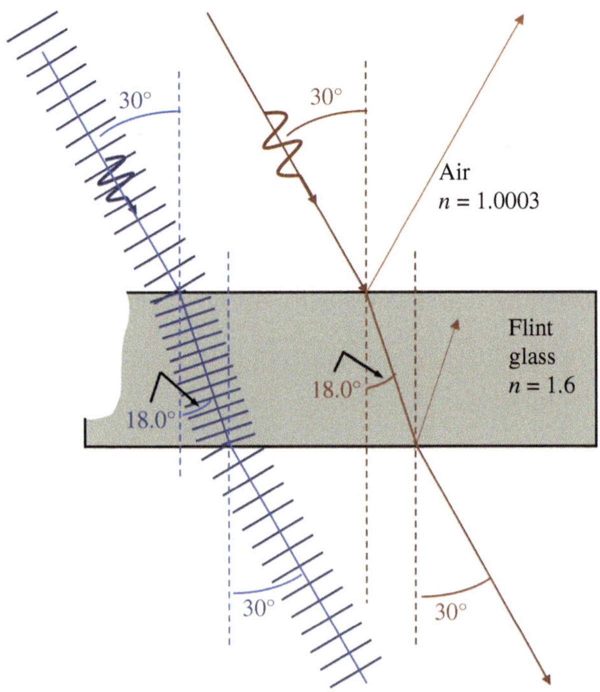

FIGURE 10.1 The refraction of a plane wave by a transparent material like glass is described by Snell's law (see Eq. 10.2).

in general materials respond differently to different colors of light, the index of refraction will depend on the wavelength; this effect is called *dispersion*. Customarily, the index of refraction of optical glasses is specified for yellow light at 589.2 nm, corresponding to the D-line of the spectrum of sodium.

From Snell's law, if $n_1 < n_2$, the angle of refraction will satisfy $\theta_2 < \theta_1$; if $n_1 > n_2$, on the other hand, $\theta_2 > \theta_1$ so the refracted angle will reach 90° when $\theta_1 = \sin^{-1}(n_2/n_1)$. In this latter case, the refracted light will be grazing the surface separating the optical media; for angles greater than $\sin^{-1}(n_2/n_1)$, *total internal reflection* will occur.

Snell's law can be derived from Maxwell's equations in material media, but it is also the consequence of the conservation of the component of light *momentum* parallel to the surface dividing the media, and the photon *energy* (directly proportional to frequency). Snell's law also implies that the speed of light, denoted by c in vacuum, is changed from c/n_1 in the first medium to c/n_2 in the second medium. In principle, Snell's law is sufficient for calculating ray trajectories through any refracting media; in fact, Snell's law is used to do detailed designs of optical systems for all applications, from cameras to microscopes and telescopes. For first-order calculations, however, the *paraxial approximation* is used: it is assumed that the rays are close to the optical axis, that is, the symmetry axis of the optical system, and that the angles relative to this axis are small so that $\sin\theta \approx \theta$. In addition, the rays are assumed to lie on a *meridian* plane, that is, a plane containing the optical axis.

The simplest lenses and mirrors use flat and spherical surfaces. If r is the radius of curvature of a spherical surface separating two media of indices n_1 and n_2, and s_o is the distance from an object to the surface, one can derive the following (paraxial) Gaussian formula for the image distance s_i inside the material of refractive index n_2:

$$\frac{n_1}{s_o} + \frac{n_2}{s_i} = \frac{n_2 - n_1}{r}. \tag{10.3}$$

A typical lens is made of two spherical surfaces of radii of curvature r_1 and r_2. From the Gaussian formula just given, we can obtain the *lens maker's formula*:

$$\frac{1}{s_o} + \frac{1}{s_i} = (n-1)\left(\frac{1}{r_1} - \frac{1}{r_2}\right), \tag{10.4}$$

where we have assumed that the lens has an index of refraction "n" and that it is surrounded by air. If the object is placed very far from the lens, at "infinity," the image distance becomes the lens's focal length f; thus, from the previous equation we obtain the lens formula:

$$\frac{1}{s_o} + \frac{1}{s_i} = \frac{1}{f}. \tag{10.5}$$

If the focal length is large compared to the lens thickness, the lens is a *thin* one and all distances are measured from its center. If the focal length is comparable to the lens thickness, the lens is called *thick*; the lens formula is applicable in this case too, but with the understanding that the distances are measured from special planes called the *principal planes* of the lens. See Advanced Concept AC10.1 for additional discussion and an example.

By convention, focal lengths are positive for converging lenses and negative for diverging lenses; further, object distances are positive when measured to the left of the lens and negative otherwise. Similarly, image distances are positive when measured to the right of the lens and negative otherwise. The image transverse magnification M_T of a lens can be easily derived from the geometry in Figure 10.2; from the triangles OPQ and $O'P'Q$, we find that the ratio $O'P'/OP$ is

$$M_T = -\frac{s_i}{s_o}, \tag{10.6}$$

the negative sign indicating that the image is inverted. In the figure used to derive the previous equation, both s_o and s_i are positive, as well as the focal length f of the convergent lens illustrated; the resulting image is *real* as it can be projected on a screen. By contrast, the image distance s_i would be negative if the image appears on the same side as the object, which would occur if $s_o < f$. The image in this case would be upright but *virtual*, as in the image produced when examining an object with a common viewing glass.

The *power P* of a thin lens is defined as the reciprocal of the focal length *in meters*; the resulting units are called *diopters* and denoted by "D." The power, as the focal length, can be positive or negative and is commonly used to characterize lenses for eyeglasses. The power of the eye itself comes mostly from the cornea at $P = 43$ D, so $f = 0.023$ m (see Chapter 15). Another important quantity used to characterize simple or compound lenses is the f-number, denoted by $f/\#$ and defined

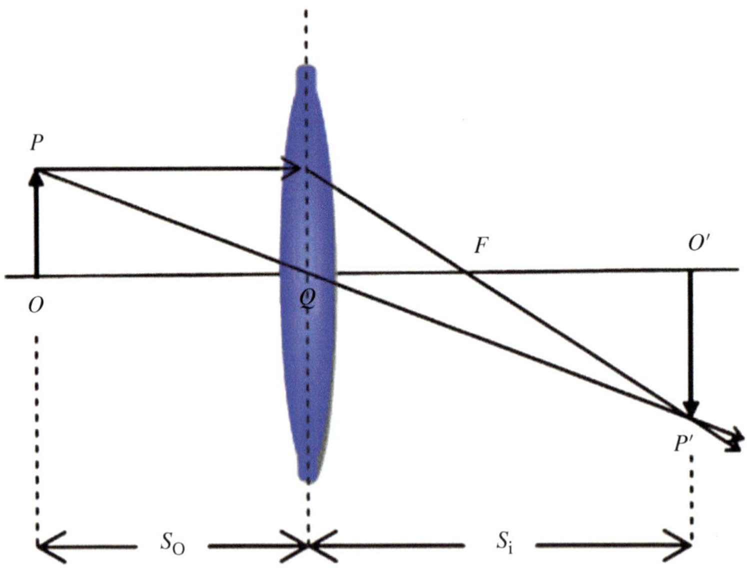

FIGURE 10.2 Ray tracing for calculation of transverse magnification (see Eq. 10.6).

as the ratio f/a, where f is the lens focal length and a is the diameter of the lens aperture. The amount of light energy transmitted by a lens (assuming negligible absorption and reflections) is proportional to the square of $1/f/\#$, so doubling $f/\#$ will decrease the light falling on the detector by a factor of 4. Thus, $f/\#$ is also a measure of the *speed* of the lens: smaller $f/\#$ indicates a lens capable of collecting more light, that is, a *faster* lens. However, there is always a tradeoff between speed and image quality as aberrations become more of an issue for larger lens apertures (Section 10.5).

INTRO PHYSICS FLASHBACK FB10.1

Refracting Telescope

A refracting telescope consists of two lenses, the objective and the eyepiece. The objective produces a real inverted image that the eyepiece magnifies, that is, transforms into a virtual image that the eye can focus to (see Fig. FB10.1) In the Keplerian astronomical telescope, the two lenses are convergent and such that the image focal point of the objective coincides with the object focal point of the eyepiece. This configuration yields a transverse magnification equal to the ratio of the focal lengths of the objective and eyepiece lenses: $M_T = -f_O/f_E$. The resulting image in the retina is inverted, but this is of little consequence for astronomical use. In another configuration, the Galilean telescope, the objective is a planar-convex lens and the eyepiece is a planar-concave lens; the image is erected in this case. The Galilean configuration is no longer in use, except as a laser-beam expander.

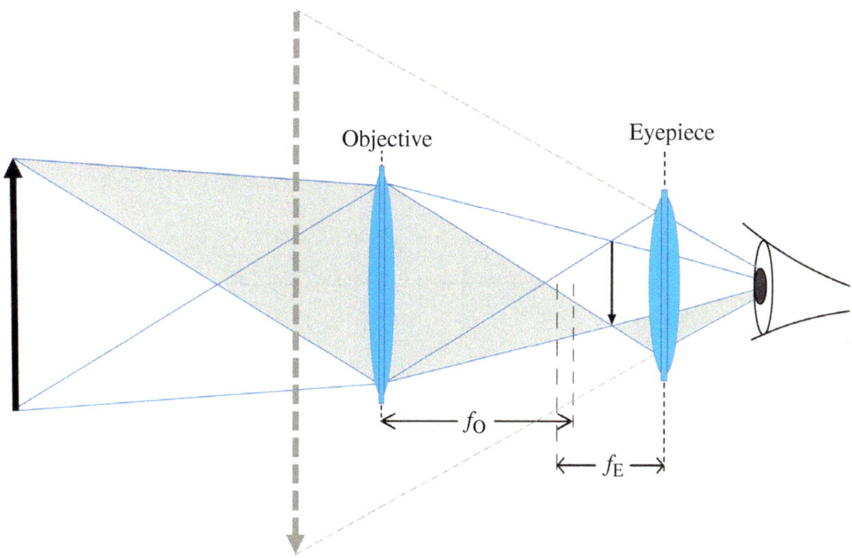

FIGURE FB10.1 Principle of the Keplerian refracting telescope.

10.2 POLARIZATION AND BIREFRINGENCE

10.2.1 Law of Malus and Brewster's Angle

Transverse waves can be *polarized*, meaning that the oscillating "medium" has a well-defined plane of oscillations (*linear* polarization) or the plane is rotating in a uniform way (*circular* polarization). This can be easily visualized with waves that travel down a rope and intercept fences having different orientations. As shown in Figure 10.3, the fences act as *polarizers* that modify the plane of the oscillations in the rope. For light and other EM waves, on the other hand, there is no "medium," but the electrical and magnetic fields can have well-defined planes of oscillation.

Plane polarized light has a well-defined plane of vibration for its electrical field, which can be represented by a double arrow subtending and angle θ to the horizontal axis, as in Figure 10.4; unpolarized light would be represented by similar arrows with angles θ distributed at random. A more realistic representation of ordinary unpolarized light (e.g., light from an incandescent light bulb) would also have a random variation of the length of the arrows, that is, the wave amplitudes at a given instance. The angle θ is the angle of polarization; it is also the *transmission axis* of a hypothetical *plane polarizer* used to obtain the plane polarized beam out of unpolarized light.

In general, light is partially polarized, meaning that it contains both polarized and unpolarized components. The *degree of polarization* of a light beam is defined as the ratio of light intensity from its polarized component, I_p, and the total intensity from polarized *and* unpolarized components:

$$V = \frac{I_p}{I_p + I_u}. \tag{10.7}$$

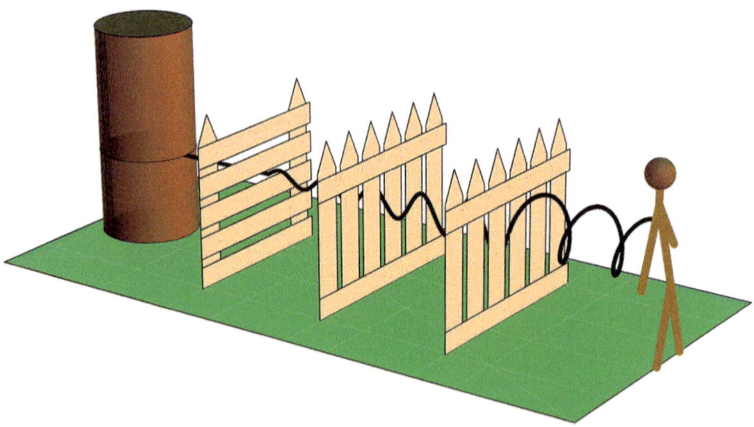

FIGURE 10.3 Polarized waves on a rope: an initially circularly "polarized" rope wave traverses a fence (vertical polarizer) to become vertically polarized and then crosses a horizontal fence (horizontal polarizer) and is "absorbed." Source: From "Understanding Polarization with an Analogy" from the Wolfram Demonstrations Project http://demonstrations.wolfram.com/UnderstandingPolarizationWithAnAnalogy/. Contributed by: Enrique Zeleny.

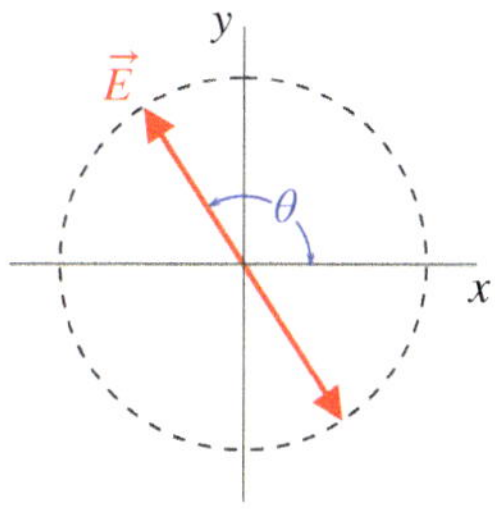

FIGURE 10.4 The double arrow represents the plane of vibration of the electrical field of the plane-polarized EM wave moving out of the plane of the figure.

Thus, completely polarized light would have $V=1$, while completely unpolarized light has $V=0$. (Naturally, however, these two extremes are only approximately approached in practice.) As an example, the degree of polarization of the skylight observed directly overhead (zenith) at sunrise is around 84%.

Polarization can be produced through reflection, transmission, dichroism, double refraction, and scattering. The French scientist Étienne-Louis Malus discovered around 1810 polarization by reflection and also the phenomenon of birefringence or double refraction (see later.) The *Law of Malus* establishes that if two polarizers have transmission axes at an angle θ relative to each other, the intensity of the light transmitted through the second polarizer ("analyzer") is given by

$$I = I_0 \cos^2 \theta, \tag{10.8}$$

where I_0 is the transmitted light intensity when the two polarizers are aligned, that is, when $\theta=0$.

In reflection and refraction of light, the *plane of incidence* is the plane containing the vibrations of the electrical field in the beam and perpendicular to the surface dividing the two media; the incidence plane defines *parallel* and *perpendicular* directions of polarization, as shown in Figure 10.5. Malus discovered that at a special incident angle, the "polarizing angle," the reflected light has maximum polarization in a plane perpendicular to the plane of incidence and essentially zero polarization parallel to that plane. In addition, at this special angle the refracted and reflected rays form a right angle (Fig. 10.5b), making it possible to connect polarization to the index of refraction. Malus had made his original observations in air–water, but Sir David Brewster extended them to glasses around 1815. From Snell's law and assuming light incident from a medium with index of refraction n_1 into a medium of index of refraction n_2, it is easy to derive an expression for *Brewster's angle* ϕ_B:

$$n_2 = n_1 \tan \phi_B. \tag{10.9}$$

As an example, if $n_1=1.00$ (e.g., air) and $n_2=1.53$ (e.g., glass), we find $\phi_B=57°$, as illustrated in Figure 10.5b. Polarization by reflection at the Brewster angle has found an important application for orienting the mirrors in laser cavities; in this way, the laser beam coming out of the cavity has a high degree of polarization. In another

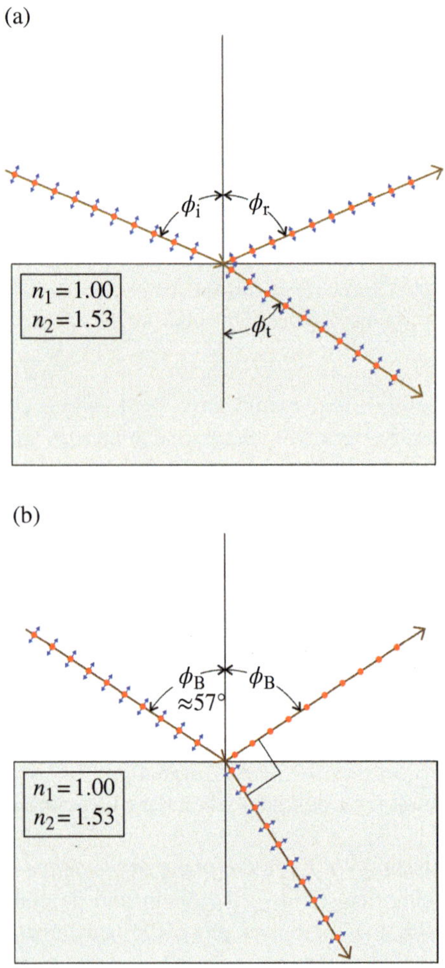

FIGURE 10.5 (a) The polarization of reflected and transmitted light is not affected much at a general angle of incidence $\phi_i = \phi_r$. (b) At Brewster's angle of incidence, the reflected light is polarized in a direction perpendicular to the plane of incidence (plane of the figure). Parallel and perpendicular directions of polarization are indicated by arrows and solid circles, respectively.

example, partially polarized light from level surfaces, for example liquid water or snow, can be filtered out with polaroid (see later) sunglasses.

10.2.2 Dichroism and Birefringence

Some substances such as the mineral tourmaline absorb one component of polarization and transmit the other. The phenomenon, called *dichroism*, is the basis of the original polarizing filters invented by Edwin. H. Land in 1932. The *J-sheets* of the original polaroids contain ultramicroscopic dichroic crystals whose polarizing axes are all parallel; in effect, the sheet is like a gigantic dichroic crystal. Problems arising from light scattering in J-sheets led Land to invent the *H-sheets* in 1938. Unlike J-sheets, H-sheets do not contain dichroic crystals; instead, they are made of long polymeric carbon molecules that act like a grid of parallel wires. Light whose polarization direction coincides with the direction of the molecules is absorbed. Polaroids of the H variety have found many applications in optical instruments, photography, sunglasses, visors and displays (see Chapter 15), automobile headlights and mirrors, and others. In Interesting Tidbit TB10.2, we quote Land on the interesting story of how he came up with the idea for his invention.

INTERESTING TIDBIT TB10.2

Dogs, Kaleidoscopes and Synthetic Polarizers

The early history in the development of sheet polarizers or polaroids is surrounded by very interesting circumstances. Edwin H. Land, the founder of Polaroid Corporation, recounts some of these in the following words (from Edwin H. Land, 1951):

"In the literature there are a few pertinent high spots in the development of polarizers, particularly the work of William Bird Herapath, a physician in Bristol, England, whose pupil, a Mr. Phelps, had found that when he dropped iodine into the urine of a dog that had been fed quinine, little scintillating green crystals formed in the reaction liquid. Phelps went to his teacher, and Herapath then did something which I think was curious under the circumstances; he looked at the crystals under a microscope and noticed that in some places they were light where they overlapped and in some places they were dark. He was shrewd enough to recognize that here was a remarkable phenomenon, a new polarizing material ...."

"Herapath's work caught the attention of Sir David Brewster, who was working in those happy days on the kaleidoscope. Brewster thought that it would be more interesting to have interference colors in his kaleidoscope than it would to have just different-colored pieces of glass. The kaleidoscope was the television of the 1850s and no respectable home would be without a kaleidoscope in the middle of the library. Brewster, who invented the kaleidoscope, wrote a book about it, and in that book he mentioned that he would like to use the herapathite crystals for the eyepiece. When I was reading this book, back in 1926 and 1927, I came across his reference to these remarkable crystals, and that started my interest in herapathite."

Other materials such as calcite and quartz crystals exhibit optical properties that depend on the direction of light propagation. This *anisotropy* can be taken advantage of for the production of polarized beams over a wider range of wavelengths than is possible with polaroid films. When unpolarized light enters a calcite or quartz crystal, two refracted rays instead of one are observed. One of the rays, the *ordinary* or *O-ray*, obeys Snell's law and stays on the plane of incidence, while the other ray, the *extraordinary* or *E-ray* does not stay, in general, on the incident plane. The phenomenon is called double refraction or *birefringence*.

Naturally, light propagation in anisotropic crystals requires additional definitions. Calcite and quartz are examples of *uniaxial* crystals, that is, crystals with an axis of

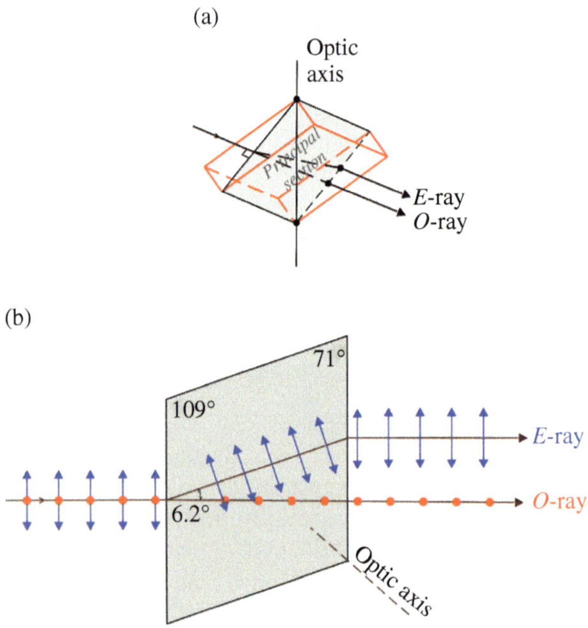

FIGURE 10.6 (a) Optic axis and principal section in calcite crystal, and (b) light traversing calcite crystal on the principal section. The directions of polarization are indicated as in Figure 10.5. Source: HECHT-ZAJAC, OPTICS, 1st Edition, © 1974, p. 234. Adapted by permission of Pearson Education, Inc., Upper Saddle River, NJ.

symmetry that defines a direction called the *optic axis*. If light enters the crystal along this direction, no double refraction occurs. The plane containing the optic axis and perpendicular to a *cleavage* surface, that is, a surface along which the crystal can be cut to yield a similar crystal, is called a *principal section*. In addition, principal planes for both the ordinary and extraordinary rays are defined such that they contain the optic axis and the respective rays. These principal planes do not coincide in general; they do only when the plane of incidence is also a principal section, as shown in Figure 10.6a for calcite.

When unpolarized light enters a crystal such as calcite in the manner depicted in Figure 10.6b, the *O*-ray has a polarization perpendicular to the principal section (also the plane of incidence in the figure), while the *E*-ray has a polarization parallel to the same plane. Polarization by birefringence was discovered by Huygens in 1678 and is the basis for birefringent polarizers that work over a broad spectral range. Several types of prisms, some of only historical interest, work as birefringent polarizers by isolating one of the two refracted rays.

10.2.3 Retarder Plates and Circular Polarization

The propagation of light in uniaxial and the more common biaxial crystals can be studied in detail using constructions based on Huygens' principle, an idea discussed in Section 10.3.1. A major result of those constructions is the existence of two principal indices of refraction for uniaxial crystals and three for biaxial crystals. At any given wavelength, over a broad range from about 200 nm (UV) to 1 μm (near-infrared), the index of refraction of the *O*-ray in calcite, n_o, is *larger* than the index of refraction of the *E*-ray, n_E, for propagation of the *E*-ray perpendicular to the optic axis. The situation for quartz is reversed, that is, the index of refraction of the *E*-ray is larger than that of the *O*-ray. Geometrically, the difference between the two indices of refraction determines the ellipticity of the *E*-wavefront surface that can be constructed using Huygens' principle.

The difference in the principal indexes of refraction can be taken advantage of to construct wave *retarder plates*. In a uniaxial positive crystal like quartz, for example, the *O*-ray travels faster than the *E*-ray because $n_O < n_E$. Thus, if the distance of penetration into the crystal is *d*, then a phase difference between the *O* and *E*-waves is established at the exit:

$$\delta = \frac{2\pi}{\lambda} d\left(n_O - n_E\right), \tag{10.10}$$

where λ is the wavelength. If δ is an integer multiple of 2π, then the direction of the incident wave vibration (assumed to correspond to linear polarization as in Fig. 10.4) will be unaffected; however, if δ is a multiple of π, the initial direction will have rotated by twice the initial angle with the vertical (Fig. 10.4). This latter case corresponds to the *half-wave plate* shown in Figure 10.7a. For any other values of δ, the light will be *elliptically* polarized. This general type of polarization has an electric field vector whose terminal point traces an ellipse at any plane perpendicular to the direction of propagation, and with a rotation frequency equal to that of the wave. In Figure 10.7b, we show the special case of *circularly*

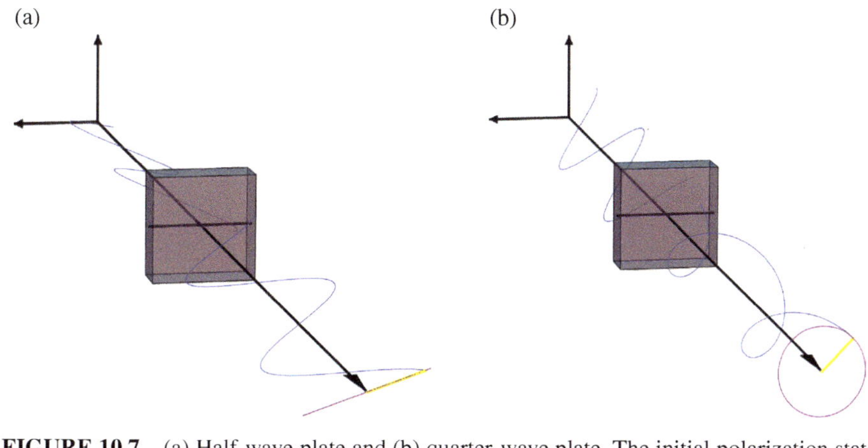

FIGURE 10.7 (a) Half-wave plate and (b) quarter-wave plate. The initial polarization state is linear in both cases. Source: From "Polarization of an Optical Wave through Polarizers and Wave Plates" from the Wolfram Demonstrations Project http://demonstrations.wolfram.com/PolarizationOfAnOpticalWaveThroughPolarizersAndWavePlates/. Contributed by: Fred E. Moolekamp and Kevin L. Stokes (University of New Orleans).

polarized light produced by the action of a *quarter-wave* plate ($\delta = \pi/2$) on an initially linearly polarized wave. In addition to the phase requirement, the O- and E-ray electrical vectors must have the same amplitude to yield $\theta = 45°$ initially. If the phase ($\pi/2$) and amplitude conditions for circular polarization are met, a polarizer made of calcite produces left-handed rotation when looking in a direction opposite to the light propagation because the E-wave travels faster than the O-wave. In quartz, the circular polarization would be right-handed, as in Figure 10.7b. The sense of rotation is reversed if $\delta = 3\pi/2$, because the situation would be equivalent to adding a half-wave plate to a quarter-wave plate.

10.3 DIFFRACTION

The imaging produced by lenses and mirrors depends on refraction and reflection of light, but the finite extent of these elements introduces limitations from *diffraction* effects, that is, effects from the encounter of the light wavefront with the edges of the lens or mirror. These effects are similar to those produced by apertures of the same size as the lens or mirror. A distinction is made between diffraction at long ("far-field") and at short ("near-field") distances; the first case, called *Fraunhofer* diffraction, involves plane wavefronts and is relatively easy to describe mathematically. For short distances, on the other hand, the diffraction is called *Fresnel* diffraction and requires more elaborate mathematics (*Fourier* optics) as it deals with curved wavefronts.

10.3.1 Huygens' Principle and Diffraction from a Single Slit

The physics of wave propagation is embodied in *Huygens' principle*, which is particularly useful to understanding diffraction: a wavefront can be thought of a collection of point sources each emitting spherical wavelets; the progression of the wavefront is then obtained from the envelope of the individual wavefronts, as shown in Figure 10.8. With this picture, it is easy to see how light and other waves can move

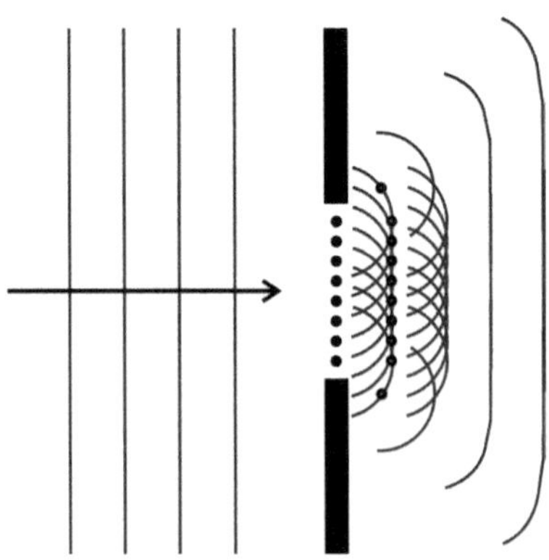

FIGURE 10.8 Illustration of Huygens' principle: 2D construction of a diffracted wavefront through a rectangular slit.

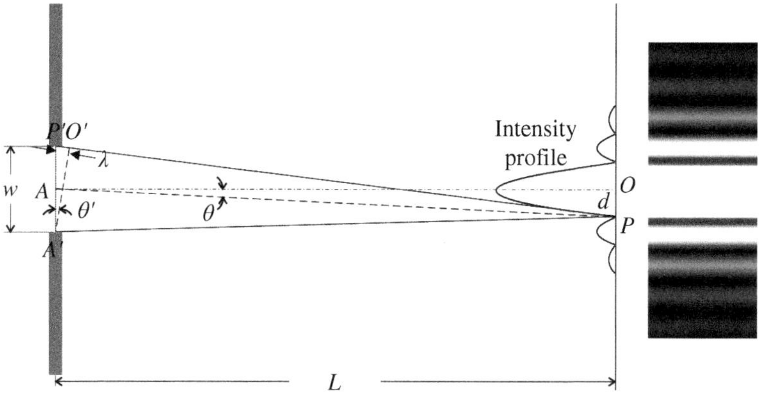

FIGURE 10.9 Fraunhofer diffraction by a rectangular slit: the point *P* indicates the location of the first minimum of intensity. For a large slit-to-screen distance *L*, comparison of triangles *OAP* and *O′A′P′* show that the angles θ and θ' are essentially equal.

around obstacles. To make Huygens' principle agree with experiment, the amplitude associated with a given secondary wavelet has to be corrected by the introduction of an *obliquity* factor, $1+\cos\theta$, where $\theta=0°$ is the forward direction and $\theta=180°$ is the backward direction. With this proviso, the wavelets do not produce wavefronts in the direction opposite to the propagation direction.

Diffraction of light of wavelength λ by a single rectangular slit in the Fraunhofer limit can be easily described mathematically. With reference to Figure 10.9, we have $\sin\theta' = \lambda/w$, and $\sin\theta = d/L$. Then, for $L \gg w$ the angles θ and θ' are indistinguishable, and we obtain the following equation for the linear distance between intensity minima on a faraway screen:

$$d = \frac{L\lambda}{w}, \tag{10.11}$$

where L is the distance between the slit and the screen, λ is the wavelength of light, and w is the slit's width.

If a circular aperture is used instead of the slit, we observe on the screen a central spot surrounded by rings. The dark rings separating the bright ones are not equally spread in angle; the first dark ring happens at an angle equal to

$$\theta = 1.22\frac{\lambda}{D}, \tag{10.12}$$

where D is the diameter of the circular aperture. Note that in the equation for the rectangular slit the quantity d/L would be the corresponding angle and that 1.00 would be the constant in place of 1.22. Figure 10.10 shows the diffraction pattern of a circular aperture; the central spot is known as the *Airy's* disk.

A smaller lens (larger *f/#*) is desired for better depth of field, but at some point diffraction effects will blur the images beyond the desired resolution of the imaging

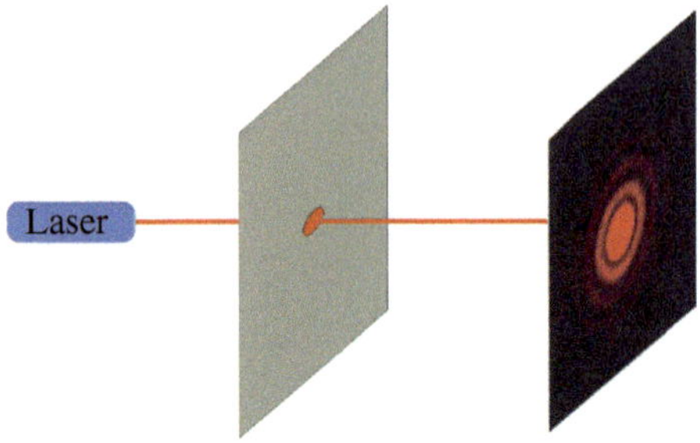

FIGURE 10.10 Fraunhofer diffraction by a circular aperture. The airy's circle is the large central region in the diffraction pattern.

device. When that occurs, we say that the lens or imaging system is *diffraction limited.* An example is provided by a telescope: if the light source is a distant star, the image will consist of a central spot surrounded by diffraction rings. If, for example, the telescope has a 6 inch (15.2 cm) objective lens or mirror, the angular resolution for white light (average $\lambda = 560\,\text{nm}$) is, from Equation 10.12, $\theta = 4.5 \times 10^{-6}$ rad or 0.9 seconds of arc. This is the diffraction limit of the telescope, as diffraction would not allow resolving two stars separated by less than 0.9 seconds of arc. Naturally, other effects like aberrations (Section 10.5) and the atmosphere can affect the effective resolution of a telescope.

10.3.2 Fresnel Zone Plate

Important additional insights to diffraction and its applications can be gained through a construction devised by Fresnel and first realized in practice by Lord Rayleigh. If concentric circles are drawn with radii proportional to the square roots of whole numbers and every other ring is darkened, a photocopy of this construction made on a slide transparency will act as a lens. This device is called a *Fresnel zone plate* (FZP) and an example is shown in Figure 10.11.

To understand semiquantitatively how the zone plate works, we imagine a spherical wavefront from a faraway point source. The wavefront can be divided into concentric *zones* such that successive zones contribute to the wave amplitude at a (on-axis) point in the forward direction with phases that differ by π. Therefore, successive zones make wave amplitude contributions of opposite signs. It can be shown that with this construction the net amplitude at the observation point is essentially equal to one-half the amplitude contributed by the first zone. The obliquity factor mentioned earlier plays a fundamental role, as it makes the effect of the last zones, that is, the ones behind on the spherical wavefront, very small.

The focal length of the FZP is related to the inner radius r_m of the *m*th ring and the wavelength λ through the following equation:

$$f = \frac{r_m^2}{m\lambda}. \tag{10.13}$$

FIGURE 10.11 Fresnel zone plate (FZP): example of zone plate with 50 rings and a focal length of 10 m for light with a wavelength $\lambda = 550$ nm. See Equation 10.13.

FIGURE 10.12 Fresnel lens concept: a lens is replaced by a series of ridges in a "collapsed" geometry of the original lens.

As an example, if $\lambda = 550$ nm, and $f = 10$ m, the radius of the first zone is $r_1 = 2.3$ mm, and that of the 50th zone is $r_{50} = 1.7$ cm, as in Figure 10.11.

By construction and unlike a standard lens, the FZP only works at a single wavelength, that is, FZP suffers from chromatic aberration. The fact that the rings on a FZP are either black or white makes it an example of a *binary optics* element. Other devices, called *Fresnel lenses* (FLs) should be distinguished from FZP lenses. FL's are essentially collapsed or "flattened out" lenses as indicated in Figure 10.12.

Inexpensive plastic Fresnel lenses find many applications where light weight, low profile and/or large area are important considerations: magnifier lenses for old TV screens and other displays, condenser lenses for projectors, collimators for light-emitting diode (LED) sources, radiation concentrators for solar cells, promotional credit-card-sized lenses, etc. FZP or lenses made of aluminum or even cardboard can also be used to focus and collimate UV radiation (for which plastic is opaque) and microwaves.

10.3.3 Diffraction Gratings

We considered before the diffraction from a single narrow slit (Fig. 10.9). Important devices called *diffraction gratings* are based on the combined action of many narrow slits equally spaced and drawn on, for example, a photographic plate, or on the reflection of light from many closely spaced (and properly shaped) grooves ruled on materials like aluminum. The wavefronts from the different slits or grooves interfere constructively or destructively to produce many sharp intensity maxima and minima, as shown in Figure 10.13. The distribution of light intensity for *Fraunhofer* diffraction as a function of the angle θ of light direction relative to the normal to the screen (as in Fig. 10.9 for one slit) can be written as

$$I = I_1 \frac{\sin^2(N\delta/2)}{\sin^2(\delta/2)}, \quad I_1 = I_m \frac{\sin^2(\pi w \sin\theta/\lambda)}{(\pi w \sin\theta/\lambda)^2}, \tag{10.14}$$

where N is the number of slits, w is the width of a single slit, I_1 is the intensity from one slit alone, λ is the wavelength, and δ is the wave phase shift from one slit to the next. The latter can be expressed as

$$\delta = \frac{2\pi d \sin\theta}{\lambda}, \tag{10.15}$$

where d is the spacing between slits. The angular positions of the principal maxima of the diffraction pattern are obtained by setting $N\delta/2 = m\pi$, where m/N is an integer and $m = 0, 1, 2, \ldots$ is called the *interference order*. Therefore, from Equation 10.15 we find

$$\sin\theta = \frac{m\lambda}{Nd}, \tag{10.16}$$

for the angular location of intensity maxima of order m. This condition is called the *Bragg* equation.

From Equation 10.14 and Figure 10.13, we can see that the diffraction pattern of one slit modulates the main intensity pattern from the grating. In other words, the diffraction pattern of one slit is the envelope of the intensity maxima of the *interference* pattern from the slits or grooves. In Figure 10.13, we have plotted the normalized intensity distribution of Fraunhofer diffraction for two cases with six slits: width $w = 0.03$, 0.3 mm, spacing $d = 3$ mm, and light of wavelength $\lambda = 550$ nm. When the slits are broader (0.3 mm), the intensity envelope is clearly seen; the sharpness of the intensity maxima, on the other hand, increases with the number of slits. In fact, from the equations above it is straightforward to show that the *resolving power* of the grating, that is, the smallest separation of wavelengths, $\Delta\lambda$, that it is possible to measure in a region around a wavelength λ, is proportional to N:

$$\frac{\lambda}{\Delta\lambda} = Nm. \tag{10.17}$$

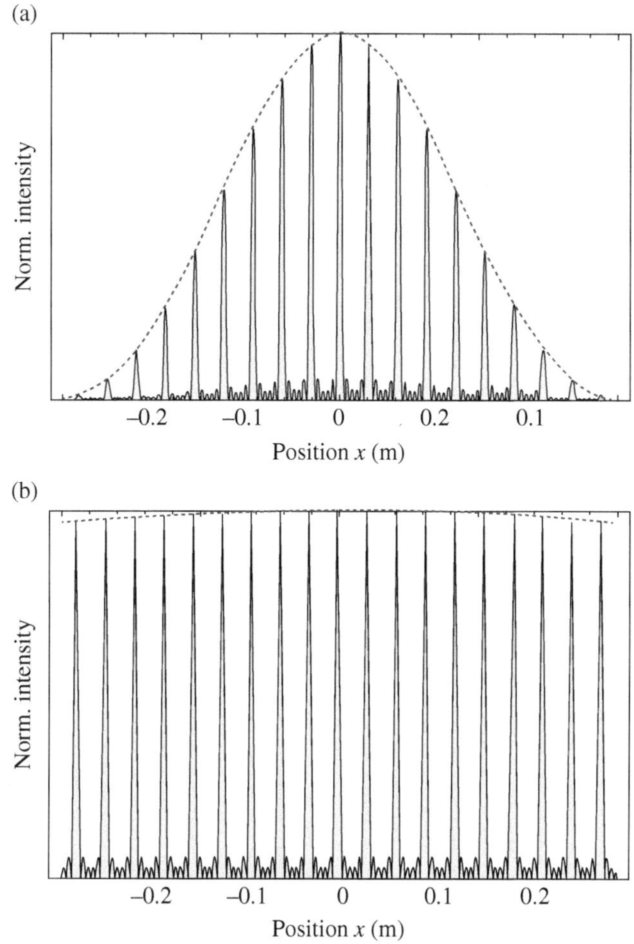

FIGURE 10.13 Fraunhofer diffraction profile from six slits (diffraction grating): (a) broad slit (0.3 mm) and (b) narrow slit (0.03 mm). The intensity is normalized to the maximum value; the dashed curve represents the diffraction intensity from one slit. Source: From "Multiple Slit Diffraction Pattern" from the Wolfram Demonstrations Project http://demonstrations.wolfram.com/MultipleSlitDiffractionPattern/ Contributed by: Peter Falloon.

As already mentioned, diffraction gratings can be based on the transmission or reflection of light. In the latter case, the gratings commonly have a sawtooth-shaped groove profile at an angle called the blaze angle. The angle is optimized for maximum efficiency at a given wavelength, that is, maximum diffracted power in first order ($m = 1$).

The resolving power of diffraction gratings is orders of magnitude better than the corresponding quantity for prisms. As an example, a 15 cm wide grating with 5700 lines/cm has a resolving power of 85,500 in first order (Eq. 10.17), which translates into a wavelength separation of just 0.0058 nm at $\lambda = 500$ nm. Diffraction gratings made possible the detailed study of atomic spectra in the late nineteenth and early twentieth centuries. This work in turn laid the foundations of present-day quantum mechanics. We discuss in Chapter 12 the application of ruled and holographic gratings to spectroscopy.

10.4 HOLOGRAPHY

Ordinary photography and TV render flat images of objects because the pictures contain information only on the amplitude of the original light waves scattered by the objects. Stereographic photography and display go farther and make possible the illusion of depth by the creation of two flat pictures that are slightly separated in space. When the two pictures are perceived with separate color or polarizing filters, the visual system process the images as coming from different angles in a way similar to the perception of a real object. However, stereographic images are not truly 3D, as very limited or no exploration around the object is possible. Holography, on the other hand, permits the display of fully 3D images whereby the viewer can perceive both depth and parallax effects.

10.4.1 Basic (Absorption) Holography

Holography, which means "entire or whole picture," was invented by Dennis Gabor (1971 physics Nobel Prize) in 1948, but several other people were involved over the years for its practical development. A holographic recording, or "hologram," unlike a standard photograph contains information not only on the amplitude but also on the phase of the light originally scattered by an object. In order to record both amplitude and phase information, an *interference pattern* is imprinted on a high-resolution photographic film (or other light-sensitive medium) by combining the object light with "reference" (unobstructed) light derived from the same source that illuminates the object. To "play back" the hologram, the developed film is illuminated with the same reference light used to record the hologram. The 3D holographic image then appears as a virtual image occupying the same space as the original object. In effect, the light wavefronts from the original object can be reconstructed because all the information, amplitude and phase, is stored in the hologram. Figure 10.14 illustrates the basic principle of holographic recording and reproduction.

Additional insights on the physics of holography are possible in terms of zone plates. In the original experiments of Gabor, quasi-monochromatic light was scattered from a small object and made to interfere with the unobstructed light on a photographic plate. The resulting interference pattern or *Gabor zone plate* is similar to a *Fresnel zone plate* (Section 10.3), except that the fringes have a radially varying intensity. When the developed hologram is illuminated with the same light used to record it, real and virtual images of the original small object appear on either side of the plate, but these are difficult to see because of the colinear geometry. The Gabor zone plate is equivalent to a combination of two lenses, one positive (convergent) and the other one negative (divergent), and an attenuator. Thus, the appearance of real and virtual images can be understood as a simple "lensless" imaging process. For an extended object, each point will produce its own Gabor zone plate, so the hologram

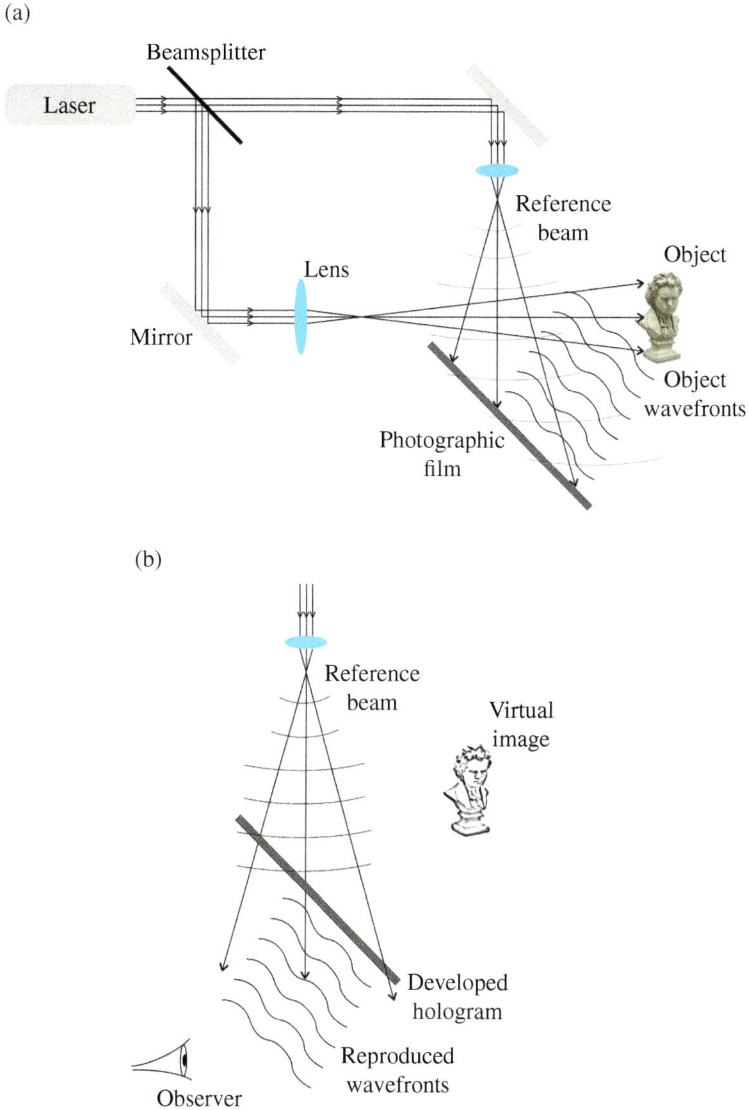

FIGURE 10.14 Principle of holography: (a) recording of hologram by mixing of mutually coherent reference and scattered beams and (b) reproduction of hologram. Only the virtual image is shown in (b). Source: O'SHEA, DONALD C.; CALLEN, W. RUSSELL; RHODES, WILLIAM T., AN INTRODUCTION TO LASERS AND THEIR APPLICATIONS, 1st Edition, © 1977, p. 186. Adapted by permission of Pearson Education Inc., Upper Saddle River, NJ.

will contain a superposition of interference patterns capable of diffracting the light to reproduce the original object wavefronts.

Basic holography can also be understood in the language of communication technology: the reference wave is similar to the carrier signal, while the amplitude and/or phase modulation resulting from the interference of object and reference waves is similar to wave mixing in a heterodyne detector. The hologram then plays the role of both detector and storage device.

10.4.2 Temporal and Spatial Coherence

A key requirement of holography is that light with a high degree of *temporal and spatial coherence* be used to record and reproduce the holograms. With high degree of coherence, the interference pattern recorded will have well-defined fringes, that is, good contrast, necessary for efficient diffraction when the hologram is illuminated.

The concept of coherence is best understood in terms of correlation of events or occurrences. Two events separated in space are said to be correlated if one leads to the prediction of the other; for example, in a parade the location of one marcher is correlated to the location of any other marcher. The same idea applies to events separated in time: the events are correlated if they follow a predictable pattern. If we take a sequence of pictures of a parade without moving the camera, a comparison of the pictures shows a time correlation. For waves, including EM radiation, good *transverse spatial coherence* means that the amplitude at any point on a wavefront (i.e., a surface of constant phase) is correlated to the amplitude at any other point on the same wavefront. In contrast, *temporal or longitudinal coherence* refers to a measure of the correlation of amplitude values at two different instances along the direction of propagation. Figure 10.15 illustrates the concepts. We can see that spatial (transverse) and temporal coherence are independent of each other.

Spatial coherence of a light beam can be measured by the largest area in a wavefront that leads to visible fringes in an interference experiment. Spatial coherence is

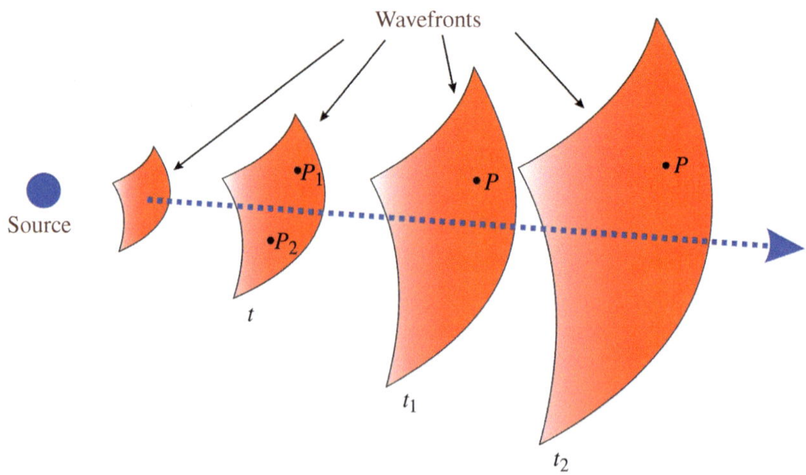

FIGURE 10.15 Spatial and temporal coherence: transverse spatial coherence refers to correlation of field values between separate points in a wavefront at one instance, while temporal coherence refers to correlation of field values at the same point in a wavefront at two separate times.

determined by the size of the source as well as the distance from it. Temporal coherence, on the other hand, is defined by the *bandwidth* of the radiation. In turn, the latter is directly related to the emission characteristics of the source. Interference experiments can also be used to quantify temporal coherence: no fringes are visible if the path difference of beams in an *interferometer* greatly exceeds the coherence length.

Only monochromatic light, that is, light with a single wavelength, is perfectly coherent in both space and time. Laser light is the closest to such a form of EM radiation. In contrast, light from an incandescent light-bulb is very incoherent in both space and time because of the multitude of wavelengths present as well as the random character of the emission of light in a thermal source.

Two measures of coherence that are often employed are the *coherence time* (t_c), or *coherence length* (l_c), and the *coherence area* ΔA_c. If the bandwidth of the radiation is $\Delta\nu$, then we have

$$t_c = \frac{1}{\Delta\nu}, \quad l_c = ct_c. \tag{10.18}$$

If the wavelength is λ, and the light source subtends a solid angle $\Delta\Omega$ in steradians (see Advanced Concept AC10.3), the *coherence area* ΔA_c is defined by

$$\Delta A_c = \frac{\overline{\lambda^2}}{\Delta\Omega}, \quad \Delta\Omega = \frac{A}{r^2}, \tag{10.19}$$

where r is the distance to the source, A is the cross-sectional area of the source, and the bar indicates average. Highly incoherent light can be made much more spatially coherent by selecting a small section of a wavefront (a process called *spatial filtering*) by using, for example, a pinhole. Temporal coherence can be significantly enhanced by means of a color filter (temporal filtering) to reduce the bandwidth.

In holography, the temporal coherence length determines the maximum path difference between object and reference beams necessary for good contrast, and the spatial coherence length determines the lateral size. See the Concept Verification CV10.2 on temporal and spatial coherence for additional discussion and numerical examples of t_c and ΔA_c.

10.4.3 Other Methods of Holography and Applications

The basic holographic process described earlier and illustrated in Figure 10.14 corresponds to what is known as *absorption holography*. Furthermore, the recording and reproduction (diffraction) was supposed to take place on the surface of the hologram, that is, the hologram was a *thin hologram*. But other types of holograms are possible. In a *phase hologram*, for example, the phase instead of the amplitude variation of wavefronts is recorded; for this, special transparent materials are used. In other, *thick holograms*, Bragg's diffraction allows the use of white light for both recording and reproduction. The *white light hologram*, also called Lippmann–Bragg or Denisyuk hologram after its inventors is a volume hologram where Bragg's diffraction acts like a filter to introduce the necessary coherence for the different wavelengths involved. Good spatial coherence, as from a small light source or the sun is still required for viewing these holograms.

Holography has many applications other than displays for commercial or entertainment purposes. A few examples are: (i) holographic interferometry for non-destructive testing of materials or live specimens, (ii) holographic optical elements for imaging and optical data processing, and (iii) holographic optical memories for massive, ultrafast, and redundant digital storage. Holographic interferometry and optical elements have been used for many years in research and industry, but the use of holography for computer memories is still in the R&D phase. Another area of strong R&D efforts is holographic displays, described in Chapter 15. The possibility of true holographic TV, for example, may depend on the realization of *real-time* computer-generated holograms (CGHs). See Interesting Tidbit TB10.3.

INTERESTING TIDBIT TB10.3

Computer-Generated Holography

Since holograms are interference patterns, they can be computed or synthesized in principle if a mathematical description of the object and the "recording" geometry are given. Thus, it is possible to create holograms in a computer without any physical object or laser light. The computed holograms can then be printed out on a transparency and then illuminated with laser light to generate a 3D image of the fictitious object. Computer programs and detailed instructions on their use are available on the Internet. See for example, "The CorticalCafe Computer Generated Hologram (CGH) Construction Kit" at http://corticalcafe.com/prog_CGHmaker.htm.

INTERESTING TIDBIT TB10.4

Diffractive Optics and Applications

The computer-generated holography (CGH) described in Interesting Tidbit TB10.3 is but one example of a broad area called Fourier optics. Ordinary optics deals with elements based on refraction and reflection, whereas Fourier optics is the underlying science of optical devices that depend on diffraction. Another application of Fourier optics that we already mentioned (Section 10.3) is the Fresnel zone plate (FZP), which can approximate a standard lens by means of diffraction by a series of concentric rings (Fig. 10.11). With diffractive optics, it is possible to manipulate the amplitude and phase of light wavefronts for applications not only in modern imaging but also in optical communications (e.g., fiber optics), data storage, networking, and sensors. Furthermore, the lithography techniques used in the microelectronics industry have been applied to "micro-optics" and "integrated-optics" to build diffractive optical components. An example of "micro-optics" is the microlens, a lens of diameter of several micrometers to

1 mm. Arrays of microlenses have found applications in digital projectors, CCD sensors, photovoltaic cells, 3D displays and others.

For signal processing, optical elements have advantages over electronic ones: extreme bandwidth (i.e., more communication channels), miniaturization, broad choice of materials, and immunity from EM interference. Although an all-optical computer has not been built yet, there has been recent progress in the development of an optical transistor.

10.5 PRIMARY ABERRATIONS

Departure from paraxial conditions in an optical system occurs when rays traverse on the periphery of the components or at steep angles to the optical axis, or both. In this case, the approximation $\sin\theta \approx \theta$ (which led to the simple lens formulae above) breaks down. The use of the exact form of Snell's law would be required for accurate ray tracing; alternatively, a first correction to the paraxial approximation can be made by using $\sin\theta \approx \theta - \theta^3/6$. The corrections introduced by this third-order theory relative to the paraxial approximation are called the *primary* or *Seidel* aberrations. There are five of them: *spherical aberration, coma, astigmatism, field curvature,* and *distortion*. It should be noted that these are *monochromatic* aberrations, as they are evaluated for light at a particular wavelength. With white light, however, the dependence of the index of refraction of the optical material on wavelength or color, that is, *dispersion*, will lead to *chromatic aberration,* very often a more serious effect than any of the primary aberrations. Although each monochromatic aberration corresponds to a specific defect in the image, it is not possible to correct the aberrations independently. Thus, a condition exists for the simultaneous correction of spherical aberration and coma, and a similar condition for correcting astigmatism and distortion.

Spherical aberration (SA) is inherent to any lenses or mirrors whose surfaces are spherical sections. As shown in Figure 10.16, the focal length for rays on the periphery of the lens is smaller than the paraxial focal length. The distance PP′ is a measure of

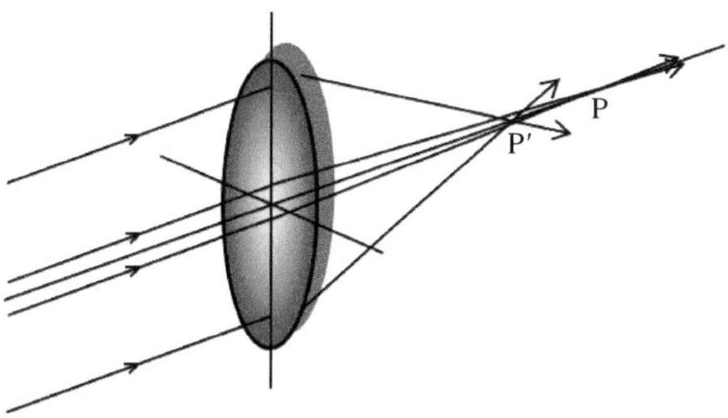

FIGURE 10.16 Spherical aberration: the focal length for rays on the periphery of the lens is smaller than the paraxial focal length.

SA, but depends on the aperture of the lens, the distance to the object and the lens shape. SA in mirrors can be easily seen in the light reflected inside a cup, as shown in Figure 10.17; the envelope of the rays crossing the optical axis forms a curve called the *caustic*. SA is eliminated completely by using parabolic mirrors, as illustrated in Figure 10.18. Similarly, *aspherical* lenses, that is, lenses whose surfaces are not spherical, can be manufactured for reducing the aberration, but this is normally a costly and not always effective alternative because SA depends on the object distance.

(a)

(b)

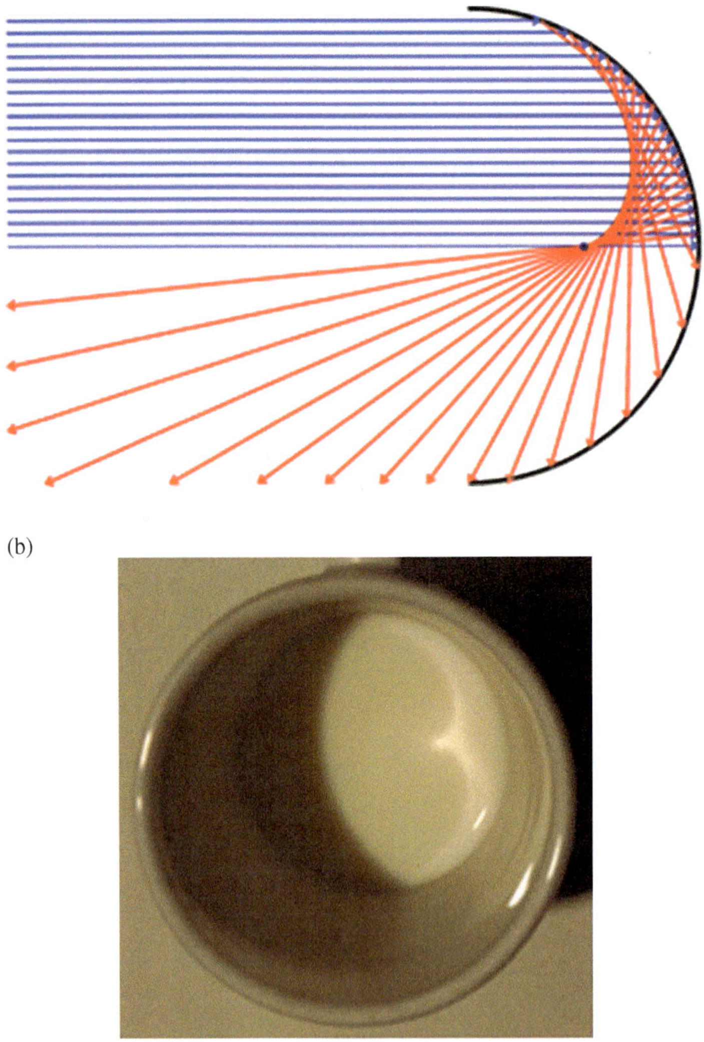

FIGURE 10.17 Caustic curve from spherical aberration: (a) spherical mirror and (b) actual realization in a cup.

Coma, illustrated in Figure 10.19, appears as a consequence of the lack of symmetry in the refraction of oblique rays, even when the rays are grouped around one crossing the optical center of the lens. The chief (center) ray determines an image point (O' in Fig. 10.19) and the rest of the rays form a displaced circle; the net result

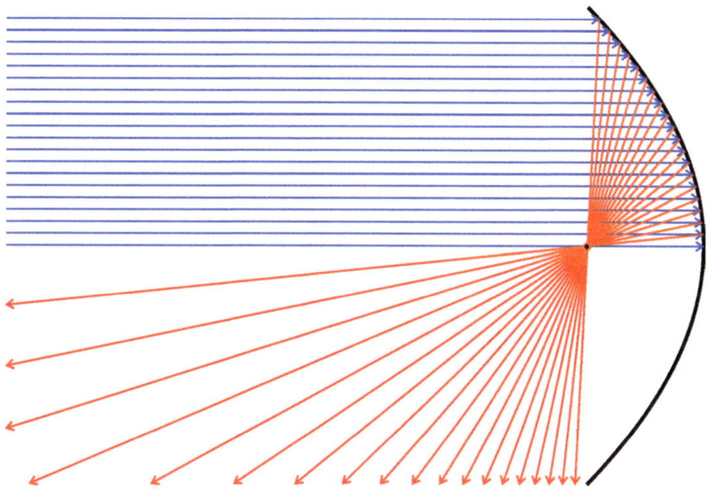

FIGURE 10.18 A parabolic mirror is free from spherical aberration.

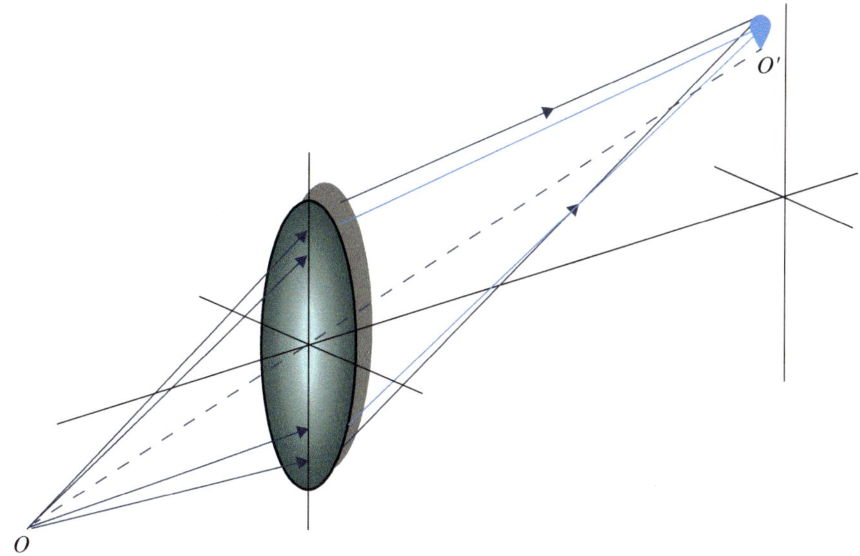

FIGURE 10.19 Aberrations, coma: in a lens with positive comma, off-axis rays form larger circular off-axis spots farther from the optical axis.

is an image that resembles a comet. Coma depends on the inclination of the incident rays and on the shape of the lens and can be corrected by the proper placement of aperture stops or by using another lens with an opposite effect. Coma, as SA, is easy to observe with a common viewing glass.

Even if a lens is corrected for both SA and coma, it will likely suffer from *astigmatism*. In Figure 10.20, rays originating from the point source *O* that traverse the lens along the axis labeled *a-b* form a vertical line image at *S* (*sagittal* focus), while those that traverse the lens along the axis *c-d* form a horizontal line image at *T* (*tangential* focus). The degree of astigmatism is measured by the separation of the *S* and *T* images, while the best image of point *O* occurs at a plane between *S* and *T* that defines the *circle of least confusion*. Astigmatism depends more on the focal length than on the lens shape; it is corrected in compound optical systems by proper separation of the elements and/or the use of aperture stops. Astigmatism is also present in mirrors, even in parabolic mirrors that are free from SA. It should also be noted that astigmatism in the eye is caused in most cases by an asymmetry of the cornea and not by asymmetrical refraction by the eye lens. When the latter occurs, it is called *lenticular* astigmatism.

The fourth monochromatic aberration, *field curvature*, consists in the formation of curved nonplanar images from relatively large planar objects perpendicular to the optical axis. The surface formed by the image points is called a *Petzval* (*P*) *surface*. Figure 10.21 shows an example for a particular location of the imaging screen or film; if the film plane is moved closer to the lens, the periphery of the image can be focused better at the expense of defocusing the central part. In the presence of astigmatism (see earlier text), two additional image surfaces are present: the *tangential*

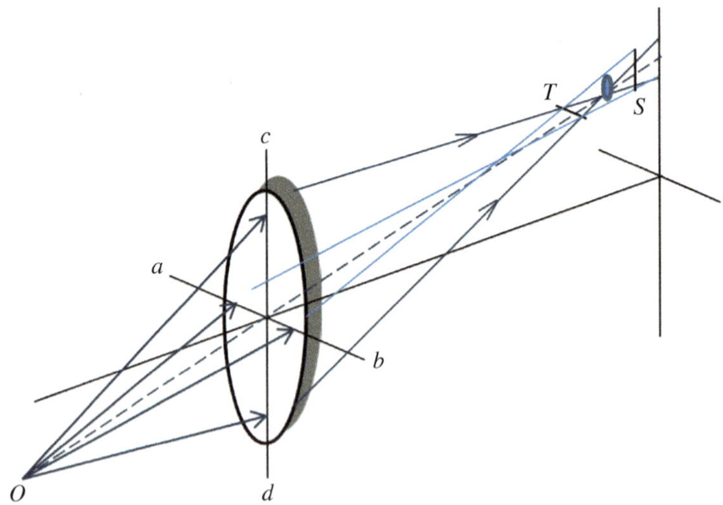

FIGURE 10.20 Aberrations, astigmatism: location of tangential (*T*) and sagittal (*S*) images of an off-axis object. The circle of least confusion falls in between.

(T) and *sagittal* (S) surfaces related to the sagittal and tangential foci shown in Figure 10.20. Astigmatism is corrected in an optical system if the T and S surfaces overlap with the P surface, but the image may still suffer from significant field curvature. However, simultaneous correction of astigmatism and field curvature is often possible in good quality photographic lens systems by proper placement of lens components and aperture stops. In contrast, field curvature is compensated in some large telescopes by curving the photographic film itself.

The last monochromatic aberration is called *distortion*. Even when the image from a lens system is planar and free from the other aberrations, a variation of the lateral magnification with the distance from the optical axis will introduce deformations near the edges. Distortion can be positive (*pincushion*) or negative (*barrel*) according to whether the magnification increases or decreases with height from the optical axis. The two types of distortion are illustrated in Figure 10.22. An aperture

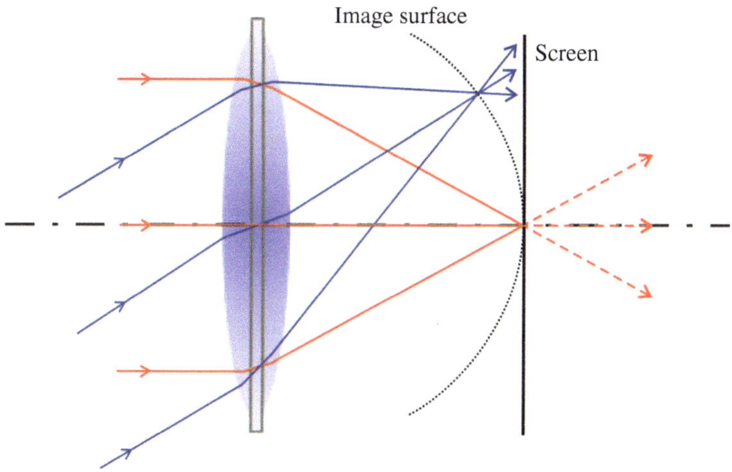

FIGURE 10.21 Aberrations, field curvature: the plane of the image is curved (Petzval surface), leading to a blurred image in the periphery for the particular location of the screen shown.

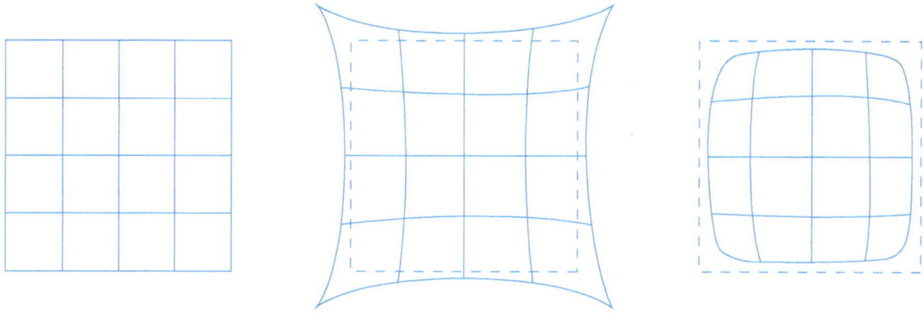

FIGURE 10.22 Aberrations, distortion: normal image (left), pincushion or positive distortion (center), and barrel or negative distortion (right).

stop placed behind a convergent lens will produce positive distortion, while placement of the stop in front of the lens will lead to negative distortion. Thus, placing stops symmetrically between the elements of doublet lens systems can eliminate distortion; the symmetry also implies that chromatic and coma aberrations are also canceled, but spherical aberration and astigmatism must be reduced by other means. This type of lens system is fairly common in photography.

Lens aberrations are not always undesirable nuisances. They can be used to introduce special effects in artistic photography and other applications. For example, a commercial lens is available that permits controlling field curvature, so peripheral focusing can be independently adjusted; in this way, it is possible to focus very closely on the center parts of an object (e.g., a person's nose), while at the same time bringing into focus the periphery of the object (e.g., the rest of the person's face). In another example, the motion of a posterior component of a special compound lens allows the introduction of diffuse halos from spherical aberration around the image; the result is to "soften" the image, but the effect works only for large apertures.

As mentioned before, *chromatic aberration* is an important consideration when designing imaging systems. The dependence of the index of refraction on wavelength implies that the focal length for colors like violet and blue is shorter than the corresponding focal length for other colors like red and orange. This is illustrated in Figure 10.23, where we have depicted a positive lens as a superposition of prisms. To correct for chromatic aberration, that is, to equalize the focal length for a pair of colors, glasses of different *dispersive powers* are employed. The *dispersive power* is proportional to the difference in indexes of refraction for blue and red wavelengths, where the exact reference wavelengths are normally taken from strong lines in the spectrum of hydrogen. The simplest example of a lens corrected for chromatic aberration is the *achromatic doublet*; it consists of a positive lens made of a low-dispersion glass

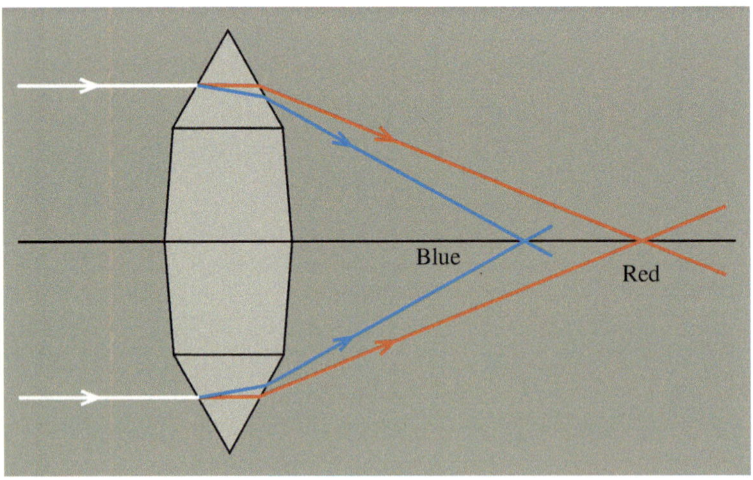

FIGURE 10.23 Chromatic aberration: a lens acts like a number of stacked prisms.

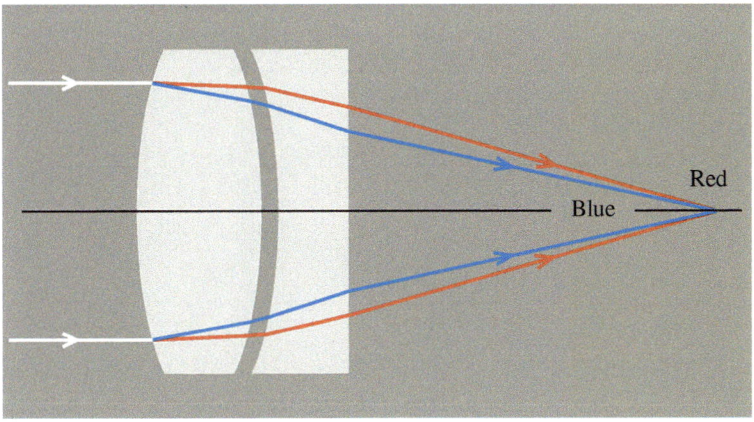

FIGURE 10.24 Correction of chromatic aberration: the achromatic doublet corrects the aberration for two colors.

like *crown* glass, and a second negative lens made of a high-dispersion material such as *flint* glass. Figure 10.24 shows the basic achromatic doublet. Correction of chromatic aberration for another pair of colors would require the addition of more lenses.

Finally, many imaging system are affected by the appearance of ghost images and glare. This is a very common phenomenon familiar to photographers and anyone that wears glasses. It can even occur inside the eye leading to problems of "double vision." The source of the ghost images is the multiple reflections that occur at the lens surfaces, both in air and inside the lenses; this is particularly noticeable with intense light sources like the Sun and at shallow angles to the optical axis. The problem can be significantly reduced, although not completely eliminated, by the use of thin *antireflection* coatings. Since the light that reaches a photographic plate can be reduced because of internal reflections in even the best corrected multiple-component lenses, a number, the *T-number*, is sometimes used (besides the *f*-number) to quantify the light losses. In single-lens reflex (SLR) cameras, however, light metering through the lens obviates the problem of knowing about light loss.

CONCEPT VERIFICATION CV10.1

Basic Lens Calculations

Problem 1: A double-convex *thin* lens has spherical faces with radii $r_1 = +10\,\text{cm}$, and $r_2 = -10\,\text{cm}$. The lens is made of glass of index of refraction $n = 1.60$, and it is surrounded by air on both sides. (Note that positive r indicates that the center of curvature is to the right, while negative r indicates that it is to the left.) Find (a) the lens focal length f, (b) the lens power P in diopters, (c) the transverse magnification M_T for an object located at $s_0 = 4f$, (d) graphical solution to part (c), and

(e) the focal length f if the lens is surrounded by oil of $n_{oil} = 1.60$ on the right side (image side) and air on the left side (object side).

Problem 2: Repeat the calculations of Problem 1 for a plano-convex lens, that is, one with $r_1 = +10$ cm, and $r_2 = \infty$.

Solution 1: (a) From Equations 10.4 and 10.5, we have $1/f = (n-1)(r_1^{-1} - r_2^{-1})$, so $f = +8.33$ cm $= 0.0833$ m; (b) $P = 1/f = +12.0$ D; (c) from Equation 10.5, $s_i = (4/3)f$, so Equation 10.6 yields $M_T = -1/3$, that is, the image is smaller than the object and inverted; (d) see Figure CV10.1; and (e) setting $s_0 = \infty$, $s_i = f$ in Equation 10.3, we obtain $f/n_2 = r/(n_2 - n_1) = 26.7$ cm, with $n_2 = 1.60$, $n_1 = 1.00$.

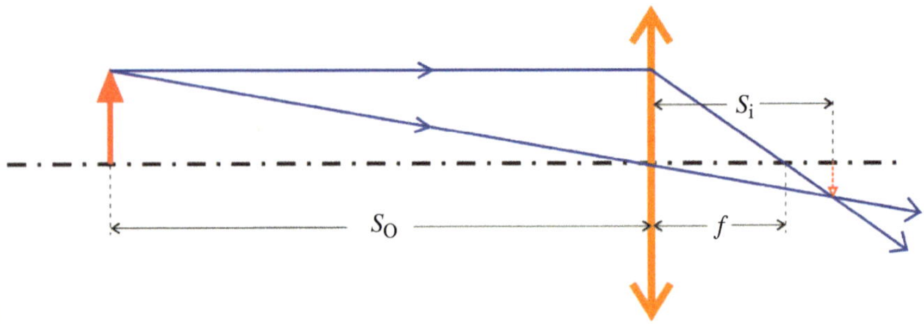

FIGURE CV10.1 Ray tracing to illustrate transverse magnification in part (d) of Problem 1 or 2.

Solution 2: (a) From Equations 10.4 and 10.5, we have $1/f = (n-1)(r_1^{-1} - r_2^{-1})$, so $f = +16.67$ cm $= 0.1667$ m; (b) $P = 1/f = +6.0$ D; (c) from Equation 10.5, $s_i = (4/3)f$, so Equation 10.6 yields $M_T = -1/3$, that is the image is smaller than the object and inverted; (d) setting $s_0 = \infty$, $s_i = f$ in Equation 10.3, we obtain $f/n_2 = r/(n_2 - n_1) = 26.7$ cm (same as for double-convex lens), with $n_2 = 1.60$, $n_1 = 1.00$.

ADVANCED CONCEPT AC10.1

Thick Lenses and Principal Planes

The actual thickness of a lens has to be taken into account for accurate ray tracing. Paraxial refraction by individual spherical surfaces can be studied separately by using Equation 10.3. With reference to Figure AC10.1, once the focal points, F_1' and F_2', of the two surfaces are determined, a ray parallel to the optical axis and starting on the object side (left of first surface) is considered. Without the second surface, this ray would cross the optical axis at the focal point F_1'. However, the second surface deflects the ray in a way that can be obtained by tracing an oblique line through its geometrical center C_2 and parallel to the

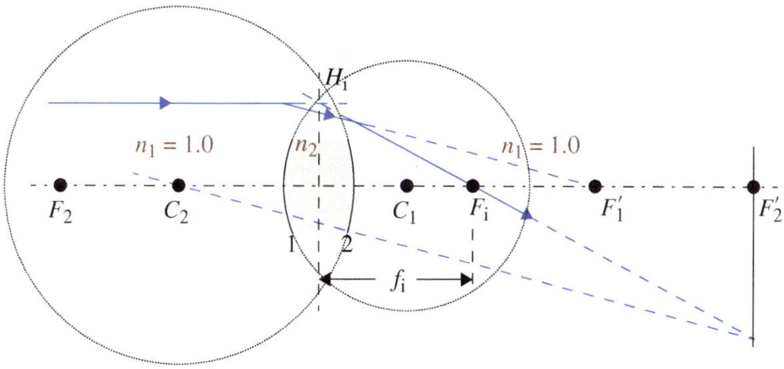

FIGURE AC10.1 Paraxial construction to determine the image principal plane $\boldsymbol{H}_{\mathrm{i}}$ of a thick lens. Equation 10.3 permits finding F_1' and F_2', the focal points of surfaces 1 and 2.

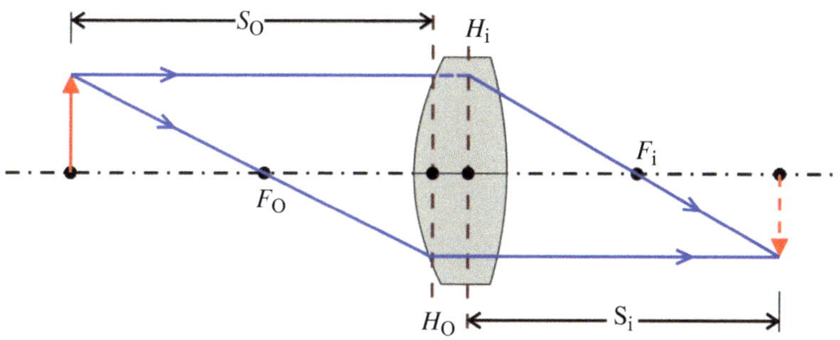

FIGURE AC10.2 Imaging with a thick lens. The principal planes are indicated by $\boldsymbol{H}_{\mathrm{i}}$ and $\boldsymbol{H}_{\mathrm{O}}$.

hypothetical undeflected ray. The construction is shown in Figure AC10.1; we determine in the way implied the location of the *image or secondary principal plane* H_i. By turning the lens around, the *primary or object principal plane* H_o is obtained.

The principal planes of a thick lens or system of lenses can be described as hypothetical planes that connect locations of unit magnification, that is, an object located at one plane is imaged at the second plane and with the same size. In Figure AC10.2, we illustrate the use of the principal planes for imaging with a thick lens. The method can be extended to systems of lenses; in effect, any number of lenses can be modeled as a single thick lens. For example, a system of two thin lenses separated by a distance d and immersed in air has a focal length f given by $1/f = (1/f_1) + (1/f_2) - (d/f_1 f_2)$, where f is measured from the object principal plane.

ADVANCED CONCEPT AC10.2

Ray Tracing and Matrix Methods

Gaussian optics lends itself to a convenient representation of rays and lenses with matrices. A ray is represented by a 2×1 matrix, that is, by a column matrix, containing the ray's transverse coordinate and its slope at a particular plane along the optical axis: $M_{\text{ray}} = \begin{bmatrix} y \\ y' \end{bmatrix}$. Thus, a ray parallel to the optical axis is designated by $M_{\text{parallel ray}} = \begin{bmatrix} y \\ 0 \end{bmatrix}$ with y=constant, the height over the optical axis. A thin lens of focal length f is represented by $M_{\text{thin lens}} = \begin{bmatrix} 1 & 0 \\ -1/f & 1 \end{bmatrix}$. Therefore, the effect of a thin lens on a parallel ray is expressed as the matrix product $M_{\text{thin lens}} M_{\text{parallel ray}}$, which yields the new ray matrix $\begin{bmatrix} y \\ -y/f \end{bmatrix}$ right at the image side of the lens. A drift space of length d is represented by $M_{\text{drift}} = \begin{bmatrix} 1 & d \\ 0 & 1 \end{bmatrix}$. As an exercise, the reader can easily show that the matrix for an interface between two media of indices of refraction n_1, and n_2 is $M_{\text{interface}} = \begin{bmatrix} 1 & 0 \\ 0 & n_1/n_2 \end{bmatrix}$. (*Hint*: use Snell's law in the paraxial approximation whereby $\sin\theta \approx \theta = y'$.) Imaging in any optical system can be studied with a matrix representation of its individual elements, which is very convenient for fast computation. Naturally, this works well in the paraxial approximation only.

INTERESTING TIDBIT TB10.1

Plastic Optics

Plastic has several advantages over glass: it is lighter, more "formable," and cheaper. But plastic has also larger dispersion (Section 10.5), absorption, stress birefringence (Section 10.2), and coefficient of thermal expansion than glass. However, plastic is perfectly acceptable for making the spherical and aspheric lenses used in many consumer applications such as cellphone cameras, scanners and light-emitting diode (LED) lighting. A typical application of aspheric plastic lenses is for correction of spherical and other optical aberrations (Section 10.5). Advances in the processing of polymers have led to the use of high-quality plastic optics in non-consumer markets such as medical equipment and sensors. Hybrid systems containing both glass and plastic optics are also more common nowadays.

ADVANCED CONCEPT AC10.3

Solid Angle and Steradians

The concept of solid angle is an extension to 3D of the standard angle of plane geometry. The solid angle is a quantitative measure of "how much of the visual field is occupied by an object?" It is denoted by Ω, or $\Delta\Omega$ and defined by

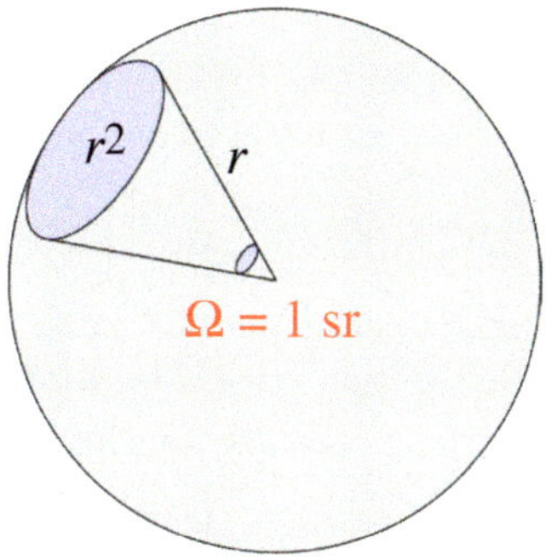

FIGURE AC10.3 Solid angle and steradian.

$$\Omega = \frac{\text{Subtended area}}{r^2}.$$

Just as 1 rad is defined in plane geometry as the angle subtended at a distance r by an arc of circumference whose length is equal to r, 1 sr is defined as the solid angle subtended at a distance r by a section of a sphere of area r^2. (Sometimes the steradian is called radian^2.) See Figure AC10.3. In spherical coordinates, the differential solid angle is written as $d\Omega = \sin\theta d\theta d\varphi$, since the area element is $dA = r^2 \sin\theta d\theta d\varphi$.

Problem 1: What is the solid angle subtended at its center by a sphere of radius R? What is the solid angle of the same sphere subtended at a distance $D > R$ from its center?

Problem 2: If the distance from the earth to the sun is 1.5×10^8 km, and the sun radius is 6.96×10^5, what is the angular diameter subtended by the sun on earth? What is the solid angle subtended by the sun on earth?

Solution 1: The total surface area of a sphere of radius R is $4\pi R^2$, so the solid angle at its center is $4\pi R^2/R^2 = 4\pi$. At a distance $D > R$, the solid angle is calculated using the differential form quoted before; the result after integration is $2\pi(1-\cos\theta)$, where $\theta = \arcsin\left(R/D\right) \approx R/D$ (the last equality valid if $D \gg R$.)

Solution 2: Sun's angular diameter $\approx 2 \times \text{Arctan}(6.96 \times 10^5/1.5 \times 10^8) = 0.53° = 9.3 \times 10^{-3}$ rad.

Sun's solid angle $\approx \pi \times (6.96 \times 10^5)^2/(1.5 \times 10^8)^2 = 6.8 \times 10^{-5}$ sr. The exact result using the equation $2\pi(1-\cos\theta)$ is essentially the same because $D \gg R$.

CONCEPT VERIFICATION CV10.2

Temporal and Spatial Coherence of Light

Problem 1: Light from a typical semiconductor (diode) laser has a peak wavelength at 655 nm and a bandwidth of 2 nm. In contrast, an incandescent source emits light in the range 400–1000 nm with an average wavelength equal to 700 nm. Calculate the coherence times and lengths for light from both sources and comment on their use for holography. *Hint*: use Equation 10.18.

Problem 2: Calculate the spatial coherence of sunlight and starlight. Also comment on the use of sunlight and starlight for interference experiments (e.g., in a two-pinhole setup). *Hint*: Before using Equation 10.19, see the previous box on "solid angle and steradians."

Solution 1: By taking derivatives on both sides of the equation $\nu = c/\lambda$, we obtain $\Delta\nu = -\frac{c}{\lambda^2}\Delta\lambda$. Then, by ignoring the sign, we can rewrite Equation 10.18 as $t_c = \frac{\lambda^2}{c\Delta\lambda}$, $l_c = \frac{\lambda^2}{\Delta\lambda}$. Therefore, the coherence time and length for the laser light are $t_c = 655^2 \times 10^{-9}/2 \times 3 \times 10^8 = 7.16 \times 10^{-13}$ seconds, and $l_c = 0.21$ mm. For the incandescent light, $t_c = 700^2 \times 10^{-9}/600 \times 3 \times 10^8 = 2.72 \times 10^{-15}$ seconds, and $l_c = 817$ nm. Since $\Delta\lambda$ is comparable to λ (average) for the incandescent light, a more careful calculation should be tried. We obtain $t_c = 2.22 \times 10^{-15}$ and $l_c = 667$ nm. In any case, the coherence length for the incandescent light is very close to the average wavelength. For the laser light, on the other hand, the coherence length is 315 times larger, but it is still too small to be used in holography. Single-mode lasers are available that produce radiation with much longer l_c, of the order of 100 m.

Solution 2: The solid angle subtended by the Sun on earth is 6.8×10^{-5} sr, so $\Delta A_c = 500^2 \times 10^{-18}\,\text{m}^2/6.8 \times 10^{-5} = 4 \times 10^{-3}\,\text{mm}^2$. The solid angle subtended by a typical star is 9.6×10^{-16} sr, so $\Delta A_c = 400^2 \times 10^{-18}\,\text{m}^2/9.6 \times 10^{-16} = 167\,\text{m}^2$. The coherence area of starlight is much larger because the star is much farther away than the Sun. The two pinholes in an interference experiment with sunlight would have to cover a very small area, of the order of ΔA_c, to yield visible fringes; starlight would work better but would require a very stable setup and a special detector due to the weak light.

11

OPTICAL COUPLERS INCLUDING OPTICAL FIBERS

There are three types of optical couplers. The first one transfers signals between electronic and photonic equipment, an important facilitating tool in the hybridization of the two types of components. These couplers enable the individual strengths of photonic and electronic technologies to be exploited. For example, powerful computers and webservers rely extensively on very-large-scale integration (VLSI) solid-state circuitry to organize and manipulate huge amounts of data. However, it is far more efficient to send these vast amounts of data over large distances via an optical fiber than via electronic signals over wires. The purpose of the first type of optical coupler is to convert back and forth between electronic and photonic signals. In the forward direction, the optical coupler consists of a constant intensity light source, followed by an electro-optical modulator. The light sources are usually either lasers or photodiodes, which have been described in Chapter 9. Currently, the optical opacity of the modulator is proportional to an input electronic voltage. In the future, engineers hope to use interference techniques to replace the electro-optical modulators, creating a pure photonic system. The forward direction ends with a light sensor, used to convert the optical signals back to electronic ones.

A second definition of an optical coupler in common usage is taken to mean the connection of two segments of an optical fiber, including a splice to separate one fiber into multiple fibers that each carries an identical signal. These couplers are used extensively in multiple applications. For example, optical couplers are necessary in many computer network systems and in fiber optical gyros (FOGs) for ground, sea, and air navigation as well as in spacecraft. The third type of optical coupler, also known as an optical isolator, introduces a small mechanical gap between two sets of electronic packages. This coupler creates electrical isolation, preventing equipment failure in one package causing a failure in the other.

Modern Devices: The Simple Physics of Sophisticated Technology, First Edition.
Charles L. Joseph and Santiago Bernal.

Companion website: www.wiley.com/go/joseph/moderndevices

11.1 OPTICAL FIBERS AND HOLLOW WAVEGUIDES

Fiber optics and hollow waveguides can be made to operate at any wavelength from $\lambda = 200$ nm to $\lambda = 10{,}000$ nm, although the telecommunications industry uses just three near-infrared (near-IR) windows extensively. These windows are centered on wavelengths: 850, 1310, and 1550 nm, and each window can transmit multiple channels (bit streams carried on a specific wavelength). The two longest-wavelength windows are particularly well suited for long-distance transmission from high-powered Nd:YAG or CO_2 lasers. (Nd:YAG stands for neodymium-doped yttrium aluminum garnet, a solid-state laser formed from a $Y_3Al_5O_{12}$ crystal doped with Nd.) Optical fibers have core diameters ranging from 8 to 400 μm, while hollow waveguides range from 100 to 400 μm, and both types can be fabricated up to a few hundred meters in length. The number of individual optical fibers contained within a cable ranges from one to a few hundred. The cross sections of a single fiber and a three fiber are depicted in Figure 11.1. Typical optical fibers are cylindrical glass or acrylic rod-in-collar designs as shown in Figure 11.2, making use of the principle of total internal reflection. A transparent and concentric rod in collar with the collar material having a slightly (~1–7%) lower index of refraction. Each optical fiber is covered with a coating or a sheath to block external light from entering the fiber from the side, potentially modifying the internal signals running down the fiber. This coating also prevents cross talk in cables having more than one optical fiber. A cable that contains a single fiber, usually has additional material to strengthen and to stiffen the cable, making it resistant to sharp bends. A fiber or hollow waveguide will loose significant amounts of signal and perhaps be damaged permanently if it is bent too sharply. The final outermost jacket of an optical cable protects the entire assembly from nicks, abrasions, strains, and stresses. This protective jacket must be suitable

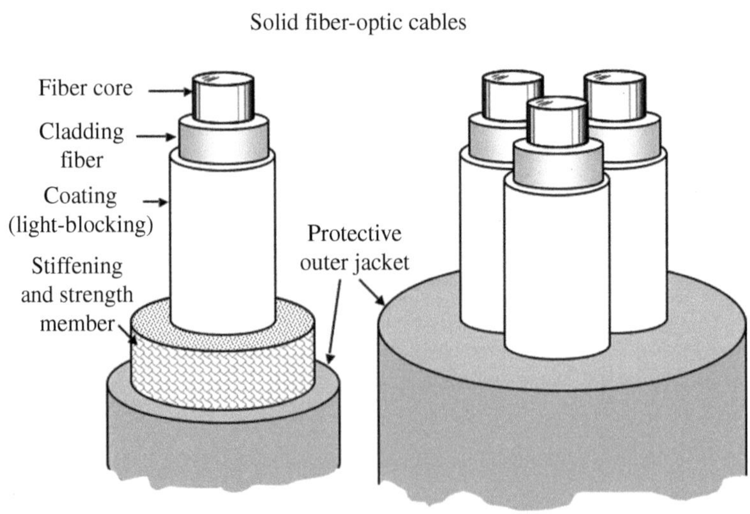

FIGURE 11.1 The anatomy of a single-fiber (left) and multi-fiber (right) cable.

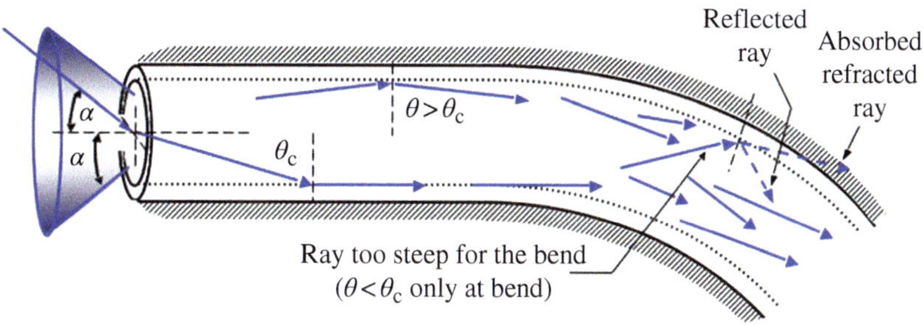

FIGURE 11.2 A cross section of an optical fiber, showing the responses to various light rays entering the fiber and traveling down the central fiber.

for its operational environment (e.g., indoor vs. outdoor, buried in a land trench or underwater, suspended on poles).

As a waveguide, optical fibers only efficiently transmit converging beams within an acceptance cone, which we define by angle, α. A light ray entering the surface at one end is refracted toward its normal as shown in Figure 11.2. This ray travels down the fiber rod, eventually encountering the boundary between the inner rod and its collar fiber. If the angle, θ, the ray makes with respect to the normal of the boundary is greater than or equal to the critical angle, θ_C, then the beam will experience total internal reflection and will continue down the rod indefinitely. This critical angle along with Snell's law at the input end of the fiber establishes the acceptance cone angle, α, where a refracted ray is equal to the critical angle for total internal reflection. Any refracted ray leading to $\theta < \theta_C$, is partially *reflected* back into the central fiber, while the other fraction of it *refracted* into the absorbing collar material. The reflected portion of this light ray experiences repeated losses down the fiber and is soon completely absorbed since the incidence angle equals the angle of reflection insures that each subsequent bounce will maintain the $\theta < \theta_C$ state. *Note*: a relatively sharp bend in the fiber cable will cause some of the rays approaching the outer fiber edge to violate the total internal reflection condition resulting in minor losses. The curvature of the fiber depicted in Figure 11.2 (right) shows one ray undergoing partial reflection and refraction into two separate, weaker rays (dashed arrows). This curvature is much sharper than that found in most practical uses.

The use of total internal reflection is highly efficient (100%), even better than the best mirrors available. A freshly coated *front-surface mirror* has the highest reflectance possible (96% for normal-incident, 98% for grazing-incident ($\theta > 85°$)) rays at visible wavelengths. This means that 2% of the light is scattered or absorbed for each bounce, resulting in a beam intensity that is only 82% of its initial intensity after just 10 reflections. While tolerable in a laboratory environment, a 2% loss per bounce is unacceptably high for long-distance telecommunications since the signal would have to be reamplified after it transverses only 1 or 200 m. For a fiber with total internal reflections, the principal signal losses are due to trace impurities in the glass or acrylic fiber materials and due to defects (e.g., microcracks), occasionally resulting in a photon being absorbed or scattered out of the fiber. Long-distance telecommunications are only possible if the fiber has sufficiently low levels of impurities and defects, which was first achieved in 1970.

Each fiber is capable of handling dozens of separate data streams simultaneously with minimal signal loss rates and no interference from adjacent fibers. Figure 11.3 shows the cross sections of a multimode, a single-mode, and a gradient optical fiber. The gradient optical fiber has an index of refraction that decreases continuously and smoothly as a function of radius. A gradient fiber produces a continuous turning of a light ray as it traverses down the fiber, causing the ray to curve back towards the central part of the fiber. The diameter of a rod-in-core architecture is of critical importance. When the diameter is approximately 10 times or less than the wavelength of light being sent down the fiber, then the acceptance angle, α, is small and it is a single-mode fiber. If the diameter is very large compared to the wavelength of light, then α is large and it is a multimode optical fiber. (Modes in this usage refer to Eigen modes, discrete resonance wavelengths from which arbitrary wave shapes can be formed from a superposition of these modes.) The speed that a light wave travels through a transparent material is slower than it is through a vacuum. Consequently, various rays of light passing through a very long fiber arrive at different times, a phenomenon known as dispersion. The paths of two separate rays are shown in the multimode fiber of Figure 11.3. Obviously, rays with significant side-to-side components (dashed-line arrows) traverse through more material to reach the end of the fiber than do rays oriented virtually down the central axis (solid-line arrows). A similar situation exists for rays with various angles of incident entering a gradient fiber. Both

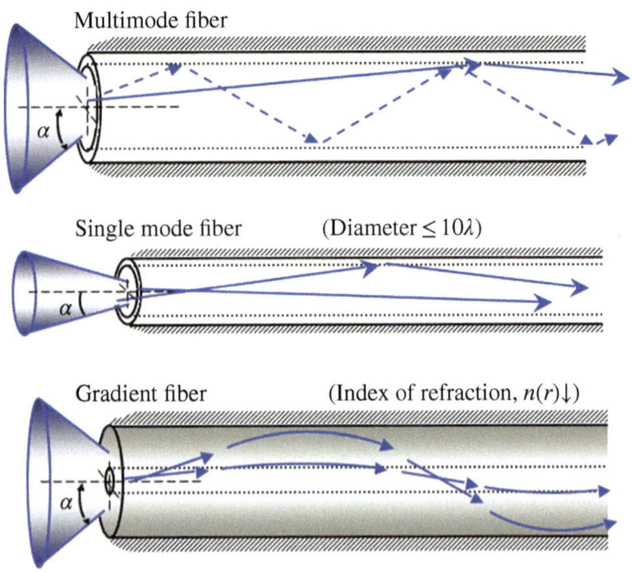

FIGURE 11.3 A few types of optical fibers. A multimode fiber (top) has a large acceptance angle, while a single-mode fiber (middle) has a small cone of acceptance. A gradient fiber (bottom) has an index of refraction that decreases continuously and smoothly as a function of fiber radius.

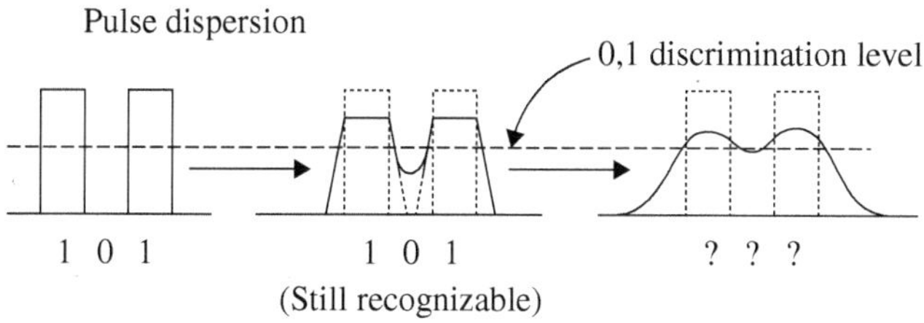

FIGURE 11.4 The dispersion of a three-bit data stream as it travels down an optical fiber.

the gradient and the multimode fibers create intermodal dispersion, which is a major disadvantage for pulse transmissions over long distances.

There are several types of dispersion within an optical fiber. A pulse of light physically spreads out and is attenuated as it propagates. Consider a sequence of three bits (1,0,1) sent down an optical fiber as shown in Figure 11.4. To distinguish between a "0" and a "1" in digital electronics, a threshold or discrimination level is used. After our sequence has traveled some distance down the fiber, the 1,0,1 is still recognizable, but has begun to blend with adjacent bits. In the presence of noise, this sequence will be recognized 99.9% of the time, depending on the signal-to-noise ratio. This is more than adequate for the transmission of pictures or voice communications, but is an unacceptable failure rate for computer programs or banking transaction data. If the sequence is allow to proceed further down the cable, the blending with adjacent bits continues and at some point the electronics cannot distinguish which bits are above threshold (i.e., a "1") from those just below (i.e., a "0"). Intermodal dispersion is the dominant form in a gradient fiber or a multimode fiber and is caused by differences in path lengths of the various light rays that have entered the end of the fiber. Intermodal dispersion is negligible for a single-mode fiber, making it and cables with this type of fiber the best choice for long-distance applications. Intramodal dispersion is the other type, consisting of chromatic dispersion and polarization-mode dispersion. High-quality lasers produce the purest single color of radiation, but the light beams still contain a small finite range of wavelengths. While photons with slightly different wavelengths will travel down a waveguide or fiber at virtually the same speed, these speeds are not identical. This process introduces chromatic dispersion. Polarization dispersion occurs due to imperfections in the cross section of a fiber. While the cross section of a fiber is always circular, it is not perfectly circular. The optical fiber will have a slightly oval shape over one length of a cable and a different slightly oval shape with a different orientation over another length. Light rays of one polarization or another will experience somewhat thicker effective fiber diameters over various portions of the cable, leading to variable path lengths traveled and the spreading out a pulse. Over very long distances, intramodal dispersion becomes the limiting process. It determines the maximum separation between repeater stations that reamplify and reshape the sequence of pulses.

Hollow-core or capillary optical fibers, working at near IR, are preferred by many telecommunication companies. Cross sections of two types of hollow-core optical fibers are shown Figure 11.5, consisting of "donut-shaped" ring of material with

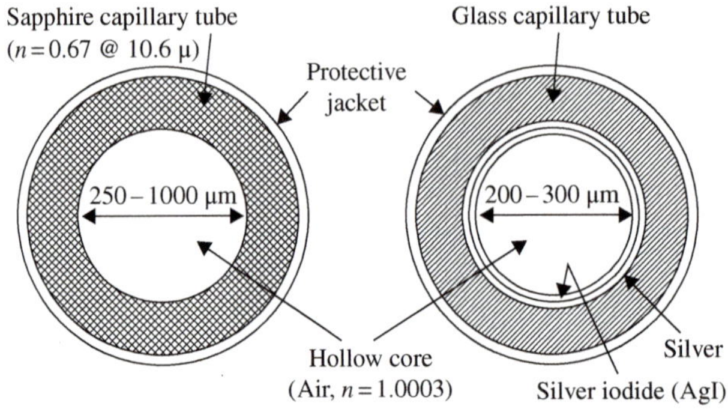

FIGURE 11.5 Cross sections of two hollow-core fiber optics (also known as a capillary tube fibers). The left one is a conventional step optical fiber using total internal reflection. The right capillary uses reflection, which is efficient at infrared (IR) wavelengths.

INTRO PHYSICS FLASHBACK FB11.1

Total Internal Reflection

In general, whenever a light beam encounters an interface between two materials with different indices of refraction, part of it is transmitted and part of it is reflected. Consider the case shown in Figure FB11.1 where the bottom portions have a higher index of refraction than the top portions. For a light beam incident from below, the portion of the beam that is refracted into the violet material follows Snell's Law: $n_1 \sin\theta_1 = n_2 \sin\theta_2$. (See Chapter 10.) The refracted angle, θ_2, is always shallower than the incident angle, θ_1, since $n_1 < n_2$. This is shown in the left two examples in Figure FB11.1. In contrast, the angle of reflection equals the angle of incidence in all reflected beams. For a sufficiently large θ_1, that of the critical angle, the refracted beam becomes horizontal ($\theta_2 = 90°$), skimming the interface. For all incident angles greater than critical, total internal reflection occurs. This is the physical principal that enables information to be transmitted through very long optical fibers.

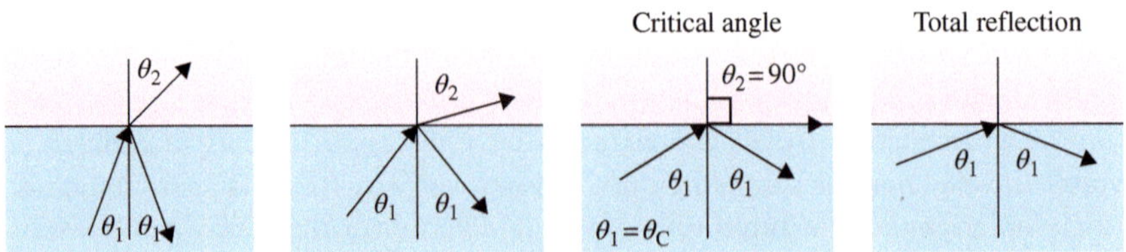

FIGURE FB11.1 For a light beam coming from a material with a higher index of refraction incident into to a material with a lower index of refraction, total internal reflection occurs when the incident angle θ_1 becomes greater than the critical angle, θ_C.

nothing in the central 200–1000 µm diameter region. The central core in this case consists of air with an index of refraction, $n = 1.0003$, virtually the same as the index for a vacuum. The sapphire capillary tube (left) is a conventional step optical fiber since the index of refraction of sapphire is less than one at infrared wavelengths ($n = 0.67$ at $\lambda = 10.6\,\mu m$), resulting in total internal reflection. A glass capillary tube uses an interior reflective surface of a dielectric over metal (Fig. 11.5, right). In our example, the inner most coating of AgI is a dielectric material used to boost the reflectivity over that of simple Ag. The AgI also serves to stabilize silver as well as be a protective overcoat. Compared to reflection off mirrored surfaces at visible wavelengths, reflective losses in the IR are minuscule, permitting the use of glass capillary tubes as efficient long-distance waveguides. From this point forward, we redefine the term "optical fibers" to include hollow-core waveguides working in the IR as well as conventional fiber optics used at visible wavelengths.

11.2 COUPLERS FOR LONG DISTANCES

Transporting data pulses over long distances presents several technological challenges. We have already discussed some of the difficulties associated with the optical fiber or hallow waveguide in the previous section including, attenuation, two types of dispersion, and the acceptance angle. In this section, we provide a systems analysis of long-distance optical telecommunications, combining the individual components discussed thus far with those necessary to complete an end-to-end long-distance system. Performance comparisons between alternative types of telecommunication networks will be discussed in Chapter 18.

Long-distance waveguides operate near-IR wavelengths and require powerful light sources at the point of origin. The intense IR beam on the focusing optical lens and on the entrance to the waveguide often produces sufficient heating to melt one or both components without some source of continuous cooling of these parts and removal of the waste heat. A hollow waveguide has an advantage over a solid-core fiber because there is no optical boundary or change in the index of refraction as the light enters the fiber. Whenever light crosses a surface boundary, two small portions of the beam intensity are lost (the reflected and absorbed fractions), and the remainder transmitted. Also, the pulse speed does not have to slow down. (The speed of light, c, down the hollow-core fiber is 3.00×10^5 km/s, whereas it is 2.04×10^5 km/s in glass. The speed in glass is the speed of light divided by the index of refraction or $v = c/n = c/1.468$). A pulse travels 300 km in 1 ms (187 miles in 1 ms) in a hollow core fiber, but requires about 1.5 ms in a solid-core fiber.

Next, the signal strength declines (attenuates) very slowly over the length of the cable. Eventually, the ends of two sections of continuous optical fiber have to be mated to each other, known as joint. A joint can be accomplished in a number of ways. In one process, the two ends are fused together by carefully raising the temperature above the melting point and then allowing the ends to solidify as a single part. It is somewhat more convenient to join the two ends using a mechanical aligning coupler, of which there a several standard types. Mechanical joints are easily

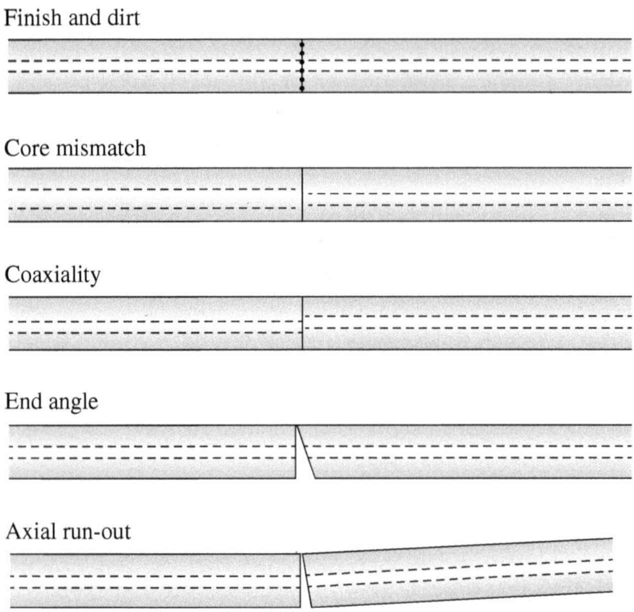

FIGURE 11.6 Some of the physical mechanisms responsible for attenuation at a joint.

disassembled, accommodating future cable expansion and rerouting needs as well as facilitating repairs. Regardless of how one fiber segment is joined to another, there are inherent physical imperfections, some of which are depicted in Figure 11.6. All of these limitations to precision in mating result in small fractional signal losses.

Branching of information flows through an optical coupler is depicted in Figure 11.7. The coupler consists of physical splices with separate fibers fused into the main trunk cable. Another possibility is a mechanical optical coupler where individual fibers with termination sockets plug into a common receptacle. Fiber-optic system diagrams are increasingly moving toward the use of standardized symbols analogous to those found in electronics circuit diagrams, although optical fiber symbols are not as universal as those found in electronics. A common symbol for an optical coupler with two branches is shown in Figure 11.7 at the bottom left. One of these symbols shows a coupler in which one of the fibers has a termination that allows it to plug into a receptacle to mate it to another device such as a detector.

Cumulative attenuations, dispersions, and other changes to profiles of the pulses require active signal restoration, including reamplification, pulse shaping, and pulse timing. Figure 11.8 shows the multiple-step process to restore the bit stream at a repeater station. A segment of the original bit-stream signal arrives at a remote distant location (second plot from the top) with pulse profiles that have been attenuated, spread (dispersed), and the individual arrival times slightly shifted relatively compared to the original bits (dotted line overlays). The degraded, raw signal is first amplified as depicted in the third plot from the top. *Note*: the peak amplitudes are

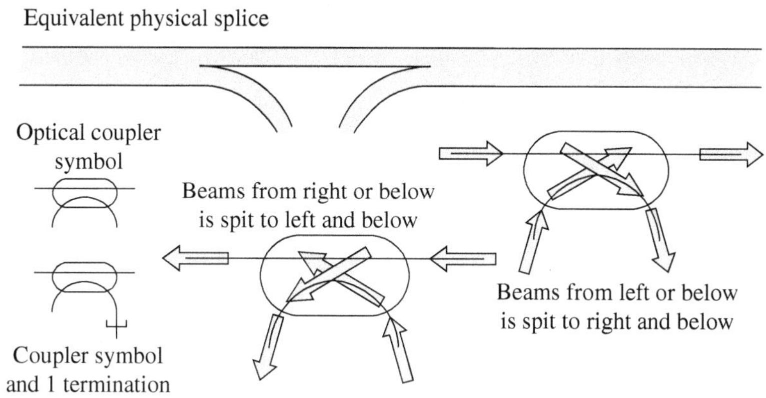

FIGURE 11.7 An example optical coupler for separating and combining signals with a fiber-optic cable. Bottom left: "circuit" symbols of fiber optics. Top: physical splice corresponding to optical coupler. Bottom center and bottom right: informational flow paths through a coupler.

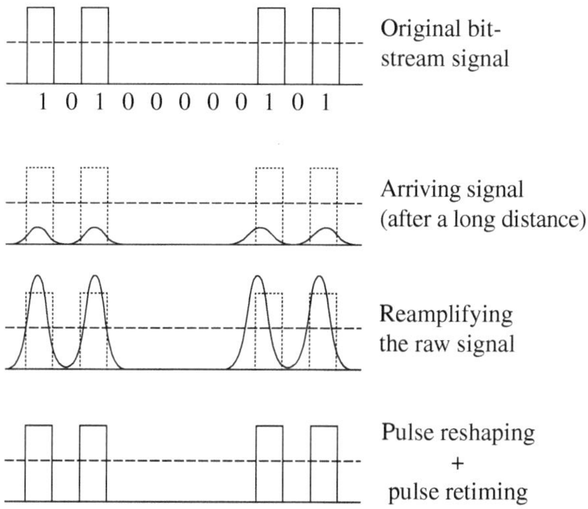

FIGURE 11.8 The multi-step process that occurs in the repeater system.

often larger than those of the original bit stream so that the areas under curves are more or less equal to the area under the original flat-topped pulses. Pulse reshaping occurs next followed by pulse retiming.

The repeater station could first convert the optical signal into an electronic signal, amplify and process the bit stream, and then convert it back to an optical signal to be sent down the next segment of optical fiber. However, this approach is very expensive and introduces significant transmission delays. It is cost effective,

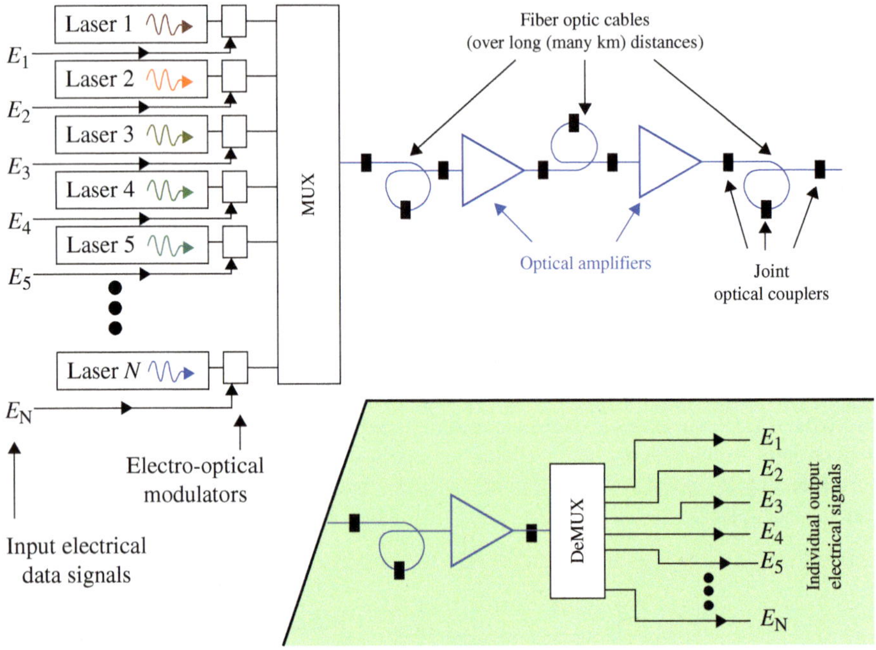

FIGURE 11.9 A basic long-distance photonic communication system using multiple channels.

faster, and more reliable to use a repeater constructed completely of photonic components. For example, optical amplification is commonly accomplished using either an *erbium doped fiber amplifier* or a *Raman optical amplifier*. Other components necessary to restore the original bit-stream signal can be accomplished photonically as well.

A complete optical-based telecommunication system is given schematically in Figure 11.9 as a block diagram. The system is depicted with multiple wavelength channels that are sent into a MUX, shorthand for multiplexer. The MUX combines all of the bit streams from the individual channels and sends the composite signal into a fiber-optic cable. Individual cable segments, spanning up to several kilometers, are mated to other cable segments using optical couplers, known a joints. The signal quality degrades slowly and continuously throughout each fiber segment as well as experiences small discrete losses at each joint and each connector, until a repeater is required to reamplified and condition the pulse shapes. Repeaters are often depicted a simple optical amplifiers in a block diagram even though it is understood that pulse conditioning must also take place. More cable segments, couplers, and repeaters are necessary over a very long distance. At the final destination, the signal again undergoes reamplification and conditioning before being "deMUXed" (undoing the

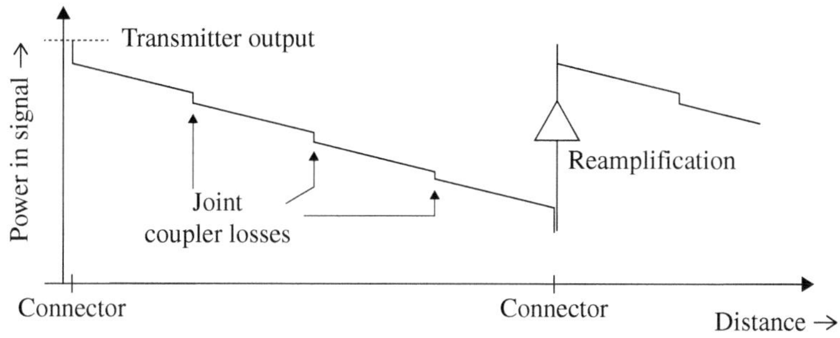

FIGURE 11.10 Plot of signal losses through a long optical fiber, including reamplification.

TABLE 11.1 Attenuation Values from the Cisco Webpages

Wavelength (nm)	Attenuation Per Fiber Length	Attenuation Per Connector	Attenuation Per Joint
1310	−0.38 dB/km (×0.957)	−0.6 dB (×0.933)	−0.1 dB (×0.989)
1550	−0.22 dB/km (×0.975)	−0.35 dB (×0.961)	−0.05 dB (×0.994)

multiplexing into individual channels). In most telecommunication systems, there are hubs, nodes, and main trunk lines. This architecture requires individual fibers to split off, sending identical signal streams to multiple locations. None of these nodes or signal splitting is depicted in Figure 11.9.

Figure 11.10 plots the signal loss as a function of distance for the first few segments of the telecom system presented in Figure 11.9. The power of the signal decreases steadily as function of cable length with small discrete jumps at the joints and connectors. Signals and signal losses are measured in decibels, a logarithmic scale defined by Equation 11.1 since signal losses are multiplicative. In other words, the final signal strength from four losses equals loss 1 times loss 2 times loss 3 times loss 4. For example, if the signal strength is half (0.5) as strong after traversing a specific length of fiber, the strength will be one fourth ($0.5 \times 0.5 = 0.25$) of the original after traversing twice that length. The decibel scale enables a numerical conversion from a multiplicative calculation to an additive calculation just as $10^3 \times 10^1 \times 10^5 = 10^{(3+1+5)} = 10^9$. Table 11.1 has the attenuations per length of fiber along with the discrete losses per joint and per connector for the devices used by Cisco Corporation. Thus, one can simply count all of the connectors, joints, and lengths of fiber, to calculate a total attenuation by adding all of the individual losses.

$$\left.\frac{S_{\text{out}}}{S_{\text{input}}}\right|_{\text{in dB}} = 20\log\left(\frac{S_{\text{out}}}{S_{\text{input}}}\right) \tag{11.1}$$

11.3 OPTICAL COUPLERS AS A MEANS OF ELECTRONIC ISOLATION

An optical coupler is used in electronics to prevent a voltage spike or a short to ground in one device from disrupting the operations of a more central electronics package. These optical couplers are also known as opto-isolators, optocoupler, or optical isolator. Optical isolators, for example, are used in NASA space missions such as the *Hubble Space Telescope* (*HST*) to protect the central spacecraft electronics from the electronic packages of the four science instruments. Operational instructions are sent to and data are transferred from each science instrument via the *HST* spacecraft. These instruments also draw their power from solar panels through the *HST* central power system, but each power supply has a fuse. All of *HST's* science instruments as well as the safety of the spacecraft itself would be vulnerable to a problem in just one instrument without optical isolators and fuses.

A schematic of an optical isolator is presented in Figure 11.11. The actual isolator is contained in the dashed-line box, consisting of a photoemitting diode and photodiode detector. However, most commercially available optical couplers come with supporting electronics, which we have represented by a simple MOS amplifier. Actual commercial products have more complicated amplifier circuits than we have depicted, including pulse-shaping electronics. (See Chapter 6 for the physics of amplifiers, their designs, and supporting electronic components.) The input and the output are electronic signals, but the two are isolated from each other with information being transferred across the gap photonically.

Originally, optical isolators used photoresistors as the light sensor, being introduced in the 1960s. These have the slowest data transfer speeds, but are the most linear, still retaining a niche market in audio applications. By the late 1970s opto-isolators had a photoemitting diode and a light sensor that was a transistor. This type of coupler, which is the most commonly used today, obtains medium rates of data transfer. The fastest optical isolators use a pair of p-type/insulating/n-type

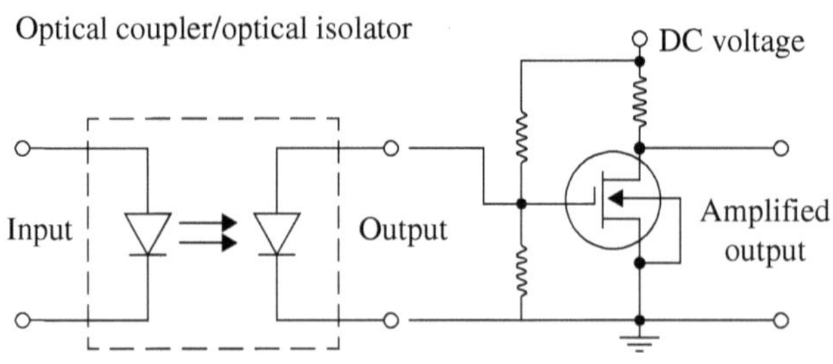

FIGURE 11.11 A simplified circuitry of an optical isolator, an optical coupler used to separate physically two electronic packages carrying data signals.

(PIN) diodes, which can be designed to work in both directions. Data rates of 50 MBd have been realized. (One Bd equals one eight-bit word per second so 50 MBd = 400,000,000 bits/s.)

COMPREHENSION VERIFICATION CV11.1

Question: Using Table 11.1, what is the signal attenuation of a 10 km long fiber optic working at a wavelength of 1330 nm? What color or bandpass is this wavelength?

Answers: 1330 nm is infrared light.

From Table 11.1, the cable attenuation is 0.38 dB/km for 1330 nm light. This implies an attenuation of −3.8 dB over 10 km or a signal strength that is 0.646 as strong. Formally,

$$\left.\frac{S_{\text{out}}}{S_{\text{input}}}\right|_{\text{in dB}} = 20 \cdot \log\left(\frac{S_{\text{out}}}{S_{\text{input}}}\right)$$

$$\Rightarrow \frac{S_{\text{out}}}{S_{\text{input}}} = \left[10^{10\left(S_{\text{out}}/S_{\text{input}}\big|_{\text{in dB}}\right)/20}\right]$$

$$= 10^{10(-0.38/20)} = 10^{-0.19} = 0.646$$

12

SPECTROGRAPHS: READING THE "BAR CODE" OF NATURE

Spectroscopy is an important tool to understand atomic properties of a substance. Spectroscopy is uniquely well suited to determine the physical properties of a gas, a liquid, or a surface of a solid, establishing the chemical makeup, temperature, pressure, and motions of the atoms. In short, a spectrograph is analogous to a checkout scanner of a retail store, reading the universal product code (UPC). Each atomic species absorbs and emits at specific wavelengths (colors) called spectral lines as described in Chapter 9. Similar to the UPC bar codes, some of the spectral lines are intrinsically strong while others are weak (UPC: thick vs. thin bars), and the separation between individual lines enables the spectroscopist to identify all of the various atoms and molecules interacting with the light as well as the physical conditions. When a spectrum is observed by eye, the instrument is a spectroscope. In contrast, a spectrograph records the spectrum using an electronic image sensor, photographic materials, or other means to make a permanent record.

High-quality spectrographs all have the following essential qualities: (i) a slit or small aperture to limit the angular bundle of light rays entering the spectrograph, (ii) a collimator optic to transform all of these incident rays into parallel ones, (iii) a dispersing element (prism or grating), and (iv) a camera optic to focus each specific wavelength of light to separate locations. The detector (image sensor) is normally considered an attachment to the spectrograph. In turn, the spectrograph is an attachment to the main instrument. Figure 12.1 schematically shows the optical layout for a pure transmissive, prism spectrograph. *Note*: to mitigate chromatic aberration, both the collimator and camera lenses are actually compound lenses, each with two or more types of glass. Any and all of the optics in the spectrograph can be replaced with reflective ones, avoiding the need for chromatic correction. Also, note that blue

Modern Devices: The Simple Physics of Sophisticated Technology, First Edition.
Charles L. Joseph and Santiago Bernal.

Companion website: www.wiley.com/go/joseph/moderndevices

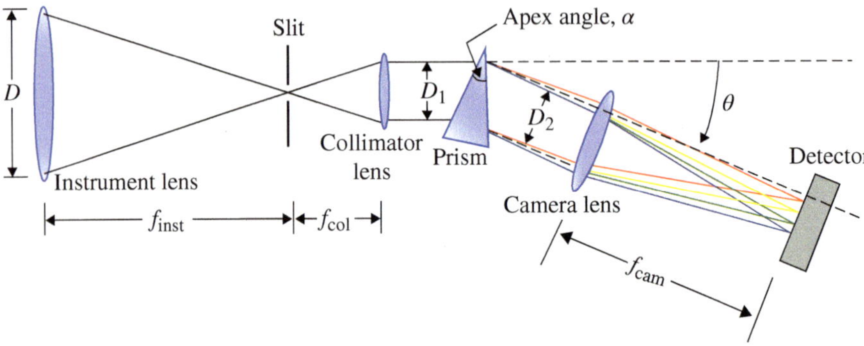

FIGURE 12.1 The basic optical components of a high-quality spectrograph: an aperture or slit, a collimator, a dispersive element (grating or prism), and a camera optic. The instrument lens and the image sensor are normally considered to be separate instrumentation attached to the spectrograph.

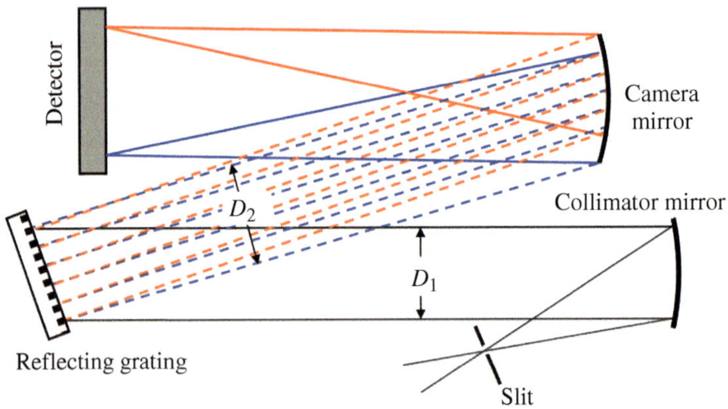

FIGURE 12.2 The basic optical components of a Wadsworth-mount spectrograph using all reflection optics: a slit/aperture, a collimator, a dispersive element (grating), and a camera mirror.

and red light rays strike slightly different portions of the camera lens, but the bundle of rays for each color almost fills it.

For a prism spectrograph, the total deflection angle, θ, depends on the apex angle, α, of the prism and the tip of the prism with respect to the collimated beam. Theta is measured for a central wavelength and various tip angles are used to sample different portions of the spectral range. The required diameter, D_1, of the collimator optic as well as the height of the dispersing optic are determined by the focal length of collimator, f_{col}, the focal ratio ($f/\#=f_{inst}/D$) and the diameter, D, of the instrument lens. Specifically, $D_1 = D\cdot(f_{col}/f_{inst})$.

Most modern spectrographs use a diffraction grating rather than a prism, making compact designs easier to implement. A Wadsworth-mount spectrograph, shown in Figure 12.2, is a common research spectrograph. The entrance slit or aperture is the first element of a spectrograph, and it is located at the focal plane of the telescope, microscope, or input optical system. The slit must be at the focal point for both the input optic and the collimating optic. This requires the focal ratio (*f*/#) of the input optic to be equal to—or greater than—f_{col} to avoid vignetting, the loss of intensity at the edges, of the collimator. *Note*: the bundle of blue-end rays falls on a slightly different portion of the camera optic than does the red-end bundle, just as these do in prism spectrograph. In a Wadsworth mount, both the camera lens and collimator are off-axis optics. That is, these optics are equivalent to a small, off-center segment that has been cut out of a much larger parabolic optic.

A common spectrograph, found in laboratories, is shown schematically in Figure 12.3. Light is focused onto the entrance slit at the bottom left. It travels up to the collimator mirror and is reflected onto a grating. Often there are three gratings residing on a rotational turret. Each grating has a different groove spacing (also known as the pitch), enabling the operator to select one of three ranges of dispersion by rotating the turret to fixed angles. Different portions of the bandpass (wavelength coverage), arriving at the detector, can be selected by fine rotational adjustments. This type of spectrograph usually has two additional ports, which can be used to pull a vacuum or for a nitrogen purge. The spectrograph can be configured by adding an internal, flat folding mirror, so that the entrance slit can be mounted at the side of the spectrograph. Similarly, an optional fold mirror enables the detector to be attached at other side port.

Another grating spectrograph is the Rowland circle type, a simple, compact design that places the slit, grating, and detector, all on a common cylindrical surface.

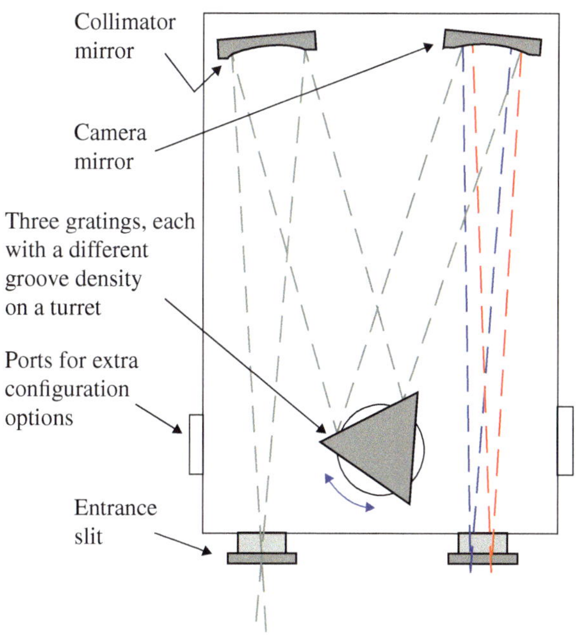

FIGURE 12.3 A common laboratory spectrograph with selectable gratings.

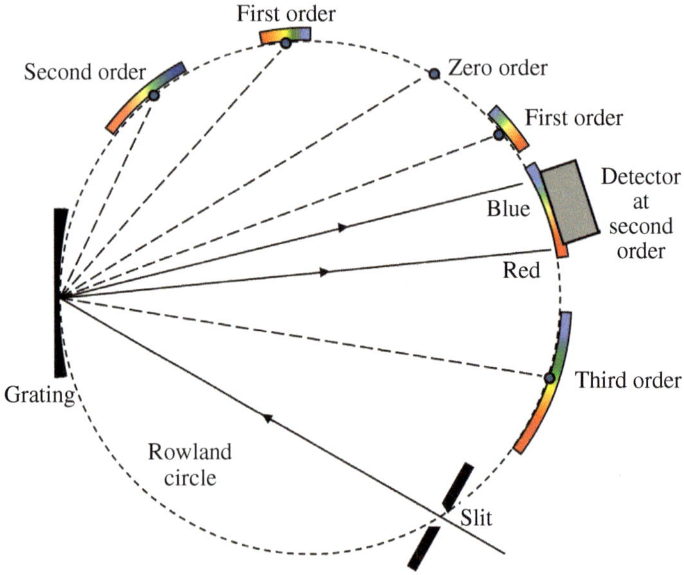

FIGURE 12.4 The basic optical components of an efficient Rowland circle spectrograph where all components are arranged on a single cylindrical surface. A single reflective surface serves as both the camera optic and the grating.

See Figure 12.4. The grating surface is convex such that it focuses a particular spectral order on one portion of the Rowland circle, combining the dispersion element and the camera mirror into one component. *Note*: the radius of curvature of the grating surface *is not* the radius of the Roland circle, and it can only be fabricated to focus one of the individual spectra shown in Figure 12.4. Ideally, the focal plane detector should have a cylindrical curved surface, but most image sensors are planar, causing the spectrum to be somewhat out of focus at the edges of its free spectral range. In addition, spatial information about how the spectra differs (in and out of the page) is limited compared to a long-slit spectrograph to be discussed later. However, a Rowland circle is a particularly efficient design for photon-starved applications, especially for ultraviolet (UV) applications where there are substantial reflection losses at each surface. Mirrors typically have reflective efficiencies of 80 and 60% at near-UV and vacuum-UV wavelengths, respectively. A near-UV spectrograph that includes a separate camera mirror and a collimator mirror is only 64% as efficient as a Rowland circle design, while in the vacuum UV it is only 36%. (The transmission efficiencies for the UV are generally much worse.) The UV efficiencies are 0.1% per optical component for glass, although UV efficiencies of approximately 80% can be obtained for various highly polished, high-quality crystal components such as sapphire or magnesium fluoride (Mg_2F), which become opaque for wavelengths shorter than 160 and 120 nm, respectively. These crystal optics are generally more expensive to fabricate and more difficult to polish.

Several general properties of all diffraction-grating spectrographs can be seen in the Roland circle. Gratings produce multiple copies of the spectra simultaneously since these diffraction devices produce alternating bands of constructive and destructive interference as a function of angle. The individual spectra are called orders, and these radiate outwardly in pairs from zero order: the point where the incident light simply reflects as if it encountered multiple flat mirrors. The blue end of each individual spectrum is closest to zero order, while the red end is furthest. Both the dispersion and the size of each order are proportional to its order number, m, with second and third orders having 2× and 3× of that of first order, respectively. See Figure 12.4.

If the spectrograph uses a collimator and camera optic to display the spectra on a flat-surface detector, the individual spectra appear along a single strip as shown at the top of Figure 12.5. The size and dispersion of the spectral component are proportional to the order number with the left group of spectra being the mirror image of the right, including blue ends toward $m=0$. In the example depicted in Figure 12.5, order 1 has no overlap with any other order, while order 2 is nearly unblended. For order numbers $m=3$ and above, there is serious blending from adjacent orders in this particular example. These overlapping spectra would produce impure colors such as purple (top) where the red portion of $m=3$ mixes with the blue of $m=4$. *Note*: the lowest orders to overlap differ from one spectrograph to another and depend on various combinations of slit widths and dispersions. In the bottom three-fourths of Figure 12.5, the spectra from the individual orders have been displaced vertically to show the contributions to the spectrum at the top. To acquire high-dispersion, one uses spectra from sufficiently high orders where the portion of spectra that can be captured at one time by the detector is often less than the free spectral range. To obtain spectra at the red or blue ends of a spectral order, a wavelength-blocking filter

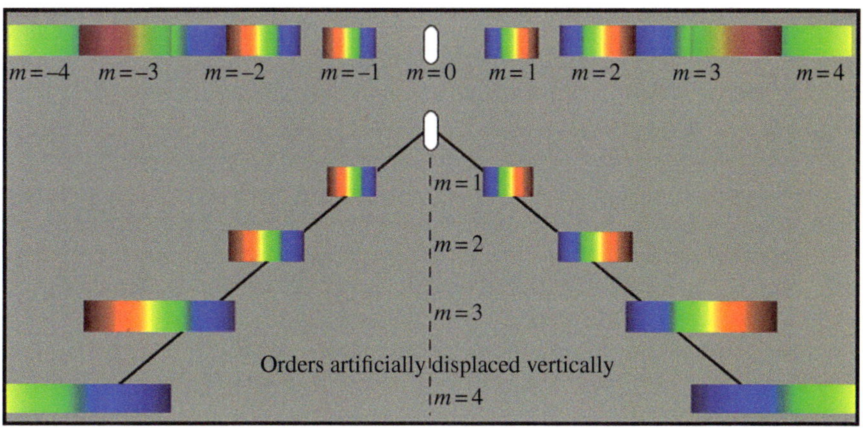

FIGURE 12.5 Multiple spectral orders ($m=0$ to $m=4$) for one spectrograph as these appear (top) on a flat screen. The dispersion of each order is proportional to m and the orders overlap for sufficiently large values of m. In this example, the first overlap occurs between $m=2$ and $m=3$.

is employed to eliminate the unwanted portion from the adjacent order. Most blocking filters transmit 80–90% of the desired spectra while reducing the contaminating spectra by a factor of approximately 1000.

The overall intensity within each spectral order diminishes as m increases. (This property is not displayed in the figures.) Moreover, a large portion of the light is not dispersed at all (found in the zero order) and most of the incident light that is dispersed appears in the lower orders where the dispersion is low. (This same phenomenon can be seen in double rainbows where the second arc is less intense than the first.) Most high dispersion ($R \sim 10^5$) spectra require the use of large spectral orders ($50 < m < 300$), which are inherently very inefficient for flat planar-ruled gratings. To circumvent this shortcoming, a blaze can be introduced to concentrate most of the light into a few large spectral orders. Blazed gratings, which will be described later, have efficiencies approximately 90% in a few high spectral orders.

All spectrograph designs require significant tradeoffs between various parameters (e.g., throughput vs. resolution, spectral range vs. dispersion, and optical efficiency vs. image quality), and it takes knowledgeable individuals with good computer tools to obtain optimal design and performance. Fortunately, many spectrographic design trades can be incorporated into more than one of the optical components within the spectrograph (e.g., into the collimator and camera optics instead of the dispersing optic), enabling the use of a few standard grating or prism configurations for many applications. Most prisms used in spectrographs, for example, either have side profile of an equilateral triangle profile ($\alpha = 60°$) or a right triangle with equal sides (45° apex angle). Also, these prisms are made of either crown or flint glass, although any clear material can be used at visible wavelengths. The two choices of apex angle have the added manufacturing advantage of minimizing waste material during fabrication. Similarly, gratings usually have groove frequencies (measured in line pairs/mm) of 300, 600, or 1200 LP/mm and are rarely manufactured with alternative values. Holographic gratings, however, do have more freedom, since these are manufactured via photolithographic processes. Some vendors offer holographic gratings with line pair densities that are more than three times the maximum of ruled gratings.

As discussed in Chapter 10, the diffraction of light is the process by which a spectrum is formed via a dispersing optic such as a prism or grating. Three important parameters of a spectrograph are (i) the free spectral range (ii) the dispersion, and (iii) the spectral resolution. Free spectral range is the total amount of spectra that can be obtained (i.e., the difference in wavelengths) without contamination from adjacent orders. Dispersion is the change in wavelength sampling per detector pixel ($\Delta\lambda$/pixel) or per spatial distance ($\Delta\lambda$ / mm), while spectral resolution ($\lambda_2 = \lambda_1 + \Delta\lambda$) is a measure of how close two spectral features can be to each other and still be identified as separate components. The spectral resolution of a spectrograph is usually the convolution of four factors, the resolving power of the grating, the projection of the slit width at the image sensor, the aberrations from the spectrograph's optics, and the detector pixel size. Specifically, the spectral resolution is $R_{\text{spect}} = \lambda / \Delta\lambda$, where the $\Delta\lambda$ term is the addition in quadrature of the various elements:

$$\Delta\lambda = \left[(\Delta\lambda)^2_{\text{Slit}} + (\Delta\lambda)^2_{\text{Disp.}} + (\Delta\lambda)^2_{\text{Optics}} + (\Delta\lambda_{2\text{p}})^2 \right]^{1/2} \tag{12.1}$$

Here, $\Delta\lambda_{2\text{p}}$ is the small wavelength range captured by two pixels of the detector—the Nyquist sampling limit. All other values are root-mean-squared (RMS) values, although the full-width-at-half-maximum (FWHM) values can be used for spectrographs producing Gaussian-shaped spectral features. The contribution from aberrations ($\Delta\lambda_{\text{Optics}}$) to the spectral resolution normally is only significant in very fast optical systems, especially very small, compact spectrographs. For most high-quality spectrographs, the third and fourth terms in Equation 12.1 can be ignored and spectral resolution depends only on the resolving power of the grating ($\Delta\lambda_{\text{Disp.}}$) and the slit width as projected at the focal plane ($\Delta\lambda_{\text{Slit}}$). Formulas for these two components are provided in Section 12.1. In most cases, the two easiest ways to increase spectral resolution are to reduce the slit width or to switch to a grating with a higher groove frequency. In either case, the efficiency of the spectrograph is reduced. In the former case, the amount of light passing through the slit is reduced in proportional to the decrease in slit width. In the latter case, light is effectively spread out (dispersed) to a greater degree with the detector capturing a smaller portion of the free spectral range.

It is worth noting that if the inherent spectral resolution is insufficient, two spectral features with slightly different wavelengths will remain blended regardless of the choices of groove frequencies and slit widths. A common, but arbitrary criterion is the Rayleigh resolution ($\lambda_2 - \lambda_1 \geq$RMS widths of the individual lines), used to separate two Gaussian spectral features. *Note*: the line spread function (LSF) rather than the point spread function (PSF) is the relevant parameter since the optical response of a grating is one-dimensional (1D), whereas it is 2D for imaging. We will return to the grating parameters quantitatively in the following section.

Finally, spectrograph designers have several other important concerns. One critical design consideration for the development of any new spectrograph are the locations and sizes of optical baffles used to minimize unwanted light reflecting back into the optical train from the several polished surfaces, creating spurious features. The optical surfaces inside spectrographs require polishing and for transmissive elements, antireflection coatings to maintain transmission efficiencies. UV and X-ray reflecting surfaces also require dielectric coatings to maintain efficiencies. Residual micro-roughness on any surface not only increases the amount of scattered light, these also reduce resolution and the efficiency of the spectrograph. Changes in ambient temperature and relative humidity also influence a spectrograph's performance. Different operating temperatures can influence the relative positions of the internal optics due to the various coefficients of expansion if a variety of materials that are used to hold the optics. Potential expansions and contractions can influence the relative angles the light rays transit through the spectrograph. Changes in the relative humidity introduce minor changes to the index of refraction of the air, which become particularly important at near-UV and shorter wavelengths.

INTRO PHYSICS FLASHBACK FB12.1

Review of Spectral Dispersion

A common childhood spectral experience is observing the rainbow from sunlight passing through a prism or, for example a dangling crystal on a chandelier. It is necessary that the light be concentrated from one direction to create a rainbow. When the Sun is the light source, its extent is only about 0.5° in diameter and it cast shadows or nearly bright rays that are essentially parallel. For prism spectrographs, the light from an extended source has to be restricted by a slit as depicted in Figure FB12.1. The slit or mask prevents numerous interfering (contaminating) spectra originating from various angles of incidence and various offset positions.

An incandescent light bulb produces a thermal, blackbody spectrum with an intensity that is smooth and continuous as a function of wavelength. If a gas, contained in a clear box, is placed in the beam, it will produce a set of emission or absorption lines, depending on the relative locations of the gas and light source. Figure FB12.2 shows two possible positions in a spectrograph. In the example pictured in the top figure, the molecules in the box remove two colors, a red and a yellow-green, from the continuous spectra and the missing light escapes in all directions. If the same box of gas is repositioned as in the bottom figure, the red plus yellow-green light (combined orangish color) refracts in the prism into the two emission lines. The corresponding spectra are shown in Figure FB12.3 along with a graphical representation of the intensity. Depicted at the top-left are two dark lines at wavelengths of approximately 665 nm (red) and 555 nm (yellow-green). *Note*: in this example, neither absorption feature is sufficiently strong to produce zero intensity (top-right plot), corresponding to a black color (top-left). Emission lines from this same gas are shown at the bottom.

Light can also be dispersed into a spectrum using multiple slits known as a grating, which can be operated in transmission as depicted in Figure FB12.4 or in reflection where a planar mirror with ruled grooves. *Note*: the density of slits or grooves needs to be high (hundreds to thousands per millimeter), far more than can be depicted schematically.

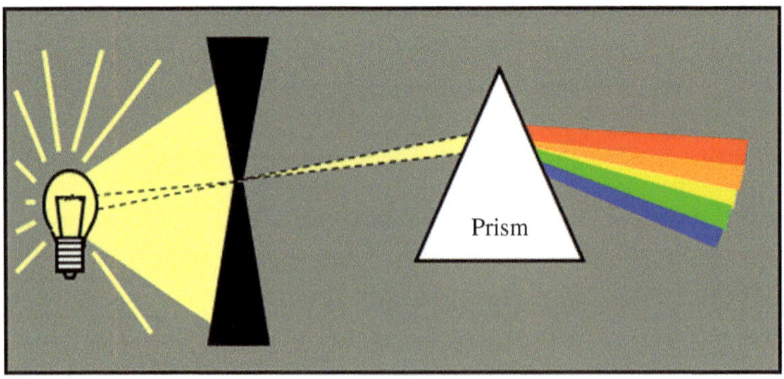

FIGURE FB12.1 The creation of spectrum with a prism.

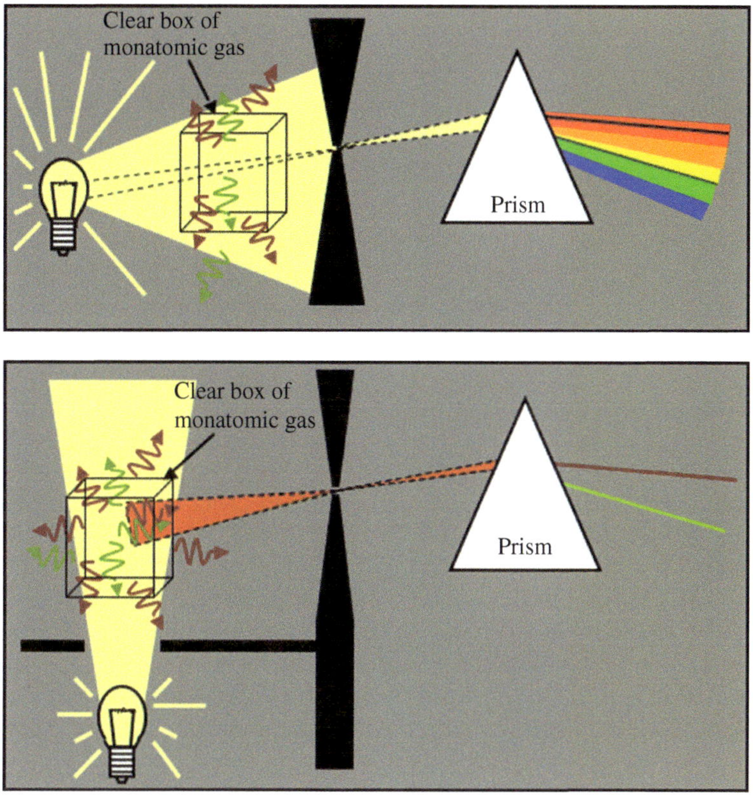

FIGURE FB12.2 Spectra from a fictitious monatomic gas.

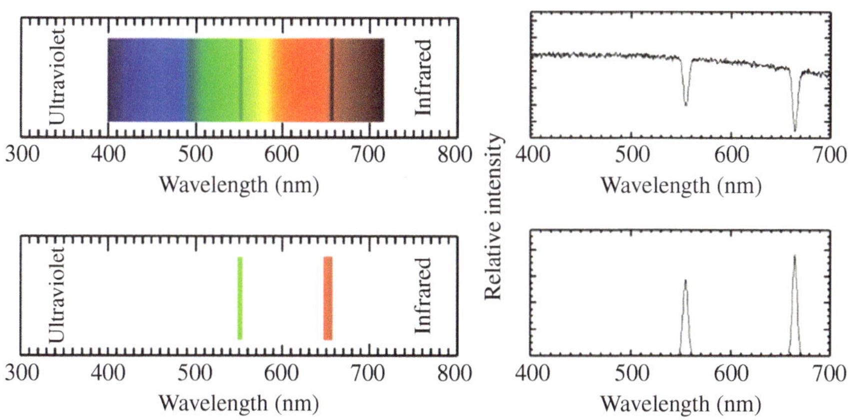

FIGURE FB12.3 Graphical representations of spectral features when a continuum of light interacts with a fictitious monatomic gas. Depicted on the left are the color intensities as a function of wavelength (absorption on top and emission on the bottom). Corresponding plots of these relative intensities are presented on the right.

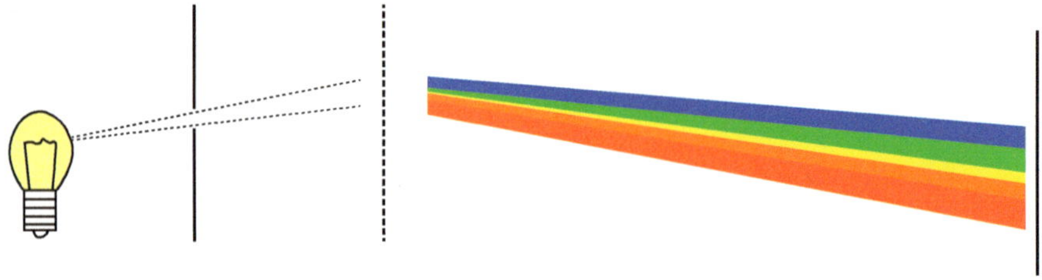

FIGURE FB12.4 Light can be dispersed into a spectrum using multiple slits known as a grating, which can be operated in transmission (as depicted) or reflective (e.g., a planar mirror with ruled grooves).

COMPREHENSION VERIFICATION CV12.1

Assume that you are to acquire a third-order spectrum from the example spectrograph that creates the spectra shown in Figure 12.5. What range of wavelengths would you need block to obtain a clean spectrum over $450\,\text{nm} < \lambda < 550\,\text{nm}$? Explain.

Answer: None. Red contamination from $m=2$ impacts only the very blue $\lambda < 420\,\text{nm}$ in $m=3$. Blue contamination from $m=4$ only impacts wavelengths longer than yellow-green. The colors as a function of wavelength can be obtained from Figure 9.4, indicating yellow-green occurs around $\lambda = 575\,\text{nm}$. To obtain clean, $m=3$ spectra at the wavelength extremes, blocking filters are required. For $\lambda < 420\,\text{nm}$, a short-pass filter with cutoff of $\lambda_C = 575\,\text{nm}$ will eliminate light from $m=2$, while a long-pass filter with $\lambda_C = 550\,\text{nm}$ will prevent contamination from $m=4$. In the later case, one can only obtain a spectrum over $550\,\text{nm} < \lambda < 700\,\text{nm}$.

INTERESTING TIDBIT TB12.1

In the days of Galileo (1564–1642) to Newton (1642–1727), astronomical objects such as stars appeared through the telescope as tiny almost point-like prisms due to the chromatic aberrations from simple glass optics. Newton invented the first all-reflecting telescope with an eyepiece coming out the side of the telescope near entrance aperture. The Newtonian telescope eliminated the chromatic aberrations and is a common telescope configuration used by amateur astronomers today.

12.1 PRISMS, RULED GRATINGS, AND HOLOGRAPHIC GRATINGS

We now examine the dispersing element in a spectrograph in greater detail. Figure 12.6 schematically depicts the light rays passing through a prism. Although the light rays in a prism are wavelength dependent, a simplifying convention is to define the

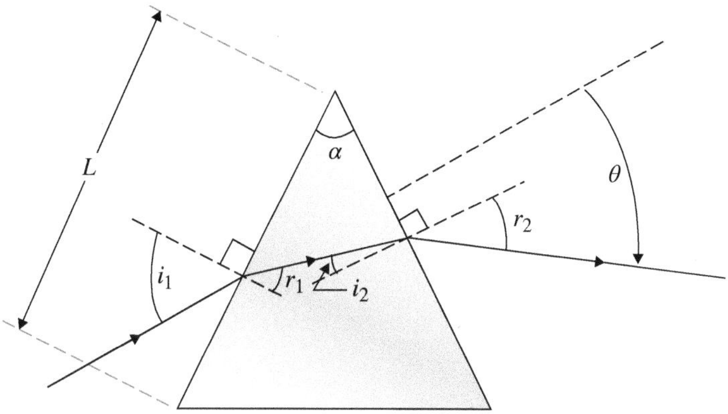

FIGURE 12.6 The light rays passing through a prism obey Snell's law with a total deflection angle, θ.

angles for the central operating wavelength and ignore any wavelength dependency. Using this approximation, it is easy to see that ray paths are simply obeying Snell's Law at the two surfaces (see Eq. 10.1).

Critical to the performance of a prism is the apex angle, α, which as noted is typically 60°. As a result, the spectral resolution can be approximated by

$$R_{\text{Disp.}} = \frac{\lambda}{\Delta\lambda} \sim 10^5 L, \tag{12.2}$$

where L is the length of a side of the equilateral triangle of the prism. Taking the material into account, $R \sim 6 \times 10^4 L$ for crown glass and $R \sim 1.5 \times 10^5 L$ for flint glass. While the actual values of R require the optical constants of the material and the detailed formula has a wavelength dependency, Equation 12.2 is a good benchmark number to remember for prisms with a 60° apex angle. In other words, very few of these prism spectrographs will have $R_{\text{Disp.}} > 3 \times 10^5 L$ or $R_{\text{Disp.}} < 3 \times 10^4 L$, and one should be suspicious of claims outside of this range for a standard-shaped glass prism. One can increase the spectral resolution easily by simply choosing a larger prism, similar to the way large camera optics produce higher spatial resolution compared with smaller optics. This statement assumes the rest of the spectrograph is scaled up so that the entire length of the prism is used to disperse the light.

A grating is simply a multi-aperture diffraction mask where the number of identical apertures is large (typically 500–50,000), creating spectra as a function of angle. See Section 10.3 for a discussion on the principles of diffraction for a small (1–10) number of apertures. A grating can be configured for use as a transmission or reflection device. We will only discuss reflection-mode gratings here since these are far more common. It can be blazed as depicted in Figure 12.7 or not. A blazed grating means its cross-sectional profile is a saw tooth. The blaze is impressed into an epoxy resin, which then has a thin-film reflecting overcoat made of aluminum, gold, or

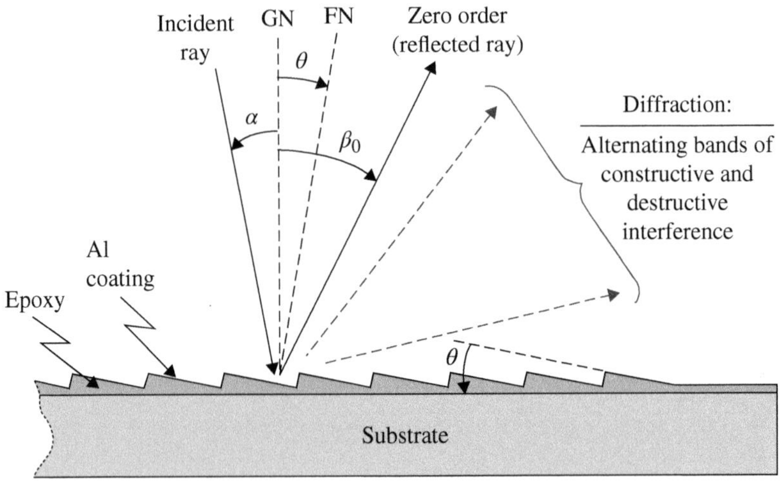

FIGURE 12.7 The angles of incidence and reflection for a grating with a blaze angle, θ. GN is the normal to the grating surface and FN is the facet normal of the reflecting surfaces. The angles α and β are defined with respect to GN, which are also the angles used by optical designers.

platinum. The blaze angle, θ, is shown in Figure 12.7 changes the path distances of individual rays and alters the angles for constructive and destructive interference of the diffracted light.

For gratings, it is all about the angles of the incident and diffracted rays. There are several diffracted angles for which the light of a particular wavelength, λ_k, experiences constructive interference. (See Section 10.3 on multi-slit diffraction.) Moreover, light at wavelengths from other spectral order numbers appear at the exact same angle as depicted in Figure 12.8. For instance, the angle of constructive interference for $\lambda = 600$ nm (order $m = 1$) is equivalent to that for $\lambda = 300$ nm ($m = 2$) and to $\lambda = 200$ nm ($m = 3$). The ratios of repeated wavelengths is equal to the ratio of order numbers. Mathematically, $\lambda_k / \lambda_n = m_n / m_k$, where n and k are integer numbers. As noted, the blazed grating enables very-high-dispersion spectrographs by concentrating most (90%) the amount of the power into a few, large, spectral order numbers ($m > 50$).

It is useful to conceptualize these various angles from a single ray as depicted in Figure 12.7 or 12.8. However, the incident flux arrives as a circular spot of collimated rays (all parallel to each other) with spatial extent, D_1, and leaves the grating with another collimated spot of diameter, D_2. The collective impact on the incident and diffracted wavefronts is simply the changes experienced over a single blaze cycle, multiplicatively repeated over the total number of blazed grooves. Figure 12.9 schematically shows a portion of these entrance/exit wavefronts over d, the spatial extent of one complete line pair of grooves. The right most incident ray must travel an extra distance of ($d \sin \alpha$) before reflecting, while the left most ray must travel an extra distance of ($d \sin \beta_m$) after reflecting, each compared to the other ray. This geometry

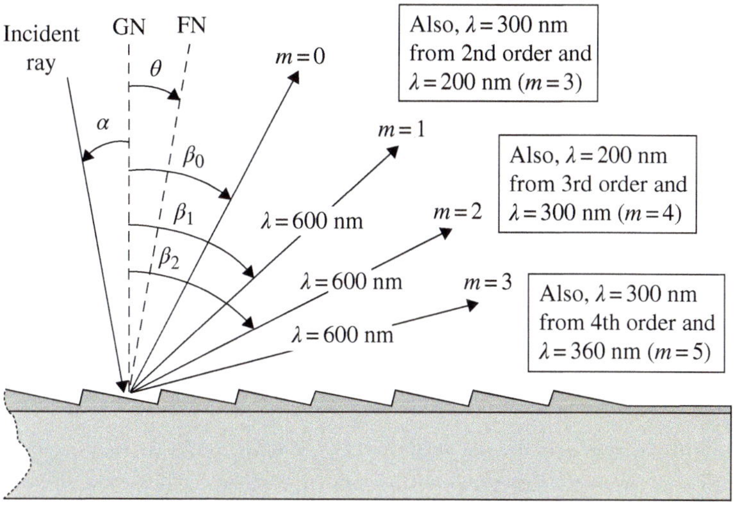

FIGURE 12.8 Constructive interference occurs for wavelength ratios of λ_k/λ_n that have a ratio of order numbers m_n/m_k, where k and n are integer numbers.

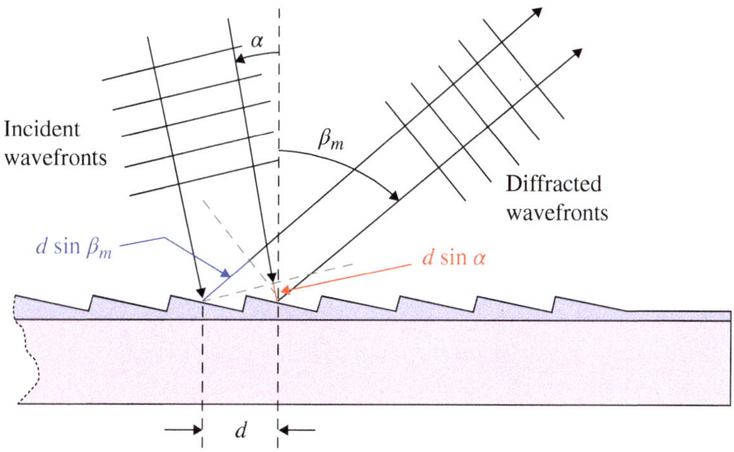

FIGURE 12.9 The projected portion of the wavefronts bounded by two rays separated by the pitch or single groove spacing, d. It shows the difference in path lengths between the two bounding rays.

leads to the *grating equation*, which very much resembles the equation found for double slit or multi-slit diffraction.

$$m\lambda = d\left(\sin\alpha + \sin\beta_m\right). \tag{12.3}$$

It is important to note that m/d is not a ratio that can be chosen independently of other parameters. A more useful form of the grating equation is obtained by substituting $G = 1/d$, the groove frequency or groove density (LP/mm).

$$Gm\lambda = \sin\alpha + \sin\beta_m. \tag{12.4}$$

The ratio of cos α/cos β_m occurs in many formulae used to calculate various properties and is known as the anamorphic factor. The ratio of diameters of entrance-to-exit collimated beams (D_1/D_2), for example, is simply cos α/cos β_m, while the anamorphic magnification resulting from the beam reflecting off the diffraction grating is the reciprocal, cos β_m/cos α, which is often less than 1 (a demagnification). In fact, the slit width projected at the detector is simply its physical width multiplied by the anamorphic magnification. Specifically, the term to be entered in the resolution calculation (Eq. 12.2) is as follows:

$$R_{\text{slit}} = \frac{\lambda}{\Delta\lambda_{\text{slit}}} = W_{\text{slit}}\left(\frac{\cos\beta_m}{\cos\alpha}\right) \tag{12.5}$$

The angular dispersion from a grating is $D_{\text{ang}} = Gm/\cos\beta_m$, indicating the angular dispersion increases as the groove frequency increases. The linear dispersion is simply the product of D_{ang} and the effective camera focal length, f_{cam}. The most useful and most often used parameter is the reciprocal linear dispersion, also known as the plate factor, P. This parameter is the change in wavelength (in nm) per pixel along the spectrum, which is the measured spectrum as sampled. The plate factor "dispersion" is

$$P = 10^{-6}\, p\cos\beta_m / G\, m\, f_{\text{cam}}\,(\text{nm/pixel}), \tag{12.6}$$

where the camera focal length and groove frequency are measured in mm and LP/mm, respectively, and p is the pixel size in mm. *Note*: most pixel sizes are expressed in micrometers (i.e., microns, μ). A detector with 15 μ pixels would have $p = 0.015$ mm in Equation 12.6. (If photographic film is used the characteristic size of the photosensitive grains is 3 μ—*c/f* electronic detectors with 5–20 μ pixels.)

As already indicated, the spectral resolution of a grating spectrograph is a measure of its ability to separate two features with a small difference in wavelength (λ_1 and $\lambda_1 + \Delta\lambda$). It depends not only on the spectral resolving power of the grating, $R_{\text{Disp.}}$ but also on the positions and performance of the other optical elements and the detector. The spectral resolution calculation for the most important parameters was given in Equation 12.1. The resolving power of a planar diffraction grating (blaze angle = 0) given in elementary textbooks is simply $R = mN$, where N is the total number of grooves and m is the spectral order number. The actual value for the second term in Equation 12.1 is given by

$$R_{\text{Disp.}} = \frac{\lambda}{\Delta\lambda} = \frac{Nd\left(\sin\alpha + \sin\beta_m\right)}{\lambda}, \tag{12.7}$$

where α, β_m are the angles shown in Figure 12.2. The product $N\,d$ is simply the length of the ruled portion of the grating, analogous to the length, L, of a prism in its resolving power calculation. To maintain optimal performance of the spectrograph, the size of the ruled portion of the grating remains constant when switching to a grating with a different groove density. Consequently, the two easiest means to increase spectral resolution are to switch gratings to one with a higher groove density or to decrease the slit width at the entrance. Both changes result in a reduction of light flux at the detector. For the former, the light is spread out, resulting in the detector capturing a small range of wavelengths. For the latter, a narrow slit passes fewer photons than a wide one does.

The fabrication of a ruled grating is a slow painstaking process, requiring a very-high-precision ruling engine that has been vibration isolated and in a tightly controlled environment. A diamond, which has its crystal axis oriented for optimal cutting behavior, is used to shape the exact profiles. A grating with a ruled area of 100×100 mm and a groove density of 1000 LP/mm requires the diamond to cut 10 km (more than 6 miles) of glass and takes about 6 weeks to rule. (Gratings used at large astronomical telescopes take a good fraction of a year.) During the ruling process, the ambient room temperature must be held constant to within 0.01° or to 0.005°C for small and large ruling engines, respectively. Moreover, large ruling engines must compensate for changes in pressure of only 2.5 mm Hg or the groove spacing will be uneven. While ruling engines are vibration isolated, small earthquakes can still render a grating useless.

Fortunately, once a master grating has been successfully ruled, it can be replicated with relative ease by pressing its groove profiles into an epoxy resin. The replication process can be repeated 5–10 times, creating 3–8 submasters. (Vendor's process yields are rarely 100%.) These submasters in turn can be used to fabricate 10–60 sub-submasters from which the commercial gratings are produced. At each fabrication stage, careful handling, inspection, and evaluation must be made with flawed gratings being discarded. Gratings, similar to many other high-quality mirrors, are front-surface reflecting optics that can be damaged significantly and irreparably merely by touching the optical surface.

Up to this point, we have only discussed ruled gratings, both blazed and planar ones. Holographic gratings have become a significant fraction of the market in recent years due to an extensive expansion in the applications for Raman spectrographs. These gratings represent a completely different approach. Holographic gratings are produced by the deposition of a photoresistant material onto a blank substrate and exposing the desired holographic pattern onto it. An ion etch (ion bombardment) is then employed to remove material, resulting in a sinusoidal profile. Figure 12.10 shows the resultant profiles for the three types of gratings discussed in this chapter, each having the same groove density, $G = 1/d$. There are differences in optical performance between ruled and holographic gratings, including the energy distribution (grating efficiency as a function of wavelength) and scattered light properties. Holographic grating masters can be fabricated more precisely than ruled gratings can. Moreover, holographic gratings can be easily placed on virtually any flat or curved (canonical) surface, enabling the dispersing element to be combined with another optical element more readily than ruled ones. Ruled gratings produce ghost features, fixed artifacts in the data caused by the ruling engine's inability to cut grooves that are precisely evenly spaced. Holographic gratings do not produce ghosts, but scatter significantly more light than do ruled gratings, making the suppression of stray light difficult.

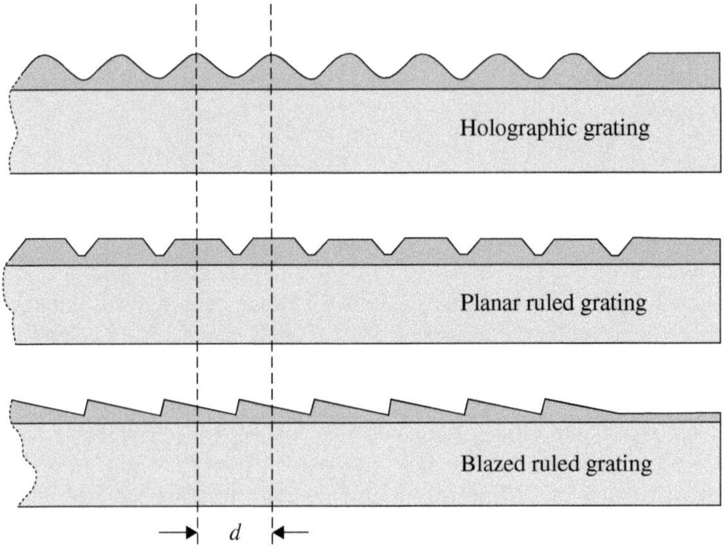

FIGURE 12.10 Various cross-sectional profiles of various gratings.

INTERESTING TIDBIT TB12.2

The first commercially available precision-ruled grating was developed by David Richardson and Robert Wiley of Bausch and Lomb between 1947 and 1950. The effort was encouraged by Professor George R. Harrison (MIT). They used a rebuilt ruling engine from the University of Chicago laboratory of Prof. Albert A. Michelson, the first American to win the Nobel Prize for Physics. Michelson was a pioneer in the measurement of the speed of light and created several novel optical systems that bear his name. The famous null result from the Michelson–Morley Experiment discredited the concept of an a either as the medium to transport light waves and was a key result supporting Albert Einstein's theory of relativity.

INTERESTING TIDBIT TB12.3

The famous German optician Joseph von Fraunhofer (1787–1826) discovered the prominent dark absorption lines in the Sun, which became known as Fraunhofer lines. He was well aware that the distribution of power among various diffraction orders in a grating depended on the shape of the grooves. It was not until 1910 that RW Wood was able to control the groove shape to a degree. Wood's gratings were observed to light up or "blaze" when viewed at the correct angle. Hence, the term "blazed gratings" with blaze angles came into being.

COMPREHENSION VERIFICATION CV12.2

Plot qualitatively the resolution of a spectrograph as a function of slit width. Put two curves on the plot: one for a grating with a groove density of 1200 LP/mm and one for 600 LP/mm. Note the minimum slit width where no additional resolution can be obtained by further reduction in the slit width.

Answer: The best way to start such a plot is to recognize the ratio of groove densities is 2, indicating all vertical values for the 1200 LP/mm will be twice that of the 600 LP/mm.

Consequently, the slope of the 1200 will also be twice that of the 600 LP/mm. The slope of the resolution versus slit width is always less than or equal to zero. We arrive at the first part of our qualitative graph, pictured in Figure CV12.1.

Note: we have left room at both ends of these plots to account for asymptotic behavior. We know from Equation 12.1 that contributions to the overall $\Delta\lambda$ term add in quadrature, which tends to be strongly dominated by one or two terms. Thus, resolution increases for narrower and narrower slit widths until the grating

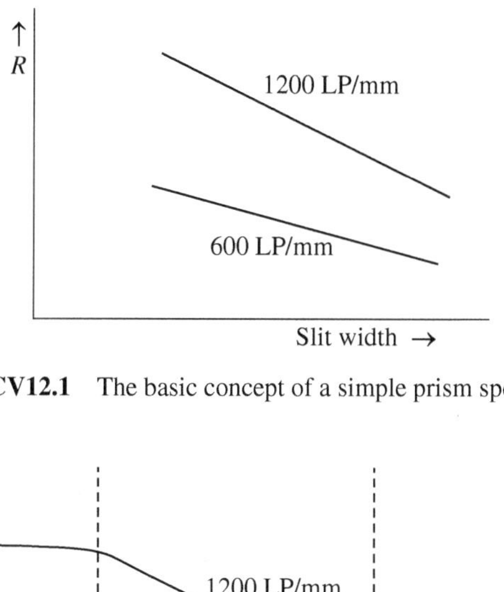

FIGURE CV12.1 The basic concept of a simple prism spectrometer.

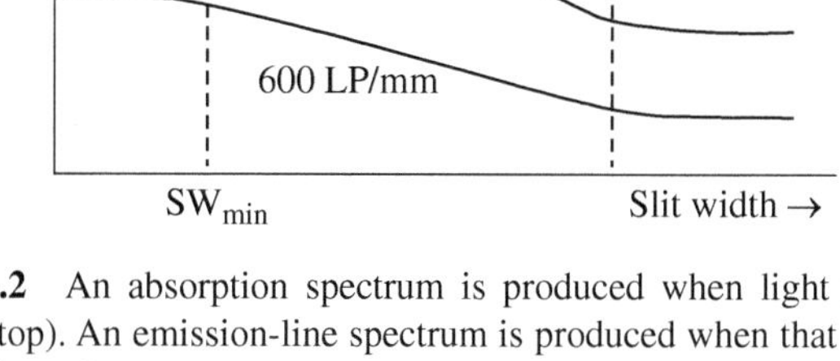

FIGURE CV12.2 An absorption spectrum is produced when light passes through a monatomic gas (top). An emission-line spectrum is produced when that same gas scatters light towards a dispersion element such as a prism (bottom).

and optical aberrations begin to dominate. At that point, no further increase in spectral resolution is possible, and the curve rolls over to a constant value. At the other extreme, the slit width has become so large that it no longer constrains the angles of incident light and again the resolution is arrives at a very low plateau. We complete the graph as pictured in Figure CV12.2.

12.2 LONG-SLIT SPECTROGRAPHS

Long-slit spectrographs are used to provide significant spectral information for a large number of spatial samples in 1D. Long-slit spectrographs are most commonly applied in astronomical and solar observatories to study extended objects such a galaxies and the solar atmosphere. Other long-slit applications might include, for example, the interactions of gas and soot rising from industrial smoke stacks and mixing with the atmosphere. Figure 12.11 combines a *Hubble Space Telescope* image and a long-slit spectrum of the two shells of gas surrounding the star, eta Carina. The dark line running through the image depicts the projection of the long slit on the gas clouds. The resulting spectrum along the slit shows differences in the iron, argon, and nickel concentrations along the length of the slit. The spectra also reveal differences in the line-of-sight velocities of these elements as a function of distance

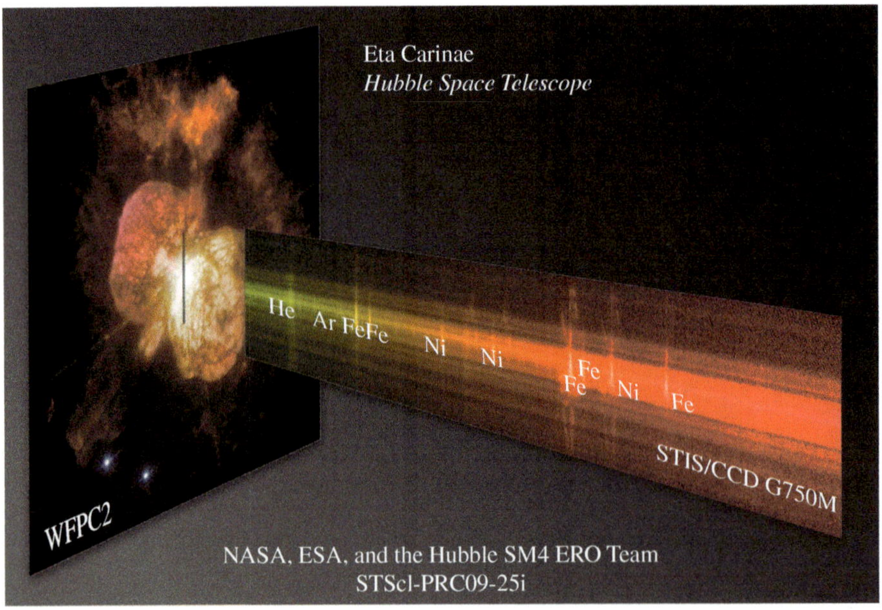

FIGURE 12.11 A composite picture, showing the image and spectra of the gas surrounding the star, eta Carina. Source: The image is courteous of NASA and the instrument team for the Space Telescope Imaging Spectrograph, an Early Release Observation.

from the center. The size in terms of total pixels (image samples) of the detector is a primary limitation of this type of spectrograph.

12.3 INTEGRAL FIELD UNIT AND FABRY–PÉROT

A Fabry–Pérot (FP) and an integral field unit (IFU) provide highly complementary 2D spectral information. FPs obtain a small amount of spectrum (typically a single spectral feature) sampled over a very large field of view (FOV). In contrast, IFU spectrographs obtain moderate amounts of spectral information, but over a very limited FOV. An IFU segments an image and separately sends each individual part though a different portion of the same spectrograph. The easiest IFU method to conceptualize is a bundle of fiber optics as depicted in Figure 12.12. A 2D image is focused onto one end of these fibers, which in this example are tightly packed. The other end of the fiber optic bundle is then arranged in a vertical column, orthogonal to the dispersion direction, at the input to a spectrograph. *Note*: fiber-optic bundle designs often leave small gaps between each output fiber to enable better separation during data processing. Normally, a dense pack spectrograph would be designed to sample, say, the inner most core of a galaxy rather than used as a very coarse 2D sampling over a large FOV.

Two other popular IFUs are an image slicer (Fig. 12.13) and an array of microlenses. Pictured is an image slicer for the mid-infrared instrument (MIRI) for the *James Webb Space Telescope*. Developed at Cranfield University, UK, it consists of 20 rectangular mirrors each with a different tip angle. A 2D image becomes segmented into 20 parts and realigned by a set of pupil mirrors that are fed into a

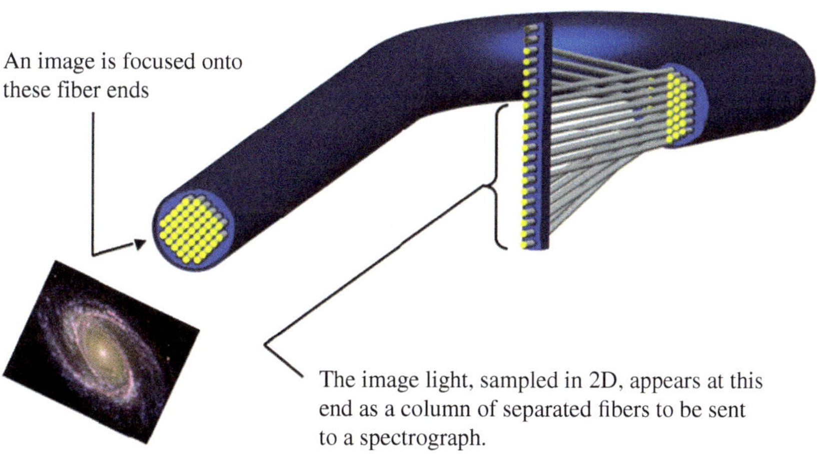

FIGURE 12.12 A dense-pack fiber-optic bundle used as a integral field unit. An image (in this case a galaxy) is focused onto one end of the fibers, while the other end has the fibers spread out vertically, effectively creating a long-slit input to a spectrograph. Differences in chemical compositions of the stars throughout the galaxy as well as rotational velocities can be sampled.

FIGURE 12.13 An image slicer, used to arrange rectangular portions of an image along a single column. Source: Cunningham and Russell (2012). Reproduced with permission of The Royal Society.

long-slit spectrograph. Image slicers have been available for almost a century and are still used extensively at astronomical ground-based telescopes. These optical slicers can be made from multiple-stacked glass elements or diamond machined from a single piece of aluminum for use at infrared wavelengths as displayed in Figure 12.13. Diamond-machined surfaces have a roughness less than 5 nm RMS.

IFUs incorporating an array of microlenses have become popular towards the end of the twentieth century to the present, the ability to manufacture arrays made of fused silica (quartz) have enabled near-UVS IFU applications. The basic operation of a microlens IFU is depicted in Figure 12.14. An image is first magnified onto a window containing an array of lenses, which samples and refocuses the image to a series of small spots. This process is depicted one dimensionally as a side view in Figure 12.14 and as a 2D image in Figure 12.15a. The fill factor, the portion of each segment focused in to each spot, is always less than 100%. In this example, each microlens has a square profile with rounded edges. Light striking at the corners of the microlenses is masked out, leaving small dark “diamond-shaped” black areas. In addition, some of the microlenses sample large amounts of the background. In this case, the background is white, resulting in a few samples being a pale green and the trunk of the tree being gray (black plus white). *Note*: one can retrieve the original image by compactly assembling these individual spots into adjacent pixels, albeit normally with some loss of resolution detail.

Once the image has been transformed into small, separated spots, each of these is dispersed as shown schematically in Figure 12.15b. *Note*: if the spectral range is too large, the spectra from one spot will overlap with another. A bandpass filter is required to limit the spectral range to prevent cross contamination. If the microlens array can produce smaller individual spots, the orientation of dispersion direction can

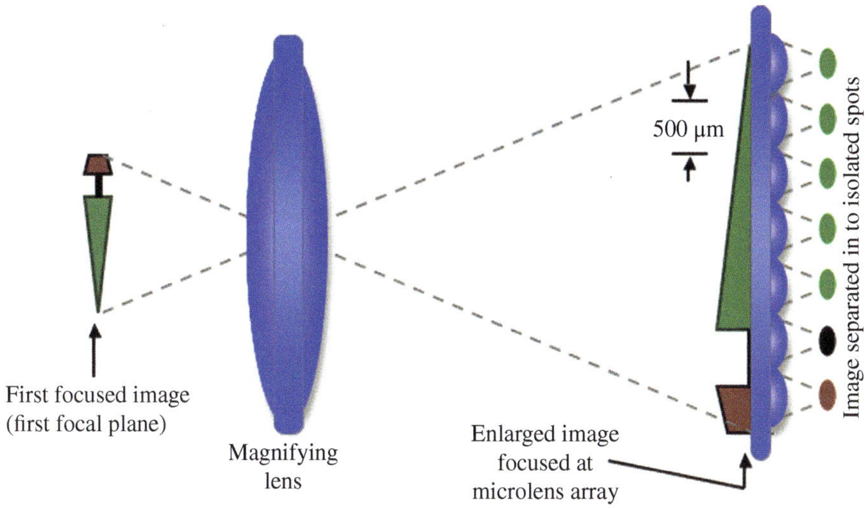

FIGURE 12.14 A schematic side-view profile of a microlens array, segmenting an image into a series of spots.

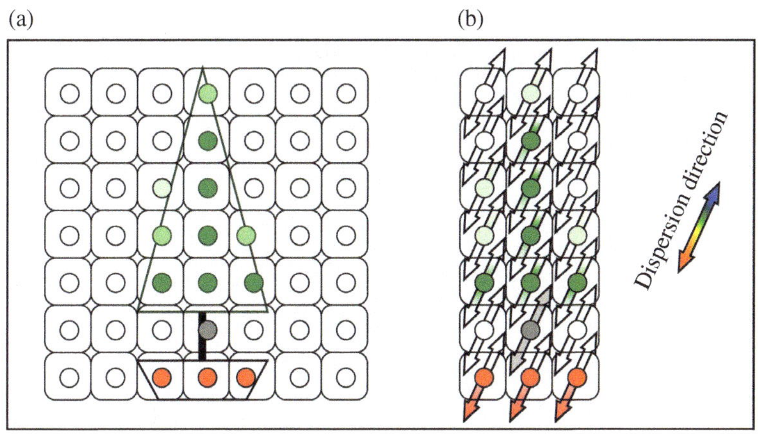

FIGURE 12.15 (a) The simple image of Figure 12.14 is overlaid against the resulting array of separated spots. (b) A portion of exposed spots from a microlens array as these would appear after dispersion.

be rotated to produce longer spectral segments without interference. Figure 12.16 shows an actual 30×30 element microlens array plus its resulting spectrum. Each microlens is 30 μ on a side and made of fused silica. Three individual spectra are outlined in turquoise to help the reader identify where one spectrum starts and another ends. The total number of microlenses and spectral range that can be obtained for each is limited by the size of the detector array.

Finally, an FP instrument provides spectral information over very large FOVs. For instance, FPs are used in astronomical and combustion (large-structure fires)

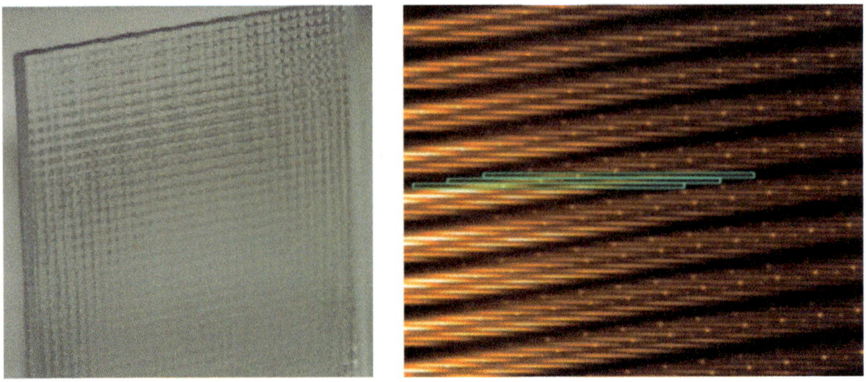

FIGURE 12.16 A 30×30 element microlens array (left) and the resulting set of spectra (right). The spectra from three separate elements are highlighted in turquoise. Source: Images courtesy of B. Woodgate, NASA/GSFC.

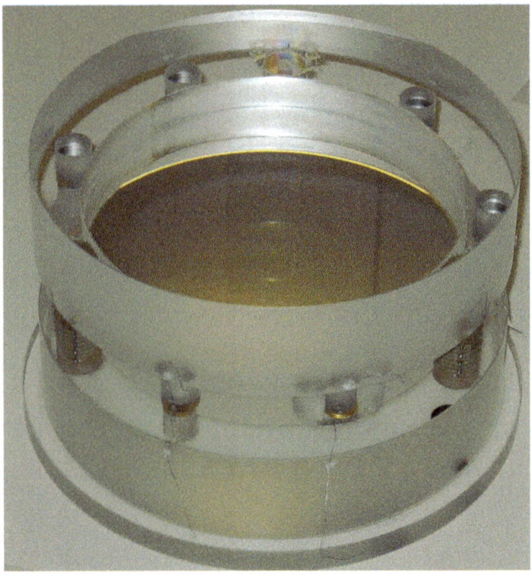

FIGURE 12.17 A large Fabry–Pérot etalon without its outer housing shell. This etalon, fabricated for the Southern African Large Telescope (SALT), accommodates a 150 mm diameter collimated beam. Source: Prof. Ted Williams, University of Rutgers. Reproduced with permission of Prof. Ted Williams.

applications. FP are also used by the *National Institute of Standards and Technology* (NIST) as well as by chemists to obtain the highest wavelength precision in the determination of atomic parameters such as oscillator strengths and energy-level transitions. The etalon, a tunable narrow-band interference filter, is the heart of an FP. The gap between the plates is adjusted by a set of piezoelectric actuators, surrounding the cylinder as shown in Figure 12.17, while a set of associated capacitors provide active feedback to keep the plates precisely parallel.

An etalon is very similar to a laser cavity discussed in Section 9.6, except both interior flat surfaces are highly reflecting (>80%) partially silvered, allowing light to enter and exit from the two ends. The exterior surface of the incident side of the etalon, however, has an antireflection coating that enables most of the light to penetrate into the cavity. The etalon, consisting of two precisely parallel plates, is placed in a collimated beam. Light entering the etalon at an angle, θ, reflects multiple times between the two plates and undergoes destructive interference for all wavelengths except those satisfying Equation 12.8.

$$2d \cos \theta = m\lambda, \tag{12.8}$$

where m is the order of interference, d is the gap between plates, θ is the incident angle relative to the plate normal, and λ is the wavelength. Essentially, any light ray with a wavelength that is a multiple of twice the gap size will pass through the etalon and all others will not. Figure 12.18 is a schematic representation of a light beam passing through an etalon that results in multiple internal reflections, causing a series of parallel reflected and transmitted interference rays. The finesse, F, of an etalon not only is a measure of the average number of reflections inside the interference filter, but it also indicates how concentrated the constructive interference is as a function of wavelength. An etalon with a low finesse ($F<5$) transmits very broad bands that often blend with the wings of adjacent harmonic transmission peaks, while an etalon with a high finesse results in narrowly peaked wavebands that are well separated from adjacent harmonic bands. For example, the transmission plot for $F=2$ (bold line) in Figure 12.18 are strongly blended. The wings of each harmonic peak are shown in dashed lines, which combine additively to form the net transmission curve plotted as a solid line. In contrast, the curve for $F=10$ has a net transmission close to zero between adjacent peaks and are well separated.

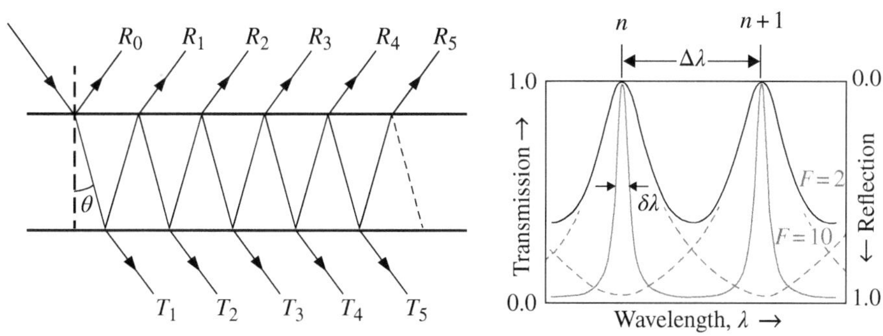

FIGURE 12.18 A graphical depiction of the interfering beams passing through a Fabry–Pérot etalon (left) and plots of the transmission interference (right) resulting from two etalons, one with a finesse, $F=2$, and one with $F=10$. The sharpness of the transmitted band, $\delta\lambda$, is proportional to its finesse. The wavelength range covers the n and the $n+1$ adjacent harmonic bands, which are separated by $\Delta\lambda$ but blended for the $F=2$ case.

Narrow bandpass blocking filters are essential to prevent contamination from the multiple (integer number of) wavelengths passed through the etalon. The nominal gap size between most etalon plates ranges from 5 to 150 μm, providing scanned spectral resolutions, $R=\lambda/\Delta\lambda$ at $\lambda=630$ nm that range from $R=500$ to $R=13{,}500$, respectively. Typically, the gap size can be changed by a factor of two (e.g., 3–6 μ for a 5 μm nominal gap system), enabling a range in wavelengths that differ by a factor of two. For the ultrahigh spectral resolution required by *NIST* and atomic transition scientists, the gap size is substantially larger and can be as much as 1 cm.

A scan of twenty steps, each 1/10 of the FWHM of the spectral feature, are necessary to scan the profile of a single spectral line with some sampling of the continuum on either side. A 2D spectral map is formed from a series of 2D exposures, each with a slightly different gap. Spectral features typically have an FWHM of 0.02 nm, requiring the etalon gap to change by 0.002 nm after every step. If only the Doppler velocity is required, nine steps of 1/3 of the FWHM are needed. (For sufficiently high signal-to-noise ratio data, the centroid of a spectral feature can be determined to 1 part in 20 of the FWHM.) A 2D velocity map can be generated from the Doppler shifts measured from the 3D spectral data cube.

12.4 ECHELLE SPECTROGRAPHS

An Echelle spectrograph is designed to provide very high dispersion ($10^4 < \lambda/\Delta\lambda < 5 \times 10^5$) over a large range of wavelengths. It is particularly well suited for applications where a large number of different atoms or molecules are to be studied simultaneously. An echelle spectrograph is a two grating system. The first is operated in a low spectral order ($m<4$) to disperse the light at low spectral resolution. The second one, the Echelle, has a large number of grooves/mm and is blazed to put much of the incident flux into high spectral orders (typically $50<m<200$). The dispersion of the Echelle is orthogonal to the first grating to separate the spectral orders. This combination of gratings creates a series of spectra, folded in such a manner, as one would scan the text of a book.

Figure 12.19 schematically demonstrates how an Echelle spectrum is formed. The light enters from out of the page at the left, where it is dispersed horizontally into a low spectral order, coming out of the page. (*Note*: the spectrum, which is represented by in the figure by five discrete colors, actually forms a continuous and smoothly changing range of colors.) Next, this dispersed light strikes a second reflection grating, the Echelle, which is ruled and blazed for high dispersion in the vertical direction. The resulting spectra are the series of high-dispersion spectra, each of limited range, running diagonally. Heuristically, if one envisions vertically collapsing down the final set of diagonal spectra to the point where the ends of adjacent orders blend together, one can create the first, low-dispersion spectra.

There are a few characteristics of Echelle spectrographs worth noting. The higher spectral orders (largest values of m), occur at the red extreme of the spectral range. The gap between spectral orders diminishes toward the blue end (lowest values of m).

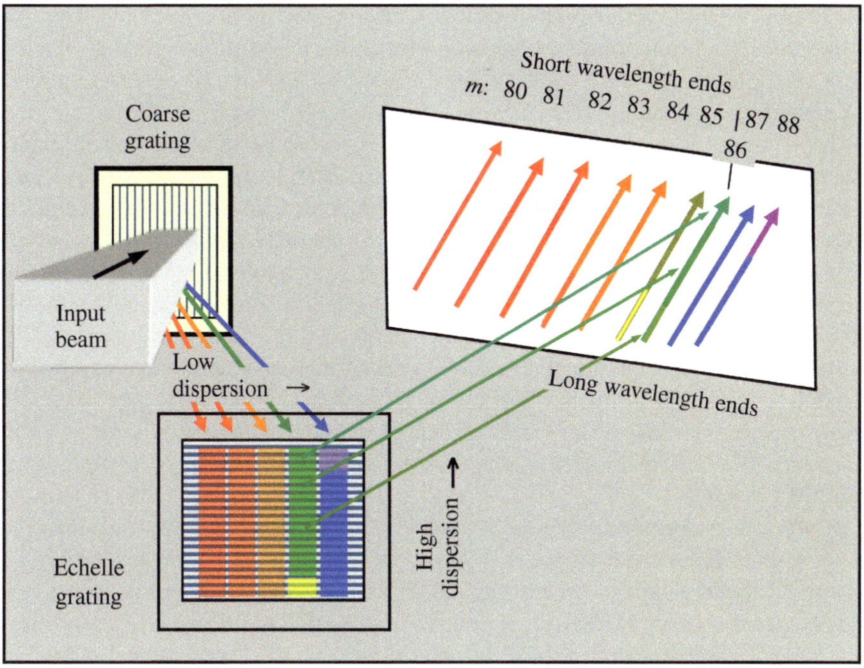

FIGURE 12.19 The formation of an Echelle-format spectra.

There is wavelength coverage overlap between adjacent orders, especially for the highest orders. For example, the short wavelength end of the $m = 84$ Echelle order also appears at the long wavelength end of $m = 85$. In the spectra shown in Figure 12.19, three shades of green, ranging from yellow-green to blue-green, are depicted for order $m = 86$. While the Echelle spectrograph gives the most amount of spectral information for a given sized format of the detector, it contains only a single spatial sampling.

12.5 RAMAN SPECTROGRAPHS

Raman spectrographs have found widespread use in molecular biology, chemistry, and environmental science applications, among others. These are specialized instruments, designed specifically to examine a very small portion of the electromagnetic spectrum surrounding a single emission line from an LED or laser, each intended to excite a certain class of molecules or specific atomic transitions. Each Raman spectrograph is intended for a particular application and some experiments may require more than one instrument, each optimized for a different wavelength. In contrast, most spectrographs are general-purpose equipment, providing access to a sizable portion of the electromagnetic spectrum. A holographic grating with its sinusoidal profile is most often used in a Raman spectrograph. Normally, the reflection efficiency from a holographic grating is not as good as it is for a ruled one, except when the sinusoidal profile is well matched to the narrow wavelengths of interest. More importantly, a grating with a sinusoidal profile scatters far less light than does a ruled one, mitigating the dominant source of contamination in Raman spectrographs.

Spontaneous Raman emission is an inherently weak process and spectrographs must induce nonlinear excitation to be practical. Explicitly, most photons scatter elastically (the Rayleigh form), while only a small portion (about 1 part in 10 million) are absorbed and re-emitted at slightly different energies (inelastically scattered), producing the Raman effect. The Raman effect has two forms: Stokes scattering where the returned photon is less energetic (red side of the Rayleigh peak) and anti-Stokes where the resulting photon is more energetic (blue side). This spectroscopic technique alters the vibrational states and bending modes of certain molecules. The Raman effect is similar to the fluorescence process, except in the latter case, the photons are completely absorbed in resonance and are emitted only after a certain resonance lifetime. Moreover, Raman scattering requires the light to be at least partially polarized.

It is worth underscoring that the Raman effect is intrinsically very weak and one of several nonlinear methods are required to enhance its signal at the expense of the unwanted Rayleigh signal. These include (i) Stimulated Raman, (ii) coherent anti-Stokes spectroscopy (CARS), (iii) Resonance Raman (RR), and (iv) surface-enhanced Raman spectroscopy (SERS). In addition to increasing the Raman signal by several orders of magnitude, each nonlinear process results in a 30–60% reduction in Rayleigh scattered light returned. Nevertheless, the central, unwanted Rayleigh peak must still be suppressed further by several orders of magnitude, using a notch filter or some other technique such as a multiple-pass spectrograph.

A pulsed laser is used in stimulated Raman spectroscopy since it provides short, polarized, very-strong oscillating electric fields compared to continuous wave laser, enhancing by four to five orders of magnitude the nonlinear Raman signal. CARS is another type of nonlinear Raman spectroscopy, which uses two co-aligned lasers having slightly different wavelengths. The third method, RR spectroscopy, is only effective for special colored substances that do not exhibit strong fluorescence, usually the primary source of contamination. A tunable laser beam is used to excite electrons from the ground vibrational states to energies that are close to the excited states of the molecules being probed. The RR effect boosts the Raman signal by three to five orders of magnitude in the so-called chromophoric group. *Note*: not all the bands of spontaneous Raman spectrum are enhanced. Finally, SERS is used to study certain metal surfaces.

For a detailed example, we now turn our attention to a particular Raman spectrograph system that is currently deployed and is under further development for the US Army. This Raman spectrograph is an instrument for the detection of aerosolized biological or chemical agents. The current field instrument is the size of a small, two-drawer filing cabinet for use on a truck platform. It houses a 6–8 inch diameter telescope and a coaligned LED light source. The *Defense Advanced Research Projects Agency* (DARPA) is developing a new replacement instrument, depicted schematically in Figure 12.20, to improve its efficiency, compactness, and robustness

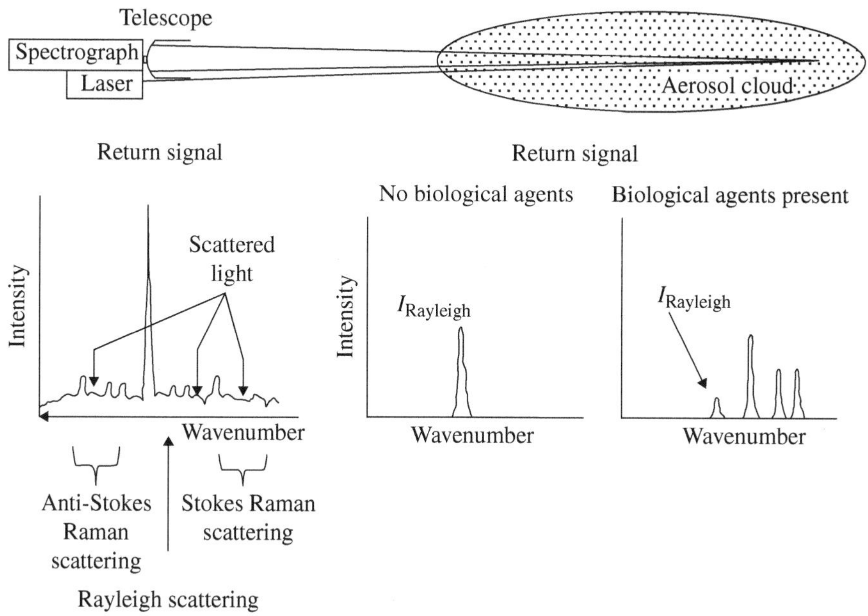

FIGURE 12.20 A schematic Raman spectrograph system being developed for the US Army. A graph (not to scale) of Rayleigh plus Raman scattering signals at visible light wavelengths is shown (bottom left), while the resulting return UV signals when biological agents are and are not present in the atmosphere (bottom right).

by replacing current components with wide-bandgap solid-state devices, operating in the near ultraviolet. The advantage of using near-UV over visible light can be seen in Equation 12.9, which gives the returned intensity from the unwanted Rayleigh scattering as a function of wavelength and direction.

$$I \propto I_0 \frac{1+\cos^2\theta}{\lambda^4} \tag{12.9}$$

I_0 is the input signal from the laser, while θ is the angle between incident and returned beams. The square of the cos θ term indicates that most of the Rayleigh-scattered light is concentrated into cones centered in the forward (θ=0°) and backward (θ=180°) directions. The wavelength (λ) raised to the fourth power in the denominator, indicates the amount of Rayleigh scattering is strongly suppressed at short wavelengths. Moreover, many biological agents (amino acids in particular) produce strong Raman scattering.

Current field systems are bulky, heavy, have a limited range of 1–2 km, are relatively fragile, and difficult to maintain and operate. These systems also are expensive and power hungry, limiting the deployment to company or battalion levels. Further, these bio/chem detection systems rely on LED light sources, operating at visible and infrared wavelengths, since UV LEDs and lasers have only been commercially available since about 2005. However, several technological advances are significantly improving the effectiveness of this aerosolized bio/chem detection system. Recent progress has been made in the development of solid-state UV LEDs and lasers with central wavelengths below 275 nm.

Future systems will be compact, rugged, and less expensive so that an individual in a platoon can carry one capable of detecting weaponized aerosols at a standoff distance of more than 2 km. Other applications for future aerosol detectors might include attaching these devices to buildings in large cities and centrally operating these for long periods unattended.

13

OPTICAL AND ELECTRON MICROSCOPY

Galileo introduced a compound microscope in 1610, only 1 year or so after aiming his first telescope at the heavens. However, the invention of the compound microscope is credited to at least two names: Zacharias Janssen (1585–1632) and Hans Lippershey (1570–1619), both of whom lived in the same Dutch city of Middelburg. The first use of a microscope to do serious research, on the other hand, is assigned to the English scientist Robert Hooke (1653–1703). The Dutch tradesman Antonie van Leeuwenhoek is also credited with conducting some of the first studies in microbiology, for which he built (around 1670) fine microscopes. Van Leeuwenhoek invented a technique for producing small glass spheres that worked very well as objective lenses. More than 250 years passed before anyone realized that particle beams, not just light, could be used to image small objects.

The first electron microscopes were built by Ernst Ruska and others before World War II, and mass-produced by Siemens in Germany, only some 35 years after the discovery of the electron by J.J. Thomson, and 10 years after the introduction of the principle of wave–particle duality by de Broglie.

In this chapter, we describe the basic optical and electron microscopes and some of the ways these instruments are used. The two devices use fundamentally different imaging mechanisms, but also have many features in common. The principles of imaging optics presented in Chapter 10 apply to both. We begin with a discussion of magnification by a single optical lens (the "simple" optical microscope), followed by a description of a two-lens microscope (the basic Galilean type), its magnification and resolution, and a brief presentation of alternative techniques of optical microscopy. In Section 13.2, we introduce the transmission electron microscope (TEM),

Modern Devices: The Simple Physics of Sophisticated Technology, First Edition.
Charles L. Joseph and Santiago Bernal.

Companion website: www.wiley.com/go/joseph/moderndevices

which is the equivalent to the standard optical microscope. We revisit several concepts and definitions that are the counterparts of those used in the context of the optical instrument; in particular, we include in that section a brief explanation of magnetostatic lenses, which are further explored in Chapter 22. In the last section, we present a summary of the various types of scanning probe microscopes (SPMs) and their applications.

13.1 OPTICAL MICROSCOPES

13.1.1 The Magnifier

The unaided human eye can focus objects at distances no shorter than about 25 cm. This distance, measured from the object to the cornea, is called the "distance of most distinct vision" or *near point*. To focus objects closer than 25 cm, a magnifier or *simple microscope* can be used. The magnifier forms a virtual image, and the eye can focus comfortably now because the *image* is located at 25 cm or farther from the eye. The magnifier is typically a lens with a focal length that is short compared to 25 cm, so a virtual image is formed when the object is placed between the lens and its focal point.

Figure 13.1 illustrates the ray tracing with a single magnifying lens. If θ is the angle subtended by an object at the (unaided) eye, and θ' the angle subtended by the *image* from a magnifier (focal length f) at the eye, we have $\tan\theta' \approx h_i/25\text{cm}$, and $\tan\theta \approx h_o/f$, if the virtual image falls at the near point of the eye and the object is near the focal plane (F) of the lens. Then, the lateral magnification is given by $M = h_i/h_o$, or

$$M \cong \frac{25\,\text{cm}}{f(\text{cm})}, \tag{13.1}$$

since $\theta' \approx \theta$. If the object is placed exactly at F, the virtual image will appear at "infinity," in which case the eye can be completely relaxed for focusing. As an example, a lens with $f = 5.0$ cm would have $M = 5\times$. Magnifications of up to about 20× are possible with a single lens.

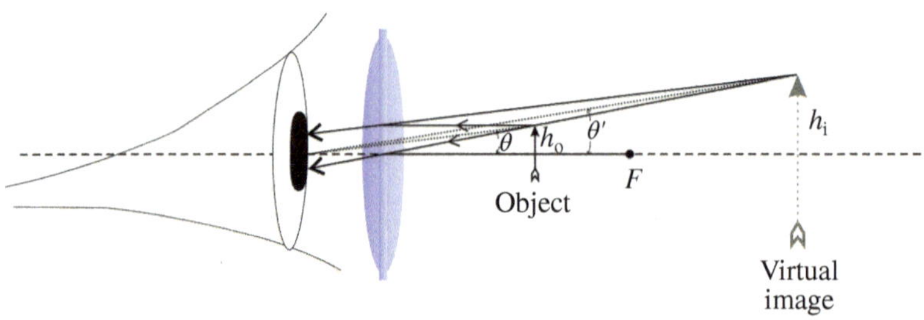

FIGURE 13.1 Ray tracing for imaging with a magnifier lens. The virtual image appears at a distance of 25 cm from the eye, the closest distance for focusing.

13.1.2 The Compound Microscope

The basic *compound microscope* consists of an *objective* lens and an *eyepiece* (ocular) lens. As shown in Figure 13.2, the specimen to be examined is placed just outside the focal point of the objective. This lens has a short focal length and forms and intermediate real (and inverted) image that becomes the object for the eyepiece lens. Further, the intermediate image is located near the focal point of the eyepiece, so the light rays entering the eye are almost parallel for comfortable viewing. Therefore, the final image to be focused by the eye or camera is virtual and located at the bottom of the microscope column, ideally at 25 cm from the eye; obviously, the image formed at the retina or photographic plate is real. Focusing is accomplished by adjusting the distance from the objective to the specimen.

The magnification of the microscope is the product of the magnifications of the objective (focal length f_O) and the eyepiece (focal length f_E). From Equations 10.6 and 13.1, we obtain

$$M = -\frac{s_i}{f_O}\frac{25}{f_E}, \tag{13.2}$$

where s_i is the distance from the objective to the intermediate (real) image (Fig. 13.2), and all quantities are in cm. The factors in Equation 13.2 can be understood easily if

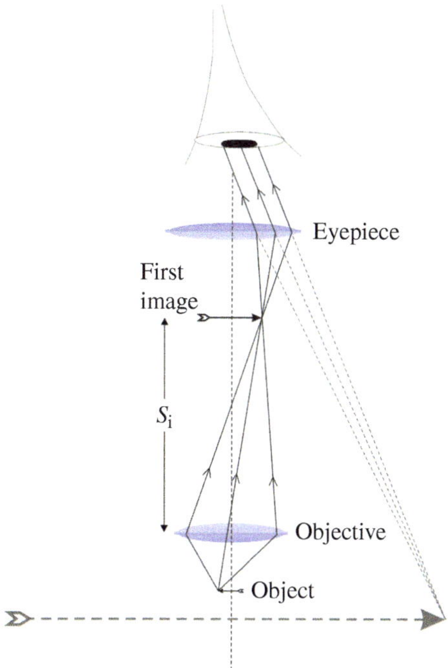

FIGURE 13.2 Basic compound microscope. This is the arrangement invented by Galileo in 1610.

we realize that the specimen is placed very close to the focal point of the objective (i.e., $s_0 = f_O$ in Eq. 10.6), and that the microscope is designed so that the image formed by the eyepiece is placed close to the near point of the eye (i.e., $s_i = 25\,\text{cm}$ in Eq. 10.6).

A real optical microscope has many components in addition to those illustrated in Figure 13.2. A major component is an illumination system consisting of a light source (e.g., halogen, mercury-arc, or xenon-arc lamps), a *condenser lens*, and a number of apertures. The function of the condenser lens is to provide uniform illumination of the specimen, while the stops help to reduce glare from the microscope column. Microscope objectives, unlike the lens in Figure 13.2, are composed of two or more achromatic elements; each achromat, in turn, consists of two lenses cemented together to correct for chromatic aberration (Fig. 10.24). In addition, the objective system is designed for correction of spherical as well as coma aberrations (Section 10.5).

In modern instruments, the eyepiece can be interchanged with CCD cameras or other detectors. Also very common is the use of binocular eyepieces for comfort viewing. Lastly, in *stereomicroscopes* two separate lens systems produce slightly displaced images for 3D viewing of samples; magnification in this case is limited to about 200×.

13.1.3 Numerical Aperture, Resolution, and Depth of Field

We saw in Section 10.3 how diffraction limits the resolution of a telescope, that is, the ability to resolve two stars or far away objects. In the same vein, microscope resolution is limited by diffraction even after corrections for spherical, chromatic, and other aberrations are implemented. To revisit the main idea, the diffraction pattern from a circular aperture consists of a central bright disk, the *Airy's disk*, and a series of concentric circles (Fig. 10.10). The *Rayleigh criterion* states that the images of two point sources are just resolvable if the central maximum of the Airy pattern of one coincides with the first minimum of the other. Thus, generalizing Equation 10.12 to include large angles and a medium (between the object and the lens) different from air, we can write the following expression for the *minimum resolvable separation* of two points by an objective lens,

$$d_{\min} = \frac{0.61\lambda}{n\sin\alpha}, \tag{13.3}$$

where λ is the wavelength of the light, n is the index of refraction in the object space, and α is the aperture angle of the lens. An important consideration here is that the object is illuminated with regular, *incoherent* light; additional interference effects would occur if the illumination is coherent (as with laser light). Figure 13.3 illustrates the imaging of a specimen with different degrees of resolution.

The quantity $n\sin\alpha$ in Equation 13.3 is called the *numerical aperture* of the lens, and is denoted by NA:

$$\text{NA} = n\sin\alpha. \tag{13.4}$$

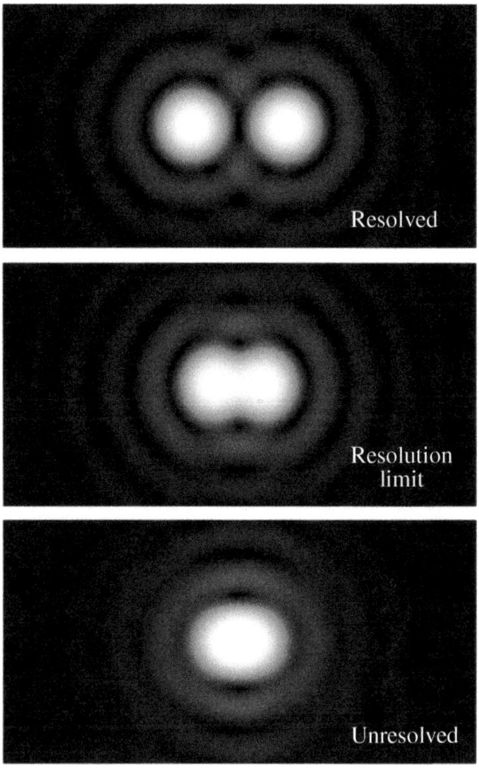

FIGURE 13.3 Airy diffraction patterns and resolution limit. Spacing between airy disks is twice the distance to first minimum (top); equal to distance to first minimum, or Rayleigh resolution criterion (middle); and half the distance to first minimum (bottom). Source: Bliven, https://commons.wikimedia.org/wiki/File:Airy_disk_spacing_near_Rayleigh_criterion.png. Public domain.

The NA of an objective lens is a measure of its ability to collect light and resolve details in a specimen. To see this, we note that $\sin\alpha = a/2f$, to a good approximation, where a is the lens aperture diameter and f is the lens focal length. But the *f-number*, introduced in Section 10.1 and defined as $f/\# = f/a$, is a measure of the speed of the lens, that is, the smaller the $f/\#$, the more light can be collected by the lens. Thus, larger $\text{NA} = n/2f/\#$ means more light gathered by the objective in a microscope. In addition, shorter f implies more magnification (Eq. 13.2) and better capacity to resolve details. However, if the lens aperture diameter a is small compared to f, the light collected by the eyepiece may not be sufficient to resolve any details; in other words, the NA would be compromised because of a small value of α.

If the space between the specimen and the objective is air ($n = 1.0$), the NA is limited to values below 0.95, corresponding to $\alpha < 72°$. If a liquid like water, glycerin, or oil is used between the objective and the specimen, the NA can be significantly increased; with oil ($n = 1.51$), for example, NAs greater than 1.40 can be achieved. We present additional details of the oil immersion technique in Advanced Concept AC13.1.

ADVANCED CONCEPT AC13.1

Immersion Oil Microscopy

The most important parameters of a microscope objective are its magnification M (Eq. 13.1) and its numerical aperture NA (Eq. 13.4). As we saw in the text, NA is directly related to the light gathering capability of the lens which in turn affects the ultimate resolution of the microscope. If the space between the objective and the specimen is just air (index of refraction $n = 1.0$), rays entering the first surface of the objective will be refracted or reflected depending on their angles of incidence. In the illustration of Figure AC13.1, some of the rays from the specimen that travel in air (right part) are either refracted outside the objective or reflected back. However, if the space between the specimen and the objective, or rather between the coverslip and the objective, is filled with a medium having an index of refraction close to that of glass, that is $n = 1.5$, the light rays originating from the specimen will not see a transition when entering the objective. This is seen on the left side of Figure AC 13.1. Therefore, more light contribute to the intermediate or first image in the microscope (Figure 13.2). Oils with $n = 1.51$ are typically used in immersion microscopy to yield NA from 1.0 to 1.4 (for magnifications in the range 60× to 100×). Note that with $n = 1.51$, the maximum theoretically possible NA would be also 1.51 (Eq. 13.4), but the aperture angle α is in practice limited to a maximum of about 72°.

Water and glycerin are common as immersion media in applications involving living cells, in which case NA is around 1.2. The use of immersion media also

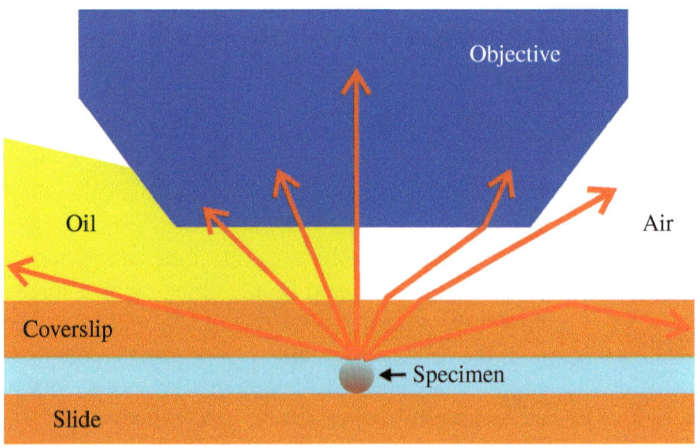

FIGURE AC13.1 Comparison of standard optical microscopy in air (right hand side) with oil-immersion microscopy (left hand side). The light gathering capacity is improved greatly with oil. Credit: Adapted from Wikipedia, http://commons.wikimedia.org/wiki/File:Immersionsvorteil.svg

eliminates problems related to uneven surfaces of the cover plates. Furthermore, properly designed immersion objectives, called *apochromatic* lenses, have minimal spherical as well as chromatic aberrations (Section 10.5). The following video from the Science Photo Library provides a dramatic demonstration of the physics behind immersion oil microscopy: http://www.sciencephoto.com/media/578356/view? utm_campaign=Clip+of+the+week%3A+17th+March+2014+K004%2F0901&utm_source=emailCampaign&utm_medium=email&utm_content=.

From Equation 13.3, we see that the resolving power of a microscope is of the same order as the wavelength of the radiation used. Thus, an alternative way to improve the resolution is to use radiation of shorter wavelength. In fact, ultraviolet (UV) microscopes use radiation in the range 200–400 nm. Because regular glass lenses absorb radiation strongly in that range, quartz lenses are required in UV microscopy. Further, special fluorescent screens or detectors are needed to make the UV images visible.

The *depth of field* is defined as the axial distance in the specimen that can be focused clearly. Objective lenses with low NA have the largest depth of field, just as camera lenses with large *f/#*. Thus, in an optical microscope, the depth of field is of the same order as the resolution, making focusing and interpreting the image relatively easy at high NA. However, the sample must be very thin in the latter case.

Table 13.1 summarizes the parameters of a typical optical microscope.

13.1.4 Alternative Methods of Optical Microscopy

In *dark field* microscopy, the image of the specimen is obtained with scattered light exclusively by eliminating the background light. The condenser lens in a dark field microscope has an annular aperture to focus a hollow cone of light at the specimen; further, the light that is not scattered passes outside an apertured objective. In this way, it is possible to discern details of very small specimens that would otherwise be overwhelmed by the transmitted light. However, the very low intensities of scattered light off a specimen require the use of very intense illumination and, at the same time, the avoidance of any spurious reflections from the microscope column.

In *fluorescence microscopy*, the specimen is irradiated with a specific band of wavelengths, normally through an *excitation filter*, and the resulting fluorescence

TABLE 13.1 Main Parameters of a Typical Optical Microscope

Microscope column length	16 cm
Objective lens magnification	5× to 100×
Eyepiece lens magnification	10× to 15×
Overall magnification	50× to 1500×
Maximum resolution (at NA = 0.95 and λ = 500 nm)	260 nm

emission from the specimen is separated out for detection through a *barrier filter*. For example, the excitation radiation can be in the range 450–500 nm, while the emitted fluorescent radiation has *longer* wavelengths, around 550 nm. The fluorescence signal is much weaker than the excitation signal by a large factor, up to a million, but allows the detection of very fine or specific details such as individual cellular components or even single molecules. If the specimen does not fluoresce on its own, it can be usually "labeled" with special fluorescent dyes called *fluorochromes*. Fluorescence microscopy has found many important applications in biology, medicine, and materials science.

Another technique, *UV microscopy* takes advantage of the improved resolving power at shorter wavelengths (Eq. 13.3), or the specific absorption of UV wavelength of some specimens (e.g., nucleic acids) but requires quartz lenses and special detectors. Mirrors, instead of lenses, have also been used because of their advantages for reducing aberrations, but at a lower NA.

In *total internal reflection microscopy*, a translucent specimen (e.g., a cell) of index of refraction n_1 is in contact with a coverslip with index of refraction $n_2 > n_1$. Thus, light incident on the interface from the specimen to the coverslip at an angle greater than the critical angle suffers total internal reflection (Section 10.1). Despite the use of the word "total" to describe the phenomenon, not all the light is reflected back. In fact, some light tunnels through the interface, the so-called *evanescent wave*, to a depth that depends on n_1 and n_2, and is of the order of 100 nm. This feeble light is collected by the objective to image the specimen, the variations of n_1 across the specimen obviously affecting the contrast and effective resolution. The technique can be used in oil immersion instruments (Advanced Concept AC13.1) and in fluorescent microscopy.

A number of modern digital techniques also exist where image processing plays a major role. An example is *digital holographic microscopy* (DHM) where the specimen is illuminated with laser light to generate a *hologram* through interference with a reference beam (Section 10.4). Since the hologram contains both amplitude and phase information, it is possible to reconstruct a 3D image of the specimen with software, and to "focus" digitally to different planes, or even make 3D movies.

13.2 THE TRANSMISSION ELECTRON MICROSCOPE

As we saw in the previous section, the wavelength of light (used to illuminate the specimen) and the numerical aperture of the objective lens are the main factors that determine the maximum resolution possible in an optical microscope. Thus, the best resolution of an optical microscope is about 200 nm, achievable with UV radiation. The best optical magnification, on the other hand, is around 1500×. In contrast, the resolution and magnification of a TEM can be orders of magnitude better (see Table 13.2). This is possible thanks to the very short wavelengths associated with electrons in the 10–100 keV ($1.0\,\text{keV} = 1.602 \times 10^{-16}\,\text{J}$), or higher, energy range. The basic relation that connects wavelength λ and linear momentum p is

TABLE 13.2 Main Parameters of Phillips CM200 FEG Transmission Electron Microscope

Voltage	20–200 kV
Electron current	1 pA to 150 mA
Diameter of beam spot	0.4–2 mm
Magnification	25× to 1,100,000×
Maximum resolution	0.1 nm
Smallest imaging area	<0.2 μm

$$\lambda = \frac{h}{p}, \tag{13.5}$$

where $h = 6.626 \times 10^{-34}$ J is Planck's constant. The momentum p can be expressed in terms of the accelerating voltage V, the speed of light c in vacuum, and the electron's charge $-e$ and mass m:

$$p = \sqrt{2meV\left(1 + \frac{eV}{2mc^2}\right)}. \tag{13.6}$$

For 1 keV electrons, for example, we obtain $p = 1.71 \times 10^{-23}$ mkg/s, and $\lambda = 3.9 \times 10^{-11}$ m $= 0.039$ nm, or about 5000 smaller than 200 nm, the shortest wavelength used in optical microscopy. Much smaller electron wavelengths are possible in electron microscopes with energies up to 4–5 MeV, but only a handful of these instruments are in operation in the world today.

Equation 13.6 includes relativistic effects and follows from the definitions of relativistic linear momentum and kinetic energy. Note that without the relativistic correction, the electron wavelength would be simply $\lambda = h/\sqrt{2meV}$, because $K = eV = p^2/2m$ is the kinetic energy written in terms of the non-relativistic momentum $p = mv$. At 1 keV kinetic energy, the relativistic correction is negligible; but at 100 keV, it amounts to a 4% correction (shorter electron wavelength), and at 1 MeV it amounts to almost 30%. A simple approximate expression for the electron wavelength can be obtained from Equations 13.5 and 13.6, without the relativistic correction:

$$\lambda(\text{nm}) \cong \frac{1.227}{\sqrt{V(\text{Volt})}}. \tag{13.7}$$

The first TEM was built by Ernst Ruska and Max Knoll in Berlin in 1931, without any knowledge of the relations presented in the previous paragraph. Ruska had been investigating for his thesis the use of the "magnetic electron lens," which had been studied before by Hans Busch and shown to do for electron trajectories as glass lenses do to light rays. Ruska had advanced the design of the magnetic lens by surrounding the coils with an iron casting thus leading to a short focal length sufficient to serve as the objective for the TEM. Then it was only natural to add a second lens to form the first electron microscope. The very first instrument had a magnification of just over 17×, but rapid progress was made and by 1933 the TEMs had magnifications vastly surpassing those of the optical instruments.

Figure 13.4 shows the schematics and photograph of a modern TEM. The main components of a TEM are (i) *electron gun* with a thermionic or field emission cathode—Figure 22.5 shows the schematics of an electron gun with a thermionic cathode; (ii) *condenser lens* system (e.g., two lenses and two aperture stops) to concentrate electrons on the sample or specimen; (iii) *objective lens* that creates an intermediate image; (iv) *projection lens* system (usually two or three lenses) that magnifies the intermediate image for projection onto a screen; (v) correction lenses such as the *stigmator* for correcting astigmatism (Section 10.5) from the objective and the condenser; (vi) *stage* for placing and manipulating the specimen; and (vii) imaging system for direct viewing, or for photography or amplification by multichannel plate (MCP) detectors. In addition, electron microscopes require that the microscope column be in a high vacuum to avoid electron scattering from the molecules of the residual gases and to prevent sample contamination. Good vacuum conditions are particularly critical at the electron gun and surroundings; typical pressures are 10^{-6} Torr, or 10^{-4} Pa. Table 13.2 summarizes parameters for a typical TEM.

The electron gun must produce an electron beam with an *energy spread* as small as possible to avoid the equivalent of *chromatic aberration* in optical lenses (Section 10.5). The best sources in this regard use field-emission cold cathodes, which produce energy spreads of the order of 0.25 eV, or only 0.00025% at 100 keV

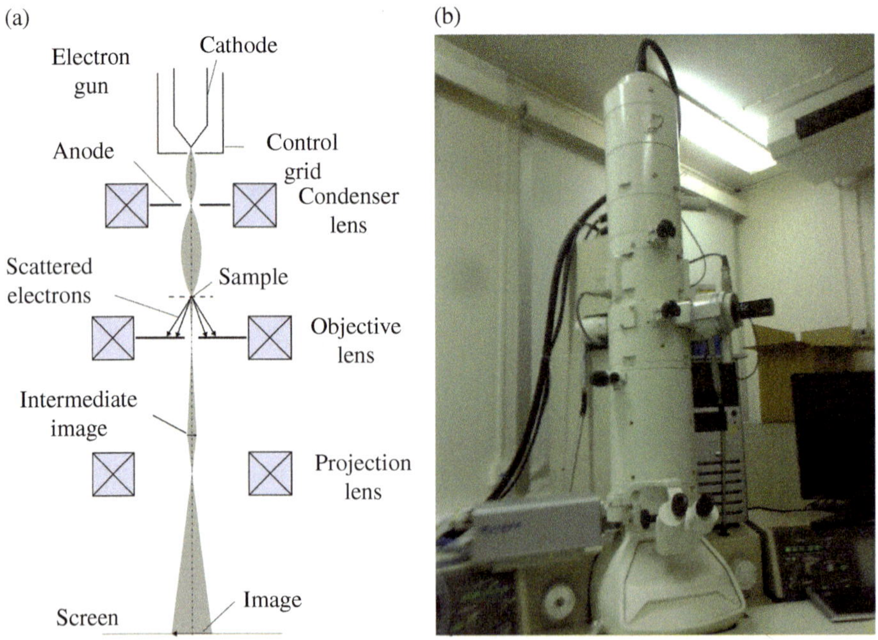

FIGURE 13.4 (a) Schematics of a basic transmission electron microscope (TEM). (b) A photograph of an electron microscope is on the right. Source: (b) Kallerna https://commons.wikimedia.org/wiki/File:Transmission_electron_microscope_Kemira.jpg. CC public domain.

nominal energy; electron guns employing these cathodes are called FEG for "field-emission gun." Since FEG cathodes are short-lived and expensive, however, heated filament cathodes are more common, with energy spreads about 10 times larger.

Modern electron microscopes use magnetic lenses almost exclusively. Magnetic lenses do not have the arcing risks of electric ones, which can compromise vacuum conditions or affect the sample; furthermore, the strength (related to the inverse focal length) of magnetic lenses can be easily adjusted and controlled by simply changing the current applied to coils. In Advanced Concept AC13.2 we give an example of the action of a TEM objective magnetic lens. (Magnetic lenses and deflectors are further discussed in connection to particle accelerators in Chapter 22.) The resolution of TEMs is limited mostly by spherical and chromatic aberrations, which are impossible to remove completely in round magnetic lenses. Unlike optical microscopes, the *depth of field* in electron microscopes is much larger than the resolution. This makes focusing and interpreting the image a much more elaborate exercise than in optical instruments. Further, focusing in TEMs is accomplished by adjusting the objective lens current instead of, as in optical instruments, by moving the objective lens or the sample.

ADVANCED CONCEPT AC13.2

Electron Optics and the Magnetostatic Lens

When a charged particle moves from a region of electrical potential V_1 to another one of electrical potential V_2, as when crossing the gap between two plates, or electrodes, at those potentials, the particle is deflected in much the same way as when light moves between regions of different indices of refraction. In fact, the square root of the potential V is the equivalent in electron optics of the index of refraction n in optics (Chapter 10). The situation is a bit more involved with magnetic forces, because particle trajectories rotate about the optical axis in coil or solenoid lenses. But all the basic concepts that apply to deflection by electrical forces and standard optics (Chapter 10) are valid also for magnetic "optics": Snell's law, paraxial approximation, Gaussian optics, lens law, etc. In Figure AC13.2a we illustrate the trajectory of an electron moving through the magnetic field of a relatively long solenoid lens. In an electron microscope, the objective lens (Fig. AC13.2b) has a much more concentrated magnetic field, with an axial field profile similar to the one shown in Figure AC13.2c. For magnetic lenses of this type (solenoidal), the trajectory equations are written in such a way that only the radial motion is left, that is by subtracting out the orbit rotation. Then, at a given electron energy E_0 the inverse of the paraxial focal length is proportional to the product of the peak field squared and the width of the magnetic field profile (a in Fig. 13.2c). For a particular field profile (see caption and broken line in Fig. 13.2c), we find $1/f=(\pi/16)\left[e^2/mE_0\right]aB_{max}^2$ for the inverse *thin-lens* focal length. As an example, if $B_{max}=0.3$ T, and $a=3$ mm, we obtain from the formula $f=11$ mm for 100 keV electrons; the plot in Figure AC13.2c shows $f=13$ mm. The difference arises from two factors: for the calculation leading to the plot, *relativistic effects* are included and the lens is treated as a *thick lens* (Chapter 10).

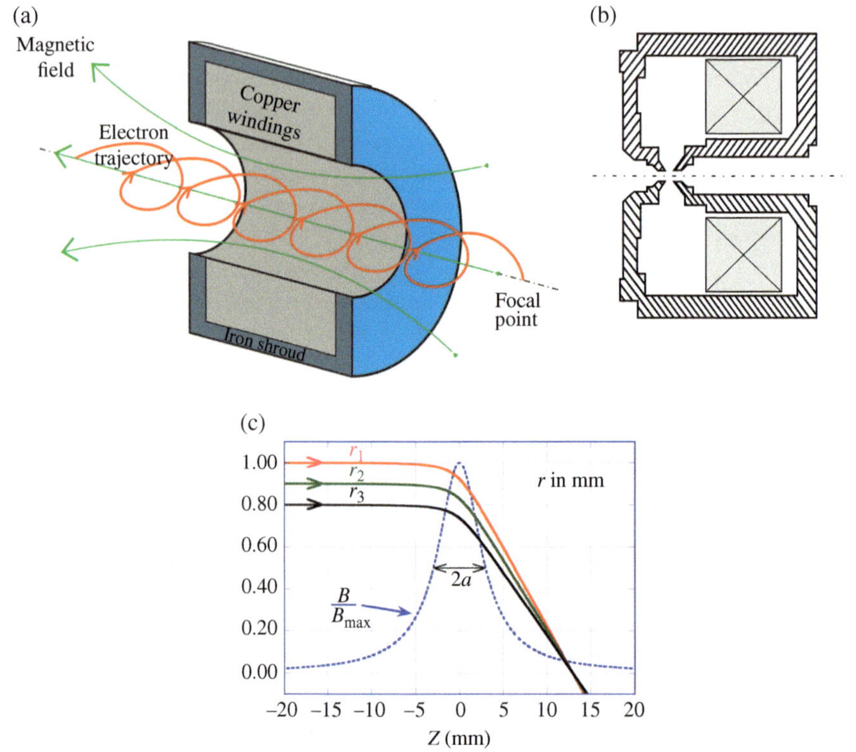

FIGURE AC 13.2 (a) Passage of an electron through a solenoid lens; (b) cross section of a magnetic objective lens used in a TEM—note the small gap at the tips of the iron shroud; (c) normalized magnetic field profile along optical axis, $B(z)/B_{max} = \left[1+(z/a)^2\right]^{-1}$, with $B_{max} = 0.3$ T, and $a = 3$ mm (dotted line), and three electron trajectories (solid lines).

Sample preparation is also much more difficult for a TEM than for an optical microscope; for example, biological samples must typically be produced as layers as thin as 5 nm. In addition, the samples are usually coated with a thin metallic layer (e.g., Au) to prevent charge accumulation. Mechanical stability of the specimen is also critical in TEMs: the sample may not drift more than the expected resolution distance during the time that it takes to record an image. Typical drifts are 1.0 nm/s, but cooling the stage/sample with, for example, liquid nitrogen can reduce this number significantly. Naturally, the entire TEM column must be isolated form shocks or vibrations and shielded to protect the operators and equipment from hazardous X-rays.

Although the magnification of TEMs can reach over 1,000,000×, the instruments are rarely used at that level because of a corresponding required increase in electron beam intensity which can easily damage or destroy the sample. Thus, TEMs are used most commonly with magnifications of about 40,000× to examine biological samples at the 0.7–1.0 nm level, or 80,000× for rugged inert objects.

13.3 ELECTRON–MATTER INTERACTIONS

In standard optical microscopes, light from the condenser lens is scattered and absorbed by the specimen before reaching the objective lens. In TEMs and related microscopes, the electron beam and the specimen can interact in several ways: (i) in *elastic scattering* the electrons bounce off without any change of energy; (ii) in *inelastic scattering,* the electrons may lose some energy by knocking out electrons from the atoms in the sample leading to *secondary electron emission*; (iii) the electrons may lose energy in collisions and produce X-rays by the process known as *bremsstrahlung*. Figure 13.5a illustrates these and some additional outcomes of the

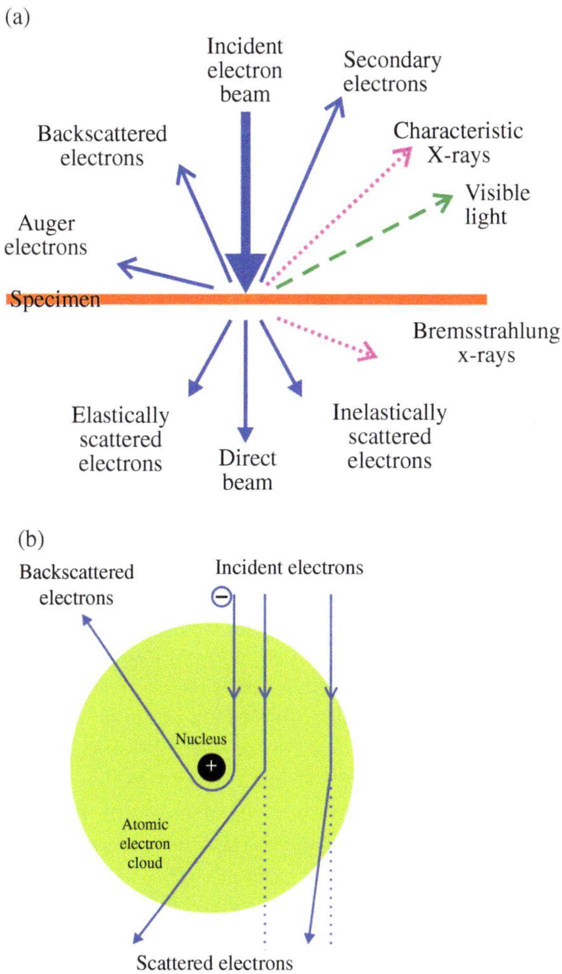

FIGURE 13.5 (a) Pictorial illustration of signals from the interaction of electrons (around 100 keV) and a thin specimen; (b) Elastic scattering of electrons by an atom.

interaction of electrons with a *thin* sample. In Figure 13.5b, elastic scattering by the internal charge of an atom is depicted.

The basic contrast mechanism in TEMs is the elastic scattering of electrons by the atoms in the specimen. The process is quantified by an *interaction cross section*, which is essentially the effective target area seen by the projectile electrons. This cross section decreases as the electron energy increases; in the words of the textbook by Williams and Carter (2009), "the target becomes smaller as the bullets become faster!" Thus, scattering at 200 keV will be less than at 100 keV. In addition, the more atoms (of a given element) per unit volume are in the sample, that is, the higher the mass density, the larger the scattering. Therefore, denser and thicker regions in the sample produce darker spots in the final image from a TEM. Furthermore, atoms of elements of higher atomic number produce more scattering because their electron clouds have more charge (Fig. 13.5b).

Most of the electron or radiation signals from the processes shown in Figure 13.5, not just scattering, depend strongly on the geometry (topography) and chemical composition of the specimen. Therefore, if the electron beam is focused to a small spot and scanned over the specimen, a detailed characterization of its surface is possible. Electron microscopes that operate in this way are called *scanning electron microscopes* (SEMs). In these instruments, imaging is not realized through lenses but by the detection and processing of a particular category of signals that originate from the interaction of a *probe* beam and the specimen. Condenser and objective lenses are still employed in SEMs, but their purpose is to focus the beam to a small spot. In addition, scan coils are employed to steer the beam in a *raster pattern* over the sample, similar to the way an electron beam scans a CRT screen (Sections 15.2 and 15.4). The first commercial SEMs appeared around 1965, but prototypes were being tested in the late 1930s.

SEMs of interest to biology operate with secondary electron (SE) emission and backscattering (Fig. 13.5). Electrons from secondary emission have energies of 50 eV or less and originate from within 10 nm, approximately, of the sample surface. Thus, SE is used to image surfaces with high resolution and contrast. Typically, the electron beam energy in SEMs is 15–40 keV, but low-voltage (1–2 keV) SEMs can be used to image surfaces of biological specimens and integrated circuits. At these low energies, heating and potential damage to the samples is minimized. Magnifications of just 1000× are typical, but with depths of field greatly exceeding those of optical microscopy. The other process, backscattering of electrons (BSEs), has much larger signal yields than SE and can also be utilized for surface imaging and characterization. In BSE mode, for example, a SEM can collect signals directly proportional to the atomic number Z of the atoms in the specimen, for Z less than around 30.

In all examples of SEMs just mentioned, the electrons or radiation induced by the probe are "reflected" off the sample, but it is also possible to detect forwardly scattered (elastic and inelastic) electrons or radiation. This is the basis for the *scanning TEM* (STEM). In *bright-field* (BF) imaging or mode, electrons from inelastic scattering events at small angles are collected; essentially, the weakening of the incident beam is measured after it passes through the sample. The same mode can be applied to regular, fixed-beam TEM instruments by inserting a small aperture near the image focal plane of the objective lens; in this way, electrons that are scattered (mostly elastically) through large angles are excluded. In the second mode of operation of STEMs, large-angle scattered electrons are collected by an annular detector; this is the *dark-field* (DF) mode of operation, which is similar in character to dark-field microscopy in optical microscopes (see earlier text).

X-rays can also be detected for imaging in SEMs. In "X-ray energy-dispersive spectrometry (XEDS)," for example, the characteristic X-rays are detected. This is an example of *analytical electron microscopy* (AEM).

In yet another type of electron microscope, the *field-emission microscope*, the specimen and the source become one. The specimen/source is a fine tungsten tip sharpened to a radius of around 100 nm. Application of a negative voltage of a few kV to the tip leads to field emission of electrons; the current, which depends on the arrangement of atoms at the tip, is used to image the atom lattice. A variation of the technique employing ions instead of electrons led to the first image of individual atoms in 1955.

13.4 BRAGG'S DIFFRACTION

Thus far, it was assumed that the specimen under study in TEM, SEM, or STEM was amorphous, that is, that its atoms were disorganized. In this case, the scattering of the electron beam probe is *incoherent*, that is, there is no correlation between the phases of the electron waves originating from neighboring scattering atoms. The contrast mechanism with incoherent scattering, as we saw above, is dominated by mass density variations over the sample under study.

By contrast, if the specimen is a single crystal, the electron beam is *coherently* scattered in a way similar to Bragg diffraction of X-rays. In effect, the ordered atoms in the crystal act as a 3D diffraction grating (Section 10.3) for the electron waves. If we consider two electron waves incident on two parallel lattice planes, as in Figure 13.6, constructive interference occurs when the difference in path lengths of the two waves is an integer number n of electron wavelengths:

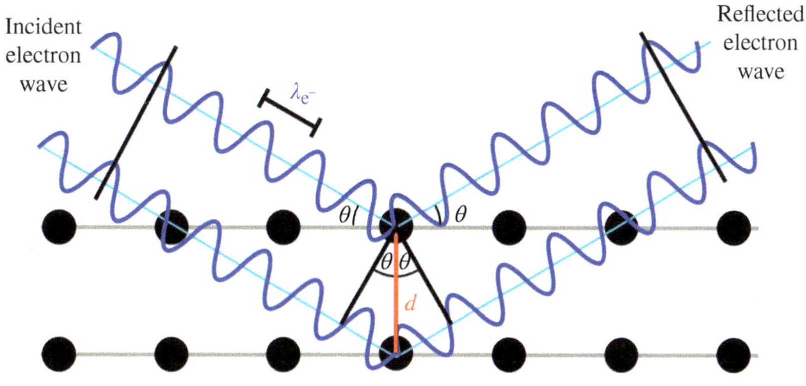

FIGURE 13.6 Constructive interference of electron waves upon diffraction off parallel lattice planes.

$$2d \sin\theta = n\lambda_{e^-}, \tag{13.8}$$

where θ is the scattering angle, d is the spacing between the lattice planes, and λ_e^- is the electron wavelength. The equation is essentially the same as the *Bragg equation* quoted before (see Eq. 10.16 with $N=2$). For example, at 300 keV and $d=0.2$ nm, we obtain from Equations 13.5 and 13.8, $\theta=0.28°$. In general, the angles in electron diffraction are much smaller than for X-ray diffraction due to the small electron wavelength.

Note from Equation 13.8 that the angle θ increases as the distance d decreases. Thus, the distances in the diffraction pattern follow an inverse relationship to the distances in real space. The space associated with the diffraction pattern is therefore called *reciprocal space*. Naturally, the diffraction pattern will depend on the orientation of the crystal planes relative to the incident electron beam. Many electrons are diffracted by the crystal for certain orientations along the so-called *zone axes*, which correspond to lines parallel to the intersection of two or more families of lattice planes. A zone axis is denoted by indices ($u\ v\ w$), while the lattice planes are denoted by *Miller indices* (hkl). Additional details can be found in any book on crystallography or solid-state physics.

The diffraction patterns constitute the coherent mechanism known as *Bragg contrast* in electron microscopes. But in general both coherent and incoherent mechanisms will be present, making the interpretation of images quite an elaborate exercise. Figure 13.7 shows an example of a diffraction pattern by a single crystal. When many crystals are present, as in *polycrystalline* samples, the diffraction patterns superimpose and give rise to ring patterns. The interpretation of images is simplified enormously by the fact that only a small number of crystalline lattice geometries exist in nature, and by the possibility of simulating diffraction in computer models.

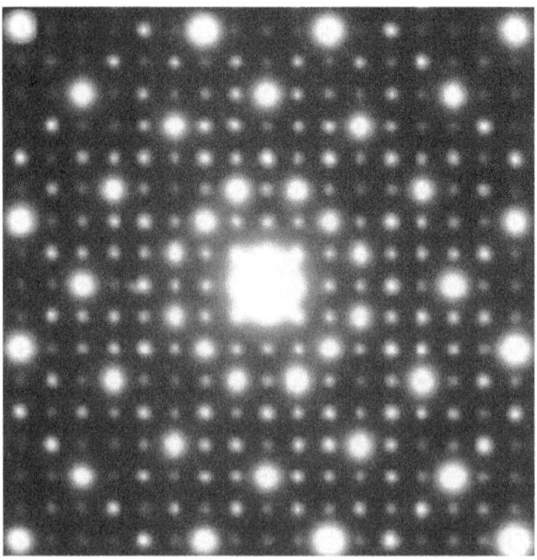

FIGURE 13.7 Example of electron diffraction pattern of a single crystal. Source: Frank Krumeich, Laboratory of Inorganic Chemistry, ETH Zurich, 2012.

The TEM plays a major role in crystallography studies or, more generally, in materials science research. TEMs cover a broad range of spatial scales from the atomic through the nano regimes and into the micrometer realm, that is from less than 1 to 100 nm and larger. The "mesoscopic" regime, which covers the many-atom structures of present-day electronics and nanotechnology, is of particular interest because of the potential for new applications and developments in mesoscopic physics.

13.5 SCANNING PROBE MICROSCOPES

A major phenomenon in the interaction of electrons and matter, not illustrated in Figure 13.5, is *quantum tunneling*. If two metal plates are connected to a battery and separated by a noticeable gap, no current will flow between them; but if the gap is made of the order of 1 nm or less, a few volts between the plates will cause a current to flow. Classically, DC current would only flow if the plates touch each other, but the wave nature of electrons makes it possible for them to traverse the gap if this is comparable to the electron wavelength. The phenomenon is the basis for the *scanning tunneling microscope* (STM), invented at the IBM Research Laboratory in Zurich by Gerd Binnig and Heinrich Rohrer in 1981. Both scientists and Ernst Ruska (see earlier text) shared the Nobel Prize for Physics in 1986.

In the STM, the probe consists of a small metal tip, usually made of tungsten, etched to a few atoms and a single one at the very end. Typically, the tip is moved over the specimen surface so a constant tunneling current flows from the tip to the electrically grounded specimen. The motion of the tip is controlled by *piezoelectric* devices that allow a raster scan and also maintain a constant vertical distance to the specimen through a feedback control mechanism. In this way, by recording the height of the probe a topographic map of the surface can be generated. Since the tunneling resistance depends on the type of atom, the STM also allows the identification of individual atoms. Although no high-energy electron beam is involved in STMs, very strong electric fields exist between the tip and the specimen: assuming a distance of 1 nm between the tip and the specimen, and a voltage of 1 V between them, the resulting electric field is 10^7 V/cm which can be potentially damaging for some materials. For this reason and because biological specimens have generally poor conductivity, STMs are primarily used for materials science studies. In these investigations, lateral resolutions of better than 0.2 nm and normal resolution of 0.01 nm can be achieved. Another drawback of STMs for 3D imaging of complex macromolecules like proteins is the limited depth (2–3 atoms) that the probe can reach.

Another major example of a scanning probe instrument is the *atomic-force microscope* or AFM. The tip in the AFM is normally a tiny diamond piece attached to a small cantilevered spring; the force between the tip and the surface is recorded as the specimen is scanned. The tip is at all times in contact with the specimen surface, similar to the way a stylus operates in a record player. Microcantilevers have spring constants of the order of 1 N/m, which implies that a deflection of 1 nm corresponds to a force of only 10^{-9} N. The sensor of the spring deflection can utilize

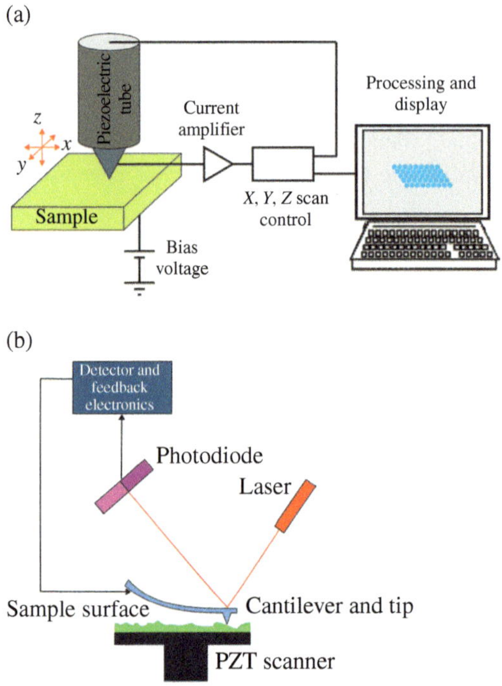

FIGURE 13.8 Operating principles of (a) scanning tunneling microscopy, and (b) atomic force microscopy. Source: (b) OverlordQ, https://commons.wikimedia.org/wiki/File:Atomic_force_microscope_block_diagram.svg. CC public domain.

FIGURE 13.9 STM image of a single-walled chiral carbon nanotube. Source: Taner Yildirim, National Institute of Standards and Technology—NIST, http://commons.wikimedia.org/wiki/File:Chiraltube.png. CC public domain.

electrical or optical signals that coupled with a feedback mechanism yield a constant deflection, similar to the constant tunneling-current operation of STMs. Thus, the AFM produces topographic images not unlike those from STMs but with reduced lateral resolution. However, AFMs can operate with non-conducting surfaces and are fairly insensitive to vibrations; this last feature follows from the high resonance frequencies typical of the microcantilever springs. Figure 13.8 illustrates the principles of both STM and AFM. Figure 13.9 shows the STM image of a carbon nanotube.

14

PHOTOELECTRIC IMAGE SENSORS

In this chapter, we consider image sensors for precision scientific and engineering applications. Many commercial and consumer grade detectors have a relatively large nonlinear response to light at the faint and bright intensity levels, have comparatively high levels of noise, and are only accurate to approximately 3% equivalent to a signal-to-noise ratio (S/N) of approximately 33. In contrast, a precision scientific or engineering photosensor routinely attains S/N = 200 and is usually capable of achieving photometric accuracies of 0.1% (S/N = 1000), if calibrated properly. *Note*: image sensors are needed for numerous scientific devices (e.g., spectrographs, scanners, and monitors for thin-film depositions) in addition to direct imaging. An image detector is normally a small fraction (2–10%) of the total cost of any apparatus requiring exactitude. It is normally, however, the component having the most significant impact on the data quality as well as being the most complex and problematic element. Moreover, state-of-the-art highly accurate detectors remain underdeveloped in bands outside of visible, never having been completely optimized in terms of sensitivity, speed, and a variety of other parameters. Despite its importance, the image sensor is frequently overlooked in the development of innovative new instrumentation since funding constraints often limit research to the advancement of a novel optical system instead. Serious detector development requires hardware specialists and substantial funds over sustained periods of time due to the sensor's inherent complexity. Detector technologies are sufficiently complex that a comprehensive treatment of these is well beyond the scope of this book.

Electronic image sensors can be broken down into two primary classes: photoemissive and photoconductive as depicted in Figure 14.1. Generally speaking, photoemissive image sensors are especially well suited for use in photon-starved situations or to capture rapidly varying signals, while photoconductive sensors can absorb large fluxes and are

Modern Devices: The Simple Physics of Sophisticated Technology, First Edition.
Charles L. Joseph and Santiago Bernal.

Companion website: www.wiley.com/go/joseph/moderndevices

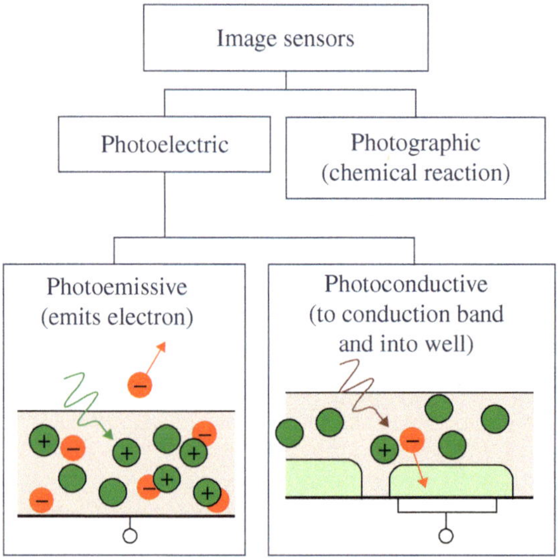

FIGURE 14.1 A diagram of the basic classifications of image detectors (sensors).

sensitive over a broad range of wavelengths. Photoconductive sensors operate by absorbing a photon in the bulk solid-state materials, which promotes electrons into the conduction band. These unbound electrons are then electrostatically moved and held in pixel wells, accumulating (integrating) charge proportional to the photon flux impinging at each pixel location. (A pixel is defined as a picture sampling element in an image sensor.) These devices usually require a mechanical shutter to determine accurate exposure times to determine flux rates. All photoconductive sensors produce various amounts of unwanted thermal backgrounds (known as dark currents), superfluous nonthermal background signals arising from materials defects, and introduce additional readout noise. The distance an electron must travel prior to being trapped in a well can introduce losses in signal as some electrons recombine with the atoms in the bulk material. Silicon-based detectors rarely experience this type of loss, but it remains an important factor for sensors made of wide-bandgap III-nitrate materials such as GaN, AlGaN, or AlN.

Photoemissive devices, which eject an electron into evacuated space, normally require some form of electron multiplication. These detectors start with a material known as a photocathode that creates the initial free-space electrons. Photocathodes usually have sensitivities that are strongly wavelength dependent. One primary amplification method electrostatically accelerates each photoelectron to strike another surface, creating additional electrons that are subsequently accelerated toward—and strike—additional surfaces, each time multiplicatively increasing the number of electrons. The intensification stage invariably requires a high voltage (2–20 kV) and a highly clean evacuated sealed tube, both requirements being inherently bulky and a survival risk. The component responsible for electron multiplication is called an image intensifier, which can be placed in front of any sensitive material (e.g., photographic film, phosphor screen, or for science-grade sensors, an electronic readout is preferred). Photoemissive devices normally respond to an impinging flux, one photon at a time known as events or photoevents. Global dynamic range (GDR), a measure of the maximum counts per second over the entire field of view, and local dynamic range (LDR), the counts/pixel, are often significant device limitations, especially for bright scenes. Nevertheless, these sensors usually have small dark event rates, corresponding to relatively smaller dark noise compared to photoemissive devices, and can be constructed to be zero-read-noise sensors. All of these qualities are especially valuable to capture high-speed transient fluxes or for photon-starved observations.

Historically, photoemissive detectors have been the premier device for use at blue, violet, and ultraviolet (UV) wavelengths (90–500 nm), while silicon-based solid-state sensors have been the preferred device over visible and near-infrared bands (500–950 nm). The wavelength division stems from the fact that the energy needed to eject an electron from most materials is a few electron volts, requiring a photon that is at least as energetic, while the energy to promote an electron to the conduction band in silicon is approximately 0.1 eV. (See Chapter 9 for energy comparisons corresponding to a photon's wavelength.)

Different applications favor one class of detector over another class with no single type of sensor being ideal for all applications. For example, if the object being imaged emits significant amounts of light over out-of-band wavelengths, then bandpass (BP) filters (i.e., wavelength rejection blocking filters) must be placed in front of photoconductive detectors to mitigate the large unwanted signal and its associated noise. Good out-of-band filters often reduce the desired in-band light transmission by about half, potentially adversely impacting the S/N ratio. Out-of-band filters might not be required for a photoemissive sensor, if its sensitivity band is well matched to the scientific needs. Conversely, bright sources often overwhelm photoemissive sensors that have to deal with the incident flux one photon at a time. Neutral density (ND) filters might then be needed to reduce the flux to accommodate the dynamic range of those detectors. ND filters decrease the transmission (increase the opacity) over a very wide range of wavelengths more or less uniformly. Regardless of sensor class, most modern detectors can be made to work well over a much broader range of wavelengths than had been used traditionally.

INTERESTING TIDBIT TB14.1

Developers of state-of-the-art image sensors have a need to promote their novel devices against strong competitors. Frequently, they adjust operating parameters such as the voltage biases to optimize one detector performance capability and then use a different set of voltages to enhance another detector function. Routinely, the list of maximum performance specifications is a compilation of multiple operating conditions. One important question for a scientist to ask is, "can the sensor obtain all of these limits simultaneously?" The answer is inevitably no.

14.1 SOLID-STATE VISIBLE WAVELENGTH SENSORS

The charge-coupled device (CCD) has been the premier scientific image sensor over visible wavelengths since the 1970s with alternative sensors (e.g., CMOS) making inroads only in niche applications during the twenty-first century. *Note*: CMOS sensors have usurped CCDs in several, non-scientific-grade consumer applications such as smartphone cameras. A cross-sectional depiction of a CCD chip is provided in Figure 14.2, sliced along a single row. Let us examine the figure for classification purposes. Photons impinge from the top and are absorbed into the bulk material, making the CCD a photoconductive device. Each photon absorbed promotes an electron to the conduction band, allowing it to drift towards the nearest well where it is held. (See Chapter 6 for conduction band physics.) Electrons accumulate in these wells proportional to the incident photon flux. The chip of Figure 14.2 is front illuminated, meaning the bias electrodes used to form the charge wells obscure a portion of each pixel, reducing its overall sensitivity. A top view of the corner of a front-illuminated CCD is shown in Figure 14.3, looking down on the image plane. The active area is depicted in green/yellow green with individual pixel boundaries denoted by dashed lines. This photo-sensitive area has electrical connections out to the edge of the chip to pins that connect to external electronics. The electrical connection for every other row is brought out to the opposite side of the chip. Normally, the nonimaging portion of the chip is masked to prevent stray light from striking it and creating unwanted extraneous signal.

After an image has been recorded, the charges in the rows of wells are shifted down the columns (out of the page as seen in Fig. 14.2 or down as viewed in Fig. 14.3) and subsequently readout. (*Note*: the horizontal electrodes in Fig. 14.3 are a simplification used to conceptualize rows of pixels. A more complicated structure is necessary to shift the contents of the wells to the readout amplifier.) Charge barriers inhibit these electrons from drifting into an adjacent column during the shifts to the readout section. Upon reaching the bottom, the charge packets encounter the readout register of Figure 14.4, which is also masked against stray light, and are shifted horizontally toward a readout

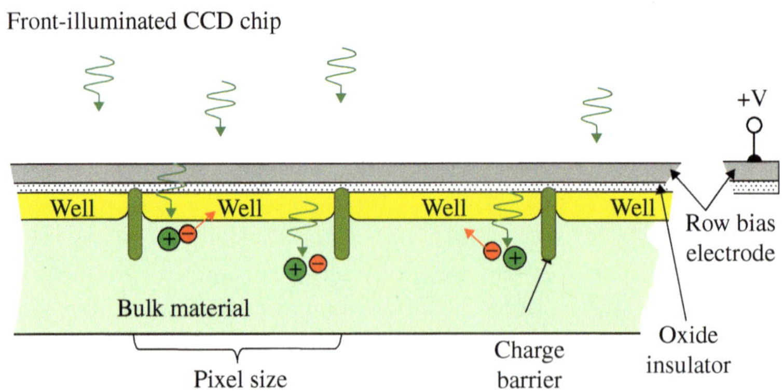

FIGURE 14.2 A cross sectional view of a front-illuminated CCD chip, cut along a single row of the active area pixels.

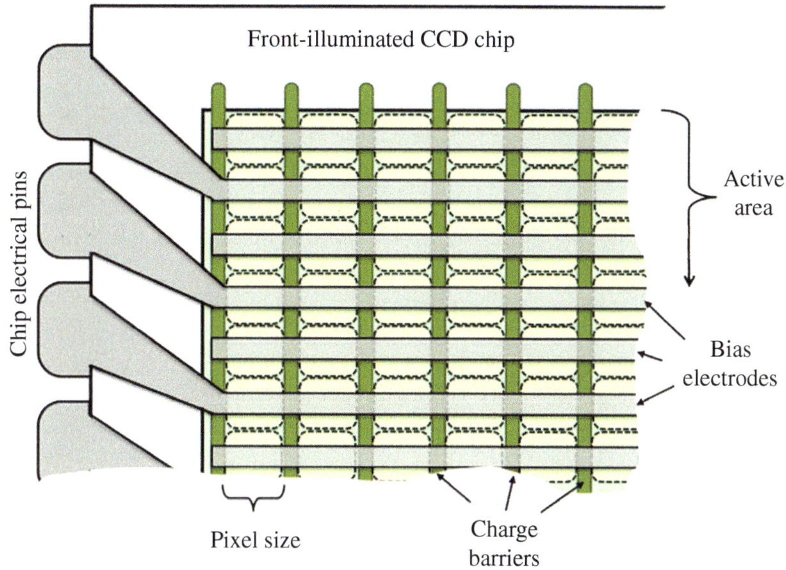

FIGURE 14.3 Corresponding top view of one corner of the chip depicted in Figure 14.2.

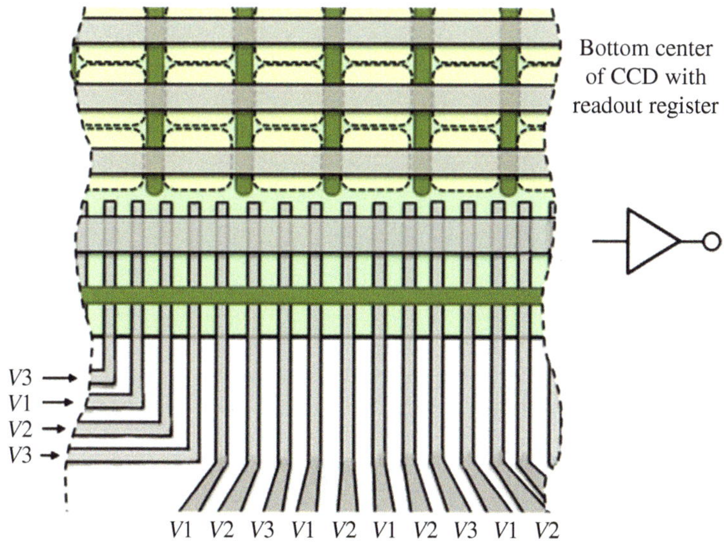

FIGURE 14.4 Top- down view of a CCD chip near a portion of the shift register.

amplifier. A three phase is a common shift register, requiring three bias voltages: $V1$, $V2$, and $V3$ per pixel to move the charge packets to the output amplifier. We return to a cross-sectional diagram in Figure 14.5 to illuminate the operation of this type of shift register. The plots at the bottom show the sequence of bias voltages that are

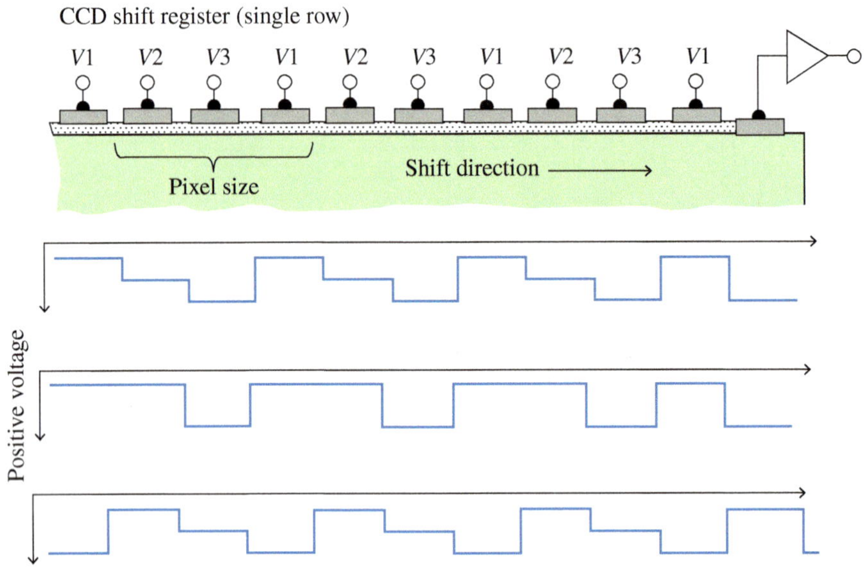

FIGURE 14.5 Cross-sectional view of Figure 14.4, showing the three phases of the shift register required to move a charge package from one pixel to the adjacent one.

applied sequentially to *V*1, *V*2, and *V*3 to move the charge horizontally. The first phase creates a "staircase" of voltages, resulting in most, but not all, of the charge in each pixel ending up in right-most third of its corresponding register element. Next, the center *V*2 potential is lowered (made a less positive voltage), isolating all of the charge packages in the respective *V*3 positions. The third phase once again creates a staircase with the bulk of each charge package now ending up in the left-most third of the adjacent pixel. (The register keeps shifting charge packages from the left-most pixels, but essentially the values being shifted are all zeros.) Once an entire row has been readout, all rows are shifted one pixel down vertically with the bottom row being moved into the shift register for horizontal readout.

There are some significant limitations for all thick, front-illuminated CCDs compared to thinned, back-illuminated CCDs. These include peak sensitivities of less than 50%, much higher thermal backgrounds generated in the bulk silicon material, and virtually no blue, violet, or UV sensitivities. If the CCD chip is simply flipped over to get the bias electrodes and oxide layer out of the way of the incoming photon flux, a large portion of the flux is still absorbed near the backside surface, away from the wells, resulting in poor pixel definition. In other words, a back-illuminated *thick* chip has significant amount of internal image smear. To improve performance, most of the bulk silicon material is removed (thinned), then the backside-illuminated CCD can be rendered scientific grade. The thinning procedure, however, is a very delicate and expensive process, typically resulting in low (~0.08) yields. Nevertheless, chips with peak quantum efficiencies (QEs) of 85–90% and with half of the thermal background as thick CCDs are routinely fabricated. Thinned CCDs also require some

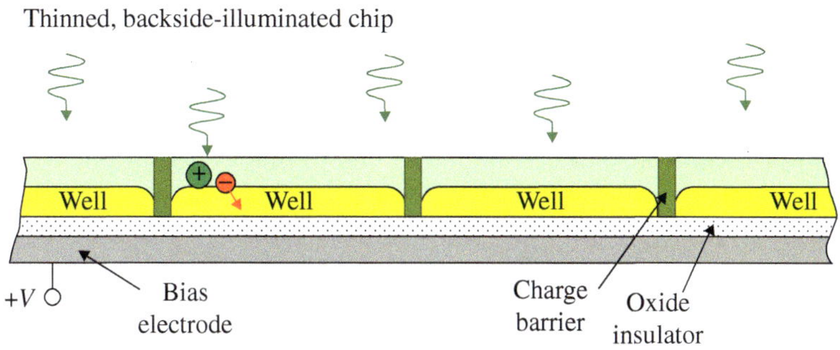

FIGURE 14.6 The best architecture for a scientific-grade CCD image sensor.

form of pacification such as a thin-film coating to prevent the formation of native silicon oxide on the surface, introducing a time variable reduction in the chip's the blue sensitivity. A cross section of a thinned chip is depicted in Figure 14.6.

Two important considerations for scientific-grade CCDs are the charge transfer efficiency (CTE) and the impact of blooming (the spilling of charge to adjacent pixels) for very bright portions of the scene that exceed the charge holding capacity of individual pixel wells. The effects of blooming can be mitigated to some degree by the inclusion of charge drains into the chip architecture. Combining two images, one short and one long exposure, is often the best strategy for scenes with an exceptionally large range between bright and faint portions of the image. The brightest levels are properly captured in the short exposure, while the faint parts in the long one. CTE, a measure of the percentage of electrical charge that can be reliably transferred to the adjacent-row pixel, limits the practical array size of a CCD. Typically, a scientific-grade CCD has a CTE of 99.999%, resulting in one one-hundredth of the charge being left behind during a 1000-row shift to the readout register. The equivalent image smear is about 150 electrons for pixel wells that are half full, compared to the typical readout noise of three electrons. (*Note*: <1 electron read noise is possible, but it requires exceptionally long readout times and high-quality electronics.)

The need for exceptionally good charge transfer efficiencies is avoided for an image sensor using a CMOS architecture where an amplifier is attached to each pixel. These detectors, which also benefit from thinning and backside illumination, have a number of advantages over CCDs such as the ability to resolve temporally any rapidly variable fluxes or to remain idle until event triggered. However, CCDs continue as of 2015 to have read noise levels that are 1–2 orders of magnitude less than CMOS sensors.

All silicon-based detectors create large thermal backgrounds if operated at room temperature. Thermally induced internal vibrations of the atoms within the crystalline silicon at room temperature have sufficient energy (kT) to promote some of the electrons to the conduction band as illustrated in Figure 9.1. This is not much of a problem for short exposure times such as those encountered in video frame rates. However, most science-grade sensors must be cooled using either a set of thermoelectric coolers or perhaps using a cold-finger attached to a cryogen coolant such as liquid nitrogen. (Astronomical imaging or spectrographic applications, which frequently have 20–50 minutes exposures, typically cool to chip temperatures between −40 and −80°C. Long-exposure images also suffer from cosmic ray hits that produce bright streaks in the image.) Despite these issues solid-state detectors based on silicon (e.g., CCD and CMOS) continue to be the sensor of choice for most precision measurements over the visible portion of the spectrum.

14.2 PHOTOEMISSIVE DEVICES FOR UV AND X-RAYS

A few important applications such as photolithography are increasingly operating at shorter and shorter wavelengths well into the vacuum UV (VUV, $90\,\text{nm} < \lambda < 180\,\text{nm}$). The move to short wavelengths increases spatial resolution, enabling very high-density electronic circuitry. Moreover, the largest number of atomic transitions, and hence the largest number of diagnostic spectral features occur at VUV wavelengths. Despite the enormous success of solid-state silicon detectors used at visible wavelengths, these devices have far fewer advantages and a few disadvantages compared to alternatives when operated at ultraviolet and X-ray wavelengths. *Note*: solid-state image sensors made of III-nitrate materials (e.g., GaN and AlGaN) might come to dominate UV and X-ray applications in the future the way silicon devices do today in the visible. The primary impediment is the quality of the crystalline materials, which need to be improved sufficiently to reduce the defect rates by several orders of magnitude.

Image sensors generally have inherent performance minimums in the near-UV, (NUV, $180\,\text{nm} < \lambda < 350\,\text{nm}$) portion of the electromagnetic spectrum. Sensitivities in particular tend to be the lowest in the NUV band due to technology limitations. For example, a CCD must either switch the thin-film passivation layer or use a technique such as ion implantation to extend its sensitivity into the NUV and VUV. Producing UV sensitivity in silicon-based devices often comes at the expense of reduced sensitivity over a portion of the visible light band. Once the incident radiation becomes hard enough (i.e., adequately short wavelength), the photons become once again penetrating with enough energy to promote two electrons to the conduction band and high QE returns. In fact, the number of electrons generated in most solid-state detectors at X-ray wavelengths is proportional to the energy of the incident photon, enabling each photon to be binned according to its energy and consequently providing spectrally resolved data within each pixel. Other issues arise with silicon detectors, however, such as radiation hardness, a measure of the number of hard photons that can be detected prior to the sensor experiencing serious degradation.

Historically, the dominant sensors in the ultraviolet and X-ray bands have been photoemissive and these continue to be essential devices in these EM bands. Photoemissive detectors begin with a photocathode, which is a material that ejects electrons in response to incident photons. The free photoelectrons typically leave the surface with kinetic energies of a few electronvolts and need some form of multiplication to trigger the detection electronics. Most metals (e.g., Fe, Ni, Cu, etc.) have QE's of approximately 2%, while some photocathodes achieve peak QE of approximately 80%. Photocathodes with maximum quantum efficiencies of 20–40%, however, are far more common with sensitivity over a relatively limited range of wavelengths. (Some detectors incorporate two distinctively different photocathodes inside the same vacuum tube to increase the range of good QE. Multiple photocathodes though can be problematic since deposition temperatures usually differ significantly requiring the tube to be processed more than once and there is a strong possibility for chemical cross contamination reducing the overall QE of each.) Photocathodes also must be deposited on a conducting surface that is connected to a low-current voltage source to replace the photoelectrons once these are ejected from the cathode. Otherwise, the photocathode would quickly become positively charged, preventing would-be photoelectrons from escaping. In many cases, the UV detector designs use a window with a thin-film conductor and a semitransparent photocathode deposited on the inside surface of the window. Semitransparent photocathodes usually have much less quantum efficiency than opaque cathodes, but other factors might prevent the use of the latter. For instance, semitransparent cathodes are sometimes required to avoid unwanted chemical interactions that might occur if deposited on the preferred surface. In addition, a photocathode on an entrance window simplifies the detector tube design, enabling proximity focusing of the photoelectrons onto the stage of the detector providing electron amplification.

As noted, photocathode-based (photoemissive) detectors must have a mechanism to increase multiplicatively the number of the electrons. This gain stage always involves high (2 kV or more) voltages, an inherent reliability risk. One widely used intensifier is the microchannel plate (MCP), shown in Figure 14.7. An MCP is a glass plate with holes (more commonly known as pores or channels) running from top face to the bottom at an angle. Some photocathodes can be deposited directly on the top surface of the MCP with some of the material being deposited into top portions of the channels. In this case, an electrostatic field is required to force the photoelectrons from the inter-channel areas back down an adjacent hole. Pore diameters run from 6 to 12 μm on 8–15 μm centers. The glass is normally doped with lead (Pb) to make it conducting but still remains highly ohmic, typically having a resistance, $R = 10$–$20\,\text{M}\Omega$. The top and bottom of the MCP are conducting surfaces to which a 1–3 kV potential is applied. Each time an electron strikes the side of a pore, it knocks off 2.2 electrons on average, which in turn, are accelerated further down the channel increasing multiplicatively the number of free electrons. The charge cloud emerging from bottom side has between 10^4 and 10^5 electrons, depending on the bias voltage and other MCP parameters such as pore size, pore length, and overall resistance of the MCP.

All photoemissive devices can be photon counting, but only some are photon noise limited. If the intensifier stage can be operated in such a manner that the number of electrons becomes self-limiting, then each emergent charge cloud has a fixed number of electrons plus or minus a small amount. The intensifier is said to be in saturation and the detector is operating in "Geiger" mode. This type of photon counter is cable of providing the theoretically highest signal-to-noise ratio (S/N) data, being limited only by the Poisson arrival statistics of the total signal collected. (*Note*: S/N = 100 requires a $S = 10^4$ counts/pixel; S/N = 1,000 requires a $S = 10^6$ counts/pixel; and S/N = 10,000 requires a $S = 10^8$ counts/pixel!) The dynamic range of the detector and the total amount of exposure time that is practical determines the S/N. Alternatively, if

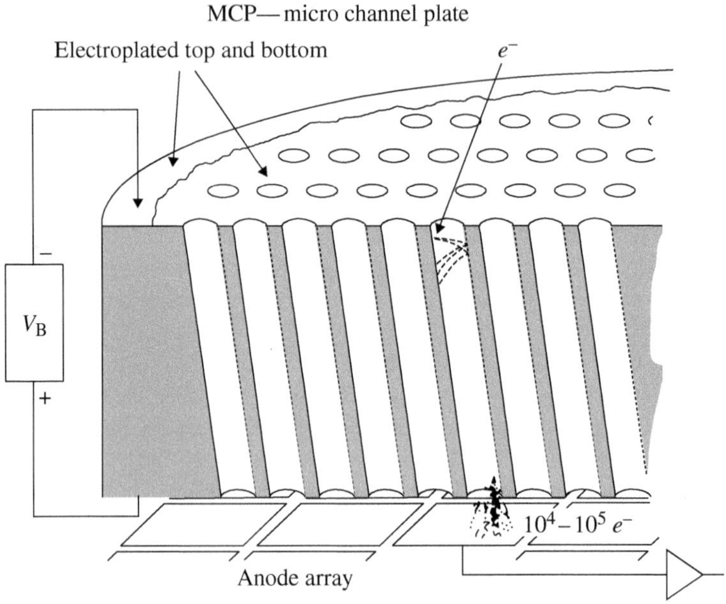

FIGURE 14.7 A schematic diagram of a microchannel plate intensifier.

the multiplication stage produces a continuous range in the number of emergent electrons, the detector is a proportional counter. In this case, any photon counting process relies to a large extent on the stability of the intensifier and the associated electronics, which even in the best circumstances have some minor drifts with temperature, relative humidity, and other environmental factors. Nevertheless, photon-counting data with an S/N of few hundred are still readily possible.

There are a number of electrode schemes to collect the charge output from an MCP. Pictured in Figure 14.7 is an array of discrete anodes, each attached to a charge amplifier. (Charge amplifiers, high-speed circuitries operating at tens of nanosecond timescales, are used to increase the electronic signals to levels needed for digital logic.) A discrete readout scheme is limited to array sizes of about 200 elements (say 20×10 pixels), the practical number of charge amplifiers that can be assembled in close proximity without significant cross talk. One readout method, the coincidence array, enables $m \times n$ pixel locations to be registered with $m+n$ charge amplifiers. The position of each emergent charge cloud is determined from two or more anodes on each axis that are triggered simultaneously by the event. (If invalid groups of anodes are activated simultaneously such is the case, for example, when two photoevents are coincident in different parts of the array, then the detector logic circuitry rejects these events.) Another readout scheme is a delay line where the emergent charge strikes a continuous anode and starts propagating in two directions toward amplifiers on opposite sides of the active area. The event location is determined from the relative arrival times, nanoseconds apart. Other readouts simply register the relative fraction of charge arriving at a few amplifiers to measure event location. Generally speaking, detectors with coincidence arrays and time-delay readout schemes dramatically out perform less sophisticated schemes such as charge sharing anode structures. MCP detectors are time-domain devices, handling each event sequentially and updating the sensor's memory. MCP-based detectors are often able to time-tag all photoevents, creating a data stream of X and Y locations along with the arrival times.

Photoemissive detectors *not* using an MCP as a gain stage include among others the electron bombarded CCD or CMOS (EB-CCD or EB-CMOS, respectively). These image sensors either electrostatically or electromagnetically focus the photoelectrons directly onto a back-illuminated chip. The EM focus in these types of photon-counting detectors also provides a several-kilovolt acceleration for each photoelectron, slamming it into the chip and creating a $3000\,e^{-}$ spot. The silicon chip is readout in frame transfer mode, identical to the method used for television. Each frame is electronically scanned for localized peak signals and the detector memory is incremented for each event found. Normally, two discriminators are used to eliminate erroneous spots with too much charge or not enough charge as invalid photoevents. In contrast with an MCP-based sensor, an electron-bombarded design is not strictly a time-domain detector. The photons, however, can be localized to an individual (several-millisecond) frame. One important advantage of an EB-CCD or EB-CMOS is its exceptionally high quantum efficiency (peak QE ~85%) for designs of an opaque photocathode on a smooth surface. Originally, this type of detector was bulky and heavy due to the use of conventional magnets. However, the volume and mass of the EB-CCD or EB-CMOS have become competitive with other high-performance detectors with the introduction of novel magnet assemblies beginning in the 1990s.

One important distinction for all detectors is the difference between QE and detective quantum efficiency (DQE). QE is usually stated as the efficiency of the photocathode, while DQE accounts for all internal losses within the sensor. Photon-counting sensors have conversion efficiencies for each stage of the detection process. For example, the photoelectrons inside an MCP-based sensor might trigger an avalanche of electrons 85% of the time and the conversion at the back end of the MCP might be 90%. Thus, a peak QE of 30% becomes a DQE of 23%, after a total conversion factor of 0.765 (0.85×0.9). For an EB-CCD device, roughly 11% of the photoelectrons simply scatter off the atoms in silicon lattice and are never detected. Thus, a peak QE of 84% with a conversion efficiency of 0.89 results in a peak DQE of 74.8%.

14.3 INFRARED "THERMAL" SENSORS AND NIGHT VISION SENSORS

Infrared (IR) radiation comprises photons having wavelengths longer than those visible to the human eye. While the IR is usually taken to begin at 700–750 nm, most people can actually see light out to 950 nm or in some cases to 1050 nm. All objects having a temperature above absolute zero (−273.15°C) produce blackbody radiation identical in nature to that described in Chapter 9. For objects at room temperature (T=25°C, 298 K) or for human bare skin temperature of 33°C (306 K), most of the emitted light occurs at infrared wavelengths, centered around $\lambda=9.5\,\mu\text{m}$ (9500 nm).

Thermal imaging sensors are generally sensitive over wavelengths of 7–14 μm (7,000–14,000 nm). Near infrared (750 nm $< \lambda <$ 2000 nm) light is generally not considered thermal radiation. Longward of approximately 2 μ (2000 nm), numerous objects emit copious amounts of IR photons, including the Earth's atmosphere, any telescope optics, and most terrestrial objects. Scientific measurements normally require measuring the small signals of interest against enormous backgrounds from extraneous sources. These thermal backgrounds, which might constitute more than half of the total flux, must be measured independently and subtracted from the total signal. IR ground-based telescopes tend to have a secondary mirror that repeatedly switches between the star or object of interest and a nearby field devoid of stars. Alternatively, some space-borne IR missions cool most of the instrumentation including the optics and detectors either using an onboard cryogen or passively radiating the heat to the 3 K temperature of deep space. Many IR spacecraft also have to use Sun screens (deployable shades) to avoid heating from the Sun, Earth, and Moon.

One common IR image sensor is a solid-state device made of mercury cadmium telluride (HgCdTe), used extensively among other applications in military remote sensing and in infrared astronomy. Most HgCdTe sensors consist of arrays of photodiodes attached to a readout integrated circuit (ROIC). In contrast with most silicon-based detectors, photodiode arrays produce a current in each pixel that must be measured in real time. Array sizes as large as 256 × 256 pixels and 1k × 1k have been used at astronomical telescopes and on NASA IR missions. Large 2k × 2k HgCdTe arrays have been under development since of 2010. The bandgap of a pure CdTe (without any Hg) material is 1.5 eV, while pure HgTe is a semimetal with essentially a 0 eV. The central wavelength of the detector's sensitivity can be set by the relative concentrations of Cd and Hg (i.e., $Hg_{1-X}Cd_XTe$ where $0 \le X \le 1$). Other competing detectors include superconducting tunnel junction (STJ) arrays and quantum-well infrared photodetectors (QWIPs) made of GaAs and AlGaAs semiconductor materials. All of these IR image sensors must be operated at cryogenic temperatures, typically at liquid nitrogen temperatures of 77 K, to reduce internal thermal backgrounds. (Even colder temperatures are required for very long wavelength IR applications.)

Finally, we conclude this section with a discussion of a common night vision binocular, which demonstrates the use of several technologies that have been discussed in this book. These are *not* science grade instruments since large amounts of noise (extraneous blips and general appearance of speckles) are present in the image. Noise is not problematic, however, because the human brain is excellent at isolating the information of interest and filtering out extraneous input. One half of a night vision binocular is schematically shown in Figure 14.8, corresponding to the optical chain for one eye. External light enters from the left, pass through a lens, and is focused onto a transparent surface with an IR photocathode. The image is very faint and upside down at this stage. The photocathode effectively converts the photon image into "an image of electrons," which is electrostatically accelerated to an MCP. Each electron entering the left side of the MCP produces a few hundred exiting it on the right at essentially the same spatial location. The MCP bias voltage is much less than that used in science a detector, resulting in such a low gain. These charge clouds are proximity focused onto a phosphor screen, converting the "electron image" back to a

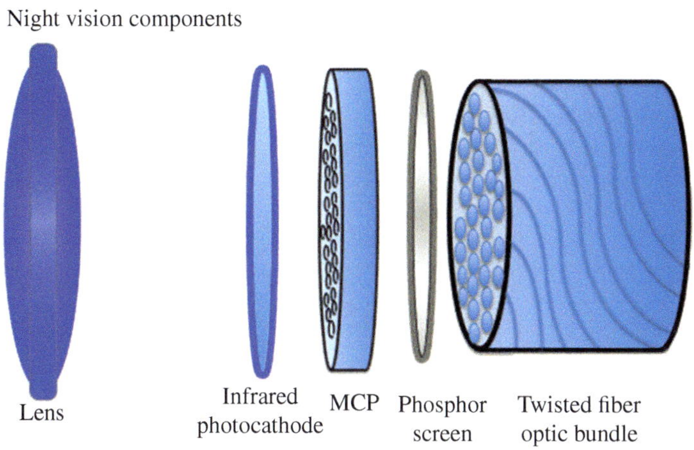

FIGURE 14.8 A schematic representation of one side of a common night vision binocular.

photon image with some additional gain since electrons striking a phosphor surface generate several photons/electron. The image, which is still upside down, is rotated 180° by a bundle of optical fibers. The bundle has been twisted and both faces of the bundle polished so that the intensified image in its correct orientation appears on the right surface of the bundle.

INTERESTING TIDBIT TB14.2

The human body emits about 100W, primarily in the infrared. Kilogram for kilogram (i.e., pound for pound), a human generates more heat than does the Sun. While the Sun is much hotter, only the inner core generates energy, making the energy/mass of the Sun less than that of a human. Recall this fact when you are stuck in a packed and stuffy auditorium.

15

IMAGE DISPLAY SYSTEMS

The first devices for projecting images relied on glass plates with emulsions and crude light sources; later, photography and incandescent light bulbs permitted a vast improvement in quality. But it was the advent of cinema and later of television followed by digital computers that revolutionized the entertainment and the technology of image displays. The rudiments of television (TV) were first demonstrated in mechanical devices like the Nipkow disk in 1926 by Baird. The invention of the cathode ray tube (CRT) by Brown in 1897 provided the primary component of the first electronic TV, and the first fully electronic color picture tube was introduced by Baird in 1944. However, the original ideas behind modern electronic TV are due to Farnsworth, as recounted later. TV technology relied for many years on *analog signals* and amplitude modulated (AM) transmission; only recently (1990s) has digital technology and associated signal processing for recording, transmission, and display largely replaced analog technology.

The history of the very first color projection, also considered as the first color photograph, is fairly obscure but worth telling. See Figure 15.1. James Clerk Maxwell was almost 30 years old and working at King's College, London. He was not yet the renowned scientist he later became for his work unifying electricity, magnetism, and optics, a contribution of such an immeasurable impact as the contributions of Newton and Einstein. Preparing for an invited lecture on color vision, Maxwell had the idea of taking three black-and-white (B&W) photographs of a tartan ribbon separately through red, green, and blue filters and then projecting them simultaneously using the same filters. Knowing that the photographic film of the time was insensitive to red light, Maxwell tried anyway, with help from his colleague Thomas Sutton. In this way, the audience at the lecture organized by the Royal Institution saw the world's first color projection in May 1861. But no one could take another color photograph for many years. It would take some 100 years before the mystery was solved. A team

Modern Devices: The Simple Physics of Sophisticated Technology, First Edition.
Charles L. Joseph and Santiago Bernal.

Companion website: www.wiley.com/go/joseph/moderndevices

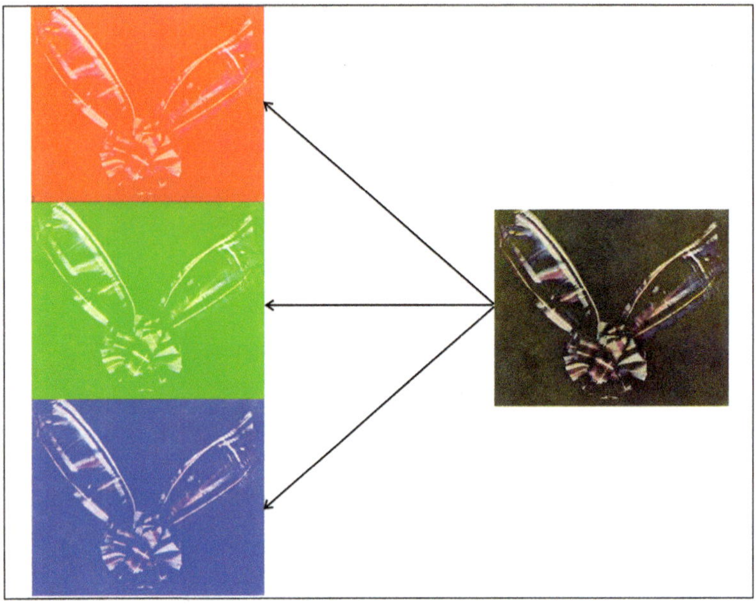

FIGURE 15.1 The first color photograph(right) was taken by James Clerk Maxwell in 1861. Its three RGB components are also shown.

at the Kodak Research Laboratories discovered that despite Maxwell having used photographic film insensitive to the red, the film was sensitive to ultraviolet (UV) wavelengths and, by an amazing series of coincidences, the red dye in the tartan ribbon happened to also reflect in the UV and within a pass-band region for UV present in the red filter's emulsion.

In this chapter, we review the main components of modern display systems. We begin with a description of the human visual system to understand better the parameters of a display system such as resolution (spatial and temporal), contrast, brightness, dynamic range, persistence, and others. We discuss the main ideas of colorimetry (i.e., the science of color measurement) and color vision. In Section 15.2, we present a short account of the invention of modern (electronic) TV, while we summarize the formats used in analog as well as high-definition digital TV (HDTV) in Section 15.3.

Display screens are based on spatial and temporal modulation of light. Both recording and displaying of images in many systems are based on *raster scanning*, that is, the image is scanned line by line; but an entire frame can be written at once in displays like liquid-crystal displays (LCDs). Another type of display is the *vector* display whereby images are drawn as contour lines. Radar, oscilloscopes, and light shows are examples of vector displays; we do not discuss those systems here. In the following sections, we describe *transmissive* displays such as liquid crystal, *emissive* displays such as CRTs and plasma displays, and *reflective* displays based on digital micromirror devices (DMDs). We also discuss the technologies behind the pervasive touch screens and electronic readers (electronic ink or paper) and conclude with sections on near-eye displays (NEDs) and 3D display technologies including holographic TV.

15.1 THE HUMAN VISUAL SYSTEM

For obvious reasons, the capabilities of the human visual system are coupled to the design of image display systems: our eyes and brains are the ultimate recipients of all image information generated and displayed by our inventions, although we can envision other recipients like computers and robots (machine vision). The human visual system itself contains all the elements of an image recording, display and storage system: optical components (lens, shutter, and iris), motion scan capabilities (via eye muscles), sensors (retina), and signal processing plus storage (retina and visual centers in the brain).

Figure 15.2 shows the schematics of the human eye. Refraction in the eye is caused by the cornea and eye lens; the liquids that fill the eye chambers and the main ocular globe, the *aqueous* and *vitreous* humour, also affect light rays. The shape of the eye lens (4–6 mm in diameter) can be modified to focus from infinity to about 10 cm, the *near point*. This capability, called *accommodation*, diminishes with age. Naturally, the image formed at the retina is inverted, but the brain centers compensate for that.

Another relevant property of the eye is the broad range of image intensities from faint to bright that it can detect, known as its dynamic range. The eye can sense light over some nine orders of magnitude in brightness, but not simultaneously over the same scene. The retina contains B&W intensity photoreceptor cells called *rods*, and color receptors called *cones*; the highest density of photoreceptors occurs at the *fovea* in the retina. Color-sensitive cones only extend over a small portion of the retina, indicating most of the field of view (FOV) is seen in B&W. The maximum sensitivity occurs near 500 nm wavelength (green light), but color perception is dependent on light intensity, being fairly degraded at dark conditions (more on color perception later).

The eye, similar to artificial light sensors such as photodiodes or CCDs, integrates temporal light variations. The eye is relatively slow, often perceiving flickering signals as steady. For example, a human will recognize image variability up to 10 Hz at 0.01 lux (e.g., quarter moon light), but at 100 lux (e.g., overcast daylight), 50–60 Hz are needed to avoid the sensation of flicker. The eye temporal response is also different for peripheral vision, which becomes more relevant at close distances from a display screen.

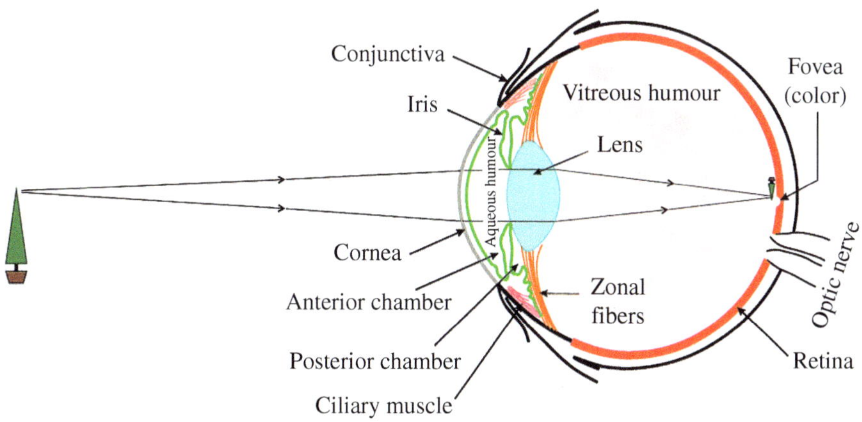

FIGURE 15.2 Schematics of the human eye.

The eye's response to brightness follows approximately a Weber–Fechner law, that is, a logarithmic relation similar to the response of other human senses like hearing:

$$S = k\log\left(\frac{R}{R_0}\right), \quad R > R_0. \tag{15.1}$$

Here, R is the magnitude of the physical stimulus intensity (brightness), S is the magnitude of the apparent brightness, that is, the subjective sensory response, R_0 is the threshold stimulus intensity, and k is a constant. This logarithmic eye response can be observed in everyday devices. For instance, the intensity of brake lights on a car must be at least 50 times brighter than the running tail lights to insure a factor of two change in the apparent brightness by other drivers. By differentiation of Equation 15.1, we can see that the logarithmic response implies that a detectable change in response is proportional to the fractional change in stimulus intensity:

$$dS = k\frac{dR}{R}. \tag{15.2}$$

Figure 15.3 illustrates the logarithmic sensitivity response of the eye. Note that a straight-line plot would result if the horizontal scale of Figure 15.3 is replaced by a logarithmic one, as often illustrated over a range of about 1–10^4 lux. Outside this range, the eye sensitivity response is relatively flat.

From Equation 15.2, we realize that contrast perception is better at higher intensities. The smallest detectable change in intensity by the eye is called *just noticeable*

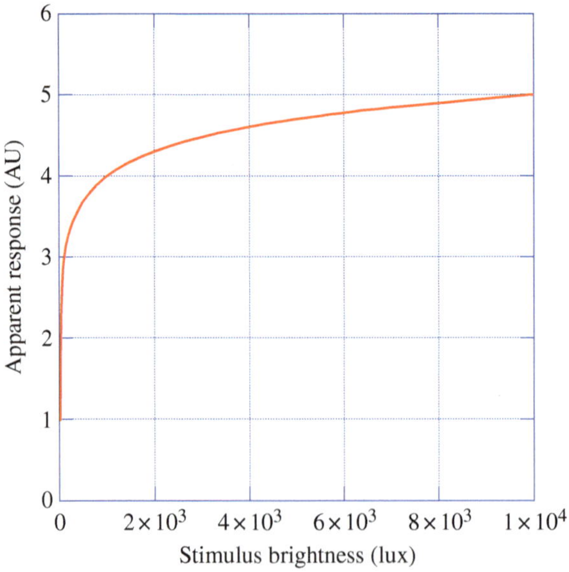

FIGURE 15.3 Approximate logarithmic response (Weber–Fechner law) of the human eye.

difference or JND. The level of JND within a single scene increases with brightness up to a certain intensity.

Another relevant parameter of the human visual system is the spatial resolution, which is determined by both physiological factors such as density of photoreceptors and physical factors such as diffraction and optical aberrations. The physics of optical systems has been discussed in Chapter 10. Using the Equation 10.12 for the angle of the first minimum in the diffraction pattern of a circular aperture, we can estimate the angular resolution of the human eye to be

$$\Delta\theta = 1.22\frac{\lambda}{a} \approx 1.22 \times \frac{560\text{nm}}{2\text{mm}} = 1.2 \text{ arcmin}, \tag{15.3}$$

where we have used $a = 2$ mm for the diameter of the iris aperture under standard illumination at an average wavelength of 560 nm. Vision physiology includes not only the densities of rods and cones in the retina but also the highly processive human brain. Physiological responses, which evolved to error on the side of seeing predatory animals at times when there is no danger present, also enable optical illusions.

While the eye spatial and temporal resolutions as well as its sensitivity are important, nothing compares to our ability to sense color. We are somewhat unique in that regard among mammals, but, strangely enough, share the possibility of color vision with other animals such as bees and other insects. As mentioned earlier, the cones in the retina are responsible for color perception in the human eye. In turn, the cones can be divided into cones sensitive to red, green, and blue, but with sensitivity curves that overlap as shown in Figure 15.4. The "mixing" of different proportions of light (not pigments!) of the *three primary colors* red, green, and blue yields most of the

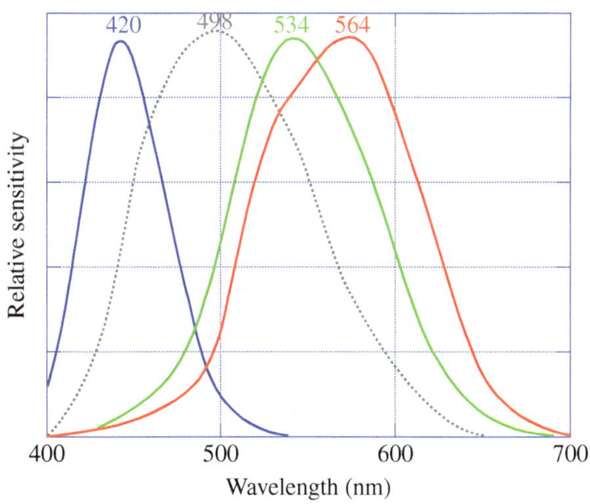

FIGURE 15.4 Relative black-and-white sensitivity curves of rods (shown in gray) and color-sensitive cones as a function of wavelength in a typical human eye.

colors that we perceive. This fact, which is far from obvious and was contended for years, was discovered by Thomas Young in a series of beautiful experiments around 1800; Young's work together with the contributions from Helmholtz, Maxwell, and others laid the foundations for the understanding of color vision. Just to see how controversial the three-color theory of vision was at some point, consider that *there are no* cones sensitive to yellow, which is the color that we perceive when light of red and green wavelengths of equal intensities impinge on our eyes; but we also see "yellow" when the light is "true" yellow, as from a sodium light. The fact is that the "red" and "green" cones are excited in the same way in both cases. Additional details can be found in Interesting Tidbit TB15.1.

INTERESTING TIDBIT TB15.1

Color Vision

From R. L. Gregoy's book *Eye and Brain—The Psychology of Seeing*, we read that the problems underlying the study of color vision have "aroused more passion than passion itself." Newton had discovered around 1692 that white light is composed of all colors. A few decades later, Thomas Young suggested the three-color model of color perception, based on the reasonable assumption that there could not be as many types of sensors in the eye as perceived colors. Young chose originally red, yellow, and blue as the "primary colors", but later picked red, green, and violet. Today we use red, green, and blue, but there is no fundamental reason to prefer this choice; in fact, the possible wavelengths of the principal colors vary over a wide range. But the important insight is that the three "primary colors" correspond to three types of sensors in the human eye. To quote Maxwell on the difficulty of studying color vision and Young's insight, "Thomas Young was the first ... who sought for the explanation of this fact (the existence of three primary colors), not in the nature of light but in the constitution of man." Yellow is observed by combining red and green lights, *not pigments*; yellow can also be produced from a sodium lamp as a *spectral* color. But many other, non-spectral colors like brown and metallic colors cannot be synthesized from the three primary colors. To complicate matters, and as E.H. Land (of Polaroid fame) showed with photography, quite good color results can be produced (brown included!) with just *two* primary colors. As it turns out, color perception also involves patterns, background, and objects, that is, context and high-level processing by the retina–brain.

"Colorimetry," that is the science of color measurement, is for obvious reasons closely tied to our perception of color. We can distinguish some 10 million different colors, spectrally resolving approximately 2 nm in wavelength. Naturally then, the design of any image recording and display system, even one with only B&W capabilities, has to take into account the nuances of our color perception.

Color quality is specified by three characteristic parameters: *hue*, *saturation,* and *brightness* (HSB). Hue refers to the dominant wavelength (peak color);

saturation corresponds to the "vividness" or "purity" of the color, with "white" being completely unsaturated, and ideal monochromatic (also called "spectral") colors being completely saturated; brightness is a measure of the light intensity, or, more precisely, *luminance*. Brightness is replaced by "lightness" or "value" when referring to reflected light. As an example, "pink" light is the result of mixing saturated red and white, the degree of saturation depending on the amount of white light mixed in. The HSB system can be graphically represented in a number of ways, the most popular being the Munsell color tree. In this tree, the trunk represents the completely unsaturated colors, from black to white, passing through grays; saturation increases from the trunk in the outward direction, while hue varies around the tree and is thus represented as an angle. The Munsell tree is very popular among artists for composing paintings; a simplified schematic is shown in Figure 15.5.

For display applications, colors are not usually quantified using the HSB system, but rather by their content of red, green, and blue, the three *primary colors*. The CIE (French acronym for International Commission on Illumination) established in 1931 the "RGB color-matching functions" based on a series of experiments on color perception. The experiments employed three arbitrary monochromatic light sources

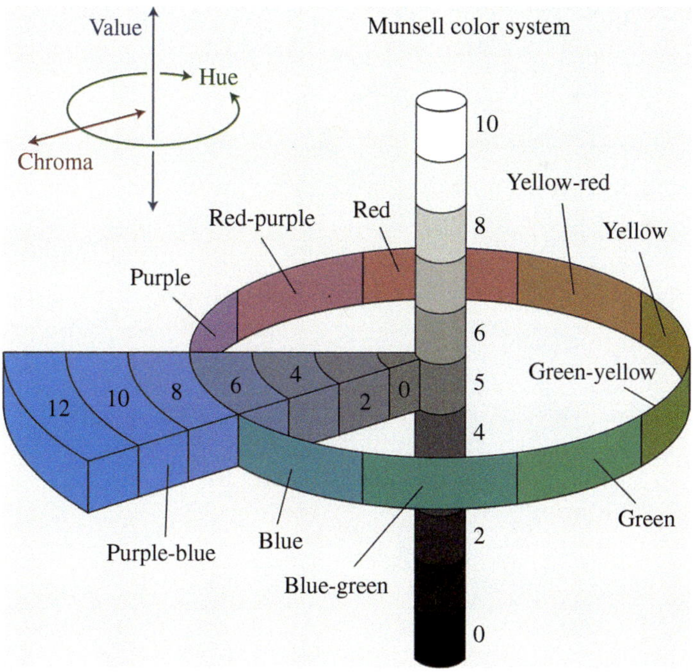

FIGURE 15.5 Simple schematic of the Munsell tree to represent hue, saturation (or chroma), and lightness (or value). Source: Rus, https://commons.wikimedia.org/wiki/File:Munsellsystem.svg. Used under CC-BY-SA 3.0 https://creativecommons.org/licenses/by-sa/3.0/deed.en.

at wavelengths of 700 nm (red), 543.1 nm (green), and 435.8 nm (blue) to synthesize other visible colors by adding the three colors with different intensities—note that these wavelengths do not coincide with those at the sensitivity peaks in Figure 15.4. The proportions $\bar{X}, \bar{Y}, \bar{Z}$ of the three *CIE* colors can be normalized by defining $(x, y, z) = (\bar{X}, \bar{Y}, \bar{Z}) / (\bar{X} + \bar{Y} + \bar{Z})$, so $x + y + z = 1$. The plot of (x,y) values that represent most perceived colors leads to the *CIE chromaticity diagram*, or horseshoe diagram, shown in Figure 15.6. The pure wavelengths, or spectral colors, are represented on the outer edge of the half ellipse. Pure white is the point $(x,y) = (1/3, 1/3)$, while all mixtures of purely blue and red, that is, the "purples" are represented by the straight line at the bottom. Similarly, the colors resulting from mixing say red at 700 nm and green at 530 nm lie along a line connecting the two points (see Fig. 15.6). Concept Verification CV15.1 further illustrates the use of the chromaticity diagram.

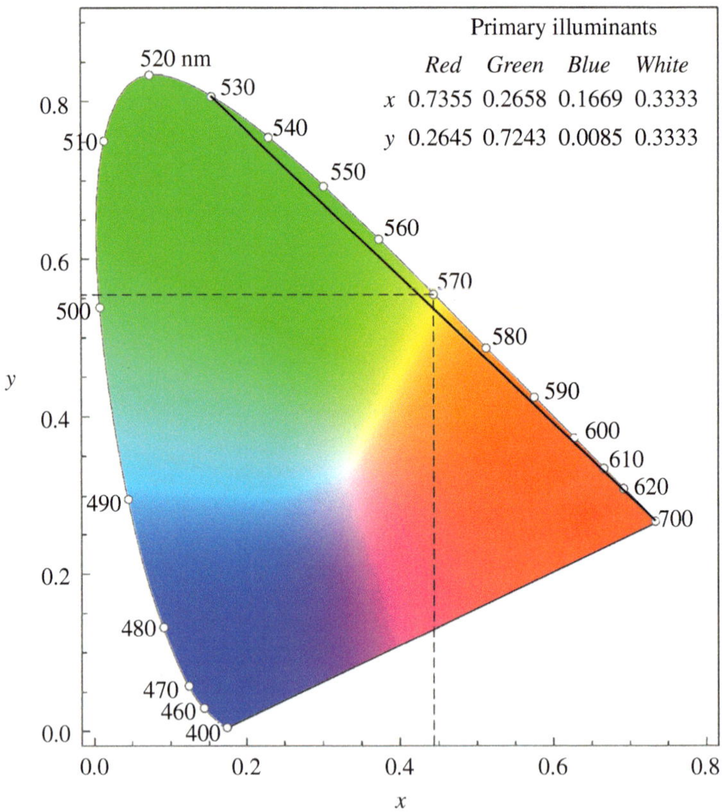

FIGURE 15.6 International Commission on Illumination (CIE) 1931 *x,y* chromaticity diagram. Source: From "CIE Chromaticity Diagram" from the Wolfram Demonstration Project http://demonstrations.wolfram.com/CIEChromaticityDiagram/. Contributed by: Yu Sung-Chang.

CONCEPT VERIFICATION CV15.1

Problem 1: Using the CIE chromaticity diagram of Figure 15.6, find the proportion "x" of red (700 nm) and green "y" (around 500 nm) that yields yellow at 570 nm. Do these numbers add up to 1? What is the value of "z" in this case?
Problem 2: The complement of a color is the color that needs to be added to produce white. Find the complement of the saturated color at 580 nm. *Hint*: the colors resulting from the addition of any two colors lie along the line connecting them.

Answers: The values of x,y,z (*tristimulus* values) in Problem 1 are (0.44, 0.56, 0). Note that "z" is equal to zero in this case because we are synthetizing a spectral color, that is, one at the edge of the diagram (570 nm). In Problem 2, tracing a line through white (1/3, 1/3) yields the point 480 nm on the opposite side of the chromaticity diagram. Many additional examples of the use of the chromaticity diagram can be found in the book *Seeing the Light* by Falk, Brill and Stork (1986).

A particular display screen can reproduce only a subset of the chromaticity diagram called the *display gamut*, typically a triangular area whose vertices correspond to the primary colors of the phosphors used, for example in CRT or plasma-based displays. Another color space, the *Uniform Color Space* or UCS (CIE 1960) was defined to better represent the color separation ability of the human eye. Figure 15.7 shows the UCS color space; the color variables (u,v) are related to the (x,y) color variables by simple relationships, for example, $x = 3u/(2u - 8v + 4)$; the white point is given by $(u,v) = (4/19, 6/19)$.

Additional considerations on chromatic variables and color spaces are given in Section 15.3.

There is a qualitative difference in the brightness, contrast, and color saturation of different display technologies. For example, a scene captured with a video camera will look different on a CRT TV, an LCD screen, and a projected display. The differences arise from particular nonlinear brightness-to-signal relations for each display technology and also from inherent differences in color reproduction. Ideally, the brightness-to-signal relations of displays should follow the logarithmic response of the human eye (see Eq. 15.1 and Fig. 15.3), but significant deviations exist in practice. Thus, a quantity called the *gamma factor* has been defined to express the relation between input and output through the following logarithmic relation:

$$\gamma = \frac{\log(V_{\text{out}})}{\log(V_{\text{in}})}, \quad \text{or} \quad V_{\text{out}} = V_{\text{in}}^{\gamma}. \tag{15.4}$$

The Vs can be understood as voltages applied to, for example, a CRT, and derived, in turn, from analog or digital signals representing an image. A $\gamma = 1$ corresponds to the ideal situation where a change in image input intensity by a given factor, leads to a change in displayed image intensity by the same factor. This ideal situation however is not always desirable because enhanced *contrast* (larger gamma) may be better in

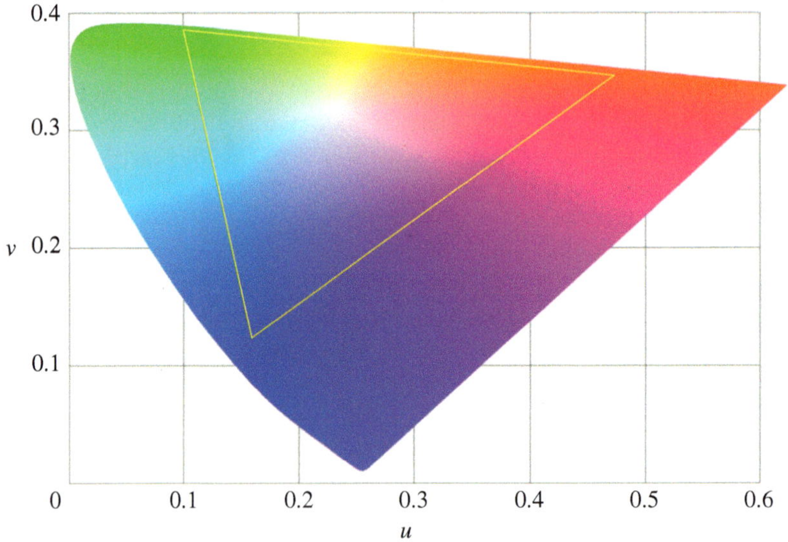

FIGURE 15.7 International Commission on Illumination (CIE) 1960 *u*,*v* chromatic diagram with NTSC (TV) 1953 standard color gamut (area inside yellow triangle). Source: Adapted from Adoniscik, https://commons.wikimedia.org/wiki/File:CIE_1960_UCS.png. CC public domain.

particular applications; for example, when reproducing line drawings or simple text as in e-readers.

CRT's have a nonlinear response corresponding to $\gamma = 2.2$, which is corrected in TV standards. Nowadays, digital image processing allows for easy gamma correction of recorded material, including the unintended changes in color saturation.

15.2 WHO INVENTED TELEVISION?

Everyone knows the electrical light bulb was invented by Thomas A. Edison; perhaps fewer know of Nicolai Tesla's many inventions, including the generation and transmission of alternating current (AC) power. But who knows the person who contributed the most to the invention of modern electronic TV? That person was Philo T. Farnsworth, pictured in Figure 15.8, an uneducated Utah boy who conceived and built the main components of modern TV. His line-by-line method for both recording and transmitting, the modern *raster scan*, was inspired while watching a field being plowed in his native Utah. The "forgotten father of television" was the first one to transmit a TV image in 1927 and whose patents (1927–1929) included camera tubes, circuitry, and CRTs for viewing, and all key components of modern TV. Other pioneers are often mentioned (e.g., Sarnoff, Zworykin, and Baird), but their roles were not as pivotal as Farnsworth's. In fact, Sarnoff in 1929 hired Zworykin, a Westinghouse engineer, to go on an

FIGURE 15.8 Philo T. Farnsworth, the "father of television." Source: Harris & Ewing, https://commons.wikimedia.org/wiki/File:Philo_T_Farnsworth.jpg. CC public domain.

intelligence-gathering mission. Zworykin visits Farnsworth and marvels at Farnsworth's invention, the world's first electronic TV camera. While several of these personalities and their companies (e.g., Sarnoff, head of the now defunct RCA corporation) were involved in lengthy litigation over patents, it was Farnsworth who won the final battle. Additional details, some truly fascinating, can be found in the references.

15.3 TRADITIONAL AND HIGH-DEFINITION TV DISPLAY FORMATS

Most analog and digital display systems have many legacy elements from the original analog TV standards such as the number of lines per frame and the refresh rate. Moreover, most digital display formats are derived from TV standards. A display screen is made of "picture elements" called *pixels*, which are only noticeable if viewed at a close distance to the screen. The resolution of the display is specified by the number of lines, the pixel size or density, and the angular resolution. Analog TV required the recording and broadcasting of two separate *fields* of *even* and *odd lines* due to bandwidth limitations for transmission. The combined even and odd fields are called a *frame,* but the two terms, *field* and *frame*, are often used interchangeably, which can create confusion. The process of tracing even and odd fields is called *interlacing*, and it is not a requirement of digital TV since the frame refresh rates are much higher than in analog TV. However, the European HDTV is interlaced because it operates at the old 50 fields/s. When all the lines are in a single field, the picture is called *progressive*; it is possible to convert interlaced pictures to progressive by using image processing; but it is more problematic to convert between different refresh rates. Figure 15.9 illustrates the interlacing technique; note that in digital interlacing there is no need to do *retrace* with blank signals as in analog TV.

Current TV standards are tabulated in Table 15.1 to show the variety. All these standards have *field frequencies* equal to the power-line frequency. For example, 60 fields/s in the United States, and 50 fields/s in Europe.

The aspect ratio 4:3 is called *full-screen*, while 16:9 is called *widescreen*. Another aspect ratio not included in Table 15.1 is "cinemascope," or 2.41:1, which is wider than 16:9 and the prevalent format of movies. However, many popular movies these days are shot in 16:9 format to match the flat screens of HDTV. When the aspect ratio of the original movie does not match the aspect ratio of the screen, black bars will appear on the screen, either on both sides or on top and bottom.

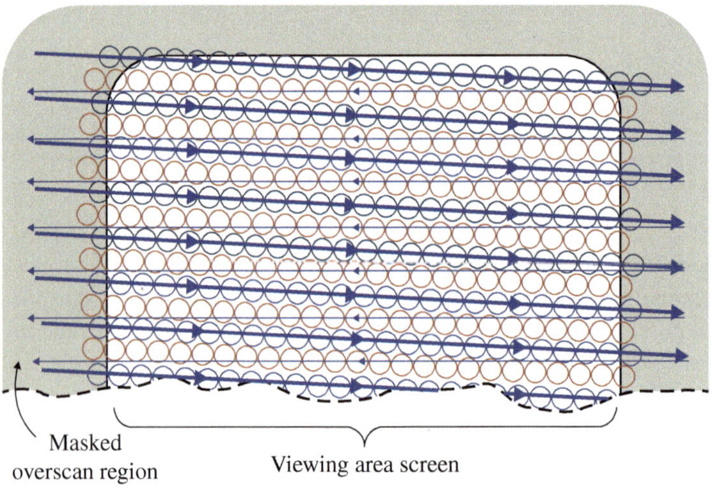

FIGURE 15.9 Interlacing in analog TV. Only the horizontal retrace is shown for clarity.

TABLE 15.1 Analog and Digital TV Standards

Name	Resolution: (H×V) Lines	Refresh Rate (Hz)	Aspect Ratio
NTSC	720×480	60 i	4:3, 16:9
PAL	720×576	50 i	4:3, 16:9
SECAM	720×576	50 i	4:3, 16:9
HDTV	1280×720	30, 50, or 60 p	16:9
Full HDTV	1920×1080	50, 60 i, 50, 60 p	16:9
QFHD	3840×2160	50, 60 p, other	16:9

Notes: "i" stands for "interlaced" and "p" for "progressive."

TABLE 15.2 Resolution and Aspect Ratio of Some Common Computer Monitor Formats

Name	Resolution: (H×V) Pixels	Aspect Ratio
VGA	640×480	4:3
SVGA	800×600	4:3
XVGA	1024×768	4:3
SXGA	1024×1024	5:4
WXGA	1365×768	16:9
QXGA	2048×1536	4:3

Modern computer monitors are primarily progressive displays, although the first computer monitors were analog TVs. Interlacing is neither needed nor desirable for computer displays. (Some of the readers may remember how text was displayed in those old monitors, with the flickering "Ts" and other annoying artifacts most coming from interlacing. Conversely, displaying old movies on modern progressive computer displays can also be problematic—in this case, the difference in refresh rates may be an issue). The first computer monitor standard, VGA, derived from the American National Television Systems Committee (NTSC) TV standard, which has 480 horizontal lines or 480 vertical pixels. *Note*: confusion is caused by the two definitions of resolution: *pixels* for computer monitors and *lines* for TVs. To clarify the situation, *vertical* resolution in TVs is quoted as the number of *horizontal* lines, and vice versa; but "H" and "V" in Table 15.1 refer to pixel *direction*, or the direction along which the lines are counted.

In Table 15.2, we give examples of resolution in *horizontal* × *vertical* (H×V) pixel counts for computer monitor formats. The resolution of the celebrated *retina* display of the third-generation iPad tablet computers and laptops since mid-2012 from Apple Computer Co. corresponds to QXGA (for "quad–XGA"). The iPad has more than three million pixels on a 9.7-inch (diagonal) screen. This resolution surpasses that of Full HDTV (1920×1080), but not that of the new QFHD (Quad HD) or 4K. In addition to the pixel counts, the diagonal size and refresh rate of the display are normally given; a typical refresh rate for computer monitors is 75 Hz.

INTERESTING TIDBIT TB15.2

Film Grain versus Pixel Size or Analog versus Digital

The spatial resolution of the best photographic films, which is limited by the grain size, is generally better than what is possible with electronic displays. However, the film emulsions with the best resolution are also fairly slow, requiring long exposure times; the CCD sensors of digital cameras, by contrast, are far better for low levels of illumination and are also not as noisy. Further, the old 35 mm film (22×16 mm) is inferior to HDTV, and electronic paper and retina displays are superior to real paper (see Section 15.9.)

CONCEPT VERIFICATION CV15.2

Problem: What size TV screen is optimal for viewing at a distance of 3 m? Compare your results to the "rule of thumb" that the ideal distance to view the big picture is 2.5 times the diagonal length of the screen. *Hint*: use the finding in Equation 15.3 for the ideal resolution of a typical human eye and in reference to Figure CV15.1 consider 480 lines for NTSC TV and 720 lines for HD TV.

Answer: We calculate the *maximum* screen diagonal required for a 1.2 arcmin resolution at a distance of $L=3$ m (10 ft) from a NTSC TV (480 lines), and a HD TV (720 lines). Referring to Figure CV 15.1, we find $D_{\text{NTSC}} = 1.2 \times (480/3600) \times 3\text{m} \approx 50\text{cm}$ or $\sqrt{2}D_{\text{NTSC}} \approx 70\text{cm}$ (27 inch), and $\sqrt{2}D_{\text{HDTV}} \approx 102\text{cm}$ (40 inch). Thus, the rule of thumb is within a factor of 2, roughly, of our calculation. Obviously, larger screens could be used if the viewing distance is increased.

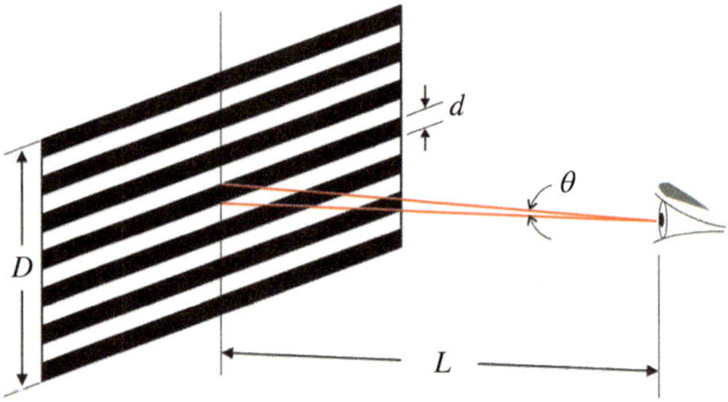

FIGURE CV15.1 Screen size and resolution of the human eye.

CONCEPT VERIFICATION CV15.3

Problem: Is a 46 inch HDTV screen with a resolution of 1920×1080 better than an iPhone 5 with a 4 inch screen and a resolution of 1136×640? *Hint*: it depends on the viewing distance; calculate the angular separation of pixels for both screens at distances of 30 cm and 3 m.

Answer: In the 46 inch HDTV screen, we have a total of 2,073,600 pixels, or 48 *pixels per inch* (ppi) along the diagonal; for the iPhone, on the other hand, we have 727,040 pixels, or 326 ppi. Let us calculate the angular separation of pixels at a viewing distance of 30 cm in both cases: we get 0.9 arcmin for the iPhone, and 6.2 arcmin for the HDTV screen; thus, the eye cannot resolve the pixels in the iPhone, but it can for the HDTV screen at such a close viewing distance. At a viewing distance of 3 m, however, the angular separation of pixels in the HDTV screen is only 0.6 arcmin!

To conclude this section, we describe briefly the implementation of color for TV displays and the encoding of color in digital formats. The details of how color is handled for recording, transmission and display in TVs, both analog and digital, are so complex that two opposite general views can be presented. On the one hand, a review article by D.H. Pritchard (1977) describes the color standards for NTSC TV as a "masterpiece of 'tradeoffs' among the pertinent psychophysical properties and electronic system techniques." On the other hand, the limitations of color reproduction in NTSC TV have been derided with the nickname "Never The Same Color." (Some readers may remember the time spent adjusting the colors in the 1970s even for the best Sony Trinitron TVs!)

As discussed in Section 15.1, a typical display cannot reproduce all the colors of the chromaticity diagram. The reproducible colors are ordinarily represented by the chromaticity values enclosed by a triangle whose vertices correspond to the dominant wavelengths of the *R*,*G*,*B* phosphors or color filters of the display. An example is the NTSC *display gamut* shown in Figure 15.7 as a light triangle. LCD displays with LED backlight (see Section 15.5), on the other hand, have a wider color gamut than old analog TV displays thanks to a combination of narrowband LEDs for the three primary colors and the color filters.

In both CRT (Section 15.4) and LCD screens (Section 15.5), the colors are obtained by addition. In both screens, the pixels are grouped in triads formed by three types of phosphors (CRTs) or filters (LCDs). The Federal Communications Commission (FCC) defined new functions $\bar{r}$, $\bar{g}$, and $\bar{b}$ based on the fluorescence of screen phosphors adopted by NTSC standards. The set $(\bar{X},\bar{Y},\bar{Z})$ and $(\bar{r},\bar{g},\bar{b})$ are connected by simple linear transformations; for example, the luminance or monochrome signal is determined by $\bar{Y} = 0.299\bar{r} + 0.586\bar{g} + 0.1146\bar{b}$. Further, the signals representing chromaticity information are generated as color difference signals, *U* and *V*, which are not identical to the (u,v) values mentioned before and used in Figure 15.7. Chromaticity (U,V) are produced by subtracting the luminance signal from the red, green, and blue signals. For backward compatibility with B&W TVs, the luminance and color signals are encoded using separate carriers. The *U* and *V* signals are encoded with the same frequency but in quadrature, that is, with a 45° phase shift. Signal processing or transmission problems can easily cause changes in this phase shift, leading to the color changes that anyone familiar with old color TVs can remember. Digital TVs also depend on processing of $\bar{Y}$ and *U*, *V* signals, but do not suffer from many of the limitations of analog signal processing.

Digital formats for basic pictures use color encoding employing 8 *bits per color channel*. The number $2^8 = 256$ represents values from 0 to 255 of different *luminance values* per primary color, that is, from no saturation to complete saturation. (See Fig. 15.5.) Thus, the number of possible colors for 8-bit RGB is $2^8 \times 2^8 \times 2^8 = 16{,}777{,}216$ colors, or more than the roughly 10 million colors that the human eye can detect. *However*, this does not mean that every color in the *visual gamut* (Figs. 15.6 and 15.7) can be reproduced with 8-bit RGB because of the *limited gamut* of any real color display system. 8-bit color is standard for the "jpg" picture format, but 16-bit is possible (even 32-bit color is available) in other formats like "png" and "tif."

For digital video cameras and video transmission, color is encoded with 10- or 12-bit versions of the YUV color space described earlier.

For printed media, a *subtractive* color system is employed; it is based on four *pigment* or *ink* colors: cyan, yellow, magenta, and black. The CMYK system needs and additional component ("K" for black) because the mixture of the other three components only approximates a true black (inks or pigments are not perfect). The RGB 8-bit colors are normally specified using hexadecimal numbers in the format #RRGGBB (e.g., red = #FF0000, since F (hexadecimal) = 15 (decimal), and FF = 15 × 16 + 15 = 255), while CMYK colors are denoted in percent levels (0–100%). Conversion between the different color spaces is normally needed as when displaying video from a digital camera in a computer monitor (YUV to RGB), or when scanning printed material (CMYK to RGB). These conversions and other digital image processing operations are covered at length in some of the references.

INTERESTING TIDBIT TB15.3

Standard Digital Versatile Disk (DVD) versus Blu-ray (BD) Disk

Standard DVD players, introduced around 1995 by Phillips, Sony, Toshiba and Panasonic, use a red laser diode (650 nm, as opposed to 780 nm for audio CDs) to read the disks, which have a capacity of 4.7 GB (single-layer). This capacity is sufficient for a 2-hour movie at standard TV definition (480 horizontal lines). Blu-ray disks (BDs), on the other hand, were developed by Sony, Matsushita, and Philips and introduced around 2005. BD players use a blue laser (405 nm) that allows reading a much higher density of information than red lasers. Single-layer BDs can store 25 GB, sufficient for a full-length HDTV movie (1080 horizontal lines). BD players have become fairly inexpensive and popular since 2014, thanks to the affordability of HDTV screens. (See the discussion of DVDs and BDs in Chapter 2.)

15.4 CATHODE RAY TUBES

CRTs are used in old TVs and computer monitors, oscilloscopes, radar screens, and other commercial and scientific display devices. "Cathode" can be translated from its Greek roots as "way down," indicating the path that electron beams follow from the negative electrodes of vacuum tubes or electrolytic cells to the positive terminals. In a CRT, the electrons from the cathode are accelerated through a potential whose magnitude depends on the application, typically from a few kilovolts to a few tens of kilovolts. The energetic electrons strike the screen and excite the atoms in the phosphor coating leading to light emission; the color and decay time (persistence) of this emission also varies according to the application. For example, the Phillips A36E CRT for use in color TVs employs an accelerating voltage of 23 kV and a maximum beam current of 1 mA. Typical decay times of emissions from TV phosphors are of the order of 1 ms, 10 times faster than persistence of the eye. The grain size of the phosphors is around 5 μm.

Commercial CRTs first appeared in 1922, but as seen in Section 15.2, the first TV transmission employing a CRT had to wait until 1927. Color TV started in 1953 and was based exclusively on CRTs for many years until 2008 when LCD-based TVs took over.

The main components of a CRT are the electron gun, a focusing system consisting of cylindrical lenses, an anode electrode to modulate the electron beam intensity, and an electrical or magnetic deflection system (made of plates or coils) for directing the beam to various locations on the screen. The screen is coated with especial phosphor(s) and grounded to avoid charge accumulation. All components, except the electrical terminals, are encapsulated in a glass tube under a low pressure (weak vacuum) of the order of a few millitorr. Further, the interior of most TV tubes are covered with a graphite layer to absorb spurious light reflections and thus increase contrast. Figure 15.10 shows a highly simplified schematic of a CRT.

The spot size on the screen of CRTs is around 0.3 mm for standard TV and 0.1 mm for HDTV. This spot size is much larger than the grain size of the phosphors because it is the result of focusing voltage variations, alignment errors of the electron gun and other elements, and the *deflection sensitivity* of the electron beam(s).

CRTs for color TV employ three electron guns, one for each one of the primary colors *R*, *G*, and *B*. The screen, in turn, is coated with vertical strips of three types of phosphors, one for each primary color. In addition, a "shadow mask" is positioned some short distance from the screen; it consists of many ($<10^6$) small holes that allow the beams to reach the correct phosphors. The holes of the shadow mask can be oblong or round (HDTV). In the celebrated Trinitron color TV from Sony, special "aperture grilles" are used instead of the shadow mask. The grilles have narrow vertical stripes etched on a thin metal plate that is positioned behind the screen; this structure improves significantly the vertical resolution and also increases display

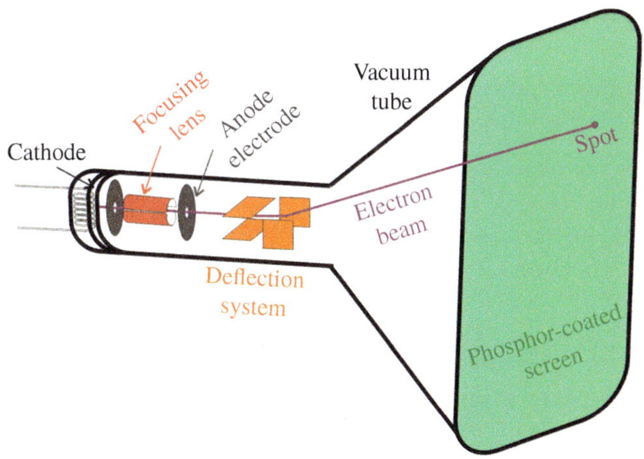

FIGURE 15.10 Cathode ray tube (CRT): simplified schematics of a typical CRT containing an evacuated enclosure, cathode, focusing lens, beam-modulating anode electrode, deflecting plates and phosphor-coated screen.

brightness because less electron beam current is intercepted than with a shadow mask. In addition, the use of a single lens for all three electron guns minimizes the effects of spherical aberration on the electron beam. The high quality of the image and construction of Trinitron TVs made them the dominant TVs in the market for decades.

Perhaps, the greatest advantage of a CRT as a vector display device is its adaptability for updating images with almost any frequency and resolution, unlike the restrictions imposed by the rastered images of LCDs and other technologies. But CRTs are limited in size, requiring relatively thick glass for large flat TVs. Moreover, CRTs emit X-rays from the electrons impinging on the phosphorous screen and being decelerated. X-ray emissions from CRTs, however, are comparable to the low-energy end of medical X-rays; the use of lead glass helps minimize this emission. Finally, there are limitations in image quality imposed by geometrical distortions, color aberrations and other factors, but these are only moderate in advanced CRTs.

15.5 LIQUID CRYSTAL DISPLAYS

Liquid crystals (LCs) were discovered in 1888 by the Austrian chemist Friedrich Reinitzer, but the name "liquid crystal" and associated concept of a new state of matter were not established until decades later. (See Interesting Tidbit TB15.4.) Standard liquids are isotropic, that is, have no preferred directions as their structures are highly disordered. In contrast, crystals have preferred directions stemming from their ordered atomic structure. Correspondingly, liquids and solids generally affect polarized light very differently. For example, bi-refringent crystals such as quartz can be used to produce linearly or circularly polarized light. (See Chapter 10 and especially Fig. 10.6.)

INTERESTING TIDBIT TB15.4

Liquid Crystal: A Substance with Two Boiling Points

The Austrian chemist Friedrich Reinitzer discovered in 1888 that a solid substance related to cholesterol appeared to have a boiling point at 145.5°C and another one at 178.5°C. The substance turned into a cloudy liquid at the first boiling point, and then into a clear one at the higher temperature. The German physicist Otto Lehmann, collaborating with Reinitzer, determined that the cloudy liquid had properties of both solids and liquids, so he coined the term "liquid crystal." It would take several years before the idea of a "liquid crystal" was accepted. Many other substances have been discovered since then that display more complex behavior (more states) than the original LC. Some of these have found applications in LCDs. The physical properties of LCs have been related to those of magnetic materials and superconductors. The French physicist Pierre-Gilles de Gennes was awarded the 1991 Nobel Prize in Physics for work related to these special materials.

LCs are partly ordered materials consisting of long molecules whose orientation can be manipulated through mechanical, electrical, or magnetic means. LCs come in two main categories or *phases*: *nematic* and *smectic*. The nematic phase approximates a liquid state; the molecules float feely but naturally adopt a preferred direction. If the molecules have long structures and also have a twist or "chirality," similar to DNA molecules, the nematic phase is called *cholesteric*. In this phase, LC molecules adopt twisted patterns that reflect light with different colors depending on the temperature; thus, cholesteric LCs can be used as temperature sensors. LC thermometers of this type can be purchased from amateur science stores and displayed as refrigerator magnets. Another type of LC similar to the cholesteric is the twisted nematic (TN) crystal.

An LC display consists of an array of individual LCs, one for each picture element (pixel). For most LC display screen applications, each pixel consists of a TN crystal placed between two crossed polarizer filters, whose linear polarization directions are orthogonal (90° apart). An "ON" state occurs when the chiral LC rotates the polarizatized light from the first filter so that it also passes the second polarizer. The "OFF" state is accomplished by destroying the chirality of the LC by applying a strong electric field, which prevents the polarized light from reaching the second filter with the correct angle. The action of the LC cell just described is shown in Figure 15.11. By arranging many LC cells with their polarizer filters and electrodes in a matrix, an LCD, or liquid crystal display, is obtained.

LCDs have limited viewing angles compared to CRT and plasma displays, but the technology has improved with ingenious geometrical arrangements of the pixels and electrodes. Another problem with LCDs has been the switching times, which can lead to artifacts in the reproduction of fast motion, but modern designs have achieved

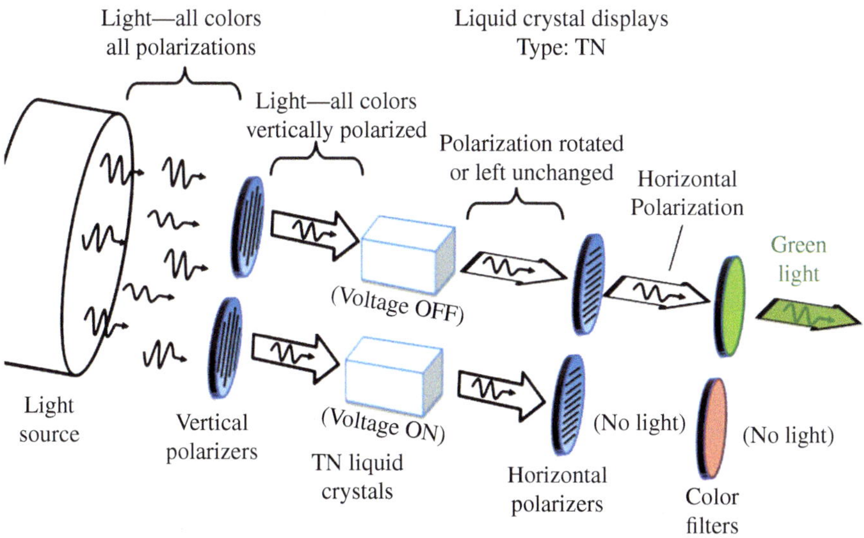

FIGURE 15.11 Principle of LCD: a twisted nematic (TN) liquid crystal is used between two polarizer filters to control the transmitted light.

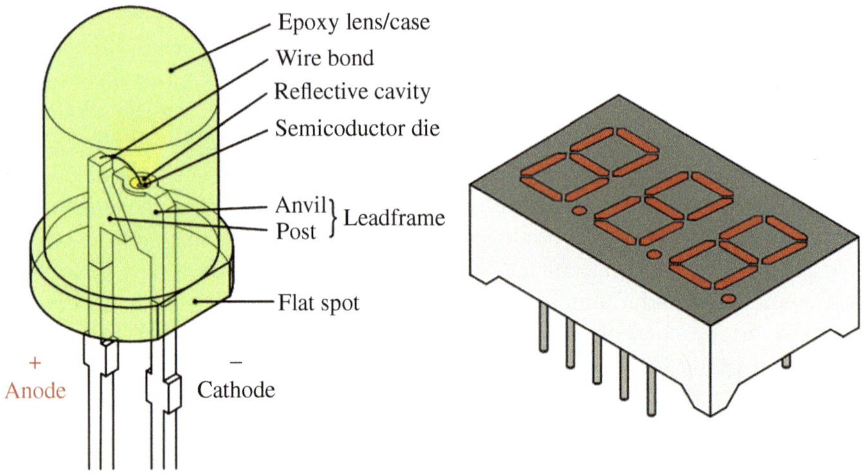

FIGURE 15.12 Schematics of a light-emitting diode (LED), left, and example of a seven-segment LED display chip, right. Source: (Left) Inductiveload, https://commons.wikimedia.org/wiki/File:LED,_5mm,_green_%28en%29.svg. CC public domain. (Right) Inductiveload, https://commons.wikimedia.org/wiki/File:7-Segment_Display,_0.36in,_Triple_%28shaded%29.svg. CC public domain.

B&W transition times of a few milliseconds, deemed suitable for most applications. A minimum frame *refresh rate* of 120 Hz is considered good for the reproduction of video. Most motion pictures were originally made with 24 frame/s.

An additional related development is that of "LED TVs," which appeared in 2007. LED TVs are not a different kind of TV, but just LCD screens that are backlit with LEDs (see Fig. 15.12). (See Chapter 9 on light sources and Section 9.4 in particular.) The illumination with LEDs can be set up directly from behind, or from the side of the screen in which case the light is redirected toward the inside front. An increased contrast is achieved, by turning on different sets of LEDs according to the darkness of the displayed image, a process referred to as "local dimming." Thanks to their energy efficiency, smaller weight, and all the improvements mentioned, LCDs are now the dominant display technology for small and large screens, from cell phones to very large TV screens.

15.6 PLASMA DISPLAYS

Plasma displays were invented in 1964 at the University of Illinois, but their development for consumer use had to wait until the development of a number of technologies related to digital circuitry and signal processing.

Each pixel in a plasma display consists of a discharge cell, which contains its own cathode and special electrodes. Figure 15.13 illustrates the basic cell. An applied high voltage to the cell initiates a plasma discharge that ionizes a gas (neon or xenon) and produces UV radiation; this radiation strikes the phosphor coating on the cell

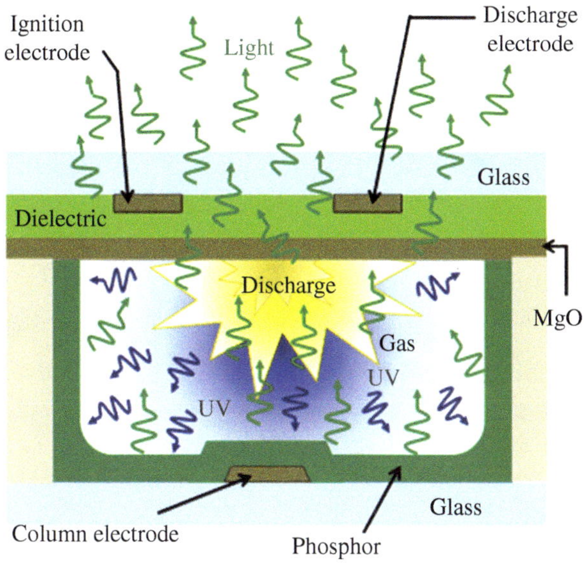

FIGURE 15.13 Schematics of the discharge cell (pixel) on a plasma display.

walls inducing the emission of visible light. The intensity and color of the light will depend on the intensity of the discharge, in turn determined by the strength of the electric field in the discharge, and on the type of phosphor.

Because the minimum size of a pixel in a plasma display is of the order of 0.3 mm, plasma displays are exclusively used for large screens. In fact, plasma screens smaller than 42 inch are very uncommon.

Plasma displays generally consume more power (up to a factor of 2) than LCD screens of the same size, but have better motion resolution because of the nature of their pulsed operation. In addition, plasma displays have better color reproduction and wider viewing angle than LCDs, but historically suffered from burn-in caused by long-term static images. This burn-in problem, which is no longer serious in the newest plasma technologies, was not caused by phosphor damage as in old CRTs, but by cathode degradation. For this reason and the difficulty in making small screens, plasma displays are not used as computer monitors. Sales of plasma TVs peaked around 2010, but have fallen significantly since then as LCD technology improved and prices became more competitive. As of 2012, Panasonic reported losses from a 40% reduction in plasma TV sales.

15.7 DIGITAL MICRO-MIRROR DEVICES

Imagine an array of two million microscopic aluminum mirrors mounted on a single silicon chip and controlled digitally to tilt in one of two orientations (+10° or −10°), with switching times of the order of 5 μs. Light from a source can be deflected in any

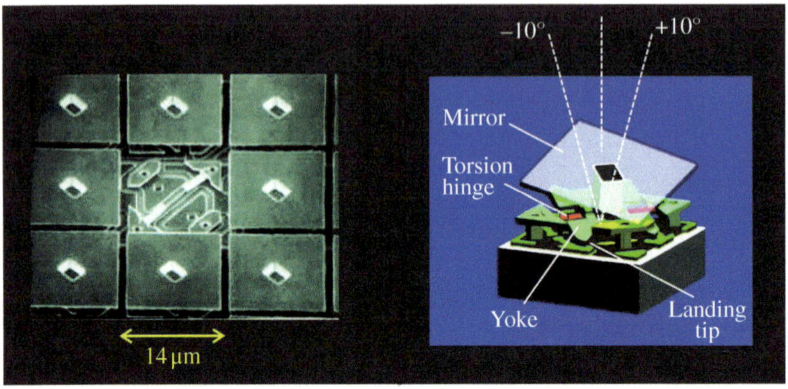

FIGURE 15.14 Left: array of digital micro-mirror devices, or DMDs (SEM photo). Right: micro-mirror architecture. Source: Courtesy of Texas Instruments.

desired pattern to create images, even fast-moving ones. DMDs are the trademark technology of Texas Instruments, invented by Dr. Larry Hornbeck in 1987. DMDs have found applications in projectors for movie theaters (more than 8000 around the world), home theater or business use, tiny cellphone-size projectors, large-screen HDTVs, and several non-display commercial and scientific uses (e.g., metrology). The technology, generally known as Digital Light Processing (DLP), is particularly suited for production of B&W pictures but has been also adapted for color projectors and displays.

Figure 15.14 shows a scanning-electron-microscope (SEM) picture of a nine-mirror DMD chip; note the small gap between mirrors, of the order of 1 μm. The center mirror in the array in Figure 15.14 is missing to illustrate the underlying spring structure. Figure 15.15 illustrates the principle of a DMD projection device. The light from a projection lamp passes through a condenser lens and a color filter system (not shown for simplicity), is reflected by the three mirrors, and projected either onto a screen or dumped to a light absorber. In this example, the two "ON" mirrors produce two white square pixel images, while the "OFF" mirror directs the light to an absorber. A complete high-resolution color picture can be projected with about one million DMD mirrors and a color filter system.

DMD-based displays have several advantages over other technologies. There is no risk of burned-in images as in plasma screens. DMD systems are more efficient than LCD systems because DMDs are reflective devices and do not depend on polarization as in LCDs where light of one polarization component is not used. DMD systems have a large fill factor (~90%) compared to LCD systems, due to the small gap between micro-mirrors (see Fig. 15.14); DLP technology can render images of very high resolution and contrast without the pixilation of, for example, LCD projectors; DLP technology is excellent for reproduction of fast video; and, finally, DMDs (and associated optics) are very reliable devices. The only downside of the best DLP projectors is their high cost relative to LCD projectors; DLP and LCD projectors divide the market share almost evenly as of 2013.

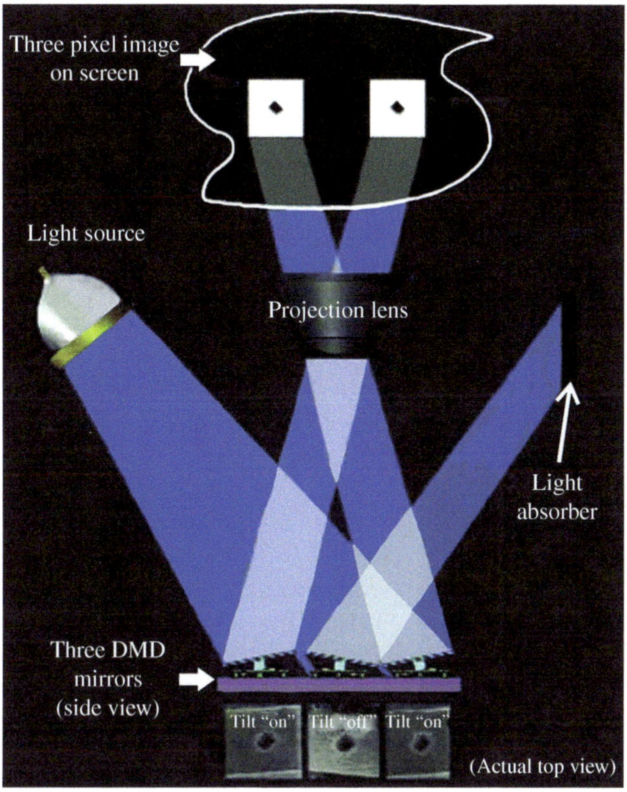

FIGURE 15.15 Illustration of DLP/DMD display technology: light is reflected off three micro-mirrors to produce a three-pixel image on a screen. In a color system, color filters and other components (not shown) are used. Source: Courtesy of Texas Instruments.

An example of a commercial, high-end DLP projector is the Digital Projection International dVision 30-WUXGA XC. It features a 1920×1200 pixel DMD, a 16 : 10 aspect ratio, corresponding to a format known as "Widescreen Ultra Extended Graphics Array" (WUXGA), and a six-segment color wheel. The contrast ratio is 7500 : 1 and the image brightness is 4300 ANSI lumens. For comparison, a typical room requires 200–300 ANSI lumens of illumination.

INTERESTING TIDBIT TB15.5

Digital Micro-Mirror Devices (DMDs)—A Miracle of MEM Technology

Imagine opening and closing a door twice every second for 20 years; it would amount to more than one trillion cycles. It is not hard to predict that because of a host of fabrication and environmental issues, the door hinges will wear out or fail long before the one trillion cycles can be completed. The miniature mirrors of DMD arrays have been subject to similar tests, but no hinge failure or "fatigue" has been observed. The fundamental reason is that the standard laws of metal fatigue in the macroscopic world do not apply to the ultrathin films of microelectromechanical devices such as DMDs. In fact, the film of a DMD hinge is only one grain thick, so no accumulation of "density dislocations" can occur that leads to the formation of fatigue cracks. Besides, the mirrors are driven electrostatically by reproducible signals, completely unlike the operation of a door. But other problems may be present in DMDs: "stiction" (or sticking), hinge memory, and contamination. Stiction can occur from water condensation and from molecular (van der Waals) forces, but it is solved by hermetic clean packaging and the use of special springs under each mirror; hinge memory is essentially a "backlash" phenomenon whereby the tilt of a mirror is not 100% reproducible, but it is eliminated by use of better materials and special drive signals; finally, contamination is avoided by manufacturing and packaging the DMDs in class 10 clean rooms.

15.8 TOUCH SCREENS

The use of touch screens is widespread nowadays, in some cases challenging the use of the keyboard as the primary input device for computing. Touch screens are used in supermarkets, ATMs, computer tablets (and even desktops), smart cell phones, GPS devices, portable video game consoles, and others. The computer operating systems themselves (e.g., Windows 10) are evolving to include touch-screen capabilities for general use.

Two main types of touch screens exist: resistive and capacitive. The resistive touch screen, shown schematically in Figure 15.16, is the simplest. Electrical current flows through two thin layers separated by tiny insulating transparent spacers. When the finger or other object presses against the top layer, the current flow is affected at that particular location, enabling the electronics and software of the device to identify the contact. One advantage of the technology is that it can be used with any object, not just the finger, but it is also less responsive than capacitive touch screens. Resistive touch screens are sensitive to only one touch at a time and no sliding (e.g., for zooming). The Nintendo portable game console uses this technology.

The capacitive touch screen relies on the electrical charge present at the tip of the finger. The action of the finger causes a change in capacitance at a particular position on the screen. In the projective capacitive touch screen, a grid of very thin wires under the glass senses the change in capacitance; the change is further processed by a fast microcontroller located under the grid. This is the technology behind the iPad, iPhone, and a myriad of other tablets and smart phones. It allows multiple touches and a host of finger motions that has made the devices so successful. The basic mechanism of the capacitive touch screen is illustrated in Figure 15.17.

Other touch-screen technologies are under development based on, for example, optical rather than electrical phenomena. For example, one technology is based on *frustrated total internal reflection* (FTIR): a finger pressed against a glass plate can destroy the (internal) light at the contact point. Cameras on the back of the screen can

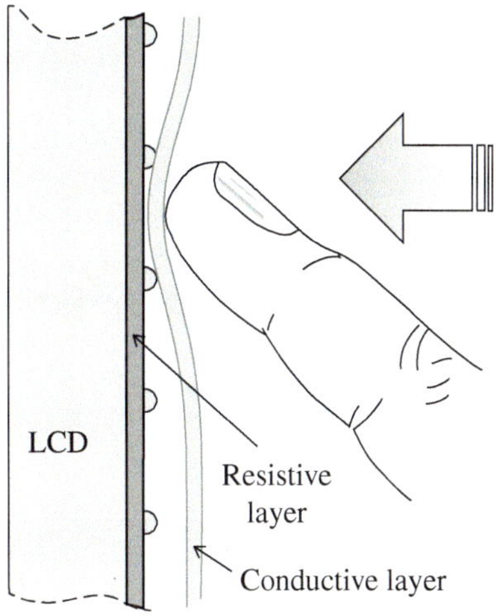

FIGURE 15.16 Schematics of *resistive* touch-screen technology.

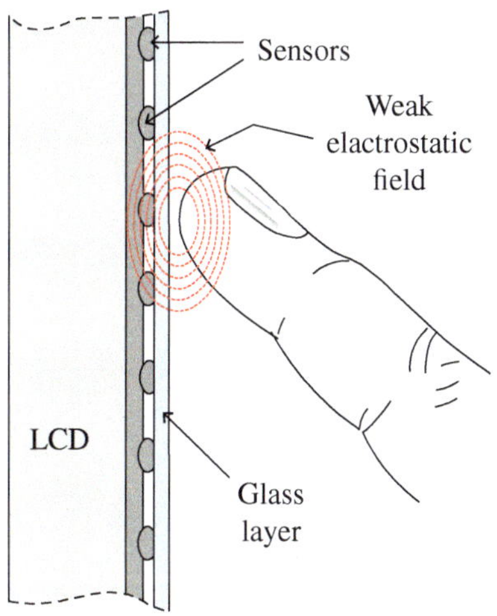

FIGURE 15.17 Schematics of *capacitive* touch-screen technology.

detect the changes for further processing. FTIR screens are sensitive to multiple touches similar to the capacitive touch screens.

15.9 ELECTROPHORETIC DISPLAYS

Electronic readers, "e-readers" for short, started with a device from Sony and later *the Kindle* from Amazon, followed by products from Barnes and Noble and other companies. Dedicated e-readers, as opposed to general-purpose computer tablets, use a special display called "electronic paper" or "electronic ink," which mimics the way ambient light is reflected by ordinary paper. Electronic paper was invented in the 1970s by Nick Sheridon from the Xerox Company in Palo Alto, CA, but was only developed into a commercial product in the late 1990s by E Ink Corp., a company founded by physicist Joseph Jacobson.

The basis for "electronic paper" is a process called "electrophoretic," which is illustrated in Figure 15.18. In one scheme, 1 μm titanium dioxide particles float in a dielectric liquid; the liquid also contains a black dye and substances that allow the particles to acquire an electrical charge. The particles and the liquid mixture are placed between two closely spaced parallel conductive plates. A pattern of B&W regions in the front panel can be created by dividing the rear electrode into small pixel elements and applying appropriate voltages between the top and rear electrodes. The electrically charged particles that move to the top electrode reflect light to give the appearance of white, while the absence of these create black regions due to the light absorbed by the dark dye.

Other technologies of electronic paper exist based on LCDs such as bistable LCDs, and cholesteric LCDs (Section 15.5). In addition, color electronic paper has been under development for years, using filters and triads of pixels to render the subtractive primary colors (see Section 15.3). The advantages of electronic paper include (i) better readability than other displays, particularly outdoors, (ii) low power consumption, (iii) low cost, and (iv) light weight.

Although dedicated e-readers have lost significant market share since 2010 to computer tablets, which can also display e-books, there is a consensus that the

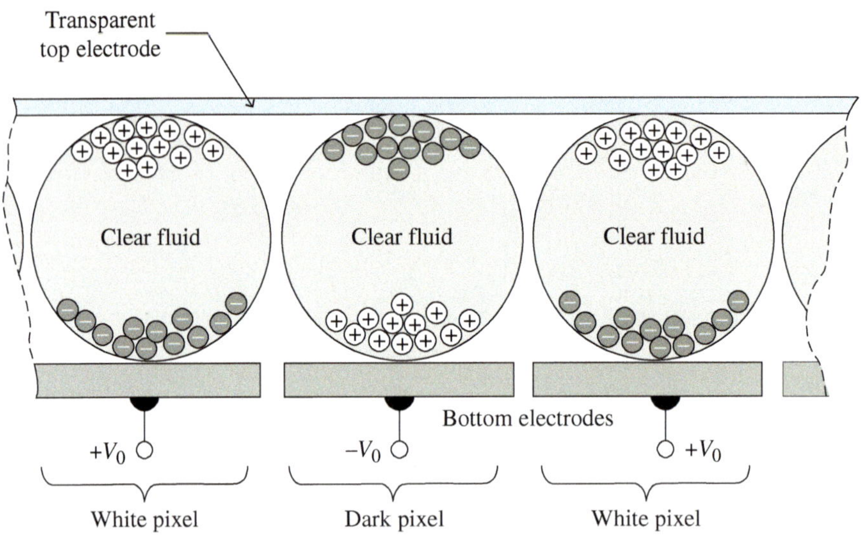

FIGURE 15.18 The electrophoretic process in one type of electronic paper: the migration of electrically charged particles determines regions of black (center) and white pixels (left and right).

technology has a future. Most consumer products based on electronic paper use rigid surfaces, but a few devices exist that employ flexible plastic. It is conceivable that in the near future entire newspapers will be available in non-disposable, flexible electronic paper. Further, the ideal digital book would have flexible pages similar to an ordinary book, except it could typeset and update itself.

15.10 NEAR-EYE DISPLAYS, AUGMENTED REALITY, AND VIRTUAL REALITY

Imagine flipping a switch in your glasses and turning the visual field into a complete virtual rendition of the screen in a movie theater, or overlaying additional information such as a map derived from GPS onto the real view of a landscape, or repairing an electronic circuit while superimposing the circuit diagram downloaded from the internet. Military aviators have used information displays projected onto their visors for a number of years. Today, there are already commercial products that "more or less" accomplish some of these tasks. We say "more or less" because most of the existing devices are prototypes, the first of their kind. There is still very intense R&D by many companies in a race to introduce something to replace smartphones and computer tablets. Nevertheless, applications other than personal use and entertainment have been developed and continue to evolve: defense (e.g., for fight pilots), industry (e.g., auto repair), and medical (e.g., surgery). We present in this section a brief overview of the physics of near-eye displays (NEDs) and augmented reality (AR).

To quote Hainich and Bimber (2011), "although the human eye is quite good at seeing, from the point of view of a classical optics design, it is a mess." In fact, a great deal of signal processing occurs at the retina and brain to compensate for many of the eye's "poor optics." Other than its color-sensing abilities, the most important capabilities of the human visual system are its large dynamic range in light intensity, some nine orders of magnitude, and the very extensive *field-of-view* (FOV), exceeding 180° and 100° in the horizontal and vertical directions, respectively. But the eye can only see clearly at the center of view due to the concentration of sensors at a region in the retina called the *fovea*—see Figure 15.2. Therefore, the eyes must scan and refocus rapidly over an extended scene. This feature can be taken advantage of for delivering a good view only in the eye's direction, which may involve eye tracking or producing a raster-scanned image directly on the retina with light coming from a device such as a DMD (Section 15.7). Eye tracking is not implemented in current NEDs, but retina beam-scanning has been developed.

There are at least three broad classes of NEDs: view-covering (total occlusion), semi-covering (partial occlusion), and complete see-through (non-occluded). Commercial view-covering NEDs are designed for watching movies, TV, and gaming in a virtual large screen. These devices can also provide virtual reality (VR), AR, 3D, or a combination of all these capabilities. NEDs with non-occlusion or partial occlusion require more elaborate optics designs that can involve not only lenses and mirrors but also prisms and other elements.

One example of view-covering NED is the Sony HMZ T3W personal 3D viewer, introduced in late 2013. The Sony viewer uses two organic LED (OLED) screens, one per eye, and provides 16:9 aspect ratio with 720p resolution and a 45° viewing angle. It is advertised as providing a virtual image size of 150 inch (!) at 20 m distance. From a Sony website, "you can enjoy not just outstanding 2D video but also the most immersive 3D experience available." A competing device, although only used for gaming, is the Oculus Rift made by Oculus VR, a company acquired by Facebook in 2014.

An example of an NED with complete see-through is Google's Glass. The Glass is essentially a "wearable" computer with camera included; it also lends itself for simple augmented-reality experiences such as overlaying a map or floor plan on the real scene. Glass was introduced to the public in 2014 and discontinued in early 2015, but Google is working on "Glass 2." The market failure of the Glass was likely due to high price and also to overwhelming competition from the capabilities and simple ergonomics of smart cellphones and tablets.

A first-generation (2004) non-occluded scanned-beam display, the "Nomad Expert Technician System" made by Microvision Inc. of Redmond, WA, employs a red laser diode and standard optics to scan a monochrome image directly on the retina. A wireless computer, worn on the belt, drives a 1.5 mm × 1.5 mm DMD on the headset; a window in front of the user's eye reflects a low-power laser light (about 1 mW) to the eye. The beam is never concentrated on a single spot for any extended period of time because the light is rapidly scanned. Moreover, an interlock circuit makes sure that the beam is "ON" only while scanning. The device has SVGA resolution and a FOV of 23° × 17°. A futuristic full-color version of the Nomad employing three color lasers is illustrated in Figure 15.19.

Second-generation AR/VR viewers are now (2015) available for mostly industrial, military, and training/educational uses. In some cases, however, the applications are not based just on NED concepts, but also as "Apps" for tablets and smart cellphones. Retailers are also beginning to use AR/VR (e.g., Lowe's Holoroom) to enhance the in-store consumer experience in a way not possible, yet, in online shopping.

To conclude, we mention Microsoft's HoloLens project. The goal of Microsoft is, apparently, to integrate AR, VR, gaming, and computing. A visor similar in appearance to Sony's HMZ T3W device would provide real-time "holographic computing." The technology relies on a "holographic processing unit" developed by Microsoft to process "terabytes of information" from a number of sensors that include cameras, accelerometers, and microphones. Ultimately, HoloLens would be a major component of the new Windows 10 or later, operating system.

There are clear advantages of NEDs over large standard displays: light and wearable, multi-use (VR, AR, communications, etc.), lower power consumption, and lower environmental impact. However, there are obvious ergonomics issues and social aspects that may preclude the easy adoption of this technology.

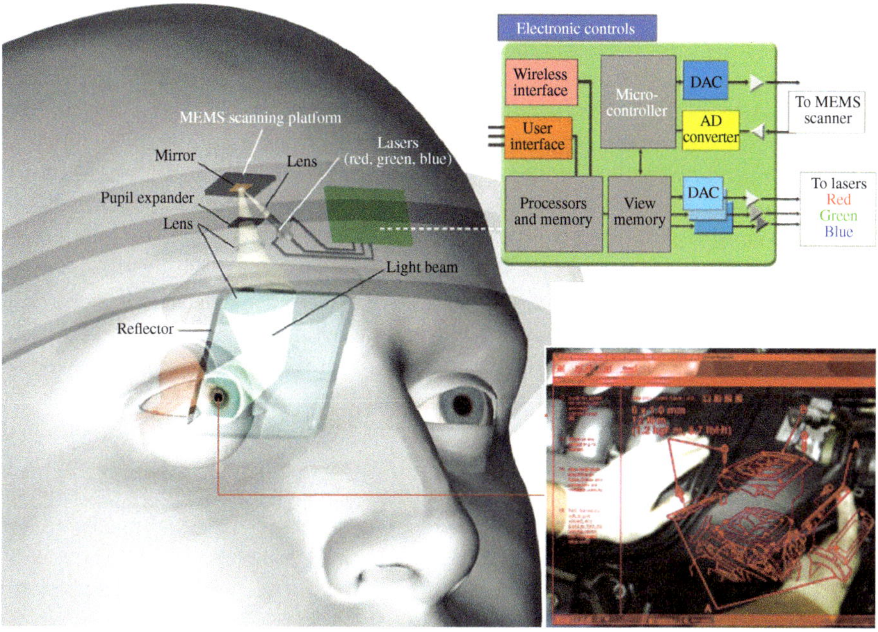

FIGURE 15.19 Main components of a futuristic color retina scan system mounted on a headset. The system is an expansion of Nomad, an existing commercial monochrome product for augmented vision made by Microvision Inc. of Redmond, WA. Source: Lewis (2004). © May 2004. Reproduced with permission of IEEE Spectrum.

15.11 STEREOSCOPIC, AUTOSTEREOSCOPIC, AND HOLOGRAPHIC 3D DISPLAYS

Paige greets you at Dulles international airport in Virginia (mid-2012) and tells you how to get through customs and about the papers that you need. Paige is a full-size "virtual assistant," similar to others at a few airports in Europe and in Dubai; she cannot answer questions but a future upgrade will have that capability. Paige is not really a fully 3D hologram, like those depicted in *Star Wars* and other science fiction movies, but a projection. She is a precursor to real holographic assistants of the not-too-distant future.

Fully 3D TV, as opposed to the current stereographic TVs, has been a dream since the invention of holography in the 1960s. As discussed briefly in Chapter 10, stereographic displays rely on the overlap of two slightly separated versions of the same scene that the human visual system perceives as one scene with depth. There are two major drawbacks of stereographic projections: the need for special goggles and, most importantly, the mismatch between *accommodation* (focusing) by the eye and *vergence* (convergence or divergence). The latter refers to the rotation of the eyes when directing the attention to different depths of the scene: the eyes rotate toward each other when focusing to closer planes, or away from each other when the planes are farther away. The mismatch results mostly from a conflict between the main *focus cue*, which is the display screen, and the intended focus depth. Many people suffer from eye strain and discomfort after watching stereographic movies as a consequence.

There are ways around the accommodation–vergence mismatch and the need for special glasses. One approach pioneered by Sharp Corp. in Japan in 2006 and more recently pursued by Apple Corporation is to use a screen with special pixels that can track the eye movement of the observer. The net effect of this technology is to create a "pseudo-holographic" image without the use of glasses. To quote from the patent issued to Apple Corporation, "By tracking movements of the eye locations of the observer, the left and right 3D sub-images are adjusted in response to the tracked eye movements to produce images that mimic a real hologram. The invention can accordingly continuously project a 3D image to the observer that recreates the actual viewing experience that the observer would have when moving in space around and in the vicinity of various virtual objects displayed therein. This is the same experiential viewing effect that is afforded by a hologram."

Technologies such as the one just described are sometimes called *autostereoscopic*, whereby light *rays* are manipulated to yield glasses-free 3D images. Other similar methods, already implemented in commercial products, involve the use of special screens to split the images into two fields, one for each eye. For example, the Nintendo 3DS portable console game employs *lenticular technology* to convert 2D images to 3D. In effect, the screen is composed of 1D vertical "lenses," actually realized with LCD elements, to project two images with electrically variable depth of field. The method works well for just one viewer at an optimal distance of about 30 cm. In mid-2013, only Toshiba has released a 3D TV based on glasses-free lenticular technology; as with the Nintendo console, the main drawbacks are the limited viewing distance and angle.

In contrast to autostereoscopic technologies, holography handles *wavefronts*. As explained in Chapter 10 and illustrated in Figure 10.12a and b, holography involves the recreation of a 3D scene from amplitude and phase information stored in an interference pattern or hologram. However, there are many vexing problems associated with generating holographic TV or *movies*, or *real-time* holography: the capture, transmission, signal processing, and display of holograms with sufficient *resolution* and *refresh rate*. Good resolution is the most important requirement of any high-quality hologram, one that can be viewed with a reasonable range of angles. From the physics of diffraction (Chapter 10), the pixel size in a hologram must be comparable to the wavelength of light that is diffracted when playing back the hologram. But as Leith and Upatnieks, the inventors of off-axis transmission holography, remarked almost 50 years ago, "The large amount of redundancy in hologram imaging does leave room for hope that considerable bandwidth reduction is possible." The situation is not completely unlike the early stages in the development of standard color TV, when many tradeoffs and very intricate analog signal processing were necessary to yield good color reproduction *and* acceptable transmission bandwidth.

In fact, several research groups around the world have pursued a number of analog techniques for the generation and transmission of simple holograms for some 30 years since their invention. In the 1990s, for example, a group in Japan designed and built a prototype analog real-time holographic TV transmission system based on LCDs. The technological focus since then has moved to digital technologies thanks to the rapid advances in computers, image processing, and communication. One subsequent approach consists in generating images with incoherent light and a number of standard 2D video cameras and then sending the information so the holograms can be *computed at the display*. This technology is also the basis for *telepresence*, the 3D version of a video conference or Skype. A group at the University of Arizona demonstrated in 2010 telepresence at 1 frame/s.

There are many applications of digital holography other than 3D TV and telepresence. One notable example is holographic microscopy: it provides the possibility of studying living cells without affecting them. Although the resolution of detectors like CCDs is not as good as that of photographic media, digital processing techniques can yield improved results. The most faithful hologram, but also complex one, is the near-field *Fresnel hologram*, but several techniques are available for reducing the computational requirements of such hologram. (See Chapter 10 on Fresnel vs. Fraunhofer diffraction.) For example, 3D holograms reduced to "horizontal parallax only" (HPO) make their computation much faster. The HPO hologram is constructed from a set of parallax views that can be obtained with several cameras or one camera in different positions (or a ranging camera) and that can be compressed with standard techniques. Using these ideas, the MIT Media Lab has produced HPOs for display at 15 frame/s.

As mentioned earlier, good resolution is a key requirement of any medium for storing a hologram. The computed or captured diffraction pattern must be imprinted on a physical medium called a *spatial light modulator* (SLM) whose properties, size, and spatial resolution must match those of the intended hologram. A SLM must be able to control the amplitude or phase, or both, of the illuminating light for recreating the original 3D scene. In practice, most SLMs can control either the amplitude or the phase, but not both. Nevertheless, controlling just the phase is sufficient for producing first-order diffracted light with good intensity. Examples of SLMs are the photorefractive materials, *acousto-optic modulator* (AOM), DMD (Section 15.7), and LCDs (Section 15.5). The DMD has been used as a phase modulator to produce dynamic Fraunhofer holograms, based on computer-generated interference patterns, by illuminating the DMD with laser light. However, DMDs have limited resolution and other problems for holographic applications, due to the mirror size and gaps between them. Figure 15.20 illustrates the process for obtaining a real-image hologram using DLP technology. In 2013, many groups in industry and academia were developing better SLMs and holographic techniques in a race to produce the first commercial holographic displays.

Another type of 3D technology is the volumetric display. The images in this case occupy a true volume, similar to the real images created by positive lenses or concave mirrors. Naturally, volumetric displays are not only glasses-free but also allow a 360° FOV. In one scheme, a large number of 2D images are projected by a high-speed video projector into a fast-rotating screen. The persistence of the eye then gives the sensation of 3D out of the fast changing 2D projections. The original object is "sliced" into many 2D images that are then mapped into a 3D matrix; the 3D data is composed of *voxels* (for "volume element"), the 3D equivalent of pixels. Figure 15.21

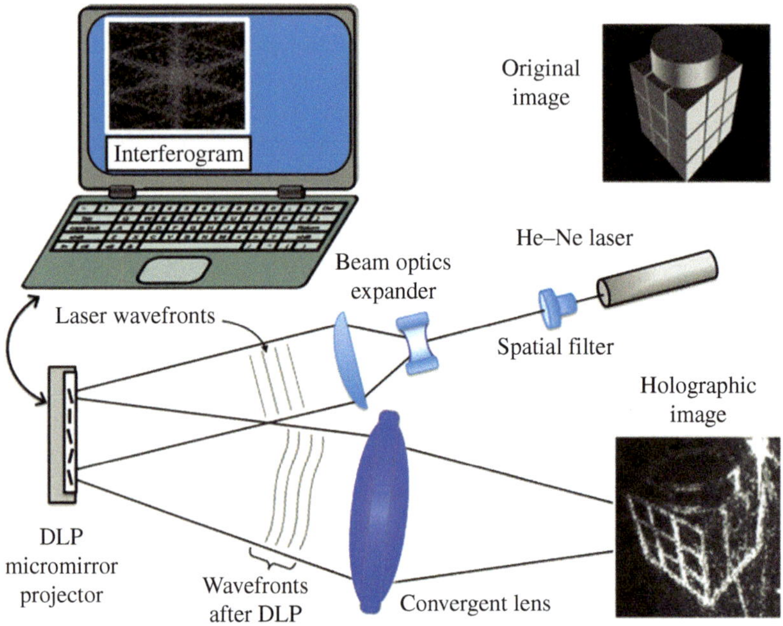

FIGURE 15.20 Digital holographic reconstruction using a laser-illuminated digital-micromirror array. Source: Adapted from Huebschman, Munjuluri, and Garner (2003). Reproduced with permission of OSA Publishing.

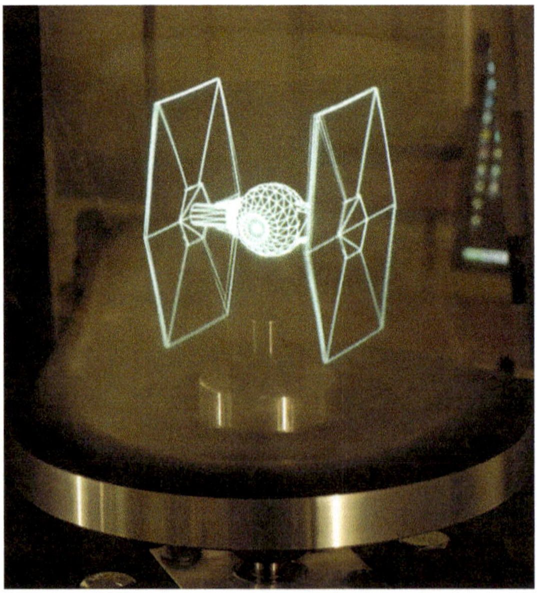

FIGURE 15.21 Example of a volumetric 3D display. Source: Photograph used with permission from University of Southern California, Institute for Creative Technologies.

shows an example of a B&W volumetric display developed by the Graphics Lab at the University of Southern California. The imaging system employs a mirror spinning at 1000 rpm and covered with a special "anisotropic holographic diffuser"; the mirror is tilted at 45° and reflects light from a high-speed DLP projector (Section 15.7). In turn, the projector is driven by a personal computer which decodes a standard DVI (video) signal and allows refresh rates of 20 Hz. The horizontal perspective obtained is accurate, but correct vertical parallax is achieved through tracking the height and distance of the viewer so that the projection can be adjusted.

Another implementation of volumetric display relies on the use of multiple stacked LCD screens that can be selected by appropriate application of voltages to scatter images from a high-speed video projector. The YouTube video at http://www.youtube.com/watch?v=RAasdH10Irg shows additional details of the 3D DepthCube, a 3D monitor made by LightSpace Technologies from Sweeden with many applications for visualization in science and technology. Finally, it is possible to generate volumetric displays without the use of screens, moving or otherwise, by the excitation of small regions in plasma by laser light. This method is the 3D version of the decades-old 2D laser light shows that are the attractions in sites like Stone Mountain Park in Atlanta, Georgia.

16

SPACECRAFT SYSTEMS

There are five essential subsystems of any manned or unmanned spacecraft: propulsion, attitude control, power, environmental control, and telemetry. Propulsion, normally in the form of a rocket, is used to get the spacecraft off the launch pad and usually into an orbit. The launch is often the most risky phase of a mission with equipment having to sustain enormous random forces in three axes due to shaking. Propulsion is also required for any mid-course corrections, for orbit insertion around a planet or moon and, if necessary, for a controlled deorbit. The pointing orientation (attitude) of the spacecraft in three axes must be known at all times. Accurate attitudes are necessary to engage propulsion systems effectively, to insure the solar panels are in a positive-power orientation, and to send or receive information via telemetry antennas. Power is necessary to operate computers and other electronics as well as to maintain the various operating temperatures throughout the spacecraft and payload. Waste heat from electronics and the power electronics itself are a significant component of thermal management with the heat being sent to various portions of the payload. Additional small heating elements are usually also required. More generally, environmental management also includes contamination control from materials that might outgas in vacuum and the enormous stresses that all subsystems experience during launch. Telemetry is the communications subsystem between the spacecraft and mission control back on Earth. Information about the spacecraft's health and performance as well as data must be transmitted to the ground and instructions must be sent to the spacecraft and payload to perform various tasks.

Modern Devices: The Simple Physics of Sophisticated Technology, First Edition.
Charles L. Joseph and Santiago Bernal.

Companion website: www.wiley.com/go/joseph/moderndevices

16.1 OPERATING IN SPACE: AN OVERVIEW

In addition to the human factor (e.g., astronauts and cosmonauts), we include in the phrase "working in space" any instrument or robotic device that must function outside of the confines of Earth. Environmental differences introduce operational difficulties that have to be mitigated. Most electronic equipment, for example, are designed to be cooled via convection, which is unavailable on orbit. Space is a particularly harsh environment for both humans and equipment even inside a pressurized spacecraft. Equipment must withstand tremendous shaking (10–15 g) during launch and subsequently function in an environment of very different temperature and pressure. The wear and tear on various apparatus operating in space is analogous to a racecar running the Indianapolis 500, where it is not uncommon for the team to have to replace or substantially rebuild the car's engine after one or two races due to the tremendous stresses. Similarly, the space environment eventually shreds the outer layers of multilayer insulator (MLI) as well as continually degrades and severely limits operational life of most equipment. There needs to be redundant backup components for critically important subsystems.

The choice of orbit is perhaps the most critical driver of all other subsystem designs and tradeoffs. For example, a satellite that must operate farther away from the Sun than the orbit of Mars probably requires a nuclear power supply since solar panels will not produce enough power. Mass (weight) is also a serious design driver since 10 kg of fuel are required for every 1 kg of payload placed in the lowest orbit. The mass budget is far more restricted if the payload is to achieve high Earth orbit (HEO) or beyond. For some or all-critical satellite components, it might not be possible to carry adequate internal shielding from this harshest radiation environment. (Mitigating long-term radiation exposure is perhaps the major challenge to putting humans on the surface of Mars. It takes 3 days to reach the moon, but several months to reach Mars where the Martian magnetic field and tenuous atmosphere provide additional shielding.)

Most orbits are closed-looped, returning repeatedly to same point in space relative to the object it orbits. There are drift-away orbits and paths designed to use the planets "to sling shot" to other planets or to leave the solar system altogether. However, all of these spacecraft paths are really in orbit about the Sun or at the very least, in a series of discrete solar orbits with transitions between each. Over the years, the vast majority of satellites and spacecraft have been launched into low Earth orbit (LEO), being 400–600 km above the surface and circling the Earth once every 70–100 minutes. As can be seen in Figure 16.1, the LEO of the Hubble Space Telescope (HST) is only about 1.1 Earth radius from the center of the Earth. There is a graphic on the left side of Figure 16.1, showing the relative scale heights and positions relative to the lunar orbit of the inner and outer Van Allen radiation belts, the range of LEOs, and two available geosynchronous orbits that fall into the class of HEOs. A geosynchronous orbit with a period of 24 hours is obtained at an altitude that is approximately one-tenth the distance to the moon, allowing a spacecraft to remain in direct contact with a ground control station if it is in a more or less equatorial orbit.

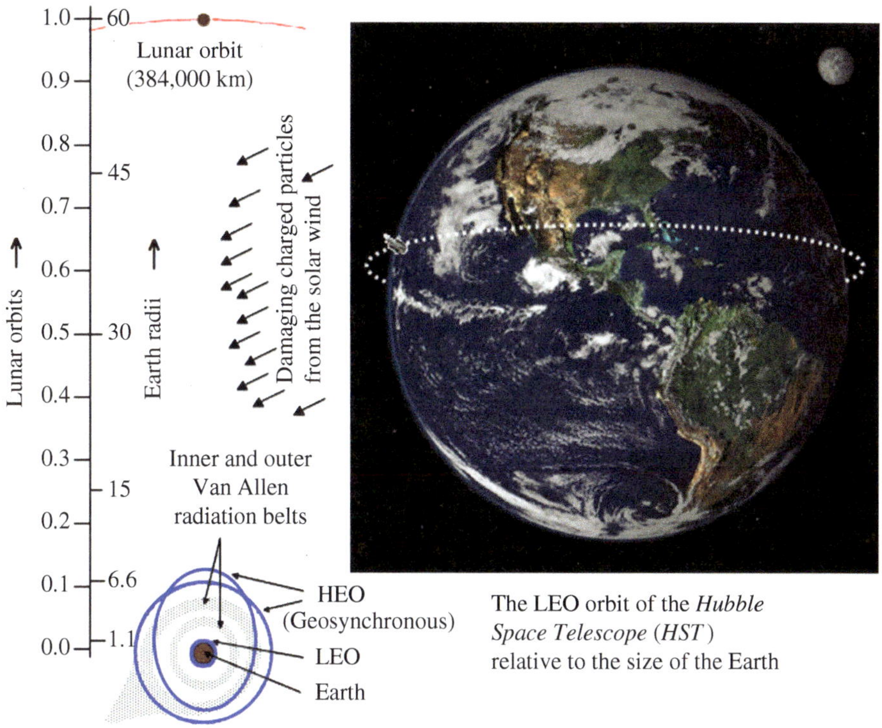

FIGURE 16.1 Low Earth orbits (LEOs) compared to the size of the Earth, the location of the Van Allen radiation belts, high Earth orbits (HEOs), and the distance to the moon. Source: NASA.

The harshest environment occurs beyond a geosynchronous orbit. Out there, a spacecraft is exposed directly to the solar wind, consisting of continuous flow of energetic charged particles (mostly protons and electrons) that are known as cosmic rays. Cosmic rays degrade solid-state circuitry and are capable of altering the digital states in a computer. This flow of charged particles is punctuated from time to time by intense, violent outbursts from the surface of the Sun, known to solar physicists as coronal mass ejections, which are capable of completely incapacitating a spacecraft or bringing down the electrical power grid on Earth. The Earth's magnetic field traps most of the cosmic rays that would impinge on the Earth or on a spacecraft in low Earth orbit. The accumulation of electrons and ions surrounding the Earth are know as the Van Allen radiation belts. Inside the Van Allen belts, the energies of these trapped cosmic rays are reduced through scattering processes and upon sufficient charge accumulation, are discharged into Earth's magnetic poles, creating the Aura Borealis (the Northern Lights seen at northern latitudes). While the Earth's atmosphere and LEO satellites are largely shielded from cosmic rays, a sufficient fraction of subatomic particles still have destructive consequences.

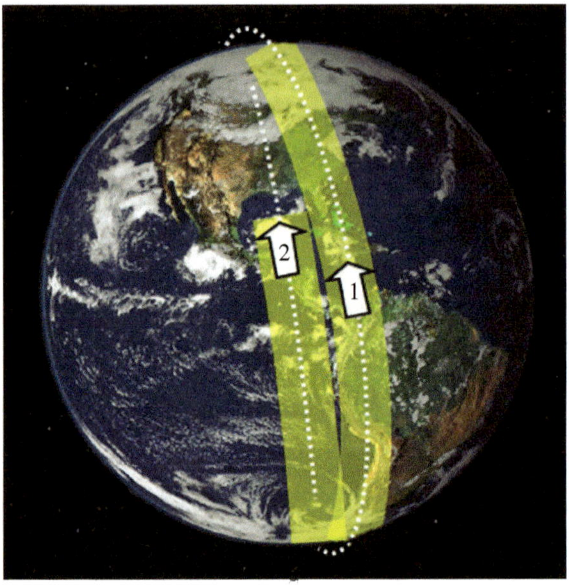

FIGURE 16.2 Spatial coverage over two orbits of a satellite in polar orbit. Source: NASA.

Many, but by all means not all, satellites and spacecraft are launched into nearly equatorial Earth orbits. A payload launched from low latitudes in an Eastward direction requires less energy to obtain a particular orbital altitude since the trajectory is taking advantage of the axial spin of the Earth. Spacecraft such as military reconnaissance or weather satellites that require full Earth coverage are often placed in a polar orbit. Figure 16.2 shows an example of a surface coverage for a satellite over two orbits. *Note*: the trajectories are not pure South to North (or visa versa) since the Earth rotates continuously beneath the satellite. Earth's rotation also causes the surface coverage of each subsequent orbit to be 1.5 time zones to the West of the previous pass, assuming a 90-minute period. There can be temporary gaps in the low-latitude coverage and usually significant overlap near the poles.

The launch phase is a particularly perilous portion of a mission. To avoid endangering the public, the United States primarily launches spacecraft destined for equatorial orbits over the Atlantic Ocean from the Eastern side of Florida (i.e., Kennedy Space Center (KSC)/Cape Canaveral) and launches polar orbiting missions Southwardly over the Pacific ocean from Southern California (Vandenberg Air Force Base). The Earth's rotation and the geographical relationship between Mexico and the United States make Vandenberg AFB an ideal launch site. The primary launch facility for the European Space Agency (ESA) is the Centre Spatial Guyanais at Kourou, French Guiana. ESA launches eastward over the Atlantic Ocean using an Ariane rocket or rather uses one from a family of French-built Ariane expendable launch vehicles (ELVs).

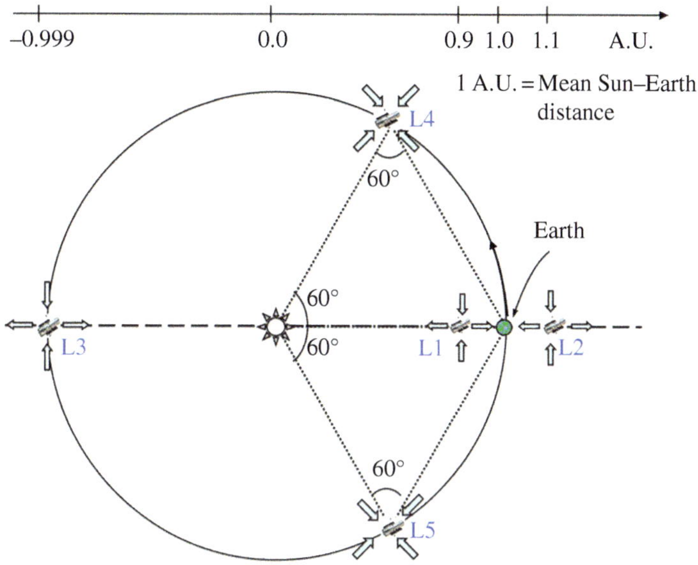

FIGURE 16.3 The locations of the five Lagrangian points in the Sun–Earth system.

Beyond the orbit of the moon are some particularly useful solar orbits, known as the Lagrangian points. These are locations where the combined gravitational pulls from two massive objects (i.e., the Sun and Earth) enable a spacecraft to maintain its relative position with minimal expenditures of energy. Figure 16.3 shows the positions of the Sun–Earth Lagrange points, L1 through L5. *Note*: the relative sizes of the Earth and satellite depictions are not to scale. L4 and L5 at the apexes of an equilateral triangle are stable. Once a spacecraft is parked near L4 or L5, the combined gravitational pulls from the Sun and Earth keep it there, counter acting most forces that perturb its orbit. Lagrangian points, L1, L2, and L3 have combined gravitational forces that create a saddle point potential. If the satellite happens to have an orbital velocity that is slightly too fast or slow, resulting in getting ahead or behind the Sun–Earth line in Figure 16.3, the combined Sun–Earth gravitational pull restores the satellite back to the Sun–Earth line. However, if a satellite near L1 or L2 drifts in the direction of toward or away from the Earth, it will eventually leave the Lagrange point region never to return. Thus, a spacecraft parked in an L1 or L2 Lagrange point must expend small amounts of energy from time to time to remain there in a process known as station keeping. The situation is the same for the L3 location, but L3 to our knowledge has never been used.

INTERESTING TIDBIT TB16.1

Approximately 10 low-energy cosmic rays (CRs) per second impinge on every person. The vast majority pass straight through not interacting. Some are absorbed by the Earth, many are not. If you have ever been awake in the dark and experienced a brief sudden flash in only one of your eyes, chances are that was a Cosmic Ray being absorbed by your eye.

16.2 ATTITUDE CONTROL SYSTEM

We define attitude control system (ACS) to be both sensing the orientation of the spacecraft with respect to an inertial reference frame as well as the mechanisms to alter its orientation. (Attitude and articulation control subsystem (AACS) or attitude determination and control (ADAC) are alternative terms used instead of ACS.) Traditionally, spacecraft used mechanical gyroscopes that are spun up to several thousand revolutions per minute (RPM) and maintained at those angular speeds by a motor. A gyro such as that depicted in Figure 16.4 is often mounted in a pair of gimbal mounts, allowing the rotational axis of the inertial reference wheel to point to a fixed location with respect to the stars. Three gyros with rotation axes arranged in a Cartesian coordinate system are required to determine completely the spacecraft's attitude via gyros alone. (Occasionally, satellites that have continued to operate long past their designed lifetimes have relied on only two gyros plus a Sun position sensor, after a gyro failure.) As the spacecraft alters its orientation, the amount of change in the two pivots are monitored for each of the three gyros to determine the new spacecraft attitude. We will defer further discussion on gyros until we revisit gyroscopes, specifically optical ones.

The three ACS axes of satellites or spacecrafts are roll, pitch, and yaw, following the tradition of naval vessels. For a ship or airplane, the centerline running stern-to-bow direction defines the roll axis, and a positive roll corresponds to the port side rising relative to the starboard side. The vessel pitches up if its bow or nose elevates relative to its aft end. (It pitches down if the front moves down relative the back, a negative value of pitch.) A positive yaw maneuver in a plane or ship corresponds to a counter clockwise rotation (i.e., turning left for the captain). Figure 16.5 shows the

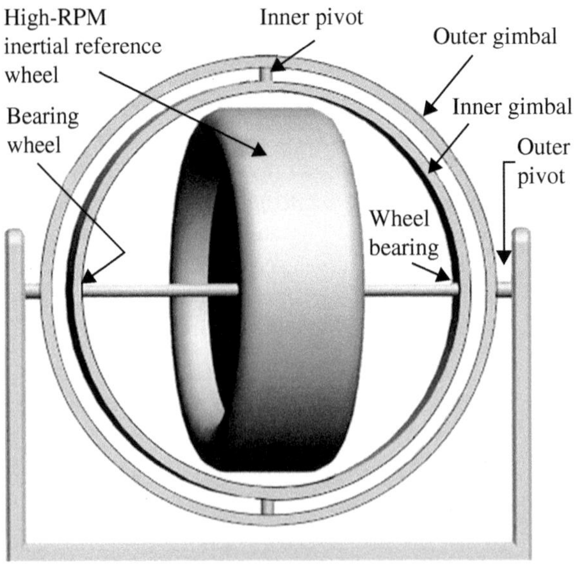

FIGURE 16.4 A mechanical gyroscope with gimbal mounts.

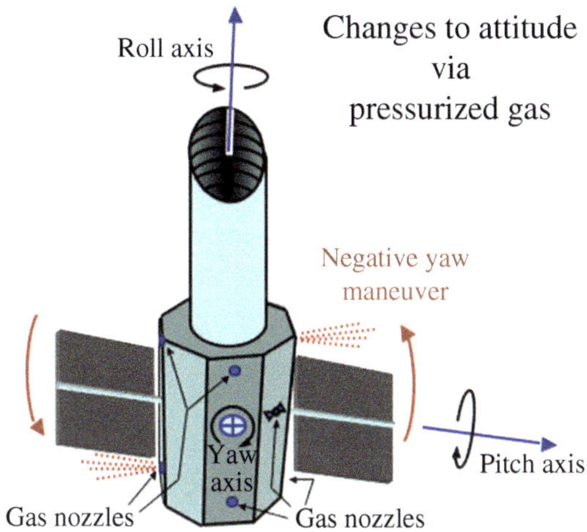

FIGURE 16.5 The roll, pitch, and yaw axes of a spacecraft carrying a telescope. The satellite is depicted in the initial stage of a negative yaw maneuver.

roll, pitch, and yaw motions for a spacecraft having a telescope. Often, a satellite will have six gas nozzles and a supply of compressed gas (usually hydrazine gas) for attitude maneuvers. The nozzles are organized in pairs, symmetrically placed on opposite sides of the center of mass of the spacecraft. Figure 16.5 depicts the start of a negative yaw maneuver where a few brief bursts of gas (shown as red dotted lines) are expelled at the lower left and upper right of the yaw axis. The satellite then rotates freely in the environment of space and continues rotating indefinitely until a second set of bursts from the opposing nozzles stops it.

A maneuver such as the one just described usually takes some trimming—weak or partial bursts to remove small residual drifts that still remain. If a satellite has a precision pointing requirement as many astronomical missions do, it usually carries a set of reaction wheels such as the one depicted in Figure 16.6 to provide smooth as well as fine attitude control. Reaction wheels operate on the principal of the conservation of angular momentum. To rotate the satellite to a new orientation, the spin rate of the reaction wheel is increased in the opposite direction and returned to its previous rate once the payload points in the desired direction. Traditionally, a reaction wheel had to have a mass that is a significant fraction of the total spacecraft mass, but mission projects may incorporate in the future a pair of high-RPM, low-mass wheels, which could reduced the mass from a few hundred kilograms to as little as 2–3 kg.

Reaction wheels are operated with some initial angular rotation rather than starting from rest for each new maneuver. The reaction wheel can be sped up to rotate the spacecraft in one direction or slowed for the opposite, which minimizes the stress on the motors and bearings. These wheels tend to accumulate excess angular momentum over time since the frictional forces are asymmetrical and maneuvers in one direction rarely exactly cancel ones in the opposite. The buildup requires a mechanism to

FIGURE 16.6 A reaction wheel assembly including drive motor and associated electronics. Source: NASA/Goddard Space Flight Center.

dump momentum occasionally to the outside universe. If the spacecraft is in LEO, it can use an electromagnet to push slowly against Earth's magnetic field to remove angular momentum. Gas propellant and nozzles similar to those depicted in Figure 16.6 can also be used, especially for a satellite far from Earth.

We now return to gyroscopes in general and optical replacements for mechanical gyros in particular. Conventional spinning-mass gyroscopes, which depend on the angular momentum generated by a rotating wheel or sphere, are highly developed and sophisticated. In fact, the *Gravity Probe B Experiment*, which had the tightest tolerance requirements of any gyro, relied on mechanically spun spheres. Nevertheless, reliability and lifetime are always issues associated with any mechanical device on orbit, especially one that is used extensively, whether it is a filter wheel, a camera shutter, or gyroscope. The first optical replacement was a ring laser gyroscope (RLG) as shown in Figure 16.6, which operates on the physical principal known as the Sagnac effect. If two beams of a single wavelength of light are sent on opposite paths around a closed loop, the two beams return with slightly different Doppler shifts, recombining to form a beating pattern that cycles back and forth between constructive and destructive interference. This is the Sagnac effect and applies to all types of optical gyros. Consider the RLG shown in Figure 16.7, which is rotating counter clockwise. The laser (top center) has partially silvered windows on both ends, allowing laser light to exit and enter both ends continuously. This laser continuously sends out two beams simultaneously, one to the left and one to the right. The beam exiting left has a longer path to complete the circuit than does the beam moving oppositely. The reason one path is longer is the mirrors, the laser, and the detector have rotated about the central point during the time it took the beams to complete the circuit. While the speed of light is extremely fast, its velocity is still finite and the interval of time that it takes for the beams to complete the loop, the mirrors, detector, and laser will have moved closer to one beam and farther away from the other.

Each of the two beams returns to the laser cavity with the same size Doppler shift, but with shifts of opposite signs (+ vs. –), creating a continuous beating in amplitude

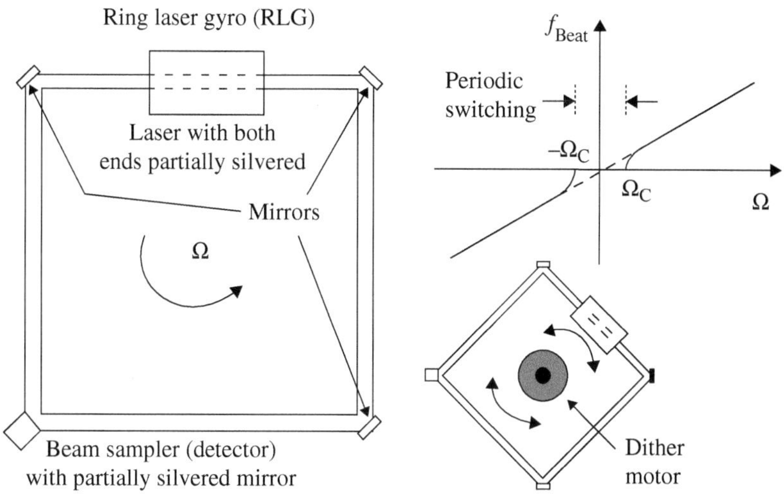

FIGURE 16.7 A Ring laser gyro (RLG), which operates on the Sagnac effect. Differences in Doppler shifts between the two interfering beams produce a beating pattern in the amplitude with a frequency, f_{Beat}, which is proportional to the rotation rate, Ω.

as the two beams sequence between constructive and destructive interference. The frequency with which the combined waves beat, f_{Beat}, is proportional to the angular rotation, Ω, and is plotted in Figure 16.7. Theoretically, an RLG is highly accurate with a rotational measurement limit of 10^{-6} degrees/h, but in practice, it is somewhat less due to fabrication limitations including cleanliness and a few inherent operational difficulties. *Note*: for extremely slow rotations ($\Omega \approx 0$), the laser cavity will first amplify one of the two beams and then the other, jumping indiscriminately and randomly between the two. The interaction between laser cavity and the two beams impinging on it, produces periodic switching and a discontinuity in the f_{Beat} versus Ω plot for rotations very close to zero. To overcome this instability for $\Omega \approx 0$, a rotational rocking back and forth of the RLG is superimposed on top of the spacecraft's rotation. The process is known as dithering and a simple motor is used to rock the RLG back and forth.

An alternative to the RLG is the fiber optic gyro (FOG). The principal advantage of FOG over RLG technology is the fiber can be 100 m to 1 km long and coiled into many loops that multiplicatively enhance the Sagnac effect while allowing FOG packages to be very small and lightweight. There are three basic types of fiber optic gyroscopes: interferometer, Brillouin, and resonance or I-FOG, B-FOG, and R-FOG, respectively. The I-FOG is the most common type in use as of 2012 with more than a dozen aerospace and electro-optical companies world wide producing various I-FOGs. R-FOGs, which offer superior performance, are primarily in the developmental stage. A generalized FOG is pictured plus an I-FOG is schematically represented in Figure 16.8 where the light source is typically a superluminescent diode (SLD). The I-FOG measures the amplitude of two interfering beams. It has a constant phase modulator that introduces a $\pi/2$ (90°) phase shift between the pair of light beams to maximize the rotationally induced change in amplitude. (See Intro Physics Flashback

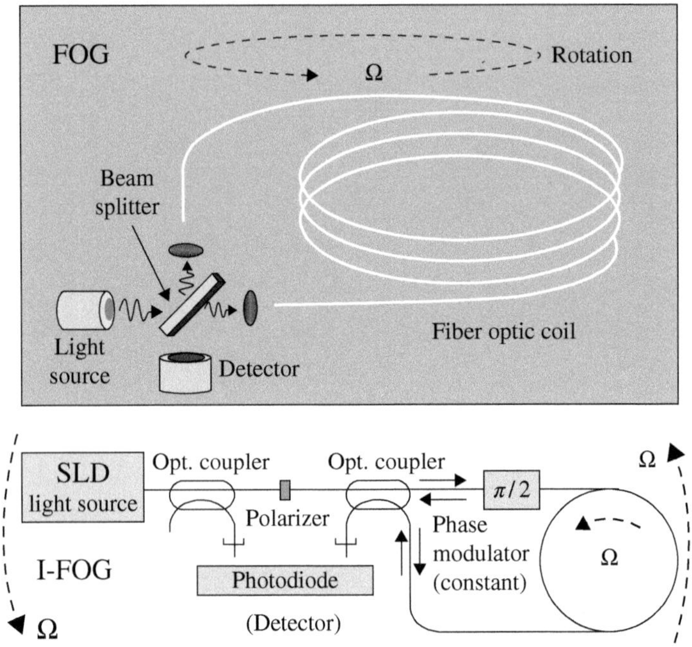

FIGURE 16.8 The basic components of a fiber optical gyroscope (FOG), consisting of a light source (normally a superluminescent diode), a beam splitter or fiber optical coupler, a 100 m to 1 km long coiled fiber cable, and a detector.

FB16.1 for the reason that small Doppler shifts centered about an induced phase shift of $\pi/2$ offers the optimal operating conditions. Also see Chapter 11 for a description of optical couplers.) It is important to realize that all components, including the optical couplers, photodiodes, and light sources, are rotating with the spacecraft. The fiber optic coil remains stationary with respect to the other parts of the FOG.

As an example of a two-axis gyroscope, the G-2000 manufactured by Northrop Grumman is pictured in Figure 16.9. This compact FOG is the smallest dynamically tuned gyroscope (DTG) produced, offering high performance, small size, and excellent reliability at a low cost. It is an excellent inertial reference system suitable for many commercial and military programs. A servo-electronics card is specifically tailored to maximize the performance accuracy of the G-2000.

An I-FOG is a simple two-pass interferometer, making it sensitive to noise factors such as any time-variant temperatures within the coil and any variations in the intensity of the light source. In contrast, an R-FOG is a multi-pass ring resonator with photons entering the ring bouncing back and forth through the resonator many times before escaping the ring. The R-FOG, which is still in the development stage, requires very few fiber optic loops around the resonator ring, making it much less sensitive than an I-FOG to noise. The average number of times that a photon reflects back through the ring resonator is known as its finesse. For an interference filter or a ring resonator, the reflectivity at both ends is approximately 80% and a high finesse typically consists of anywhere between 10 and 100 reflections before exiting. (See Section 12.3 on Fabry–Pérot etalons for

FIGURE 16.9 A miniaturized three-axis I-FOG inertial reference system. Source: Reproduced by permission of Northrop Grumman Corporation.

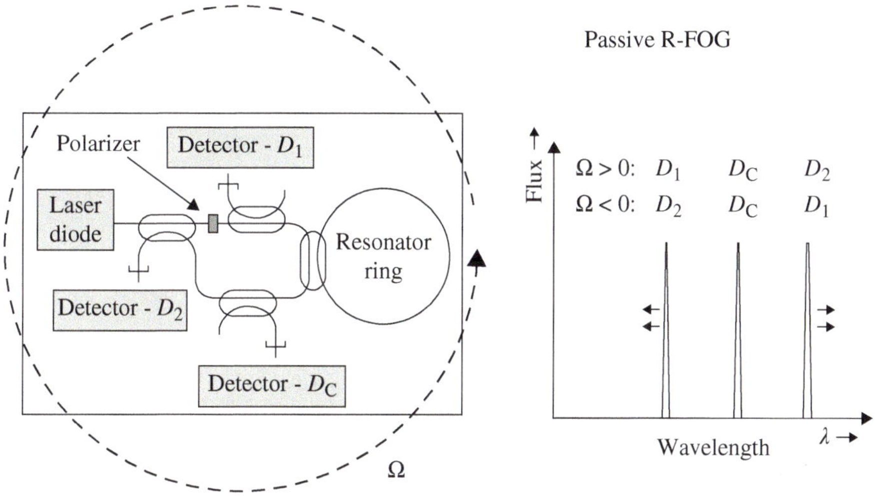

FIGURE 16.10 A passive resonance fiber optic gyro produces sharply peaked emission features with separations proportional to the rotational rate, Ω.

a more complete discussion of finesse.) Constructive interference near the resonance frequency results in three discrete narrow emission peaks, each appearing at different wavelengths at the detectors. The central detector, D_C, is the reference emission from the light source devoid of any influence from the resonator ring. Detectors, D_1 and D_2, sample the rotationally induced Sagnac effect. The wavelength separation of the emission peaks is proportional to the angular rotational rate, Ω, and the narrow width of the emission provides a very precise measurement. The direction or sign of the rotation is determined by which Doppler-shifted emission peak appears on which detector, D_1 or D_2.

Figure 16.10 depicts a passive R-FOG, while a closed-looped active R-FOG is shown in Figure 16.11. An active R-FOG employs a piezoelectric (PZT) phase shifter that is controlled by feedback circuitry including a lock-in amplifier. A signal generator is used to provide a saw-tooth or triangular wave of voltage, sweeping the phase modulator through a range of wavelength shifts. Destructive interference occurs most of the time, except for one particular point in the cycle where the phase shift leads to constructive interference. The continuous cycle results in discrete electrical pulses from the detectors with a constant period, effectively changing the angular rotation from a spatial-domain to a time-domain measurement, increasing the accuracy, precision, and robustness of the measurement.

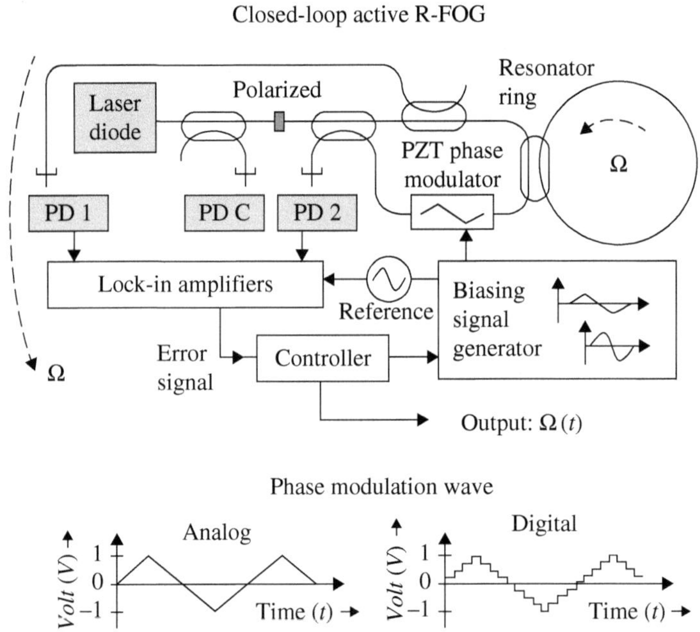

FIGURE 16.11 An active R-FOG.

COMPREHENSION VERIFICATION CV16.1

Problem: Sketch the attitude of a spacecraft as it orbits about a circular object. Place 6–8 simple arrows on all portions of the orbit to indicate the attitude pointing of the free-flying spacecraft (i.e., one that is not executing any maneuvers).

Answer:

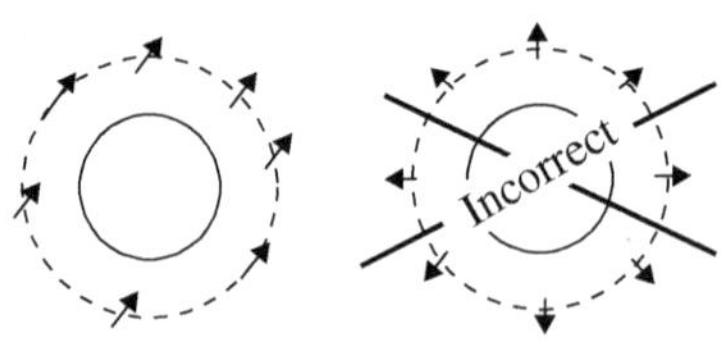

FIGURE CV16.1 Two possible solutions to the problem, a correct one and a commonly-made incorrect one.

Note: a common mistake made by students is pictured on the right side.

INTRO PHYSICS FLASHBACK FB16.1

Interference Amplitude as a Function of Phase Shift

Introductory physics textbooks usually display wave interference primarily for the case of pure constructive interference (phase difference = 0°) and for pure destructive interference (phase difference = 180° or π). (See Section 2.4.) An interference fiber optic gyroscope (I-FOG), only produces a small and a constant phase shift difference ($\Delta\phi \sim 0$) between the clockwise and counter clockwise moving beams, resulting in an interference amplitude that changes very slowly as a function of rotational speed, Ω. Figure FB16.1 plots the net wave (bold line) from the interference of two separate waves (thin lines). As can be seen from the figure, a small change in $\Delta\phi$ between the two plots, produces a tiny change in the amplitude (bold line) of combined wave when both waves are close to purely constructive interference.

If the rotationally induced phase shift occurs is added to two waves that already have a phase shift of $\pi/2$ (90°), the amplitude of the combined wave shown in bold in Figure FB16.2 becomes significant. Three interference plots are shown near $\Delta\phi = \pi/2$. The amplitude of the net wave (bold line) becomes progressively smaller as $\Delta\phi$ becomes larger than $\pi/2$, indicating its amplitude is sensitive to the rotational rate.

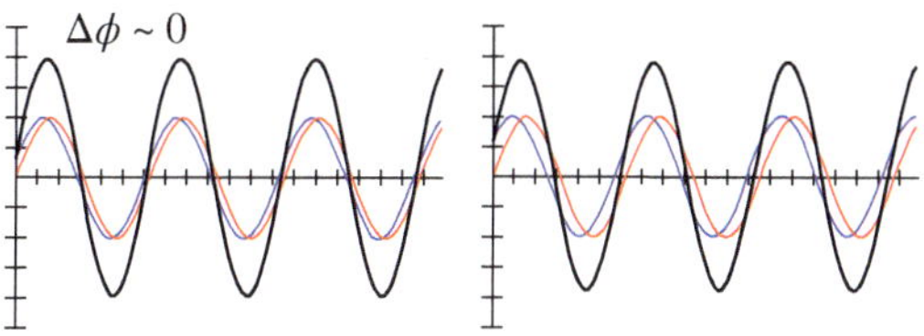

FIGURE FB16.1 A very small difference in the amplitude of the net wave is observed when there is virtually no phase shift between two interfering waves.

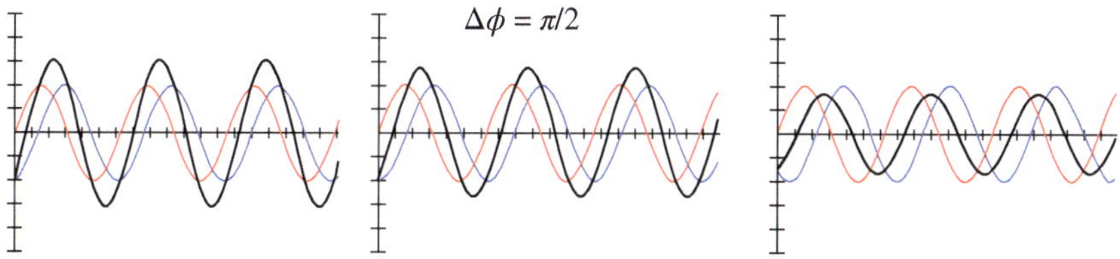

FIGURE FB16.2 A significant difference in the amplitude of the net wave is observed when the two interfering waves have a phase shift close to $\pi/2$ (90°).

16.3 SPACECRAFT POWER

DC power is usually obtained either by photovoltaic cells (solar panels) or by small nuclear reactors known as radioisotope thermoelectric generators (RTGs). Solar cells are similar to those found on rooftops on Earth to supplement the electrical power from the grid, except those used in space have a more complicated device architecture, are much higher quality devices, and made of highly pure solid-state materials. A photovoltaic cell is simply a solid-state P–N junction where incident light causes electrons to be promoted from the valance band to the conduction and then makes use of the intrinsic voltage difference near the P–N junction to induce a current. (See Section 6.3.) Cells can be arranged in a combination of series and parallel connections to create solar arrays with any desired DC voltage and current. Solar arrays can be placed on the skin of a spacecraft or unfolded into articulated panels that are more easily kept pointed in the Sun direction. The depth that incident light penetrates into a solid-state material is a function of its wavelength, enabling a simple, single P–N junction to have an absorption efficiency of only 7–10%. The efficiency can be increased to 35% by simply stacking various P–N junctions often of various solid-state materials one on top of the other to create a multijunction device in which the shortest wavelengths are captured in the top junction and longest-wavelength radiation being absorbed in the lowest P–N.

Radioisotope thermal generators generate heat by bombarding a metal surface with alpha particles from the radioactive decay of plutonium-38 (Pu38) and using the thermoelectric effect to generate a voltage across a pair of metals. The thermoelectric effect, depicted in Figure 16.12, occurs when the ends of rods (or wires) of two different metal alloys are held at different temperatures. At elevated temperatures, the electrons in both alloys easily jump from the valance to the conduction band, where these diffuse throughout the metals leading to small voltage differences between the two ends. Each metal alloy has a slightly different band gap and corresponding rate for promoting electrons to the conduction band, allowing current to flow throughout the loop. A somewhat more efficient design is to arrange most of the temperature drop (right diagram of Fig. 16.12) to occur across the alloy produces the greatest voltage.

A radioisotope thermoelectric generator, depicted in Figure 16.13, is frequently used for missions to the outer planets and moons, where the low solar flux makes solar panels impractical. Similarly, RTGs were used on Pioneer 10, Pioneer 11, Voyagers 1 and 2, all of which continue to operate after leaving the solar system. RTGs where also used on scientific instruments left on the moon by the Apollo 12 through 17 missions and by both Viking landers on Mars in the 1970s.

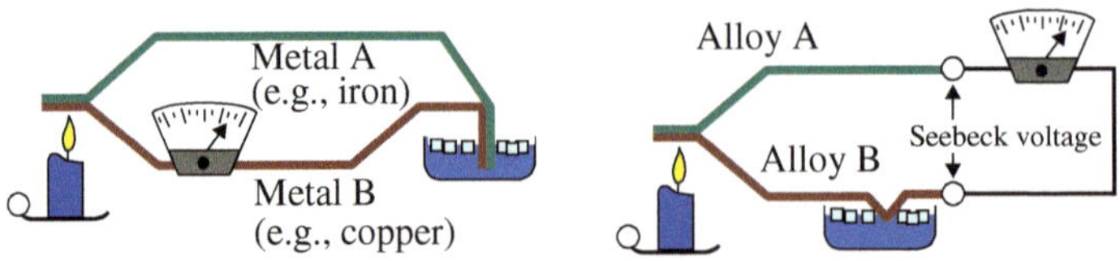

FIGURE 16.12 Two variants of the thermoelectric effect with the right diagram being somewhat more efficient than the left.

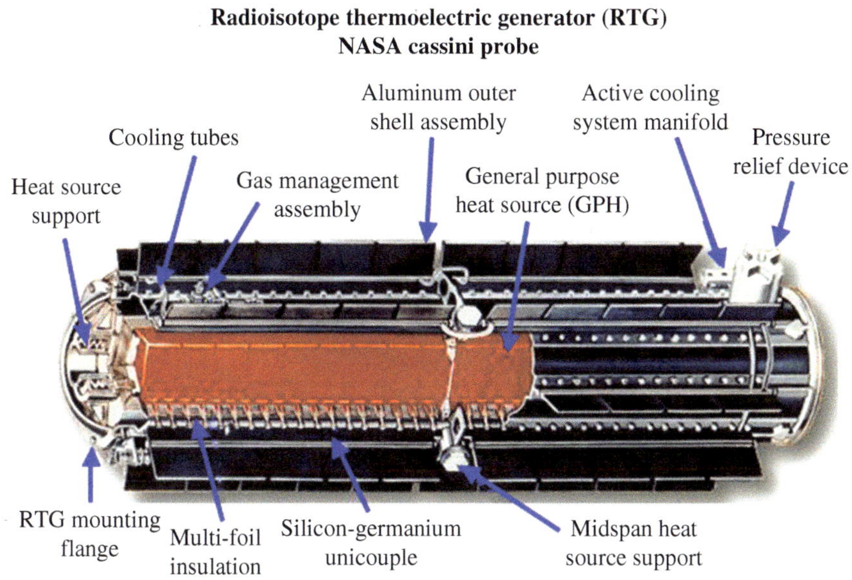

FIGURE 16.13 The schematic layout of the radioisotope thermoelectric generator (RTG) thermoelectric power system used in the Cassini mission to Saturn and its moon, Titan. Source: NASA.

16.4 THERMAL AND OTHER ENVIRONMENTAL CONTROL

As noted, operating in space is a difficult and often harsh environment. Equipment and humans must first survive a very violent three-axis shaking during launch. When in space the sunlit portion of the spacecraft faces several thousand-degree temperatures while any side exposed to deep space is facing −270°C. Oblong spacecraft can experience "hot dogging," a thermal bending that can cause stress and can cause mechanisms to jam due to structural distortion. Moreover, the alignment of optical elements, electronic packages, and mechanical devices all operate properly only over specific ranges of temperatures. The first and primary mitigating approach is to cover completely the outermost surface of the spacecraft with an MLI blanket of which the cross section one is shown in Figure 16.14. Normally, one small surface, consisting of a few metal fins, is left exposed to deep space to serve as a heat sink for cooling the spacecraft. All spacecraft undergo detailed thermal analysis to insure a balance between heat sources and cooling to deep space as well as distribute the thermal resources to maintain all subcomponents within operating tolerances.

The primary heat sources in most cases are waste heat from the inefficiencies of various electronics packages and perhaps a few small heating elements, powered by the solar panels. If the spacecraft is powered by a radioisotope thermoelectric generator, any supplemental heat at the satellite extremities is supplied by a few lightweight radioisotope heaters. Pictured in Figure 16.15 is a disassembled radioisotope heater compared to the size of a penny.

FIGURE 16.14 Multilayer insulator (MLI). Source: Rossie, https://commons.wikimedia.org/wiki/File:MultiLayerInsulationCloseup.jpg. Used under CC-BY-SA 2.5 https://creativecommons.org/licenses/by-sa/2.5/deed.en.

FIGURE 16.15 The components of a single lightweight radioisotope heater used to provide supplemental heat sources for deep-space satellites. The penny provides a size comparison.

Virtually all materials outgas in a vacuum. The surfaces of metal alloys, even if thoroughly cleaned and maintained in the cleanest environments, retain small amounts of contaminants that boil off these surfaces once in space. Most organic materials (e.g., some epoxies, carbon fiber struts, integrated circuit chips, and electronic circuit boards) slowly discharge complex molecules from the bulk materials, releasing particularly harmful contaminants and making the devices brittle and fragile. While most of the contaminants in unmanned missions eventually reach the vacuum of free space, some molecules roam around the spacecraft and can be adsorbed onto other surfaces, especially if the surface is cool or cold. Ultraviolet radiation, which is more severe in space than under Earth's atmosphere, will polymerize any contaminated surface. UV optics can be quickly degraded permanently by polymerized contaminants.

The strong presence of cosmic rays in space is another environmental factor requiring mitigation, often using shielding that adds mass. CRs damage electronics and are a significant background source for silicon-based image sensors. Any glass or crystal exposed to CRs, light up via Cherenkov radiation, causing optical couplers used for electrical isolation between the spacecraft and various subsystems to create false commands.

16.5 COMMAND, CONTROL, AND TELEMETRY

Spacecraft carry a central onboard computer, with basic software routines permanently burned into a programmable read-only memory (PROM) that automatically reloads in the event of a computer crash, power interruption, or the recovery from loss of attitude control. Microprocessors or subsystem instrument computers, depending on the complexity of the mission objectives, often support the central computer. The central computer is the primary with all other subsystems being slaves to it. For unmanned science, surveillance, or other government satellites, single commands, revised software programs, and other instructions are uploaded from the ground station from time to time and science data along with housekeeping information are downloaded back to the ground. Housekeeping information are those data from the spacecraft probes and indicators used to monitor the health and operating conditions of the various subsystems of the spacecraft. These include ACS indicators, electronic thermometers, voltage probes, and other strategic sensors with the numerical values being stored on board until commanded to download. Generally speaking, the volume of housekeeping data is very small compared to the images or other data to be downloaded. It is often inserted as part of a header, the preamble that describes the critical observational details associated with the primary image itself. Ground station commanding and control are normally carried on separate microwave frequencies within one standardized band and data telemetry link is often carried in a different band on a separate antenna. Commercial telecommunications also use one band for satellite operations and another for customer transmissions.

There are 17 microwave frequency bands defined by the Institute of Electrical and Electronics Engineers (IEEE) and 14 bands defined by the Radio Society of Great Britain. Each band is subdivided into numerous smaller communications frequency ranges, each being reserved for various uses from satellite communications, to GPS, to cell phones, to television broad casting, to maritime communications, and other applications with some being reserved exclusively for governments, for radar, and for radar ranging of satellites. Three common frequency ranges used by NASA are in the K_u (12–18 GHz), S (2–4 GHz), and X (8–12 GHz) bands. Early in NASA's history, the Mercury and Gemini missions used ultrahigh frequency (UHF) and very HF

(VHF) bands, 300 MHz to 3 GHz and 30–300 MHz, respectively. With the advent of the Apollo program to the moon, NASA began using the unified S band (USB) for voice, television transmission, commanding, and tracking. All of these manned missions required an extensive network of ground stations around the world.

NASA began replacing its multiple ground stations in the 1980s with the Tracking and Data Relay Satellite System (TDRSS, pronounced "tea dress"), consisting of six individual (three operational and three redundant contingency backup) TDRS satellites, all positioned in geosynchronous orbits. The system is designed to increase substantially the speed and volume of data transmission between the ground and collectively all spacecraft in orbit as well as increase the time intervals available for individual contact. Essentially, all US government data and information transfers now go through a pair of ground stations at White Sands Missile Range in New Mexico and the entire TDRSS is controlled at Goddard Space Flight Center in Greenbelt, Maryland. The system supports all US government departments and agencies, with highest priorities as of 2012 going to the department of defense in an emergency, the *International Space Station*, and the *HST* in that order. TDRSS also supports the telecommunications needs for the entire US government, including small NASA science missions such as stratospheric balloon flights as well as weather and climate missions by the National Oceans and Atmospheric Administration (NOAA).

A single TDRS (generation 2) satellite is pictured in Figure 16.16 after being deployed in 2002 by the space shuttle. Future Gen-3 TDRS spacecrafts will be launched on Atlas V rockets. The two large circular antennas independently articulate with one to pointing to an individual satellite or manned spacecraft and the other (if necessary) pointing to another TDRS for relaying the bit stream to the ground. Each of these two relay antennas can handle 8 data streams simultaneously, two bands (S and K_u) each with a pair of frequencies within the band and two polarization states (right- and left-hand circular polarization) for each of the frequencies. The solar arrays are seen unfolded as a set of flat rectangular objects, extending in both directions from the central portion of the satellite, and a long thin antenna extends from the bottom.

Figure 16.17 shows a schematic representation of a TDRS. There is an array of 30 multiple-access S-band antennas, 12 of which transmit and all receive, and a separate commercial K-band (small green dish). (*Note*: the Ku-band is different from—and is not a part of—the K band.) External to the TDRS body are the two large single-access antennas (black, wire mesh dishes), a space-to-ground link antenna (large green dish), and a C-band omni antenna (blue) for telemetry, tracking, and control (TT&C) of each TDRS. An omni antenna is one that transmits/receives in all directions. *Note*: there is a small solar sail (yellow) to help balance the small constant pressure of the solar wind, reducing the amount of gas that must be expended to maintain the correct TDRS attitude.

While TDRSS serves the needs of satellites in low and high Earth orbits, for missions significantly beyond geosynchronous, a Deep Space Network or DSN must be used. The US DSN is a worldwide network of large antennas with ultra-sensitive receiver electronics that are capable of receiving the very weak radio signals from interplanetary satellites and transmitting commands that are strong enough to overcome the $1/r^2$ spreading over very large distances. A NASA depiction is given in

FIGURE 16.16 Picture of TDRS (Gen 2) deployed by the space shuttle, but prior to orbital transfer to high earth orbit. Source: NASA.

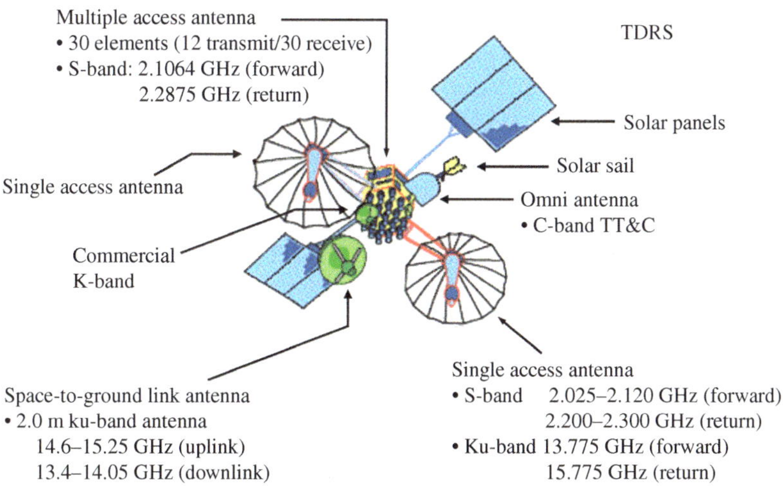

FIGURE 16.17 A schematic representation of a single TDRS spacecraft. Source: NASA/Jack Pfaller.

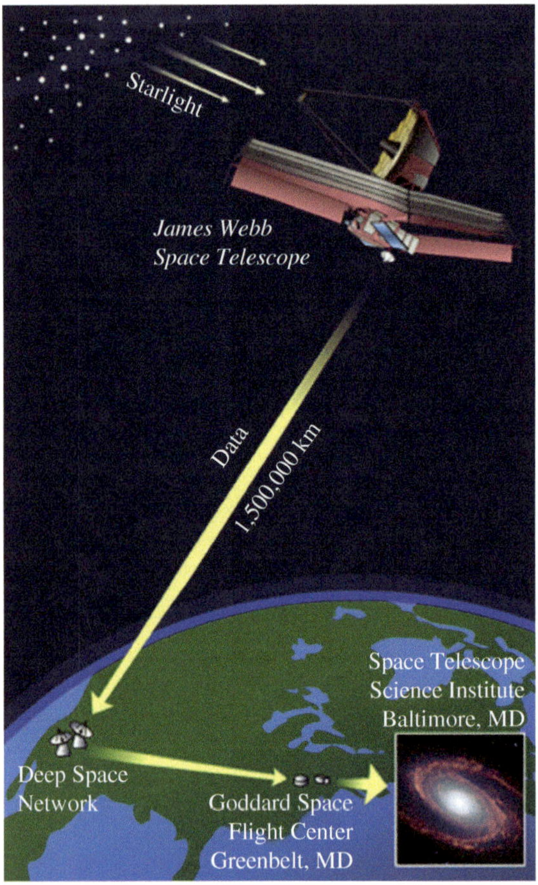

FIGURE 16.18 A schematic representation of the *James Webb Space Telescope* (*JWST*) communication system. *JWST* is scheduled to be launched in 2018. Source: NASA.

Figure 16.18 of the use of its DSN to communicate with the future *James Webb Space Telescope* (*JWST*), which will be parked six times further away than the moon in the solar L2 Lagrange point. Other similar networks include the European Space Agency's ESTRACK, the Chinese DSN, the Indian DSN, and the Russian Deep Space Network.

The United States has the oldest DSN, and it will be required to support twice as many missions in 2020 as it did in 2005, presenting some limitations and challenges in the future. First, there are four legacy missions, *Voyager* 1 & 2 plus *Pioneer* 10 & 11, which have all left the solar system and continue to operate and return scientific data well beyond their design life. Second, there is only one DSN site in the Southern Hemisphere, the Canberra Deep Space Communication Complex in Australia. New antennas, however, are scheduled to come online between 2014 and 2016. Third, the United States has been deferring major upkeep activities since the 1990s, including not replacing aging major components. Eventually, some of the DSN antennas will need to be replaced, most likely each with an array of smaller dishes tied together.

16.6 LAUNCH, PROPULSION, STATION KEEPING, AND DEORBIT

The launch phase usually carries the highest risk during any mission. Not only is there a 1 : 50 risk of a rocket malfunction, the payload experiences violent (12–15 g) shaking in three axis over a sustained period of 5–10 minutes. (Launch forces are kept just below 10 g when humans are going into space since 7 g turns in fighter jets are sufficient to cause pilots to black out.) The random forces in unmanned launches are equivalent to dropping a piece of equipment repeatedly onto a concrete floor from chest height. It is not uncommon for steel or aluminum parts to become work hardened and brittle due to the launch forces, causing parts to snap off and become ballistic projectiles capable of damaging other components. One standard procedure for space agencies such as NASA or ESA is to require quality-controlled traceability of parts, starting with material processes of purification and alloying of the raw materials, through the manufacturing of bolts, plates, and other shapes, and ending with the instillation of each subcomponent into the payload.

Rocket propulsion is the most common method use to place satellites or spacecraft into space. Future launches may one day employ an electromagnetic rail system up the side of a mountain either to throw payloads directly into space or to provide the initial acceleration prior to igniting a rocket. The primary advantage of a high-speed rail system is that much less onboard fuel must be expended initially to achieve orbit. The primary disadvantage is any rail system would require launching over populated areas. Other spacecraft propulsion systems include ion thrusters, solar sails, and cold gas thrusters such as discussed in Section 16.2 for attitude control. Cold gas jets are capable of increasing the orbital altitude of a spacecraft, circularizing its orbit, or station keeping as well as changing attitude orientations. Ion thrusters, which use either the electrostatic Coulomb force or the electromagnetic Lorentz force, create very small levels of thrust, but achieve very high specific impulses (i.e., high efficiencies with respect to the mass of the propellant consumed). Ion thrusters are only effective in space and are used primarily to provide efficient slow accelerations over prolonged periods, reducing mission costs. These systems can be used to reduce travel times to the outer planets or intermittently for station keeping. Operational ion thrusters have been demonstrated and the High Power Electric Propulsion (HiPEP) is being developed by NASA/Glenn Research Center for the Jupiter Icy Moons Orbiter (JIMO) spacecraft. Similarly, a sail can be deployed once in space, using the tenuous solar wind to provide small but sustained acceleration.

The basic physics of rocket thrust is given in the Intro Physics Flashback FB16.2. There are two basic types: liquid-fuel and solid-fuel rockets, or simply liquid and solid rockets. Both types are depicted in Figure 16.19. Cryogenically cooled liquid rocket engines, using liquefied hydrogen and oxygen, are most commonly the primary rocket. Multiple staged rockets enable more total thrust for very heavy payloads since 90% of the total weight of a rocket plus payload is fuel plus the vessel holding it. Subsequent stages only have to provide the required thrust for the payload and any remaining rocket stages, free of the deadweight of the ejected cryogenic containers and rocket skin of the previous stage. Solid rockets frequently serve as first-stage boosters, providing supplemental lift. Solid-fuel boosters supply greater thrust compared to liquid-fueled counterparts plus do not require refrigeration or

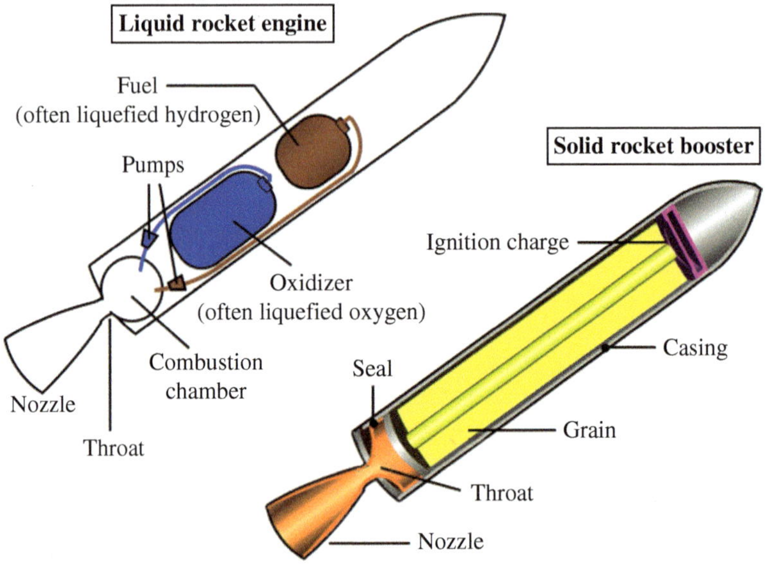

FIGURE 16.19 The anatomies of a liquid- and solid-fuel rocket. Source: Adapted from Pbroks13, https://commons.wikimedia.org/wiki/File:SolidRocketMotor.svg. Used under CC-BY-SA 3.0 https://creativecommons.org/licenses/by-sa/3.0/deed.en.

insulation. However, once ignited these boosters cannot be stopped until all of the fuel is consumed. Liquid rockets, which have fuel supplies controlled by pumps, can be ignited and shut off several times throughout a mission. These rocket systems are ideal for orbit transfer maneuvers such as leaving LEO to travel to another planet, making mid-course corrections, and then retro-firing for insertion into orbit around the destination planet or one of its moons.

INTRO PHYSICS FLASHBACK FB16.2

Rocket Propulsion and Newton's Laws of Motion

Newton's third law states for every action there is an equal and opposite reaction. We all learned in a general physics class that "action" and "reaction" of which Newton spoke are really changes in momentum. If a pair of objects such as the two examples shown in Figure FB16.3 is initially at rest and then a projectile is

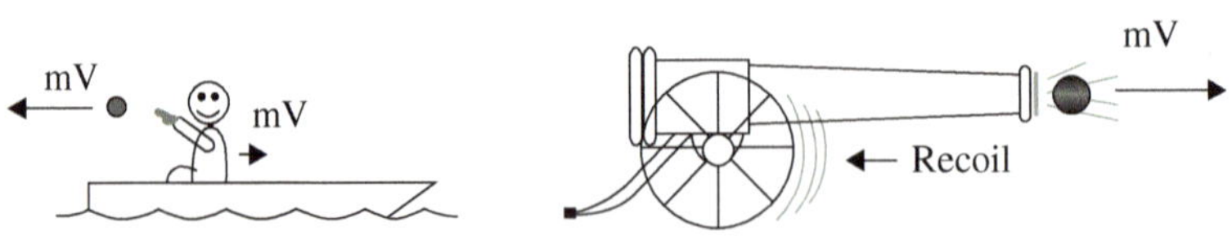

FIGURE FB16.3 Equal and opposite reactions from a horizontal projectile where all masses were at rest initially.

sent in motion, the remaining mass must recoil, moving oppositely to conserve momentum. In these examples, the change in momentum, Δp, of the projectile is simply its mass multiplied by its velocity, mv, since its initial velocity was zero. Similarly, the person plus boat or the canon move oppositely with a smaller velocity that is proportional to the ratio of projectile mass to the remaining mass.

The cases given before are examples of impulse events: a projectile is thrown or fired in a single brief event and individual momentums are abruptly changed. A rocket launch in contrast represents a continuous process of momentum change. The net mass of the rocket plus payload is modified constantly as well as its velocity. Thrust, the force from a rocket, is given by

$$\vec{F} = \frac{\Delta p}{\Delta t} = \frac{d(mv)}{dt} = m\frac{dv}{dt} + v\frac{dm}{dt}, \tag{FB16.1}$$

where the rate of expulsion of mass, dm/dt, is constant and the velocity of expelled gas is also constant. Thus, rocket thrust is simply $\vec{F} = Ru$ where R is the rate that fuel plus oxidizer are consumed and u is the speed the oxidized gas exits the nozzle. Figure FB16.4 shows the thrust forces from a rocket engine. *Note*: a continuous explosion occurs in the spherical combustion chamber. The throat enhances, u, the speed of the exhausted molecules. Also, note that total mass of the rocket is slowly decreasing at a constant rate as the fuel plus oxidizer is consumed. Consequently, the acceleration is increasing even though the force (thrust) is constant. To obtain the payload position as a function of time, one has to integrate the net forces (i.e., thrust—gravitational force—air drag) over time.

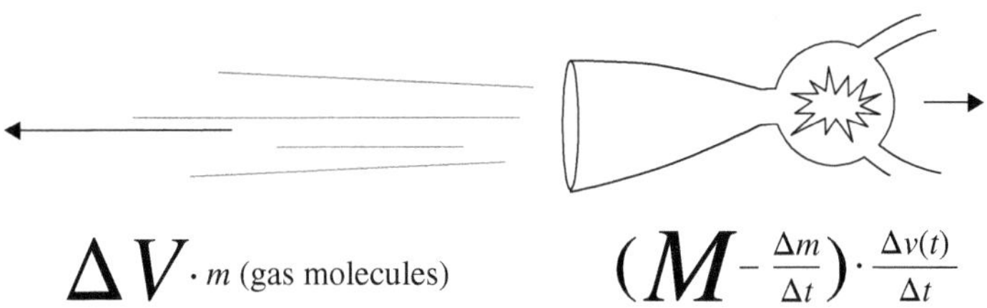

FIGURE FB16.4 Rocket thrust from a small amount of gas mass being accelerated to a very high velocity. The reaction is an accelerating rocket in which its acceleration is a function of time, $a = a(t)$.

One of the most successful, general-purpose rockets that continue to be major workhorses for NASA and commercial telecommunications is the Delta family of rockets made by Boeing Corporation. In particular, the Delta II have been used for many scientific missions and small payloads, while the Delta IV class are used for medium, medium+, and heavy payloads. The profiles of the entire family of Delta rockets are shown in Figure 16.20. Another prominent rocket includes the Atlas rocket family made by Convair, a division of General Dynamics, which carried the first US astronauts into space and were initially used for the first intercontinental ballistic missiles (ICBMs). An Atlas II was the expendable launch vehicle 63 times

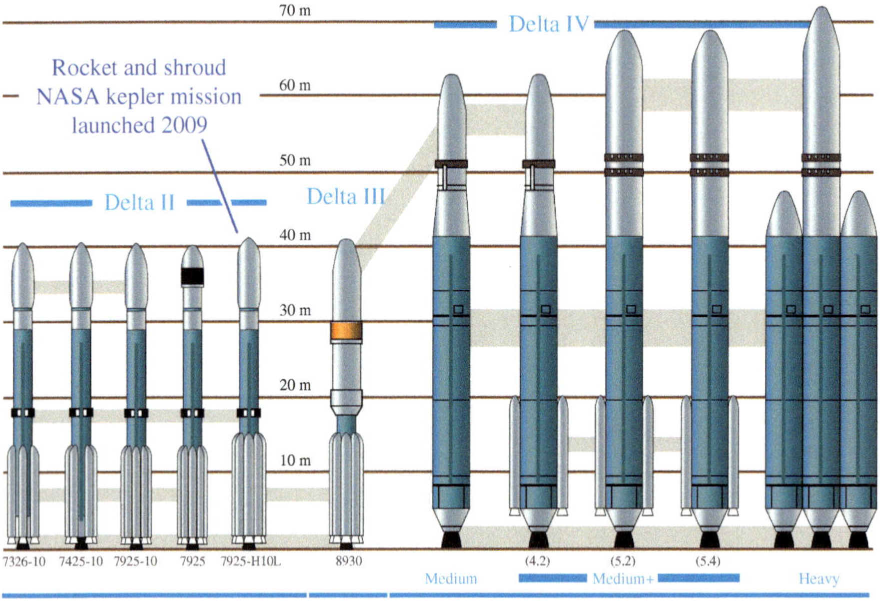

FIGURE 16.20 The Delta family of rockets designed and built by Boeing Corporation. *Note*: the small solid-rocket boosters surrounding the base and the various shrouds available. Source: NASA.

between 1991 and 2004, while only six launches with the Atlas III where made between 2000 and 2005. The Atlas V is the only rocket from the family still in service with launches planned until 2020. The third major player on the world's stage is the Ariane family of rockets used by the European Space Agency. These rockets were developed and fabricated by the French company, Arianespace SA, the first commercial space transportation company. Arianespace holds the bragging rights for providing the most rocket boosters to transfer satellites from LEO to geostationary orbits. It had more than 50% of the world market in 2004. An Ariane V will launch the *JWST*, the next great observatory and follow-on mission to the *HST*.

All of the rocket components along with the payload are integrated into a single unit on top of a mobile platform inside a very large building known as the vehicle assembly building (VAB). Launch complex 39A at NASA/Kennedy Space Center is pictured in Figure 16.21, including the VAB, a mobile platform with a Saturn V rocket used to send man to the moon, and two empty platforms in the background. *Note*: the VAB has two large roll-up doors, enabling it to accommodate the assembly of two launch vehicles simultaneously. KSC has several launch pads and the VAB has two cranes that can lift weights in excess of 200 tons.

A shroud (also known as a fairing) sits at the top of an ELV rocket, which is a protective outer casing that detaches once in space. Figure 16.22 shows the Kepler satellite, which was launched in 2009, being integrated with the shroud and rocket inside the VAB at NASA/KSC. There are only a handful of standardized shrouds

FIGURE 16.21 Launch complex 39A. Source: NASA.

FIGURE 16.22 The *Kepler* spacecraft, including third-stage booster, inside one half of the Delta II 7925-10L shroud. The second half is attached to a crane in the foreground. Source: NASA/Jack Pfaller.

available for each type of rocket. For example, there are only three distinct shrouds available for launch on a Delta II rocket and independently three for a Delta IV.

Missions often have to be designed specifically within the limits of the lift capacity of the rocket as well as the inside volume of the fairing. The largest shrouds can accommodate a maximum diameter of about 4.5 m and a length of approximately 10–15 m. These numbers are static dimensions. All spacecraft and satellites experience various amounts of vibrational flexures during launch and must remain free of contacting the interior of the fairing. Some future astronomical missions such as the 6-m *JWST*, must incorporate primary mirror segments as well as a telescope design that unfolds after space deployment to accommodate the volume restrictions of the launch shroud. Figure 16.23 displays the components of the Ariane 5 ESC-A along with the folded *JWST* design inside the shroud. (ESC-A is a cyrogenic liquid-fuel upper stage rocket that resides inside the fairing.)

After launch and a LEO is established both the *Kepler* and *JWST* missions fire their upper stage rockets, changing from a geocentric to a heliocentric orbit. For *Kepler*, the solar orbit was one that slowly drifted away from the Earth. The ESC-A stage of *JWST* in contrast is designed to transfer the observatory to the solar L2—Lagrange point, where station-keeping processes will maintain the observatory in L2 for the duration of the mission. More information, including the physics of station keeping, can be found in Section 17.4.

Finally, all missions launched in the twenty-first century have to have in place prior to launch a plan to deorbit the spacecraft safely at the end of its mission. Space debris from old, no-longer functional satellites has become a major hazard,

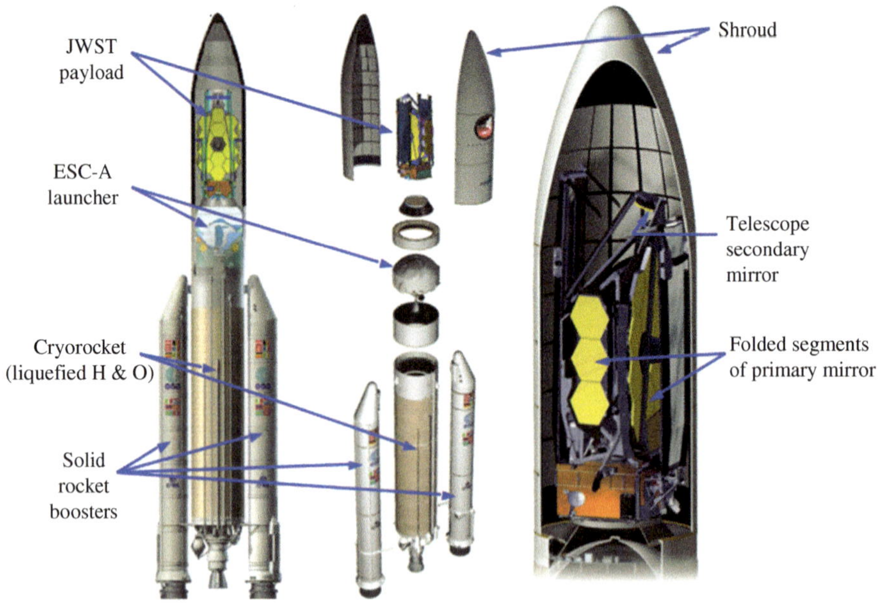

FIGURE 16.23 The Ariane 5 ESC-A configuration assemblies including the folded *JWST* design. Source: Arianespace–ESA–NASA.

especially in LEO. A single bolt from an old spacecraft, for example, that crosses the path of an operating satellite does so at an average speed of about 10 km/s. While satellites in LEOs do slowly decay, eventually falling back to Earth, not all will burn up completely in the atmosphere. The reentry point of these satellites must be controlled to insure the remnants splash down in the ocean rather than fall on to populated areas.

INTERESTING TIDBIT TB16.2

KSC's vehicle assembly building (VAB) is one of the largest buildings in the world with a volume that is 1.67 times larger than the Pentagon and 3.75 times larger than the Empire State Building. It is 526 ft high, 518 ft wide, and 716 ft long, enclosing 129,428,000 ft^3 and covering 8 acres. The VAB is large enough to have its own weather inside such as cloud formation, but contrary to popular myth, it has never rained inside the VAB except when the very tall doors were left open on a rainy day.

17

ASTRONOMICAL AND PLANETARY OBSERVATORIES

Astronomical discovery in twentieth century has gone into the history books as one of the greatest advances since the invention of the telescope 400 years ago, which ushered in the epoch of Galileo, Copernicus, Tyco, and Kepler. Large telescopes in the twentieth century revealed that our own Milky Way galaxy consists of more than 100 billion stars and subsequently, many billions of galaxies similar to our own, all flying away from each other in a great expansion of the universe. Working backward, all of these galaxies originated from a single location at a single point in time, the *Big Bang*. Later observations revealed the existence of the cosmic microwave background—the left-over remnant radiation from the still-cooling-off Big Bang, as well as the existence of cold dark matter and cold dark energy, which collectively constitute 96% of the mass of the universe. New technologies plus the appearance of observatories operating in space opened up new diagnostic tools associated with new wavelength bands. At the dawn of the twenty-first century, the largest ground-based telescopes had become six times larger than those available just a few decades earlier, novel optical designs with adaptive optics have increased resolution substantially, and larger more powerful telescopes, operating in space, are all adding significantly to the astronomers' tool box at a time when many fundamental questions have yet to be answered.

Planetary research has also made tremendous strides, especially since the middle of the twentieth century. Spacecraft have made flyby, up-close observations, revealing chemical analysis and flow patterns in unprecedented detail. Voyager flybys discovered Saturn's rings were broken into dozens of ringlets and the very existence of rings around the gaseous planets: Jupiter, Uranus, and Neptune. The surface of Venus has been radio mapped by orbital spacecraft and probes have been sent deep into the dense, hot Venetian atmosphere. Single-point landings and rovers have been sent to Mars and the Moon. Spacecraft have rendezvoused with both comets and asteroids,

Modern Devices: The Simple Physics of Sophisticated Technology, First Edition.
Charles L. Joseph and Santiago Bernal.

Companion website: www.wiley.com/go/joseph/moderndevices

including a landing on an asteroid. Perhaps, the most exciting developments of the past 50 years has been studies that exposed microbial extremophiles thriving in the most inhospitable places here on Earth. These findings have renewed the interest in searching for life or evidence of past microbial life elsewhere in our solar system.

Historically, astronomical and planetary observatories have been ground-based telescopes, operating at visible wavelengths. These observatories were classified to be either reflecting (mirror based) or refracting (glass transmitting optics) and defined by the aperture of the primary mirror of the telescope, which directly translated to the faintness of object that could be observed. (Without adaptive optics, *atmospheric seeing*, the phenomenon that makes stars dance around and twinkle, limits the spatial resolution to 0.7–1.0 arcsec—the theoretical limit of a 12 cm (~4.5 inch) aperture. In the past, a large visible wavelength telescope could see fainter, but with no more resolution than a small one.) Refracting optics also suffered from chromatic aberrations, causing point sources to appear as very small rainbows, although compound, corrective lenses largely mitigated those distortions. Throughout the twentieth century as telescope apertures continually increased, virtually all research telescopes became reflecting and the range of accessible bands expanded into the infrared (IR) and radio bands. Subsequently, the range of accessible wavelengths extended into the ultraviolet and X-ray bands with the advent of rocketry and orbital satellites. High-energy physicists, cosmic ray astronomers, and nuclear physicists began various collaborative programs, some launching satellites while others going deep underground into caves to study, for example, neutrinos. Today, observatories study astronomical and planetary objects over the entire electromagnetic spectrum from the radio to gamma rays as well as the cosmic rays of atomic and subatomic particles.

17.1 TELESCOPE DESIGNS

A telescope with a monolithic primary mirror is configured and polished to have a concave parabolic cross section. (A variant of this mirror shape is actually a hyperbola, which eliminates an optical aberration called *coma* that affects the outer portions of wide-field images.) Light from a planet or star arrives at Earth highly collimated (i.e., parallel rays) since the distances to even the nearest stars are so vast. A parabolic reflecting mirror is the perfect surface to collect and focus collimated light since all rays regardless of wavelength converge to a single focal point. Rays of light from adjacent points on the sky are focused from parabolic mirrors to form a convex spherical focal plane. In addition, any portion or segment of a concave parabolic mirror will focus to the same point as Figure 17.1 shows graphically.

The type of telescope is further defined by the location of the focal point (i.e., where an instrument or eyepiece is placed) relative to the rest of the optics. Figure 17.2 shows the four basic focus positions for reflecting telescopes. Each diagram is shown with the primary mirror at the bottom as if the telescope were pointing to its zenith. In all cases, some portion of the central region of the primary mirror is occulted by an instrument or secondary mirror. The few exceptions are specialized telescopes, usually designed around an off-axis parabola. These blocked portions, however, only

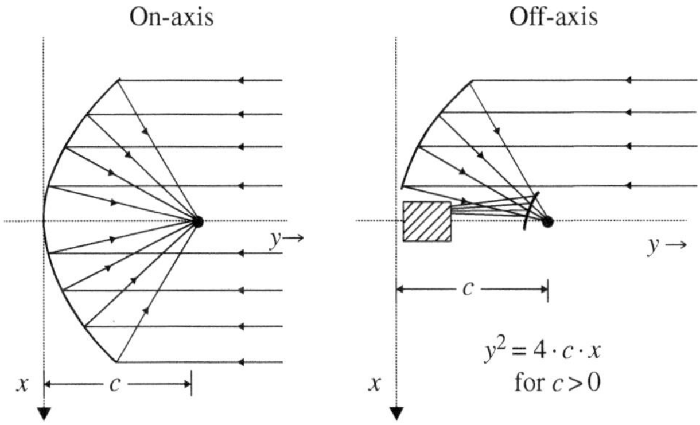

FIGURE 17.1 Two-dimensional ray tracing of a parabolic primary mirror. Right: the same for an off-axis parabola with a secondary convex parabola mirror and a detector (hashed marked box). *Note*: *X–Y* coordinate system is rotated 90° from the normal projection.

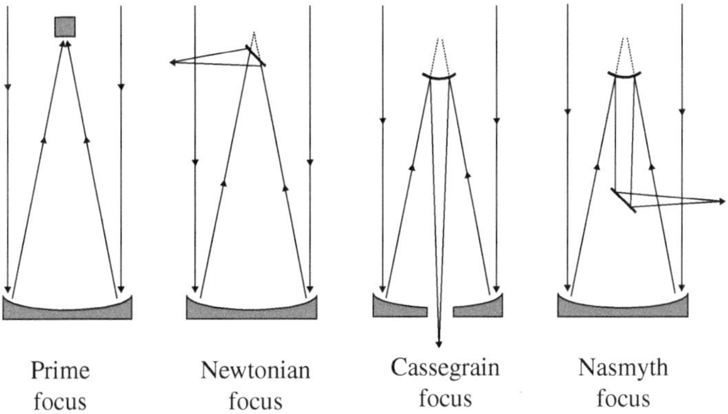

FIGURE 17.2 Telescope configurations based on focal points.

reduce the sensitivity of the telescope by small amounts, having a minor impact since the remaining portions of the parabolic primary are unaffected. *Note*: the secondary mirror is usually a convex parabola, which transforms the *f*-stop (*f*/#) to a longer, slower beam. (If the secondary is concaved rather than convex, the telescope is a Gregorian design most often used by solar observatories.)

Most astronomical research telescopes are configured for one or more instruments located at the various focal points shown in Figure 17.2. Newtonian focus telescopes, however, have been relegated primarily to portable, field-observing telescopes with apertures of 30 cm (12 inches) or less. Cassegrain focus telescopes, equipped with electronic instruments, are popular for all apertures, but especially mid-sized (45 cm to 1.5 m) telescopes used primarily for teaching and to a lesser extent research. A camera or spectrograph at Cassegrain maintains fixed with respect rest of the telescope optics. Nevertheless, gravity-induced flexure between the telescope and the

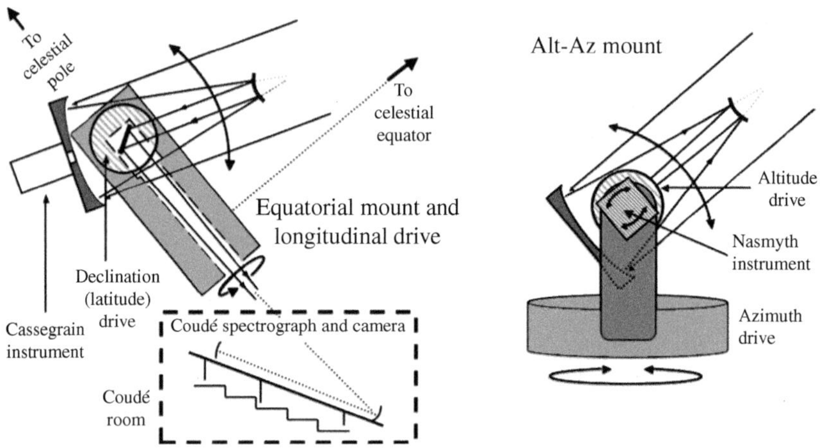

FIGURE 17.3 The two different types of telescope mounts.

Cassegrain spectrographs introduces small variable changes throughout the night, requiring periodic wavelength recalibrations. Until almost the turn of the century, it took a few hours and often required cranes to swap various instruments, making daylight hours when the telescope is otherwise not in use, the preferred time for reconfiguring the telescope operation. The largest telescopes today have four or more instruments semi-permanently mounted, and it may take only 1–2 minutes to switch between different instruments.

The telescope mount and pointing system is the third means of defining a telescope. Most permanently housed telescopes have either an equatorial mount or an altitude–azimuth (alt–az) mount as depicted in Figure 17.3. An equatorial mount is based on two orthogonal cylindrical coordinates: latitude and longitude. The axis of the longitudinal drive is arranged to point to the celestial North Pole (or South Pole if in the southern hemisphere); the angle with respect to the horizontal plane is simply the latitude of the observatory, which differs from one telescope to the next. The primary advantage of a polar mount is an object (i.e., a star) can be tracked throughout the night simply by a slow drive motor that rotates at the same rate as does the Earth. In other words, the tracking drive has to have a rate of 360° per 24 hours or 15°/h (0.004167°/s). Counter-balancing weights are added, depending on which instruments are attached, and the telescope normally has an internal counter weight that automatically adjusts its distance from the drive axis, all to insure a smooth, uniform tracking. In practice, small adjustments still have to be made continually in declination and longitude, a process known as guiding. The resolution of major research telescopes is sufficiently high that guiding is necessary to insure small amounts of image smearing are not introduced during long exposures.

Prior to 1990, most research telescopes had an equatorial mount. These telescopes usually were equipped with one permanently mounted Coudé instrument at a Nasmyth focus. Coudé instruments have very large, slow optics that offer very-high-precision measurements. Coudé instruments are usually mounted on very long optical benches that reside permanently in a dedicated room below the floor where the telescope is operated. Coudé instruments do not suffer from flexure changes since all of its optics are mounted on a fixed optical bench. Another advantage of the polar mount is the light beam from the telescope to the Coudé room remains fixed during a 30 minutes to 1-hour observation. The tip of the pick off mirror only has to be changed when a new target with a new latitude is acquired. Most major ground-based observatories were equipped with prime focus, Cassegrain, and Coudé instruments, each optimized for different types of data collection.

Most research telescopes designed and built after 1990 have altitude–azimuth mounts, which are easier and cheaper to fabricate, especially for the very largest observatories. Tracking an astronomical object with an alt-az requires both the azimuth and the altitude settings to change continuously or at least be updated every second, this requirement is no longer a disadvantage with modern computers and orientation sensors. Moreover, large bundles of optical fibers means an image from an alt–az telescope can easily transferred to another room without the orientation of the pickoff mirror having to be changed continually. Many of the very largest telescopes have 1-3 Nasmyth instruments attached to each side of the telescope mount, centered on the altitude rotational axis. These instruments rotate synchronously with the elevation of the telescope, maintaining a constant field of view with respect to the telescope optics regardless of movement. The popularity of Nasmyth instruments on very large telescopes with alt–az mounts has reduced the popularity of classical Cassegrain instruments mounted behind the primary mirror.

INTERESTING TIDBIT TB17.1

The headlights on automobiles employ concave parabolic mirrors to maximize the amount of light in its forward head beams. While the light source is a small bulb placed at the focal point, the physical size of the bulb is large compared to the mirror focal point, creating a fairly collimated beam, but with some divergence.

INTERESTING TIDBIT TB17.2

Mirrors with a parabolic cross section suffer from an aberration known as *coma* and some large telescopes have mirrors with hyperbola, which dramatically reduces coma. Similarly, a Ritchey–Chrétien is a special class of Cassegrain instrument that a has a hyperbolic

INTERESTING TIDBIT TB17.3

Specially trained technicians called telescope operators operate large ground-based telescopes. Observatories do not let astronomers operate their telescopes since each is far too valuable and the cost of operating one runs into the tens of thousands of dollars/night. Instead, the observer is relegated to giving lists of target coordinates and exposure times to the telescope operator, all of whom work in an isolated and warm control room. Astronomers often use telescopes nowadays remotely from comfort of their office, which can be a real advantage if the observatory is in a very different time zone. However, remote observing from one's office during the nighttime rarely releases the astronomer from his/her daytime duties given by spouses or university teaching assignments.

17.2 VERY LARGE, ULTRA-LIGHTWEIGHT OR SEGMENTED MIRRORS

Single monolithic mirrors have become prohibitively expensive and heavy for apertures larger than about 5–6 m. (The Palomar 5-m is the largest telescope with a conventional single-mirror primary fabricated for a US observatory. The all time record holder, is the 6-m Bol'shoi Teleskop Azimuthal'nyi (BTA) in Nizhnii Arkhyz, Russia, which went into operation in 1976. Its primary mirror without the mechanical support, weighs 176,000 kg or 40 tons!) A conventional primary mirror larger than 6 m distorts its shape under its own weight. Beginning in the 1970s, research into telescope mirror designs and methods to improve atmospheric seeing led to the development of light-weighted mirrors with honeycomb backside structures, meniscus mirrors, segmented mirrors, and multiple-mirror telescopes. New designs and innovation have produced a new generation of large telescopes with apertures between 8 and 12 m that are far less expensive than conventionally designed telescopes. There are 13 ground-based telescopes with apertures of 8–12 m and another 4 with 6–6.5 m that became operational between 1990 and 2010. All of these telescopes were built and are operated by international consortiums, representing countries from every continent (except Antarctica).

Moreover, there are three extremely large telescopes under development as of 2013. These include (i) the Giant Magellan Telescope (GMT), a 21.4 m equivalent aperture to be placed on the Chilean Andes, (ii) the 30-m telescope (TMT) to be sited on Mauna Kea, Hawaii, USA, and (iii) the European Extremely Large Telescope (E-ELT), a whopping 42-m aperture telescope being built by ESO and destined for the Chilean Andes as well.

The first of the current generation of very-large ground-based telescopes are the pair of Keck 10-m telescopes on Mauna Kea, Hawaii, operated by the California Association for Research in Astronomy (CARA). The primary mirror for each Keck consists of 36 hexagonal segments, each a unique portion of an off-axis parabola, configured to function as a single 10-m diameter primary mirror. Another duo of

FIGURE 17.4 The thin, meniscus primary mirror of the Gemini North Telescope. This mirror is only 20 cm thick, but maintains a high surface accuracy over its 8.1 m diameter aperture with numerous support points on active actuators. Source: The Gemini Observatory and the National Science Foundation. Public Domain.

large-aperture telescopes are the 8-m Gemini telescopes, one on Mauna Kea, Hawaii, and the other on Cerro Tololo, Chile. An international consortium that includes the United States, United Kingdom, Canada, Chile, Australia, Brazil, and Argentina operates the Gemini observatory. Each primary mirror, Gemini North shown in Figure 17.4, is a meniscus mirror that is only 20 cm thick over the entire 8.1 m diameter aperture. The mirror cannot support its own weight without breaking and must be cradle-supported at numerous points at all times. These mirrors have a surface accuracy of 15.6 nm (root mean squared, RMS), the equivalent smoothness of approximately 156 atoms over the entire surface. ESO operates four similarly designed meniscus-mirror 8.2 m telescopes, all on a single site in the Chilean Andes. These mirrors weigh 22,200 kg, approximately 13% of that of the Russian 6-m.

Another novel approach is one that optimizes one specific type of observation at the expense of everything else. Most telescopes are designed for high performance over a wide variety of spectrographic and imaging applications, driving costs up to achieve excellence in all areas of research. Very-large aperture telescopes such as the Hobby–Eberly Telescope (HET) and its sister the Southern African Large Telescope (SALT) achieve superb spectra of faint objects with only good-quality imaging capabilities, but at about one-fourth of the fabrication cost of a general-purpose design. Both telescopes have two innovative features that keep costs down: (i) a segmented primary mirror with a spherical (rather than parabolic) cross section and (ii) a fixed-altitude but flexible-azimuth mount with target tracking being performed at the top of the telescope. A

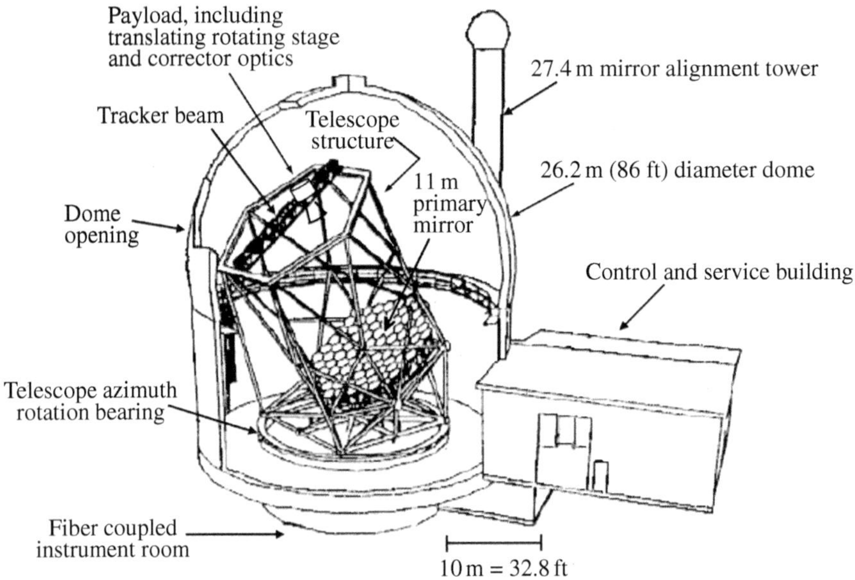

FIGURE 17.5 Diagram of the novel design of the Southern African Large Telescope (SALT). Source: Darragh O'Donoghue, South African Astronomical Observatory. Reproduced with permission of SALT Foundation Pty Ltd.

diagram of the SALT telescope is given in Figure 17.5, showing these cost-saving design features. Although these primary mirrors require substantial four-mirror corrector optics to compensate for spherical aberrations, all of the segments are identical and interchangeable. (If the primary had a parabola cross section, the segments would have various polished profiles that would have to be installed in the correct location and with the correct orientation.) The HET/SALT design also avoids the need for two sets of motors large enough for the entire telescope to track a point in the sky. Instead, a payload containing corrective optics plus a primary focus instrument moves along the tracker beam, tipping and rotating to keep the payload pointing towards the center of curvature of the primary. The telescope remains stationary through an observation.

A portion of the primary mirror is pictured in Figure 17.6. There are 91 (plus two spares) hexagonal segments, each is held in place on a set of backside actuators. When replacing one or more segments, the telescope is rotated to point to the top of a 27.4 m high Mirror Alignment Tower, which is outside of the main building and dome. (See Fig. 17.5.) The optics in this tower allow small adjustments with a precision of one-tenth of a wavelength to be made to insure the all parts of the entire mirror assembly performs as a precision, uniformly spherical reflecting surface.

There are a few limitations with the HET and SALT designs. Both telescopes can only make observations over approximately 70% of the sky centered on the celestial latitude corresponding to the fixed elevation angle of the telescope and can only observe a particular object for a few hours at a time. Neither telescope can observe the polar regions of the sky. To make an observation at a different latitude, the entire

FIGURE 17.6 A few of the segments comprising the 11-m telescope mirror for SALT. Source: Darragh O'Donoghue, South African Astronomical Observatory. Reproduced with permission of SALT Foundation Pty Ltd.

telescope structure and base are elevated by a few centimeters, rotated on large bearings, and set down at a new azimuth position. Once a star, galaxy, or other object is acquired, it can be tracked for 45 minutes to 2.5 hours as the object transits the telescope. While the primary mirror is equivalent to an 11-m diameter telescope, the effective aperture usually ranges from 8 to 9.9 m diameter throughout an observation. (The nominal average is equivalent to a 9.2 m aperture, but can be as low as 7 m in the worst case.) Despite these limitations, a high-quality large-aperture telescope can be fabricated at a fraction of the cost of making a general-purpose observatory. This type of telescope naturally lends itself to queue mode observing where individual research programs may be assigned a fraction of a night.

Finally, the University of Arizona has been at the forefront of designing and building lightweight mirrors as well as multiple-mirror telescopes since the 1970s. A multiple-mirror telescope differs from one with a segmented primary in that each primary mirror is its own independent telescope. While it is difficult to phase up and align multiple images from separate telescopes, an array of telescopes on a common mount and pointing system makes multiple aperture interferometry at visible and IR wavelengths feasible compared to the use of several separate telescopes. (*Note*: radio telescopes have had tremendous success combining multiple dish antennas such as the Very Large Array (VLA) in New Mexico, USA, consisting of 27 antennas each with a 25 m diameter. Individual antenna are moved along railroad tracks and set down on concrete pads, forming a Y-shaped configuration that spans up to 36 km across. Unfortunately, visible and IR wavelength astronomers have had very limited success with interferometric observations, especially between telescopes that are not on a common mount.) Stewart Observatory of the University of Arizona built the first multiple mirror telescope, the MMT, which it operated from 1979 to 1998. It consisted of six 1.8 m primary mirrors, the equivalent of a single 4.5 m aperture telescope. The MMT, which has an alt–az mount, also had an innovative design in which the dome and the telescope share a common platform that rotates together. By 2000, the six-mirror primary of the MMT had been replaced by a single, lightweight honeycombed-back 6.5 m primary, fabricated by the Stewart Observatory Mirror Laboratory. Another MMT-style observatory is the Large Binocular Telescope (LBT), consisting of two 8.4 m diameter mirrors on a single mount, which became fully operational in 2008. It has the same light gathering power as an 11.8 m diameter conventional telescope, while capable of providing spatial detail equivalent to a 22.8 m along one axis. The BLT is located on Mount Graham, Arizona, USA, and is operated by an international consortium of universities and government research centers from the United States, Italy, and Germany.

17.3 ADAPTIVE OPTICS AND ACTIVE OPTICS

Active optics should not be confused with adaptive optics (AO). Active optics are simple measures that compensate for the distortions that vary slowly, minutes to hours in time caused by gravity and temperature, while AO systems compensate for atmospheric seeing, varying on timescales less than 1 second. A telescope experiences subtle change in the position of one optic with respect to another caused by temperature variations or by gravity as the telescope changes its orientation. These effects are significant for large telescopes and are often automatically corrected with active optics, located on one or more mirrors. Active optics consist of a sufficient number of piezoelectric transducers (PZTs) attached directly to the back of the primary mirror, or to the mounts of the secondary, or at other strategic locations if need be. For a primary mirror, the array of piezoelectric actuators on the primary can be mounted either parallel or perpendicular to the mirror surface and are used to introduce a coordinated set of stresses that alter the mirror's shape very slightly. The primary mirror may, for example, become slightly oblate when pointed close to the horizon or may change its focal length slightly due thermal contraction as the temperature drops throughout the night. The PZTs compensate for these alterations by pulling or pushing on the optic in a systematic way to introduce an opposite set of strains. Manual periodic adjustments for a limited number of corrections such as refocusing can be made throughout the night if active optics are not available.

AO differ from active optics in that adaptive optic systems compensate for the moment-to-moment image aberrations caused by the continually changing optical properties of the atmosphere. The atmosphere as shown in Figure 17.7 distorts the wavefronts arriving from space. *Note*: the distortions are always a small fraction of a wavelength. If these uneven wavefronts strike a reflecting surface that has been deformed exactly half as much as the incident wavefronts, then the reflected waves will be restored to the original plane parallel wavefronts.

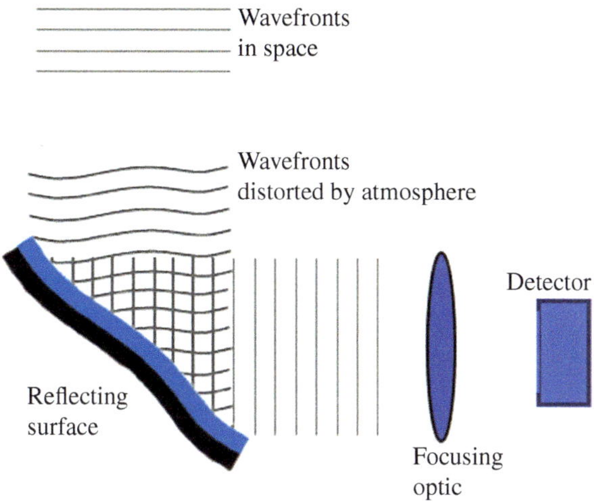

FIGURE 17.7 The basic principal of deforming a mirror to restore the wavefronts to correct atmospheric aberrations.

One AO approach most easy to conceive is a computer-controlled rubber mirror system in which a sensor measures the wavefront perturbations and a normally flat small mirror is distorted appropriately. (Rubber mirrors do not have any rubber, but in fact, are quartz crystals with a front-side coating of silver. These mirrors are attached to piezoelectric actuators that produce the necessary distortions.) One common approach to measure changes to the incoming wavefronts is a Shack–Hartmann sensor, consisting of a microlens array in front of an image detector. When the wavefronts are plane-parallel, an array of bright spots are evenly spaced in two dimensions. Distorted wavefronts result in spots that are not uniformly distributed. The spaces between any one spot and its adjacent spots is directly proportional to the amount a corresponding location on the rubber mirror must be displaced. (See Fig. 17.8.) Adaptive optics have taken hold in astronomy over the past few decades, especially the use of *rubber mirrors* for IR observations. In the late 1980s, the defense departments of several countries in the NATO alliance began declassifying technologies associated with IR imaging through the Earth's atmosphere. The declassifying has been a major catalyst in the advancing adaptive optics for astronomical applications.

The essential problem with any AO system is to measure the aberrations to the wavefronts and make the necessary adjustments in real time. Figure 17.9 shows the essential features of a rubber mirror system. The distorted wavefronts arrive from the telescope on a downward path where a deformable fold mirror sends the waves to the right. Next, a beam splitter sends part of the flux to a wavefront sensor and the rest to a camera, represented in Figure 17.9 by a focusing optic and detector. The continuous real-time image from the wavefront sensor, consisting of an array of spots, is processed by the computer, which in turn sends signals to the individual actuators to null out the distortions registered by wavefront sensor. Very quickly, the feedback loop becomes synchronized, tracking and correcting the atmospheric aberrations. In practice, the wavefront correction is not perfect and residual aberrations remain.

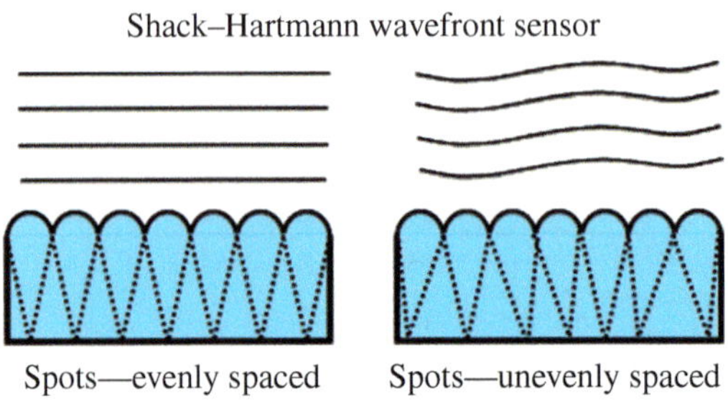

FIGURE 17.8 The basic principal of a Shack–Hartmann wavefront sensor. For undistorted parallel wavefronts incident from the top, each portion of the wavefront is focused to a spot directly behind one element of the microlens array. Atmospheric aberrations introduce localized slopes in the wavefronts, causing each spot to move proportional to the localized slope.

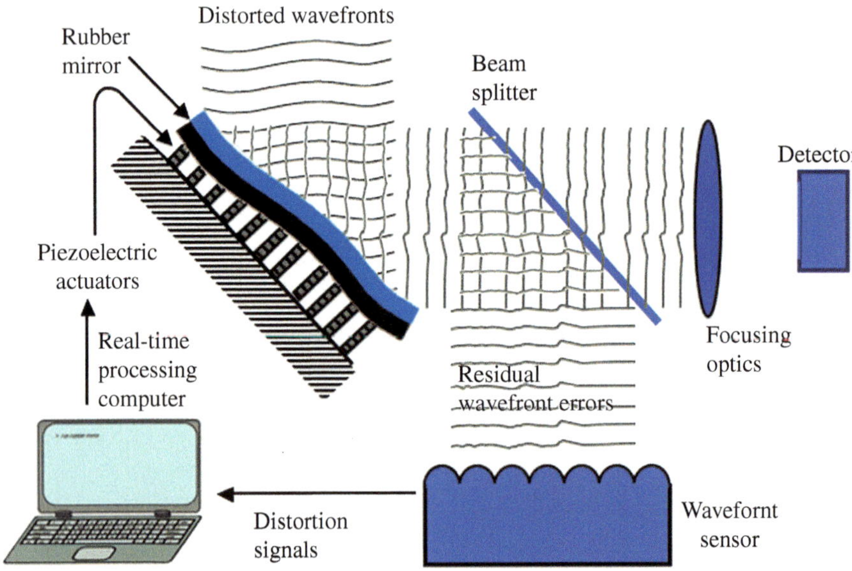

FIGURE 17.9 The basic principal of deforming a mirror to restore the wavefronts to correct atmospheric aberrations.

Atmospheric aberrations are extraordinarily more difficult to correct at visible than at IR wavelengths since the number of actuators balloons and the rate of updated changes increases. Several hundred actuators spread over the field of view (FOV) are required in a visible AO system, while only 9–16 are need for thermal IR operations. (In most IR applications, a simple tip-tilt mirror, a one-element AO, can make the bulk of the corrections necessary to retrieve the theoretically best image quality.) The atmospheric disturbances of these wavefronts change rapidly, requiring a new set of corrections every 10 ms (0.010 seconds) in visible bands to hundreds of milliseconds (0.500–0.900 seconds) in the IR.

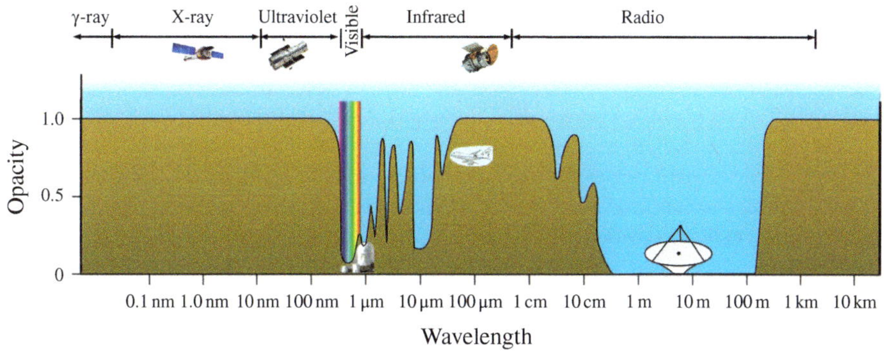

FIGURE 17.10 The opacity for light entering the top of the atmosphere and arriving at sea level. Source: Adapted from NASA illustration.

Finally, the Earth's atmosphere absorbs and scatters light over many wavelength bands, requiring some observatories to operate in orbit above the atmosphere. Figure 17.10 plots the opacity of the atmosphere as a function of wavelength for astronomical light arriving at ground level. As can be seen in Figure 17.10, the atmosphere is transparent for visible wavelength light, but even more so at radio wavelengths. (Consequently, it is advantageous to place visible light telescopes on high mountains, but not most radio telescopes.) While certain portions of the IR spectrum are also relatively transparent, the atmosphere itself becomes a diffuse light source over the so-called thermal-imaging portion of the spectrum longward of 2 μm. In fact, our own atmosphere creates a huge background signal (90% or more of the total), which must be subtracted from the total to measure the net signal from the astronomical source. One common IR method for subtracting two large numbers to obtain an accurate small number is to chop the image, alternating observations every few seconds between the astronomical object of interest and a nearby blank portion of sky.

17.4 SPACE OBSERVATORIES

As noted in the previous chapter, the selection of an orbit for a spacecraft drives many of the engineering parameters associated with each mission. There have been a number of astronomical flagship missions in low Earth orbit (LEO) such as the *Spitzer Space Telescope* (IR), the *Hubble Space Telescope* or *HST* (ultraviolet-visible), the Chandra Observatory (X-ray), and the Gamma Ray Observatory (γ-ray). The largest single-mirror ever flown is the 3.5 m telescope of the Herschel Space Observatory. The mission was built and is operated by the European Space Agency, being launched in May 2009 on an Ariane 5 rocket from the Guiana Space Center and operating at the solar L2 Lagrange Point (1.5 million km from Earth). Both Canada and the United States made contributions to the instruments. Herschel operates over IR to submillimeter wavelengths, being cryogenically cooled to approximately 1.4 K by liquid helium. The 2000-L supply of liquid He limited its operating life to about 3 years (2009 to 2013). We will restrict our treatment here to a few current and future missions that are expected to have a major impact on science and the public imagination. Similar to Herschel, all of these happen to orbit the Sun rather than the Earth. Let us revisit Figure 16.3, which identifies the locations of the solar Lagrange points, and place on that diagram the spacecraft locations. Figure 17.11 depicts the relative positions of three solar observatories as well as the Kepler Mission to look for evidence of Earth-similar exosolar planets and the future, *James Webb Space Telescope* (*JWST*), the premier observatory replacement for the *Hubble Space Telescope*. We shall briefly describe these missions and discuss some of the technological challenges faced by each.

SOHO, the Solar and Heliospheric Observatory, is an international collaborative mission between ESA and NASA to study the Sun from its deep core to the outer corona and solar wind. While launched in December 1995 with a 2-year design life, the SOHO mission has been extended several times and has now successfully observed a full 11-year solar cycle. One of the primary science objectives of SOHO has been to understand more fully the development and dynamics of coronal mass ejections (CMEs), which are huge bursts of ionized plasma released into planetary space. A single CME has the potential to destroy the electrical system of a spacecraft or to bring down the electrical grid, depending on its intensity and location on the Sun.

While SOHO has been very successful, providing numerous insights into the solar atmosphere and corona, many ambiguities in the models remained. Additional constraints on the physics leading to CMEs can only be obtained by simultaneous observations from more than one advantage point. Hence, the twin spacecrafts of the Solar TErrestrial RElations Observatory (STEREO) mission were launched in 2006 and placed into the solar orbits as shown approximately in Figure 17.11. Each STEREO spacecraft contains a variety of instruments. (See Fig. 17.12 of STEREO-B.) One instrument, the Sun Earth Connection Coronal and Heliospheric Investigation (SECCHI), contains an extreme ultraviolet imager, two white light coronographs, and a heliospheric imager. Combined with SOHO measurements, STEREO's pair of SECCHI instruments provides 3D pictures of the birth of CMEs at the Sun's surface (the heliosphere), and follows the 3D propagation through the low corona and into the interplanetary medium. Other STEREO instruments improve our measurements of the structure of the ambient solar wind.

The *JWST*, the premier replacement observatory for the *Hubble Space Telescope* the *Spitzer Space Telescope*, and Herschel is scheduled to be launched in 2018. It will look back in time to what astronomers call the *Dark Ages*, a period in time shortly after the *Big Bang* that is virtually unexplored. It is the epoch when very first stars formed and gravity began to organize these stars into precursor galaxies. *JWST*, pictured in Figure 17.13, will be a 6.5 m aperture IR telescope with an orbital location at solar L2. The project represents an international collaboration of 17 countries led by NASA with significant involvement from the European and Canadian Space Agencies.

JWST represents the most technologically ambitious and challenging mission ever undertaken and its orbit at 1.5 million kilometers from Earth renders it unrepairable.

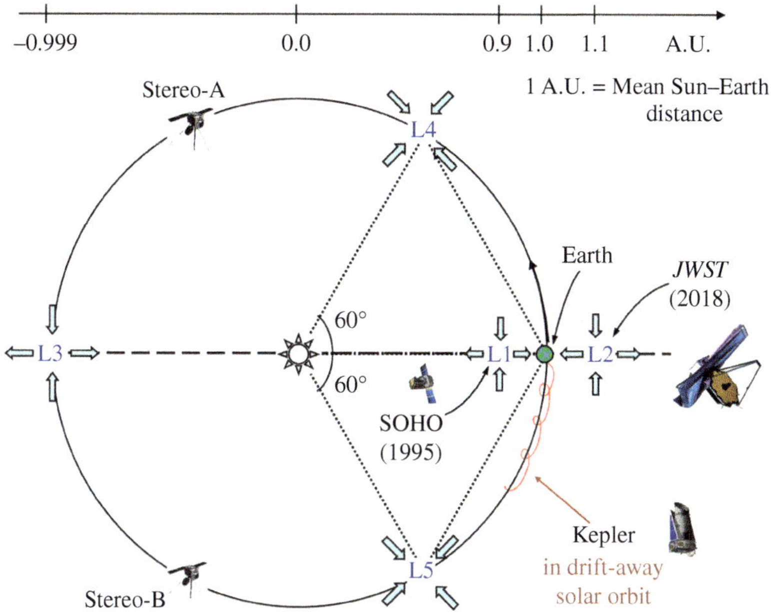

FIGURE 17.11 The locations of several missions, having high scientific impact and wide popular appeal. Launch dates are in parentheses.

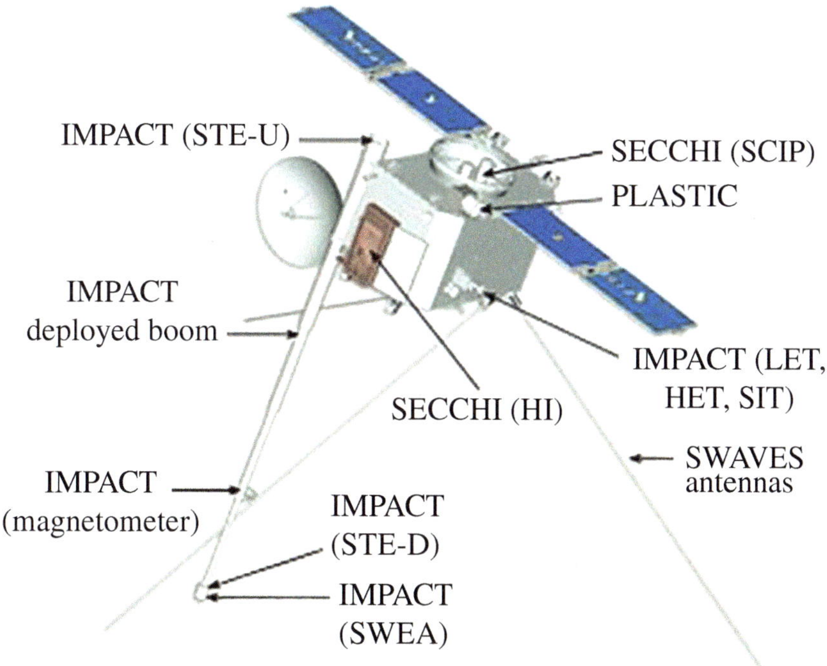

FIGURE 17.12 Mounted onto the STEREO-B spacecraft are four instrument packages: SECCHI, SWAVES, IMPACT, and PLASTIC. Source: Adapted from NASA illustration.

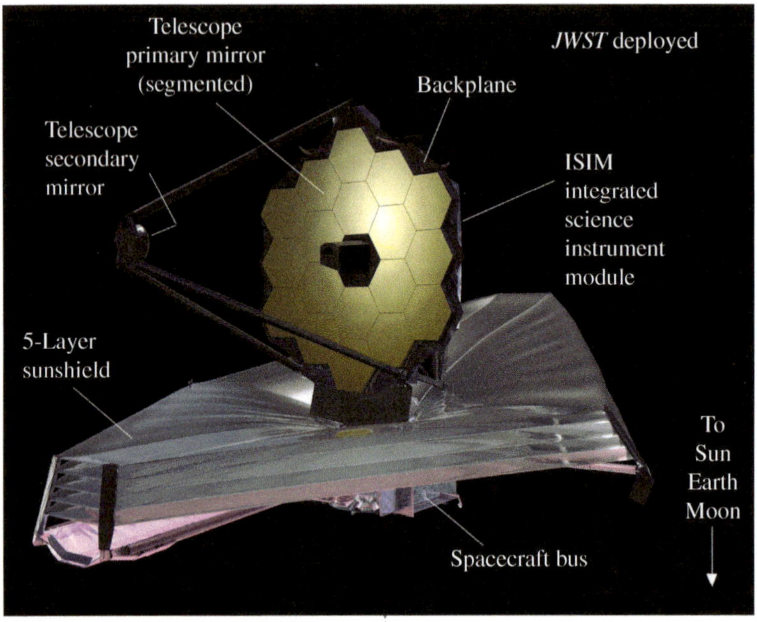

FIGURE 17.13 An artist's conception of the *JWST* observatory after being deployed. Source: NASA JWST project.

Three major challenges for *JWST* are (i) extensive unfurling of the telescope optics, the sunscreen, and other subsystems, (ii) operating with a 400°C difference between two portions of the observatory, and (iii) station-keeping *JWST* at L2 while preventing it from flipping over. In addition, the primary mirror represents novel technology with the segments being made of polished beryllium with a coating of highly pure gold. Beryllium has much less thermal contraction than glass. Nevertheless, the shape of each mirror segment will change over the very large difference between the fabrication and operational temperatures. Engineers calculate using computer models how the mirror will deform as its temperature drops. The goal is to start with a mirror that is warped in the opposite sense at room temperature such that when the mirror reaches 40–50 K, it will be distorted to the correct shape.

JWST must operate at cryogenically cold temperatures to prevent the telescope itself from creating a enormous thermal IR background that swamps the signals from deep space. The telescope optics will operate at 40–50 K (−233 to −223°C), being passively cooled by radiating to approximately 3 K deep space. The science instruments must operate even colder at 6 K, only a few degrees above absolute zero. The instrument module, which contains electronics that generates some internal heat, uses a two-stage cryocooler. (A cryocooler operates essentially on the same principles as a refrigerator or an air conditioner described in Chapter 2, but is designed to operate at the extremely low temperatures with gases that are able to remain fluid.) The spacecraft bus, which is responsible for communications back to Earth and for repositioning the observatory, will be perpetually bathed in sunlight and be as warm

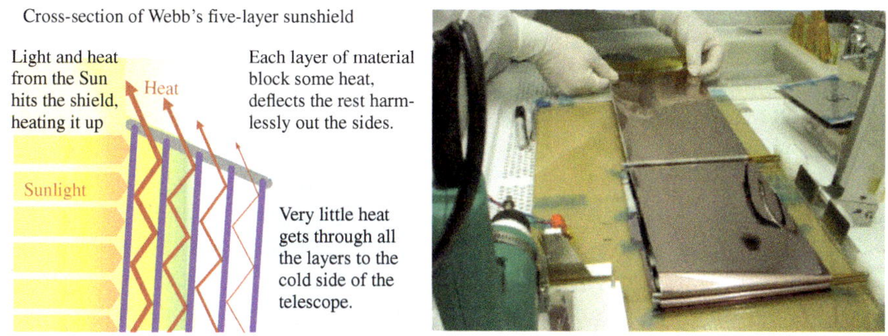

FIGURE 17.14 Left: The IR reflection mechanism used to shield the cold portion of the *JWST Observatory* from the light of the Sun, Earth, and Moon. Each sheet is colder than the previous one and consequently, creates less thermal IR. Right: Samples of the coated Kapton material used in the multilayer sunshield. Source: NASA JWST Project.

as 85°C, close to the boiling point of water on the Earth. The warm and cold portions of *JWST* are separated by a five-layer insulating shield, which is unfurled to become the size of a tennis court.

JWST is sufficiently large that the observatory has to be extensively folded to fit in its launch shroud. Compare *JWST* in its launch configuration (Fig. 16.23) to it being fully deployed (Fig. 17.13). Mechanical deployments, the unfolding and position of various subsystems, are inherently risky and *JWST* has a sizable number, any one of which could adversely impact the observatory performance. Some of the subsystems to unpack are as follows. The solar panels and telemetry antenna unfold. Six segments (three on each side) of the primary mirror are fold to the sides. These must be brought to precise alignment with the central portion of the primary. The secondary mirror attached to a boom is launched in a position over the top edge of the primary. As was pictured in Figure 16.23, the boom rotates down from the right and the third leg of the support tripod is hinged to come over the top of the primary and lock into position. The five-layer insulating shield is the most complicated deployable system on *JWST*. First, the five layers unfold on booms on each side. Then, the layers spread out orthogonally by two additional protracting booms, while still in contact with each other. Finally, each of the five layers is sequentially separated from the others, moving along the four rails at the corners of the shield.

Each layer of the heat shield is made of Kapton with aluminum and doped-silicon coatings that reflect the Sun's warming light back into space. The physics of the multilayer shield along with a picture of the sunscreen material are given in Figure 17.14. The layer bathed directly in sunlight reflects essentially all of the visible light and most of the IR. The small percentage of the light that is not reflected is absorbed, causing the temperature of that layer to raise somewhat and any material above absolute zero produces IR light in proportional to its temperature, the so-called heat light. Compared to the first layer, the next one has no visible and much less IR light to screen. Again, it absorbs a small fraction of the residual IR radiation and reflects the rest. Hence, the second layer is colder than the first and it emits less IR. Most of the residual IR light bounces between the two highly reflecting layers until it reaches an edge and is radiated into space. Five layers are required, each being colder than the previous layer, until the heating from IR radiation from the Sun, Earth, and Moon becomes negligible.

One important challenge for *JWST* will be station-keeping. As noted in Chapter 16, the solar L2 Lagrange point is a saddle point in the gravitational potential field. A saddle point implies the gravitational potential is restorable (stable) in one direction, but not in the orthogonal direction. In other words, if the *JWST* is perturbed such that it gets ahead or behind the Sun–Earth line, the combined gravitational pull of the Sun and Earth will restore it toward that line. If the perturbation is along the Sun–Earth direction, it will fly away from the region without additional station-keeping thrust from the spacecraft. Complicating matters for *JWST* is the fact that the expansive, lightweight sunshield acts like a solar sail against the solar wind. Most of the volume and much of the mass of the observatory is down wind of the sail-like sunshield, a somewhat unstable configuration against forces that will flip *JWST* to point toward the Sun. Care must be taken in making station-keeping maneuvers such that the sunshield remains squarely pointed toward the Sun to a sufficient degree.

Another high-impact mission is the Kepler Observatory, pictured in Figure 17.15 and launched in March 2009. It is in an Earth-trailing solar orbit with an orbital period of 371 days. (Nearly 30.5 years after launch the spacecraft and Earth will be on the opposite sides of the Sun and 61 years later Earth will catch back up to be once

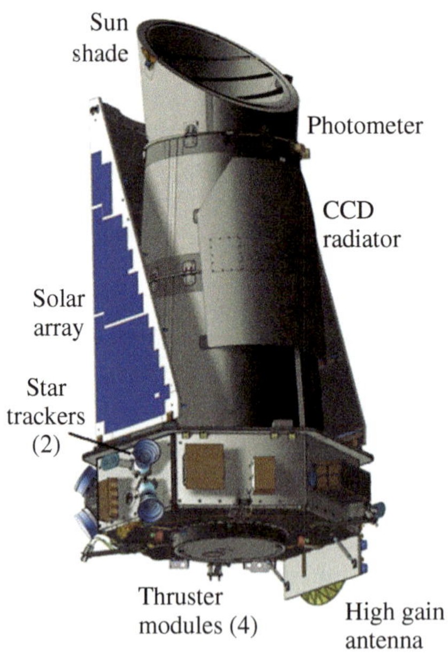

FIGURE 17.15 The Kepler Observatory to search for terrestrial sized planets in the habitable zone. Source: NASA.

again in the vicinity of Kepler.) Kepler's primary science objective is to determine the frequency of Earth-sized planets with orbits in the habitable zone around stars. (The habitable of a star is defined loosely as a region around a star where a solid planet can orbit with an atmospheric temperature range suitable for life. Cooler, redder stars than the Sun have habitable zones closer to their surface. *Note*: the life span of a star increases the cooler it is, sustaining a habitable zone correspondingly longer.) As a planet orbits its star, it periodically blocks out a small amount of the star's light. Giant planets similar to Jupiter or Saturn, dim the star by approximately 1 part in 1000, while an eclipse by solid planets similar to the Earth or Mars reduces the light by a few parts in 10^5, a hundred fold less occultation. Kepler monitors continuously more than 160,000 stars, looking for very tiny dips in the flux from some of these stars that are undergoing a planetary eclipse. The method of monitoring stellar fluxes is known as photometry and the uncertainty is a measure of its photometric precision.

To photometrically monitor approximately 160,000 stars, Kepler requires an array of large image sensors and a star field that is neither too densely or too sparsely populated, one that is also not occulted by the Earth or Sun. Kepler is a 1.4 m telescope with a focal plane that contains 21 modules, each with two 2200×1024 pixel detectors, constituting 95 million pixels. This curved focal plane is pictured in Figure 17.16. It is difficult to achieve 10^{-5} photometric precision. The absolute precision of a sensor in a commercially available camcorder is approximately 1% and scientific grade detectors are 0.1% (10^{-3}) at best. Relative photometric measurements, that is, comparing many stars simultaneously, increases the precision by more than an order of

FIGURE 17.16 The curved focal plane of Kepler, populated with an array detectors with a combined 95 million pixels. Source: NASA & Ball Aerospace Systems Group.

magnitude, especially if multiple exposures are made. Kepler increases the photometric precision further by defocusing the telescope. (Stars are so far away that virtually all of these remain unresolved point sources. Defocusing the telescope spreads the image from individual stars over several pixels, reducing any pixel-to-pixel systematic noise to gain a signal-to-noise improvement of $\sqrt{N}$ where N is the number of pixels containing flux.) Kepler has demonstrated a 10^{-5} photometric precision using these combined methods.

Prior to Kepler, several extensive ground-based surveys had found evidence of large planets around several close by stars, mostly planets with masses similar to Jupiter and orbits relatively close to the central star. While representing breakthrough science revealing the existence of exoplanets, those observations were at the very limits of what the instrumentation could achieve and remain inadequate to detect terrestrial-type planets similar to Earth. Kepler stares at a field of stars in an adjacent spiral arm of our Milky Way galaxy, stars much farther away than the previous studies. In the first 2 years, Kepler has found 47 star systems with planets and thousands more candidate planets. Photometric dips associated with a planet partially eclipsing its star are considered candidate measurements until the same dip in flux is repeated. Once an occultation reoccurs, the star becomes one with a verifiable planet in orbit. Since some planets take two or more years to complete one orbit, it will take several years of observations to confirm many of Kepler's candidates. Unfortunately, the failure of two reaction wheels, one in 2012 and the other in 2013, forced the cancellation of further observations.

INTERESTING TIDBIT TB17.4

As of March 2012, the Kepler Science Team has found 1790 host stars with a total of 2321 planet candidates. Forty-seven stars have been found with planets orbiting, most with multiple planets. On January 11, 2012, a new class of planetary system was found, a planet with two suns.

17.5 PLANETARY PROBES

Planetary probes historically have used brute force methods to land on the Moon, Mars, and other planets. For instance, the Viking Mission to Mars in the 1970s used rocket thrusters to descend to the surface. (*Note*: the atmosphere of Venus is so dense and hot that all probes have been destroyed long before reaching the surface. Gaseous giant planets such as Jupiter or Saturn do not actually have a surface.) The exclusive use of rocket thrusters is both expensive and risky. Several early missions to the Moon and Mercury simply opted to approach, slow, and impact the surface, essentially a controlled crash landing.

During the 1990s, two scientific advances renewed interest in the search for life within our own solar system. First, biological research here on Earth during this period indicated that life can evolve and survive in much harsher environments than had been previously thought. Microorganisms called extremophiles have been found in the coldest Arctic regions as well as around hot ocean-floor vents that spew out

FIGURE 17.17 The two-stage landing method of the Mars Rover. Source: NASA.

high concentrations of sulfur. Life has been found to exist in any naturally occurring liquid on Earth, even those devoid of sunlight. Flyby missions of the 1980s and 1990s as well as the 1997 launch of NASA's Cassini mission to Saturn and its moon, Titan, provided new evidence of liquid flows especially on Mars and Titan. (Saturn's moon, Titan, is similar in size and mass to that of the Earth and has long been known to have an atmosphere made primarily of ethane and methane.) Cassini revealed rivers and lakes, presumably of liquid ethane and methane since surface temperatures are −179°C, much too cold for water ice to melt, have shaped the surface of Titan. Cassini also carried a European probe, Huygens, snapping close-up pictures of Titan's surface as it parachuted down to the surface. Huygens measured wind speeds and pressures during decent and found evidence of surface erosion.

Budget constraints and some soul searching at NASA after the initial optical problem with the *Hubble Space Telescope* prompted planetary science teams to search for novel, less-expensive approaches to place probes on the planets in our own solar system. One the first innovations was the use of parachutes plus airbag technologies from the automobile industry to land the Mars rovers. Pictured in Figure 17.17 are the initial parachute stage and the bouncing plus rolling stage once the airbags had been deployed. The Martian atmosphere is far more tenuous than the Earth's and required a redesign of the parachute. After the rover had come to rest, the airbags deflated and the rover had to right itself. An artist's conception of a Mars rover on the Martian surface is given in Figure 17.18. These new landing technologies and integrated approach of the Mars Exploration Rover Mission reduce the costs in half.

It is important to keep in mind that any integrated approach to solve a safe landing on Mars, for example, cannot be applied universally to all future missions. Indeed, the SUV-sized *Curiosity* rover that landed on Mars August 14, 2012, was much too heavy for existing airbag technology. While, *Curiosity* also used a supersonic parachute initially to slow its decent and position it over the landing

FIGURE 17.18 An artist's concept of the Mars Rover operating on the surface. Source: NASA.

site, a rocket propelled sky crane did the rest. An artist's rendering of the finally landing stage is given in Figure 17.19. The use of a sky crane rather than a common legged lander offers the advantages of keeping the center of gravity (cg) low with respect to the thrusters while preventing the thrusters from kicking up dust and debris that could damage the rover's instruments during final approach. A conventional landing would have required long, massive legs to accomplish a low cg plus keeping thrusters well above the surface. Moreover, the latter would have also required a drive-off ramp, adding additional risk if *Curiosity* landed on a steep slope or a sizable rock. Upon touch down, the sky crane hovered for 2 seconds, waiting for confirmation from the sensors on the wheels that these were detecting weight. Then the sky crane fired several pyros (small explosives) to cut the cables and flew off, crashing about 650 m away.

The *Curiosity* rover is much larger, more mobile, and contains many more scientific instruments than did its predecessors, the two *Mars Exploration Rovers*. *Curiosity* has a radioisotope power source as described in Chapter 16, allowing it to move faster, dig/drill/scoop samples more quickly, and have greater experimental flexibilities than the solar-panel-powered Mars Exploration Rovers. It also has a rock-vaporizing laser and numerous cameras.

A particularly interesting destination to look for life is Europa, one of the four large moons orbiting Jupiter. Europa consists of one giant ocean devoid of any known landmasses. Its surface is frozen ice since Jupiter and its moons are five times further away from the Sun than is the Earth. An envisaged mission would land on Europa along one of the major cracks in the ice, drill a hole, and drop a tiny submarine to

FIGURE 17.19 A sky crane with three 7.6 m nylon tethers was used in final decent to lower the *Curiosity* rover to the Martian surface. The final landing occurs with wheels down. Source: NASA.

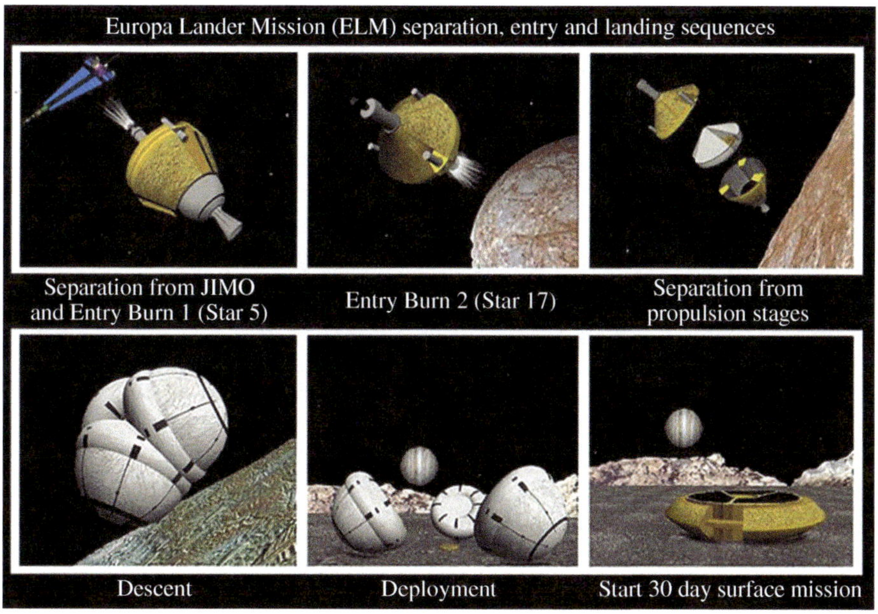

FIGURE 17.20 An envisaged Europa Landing Mission (ELM). Source: NASA.

FIGURE 17.21 A deployed microsubmarine to search for evidence of microorganisms as well as cameras to capture swimming creatures if these exist. Source: NASA.

explore its ocean. The concept mission is depicted in Figures 17.20 and 17.21, showing the landing methods and an artist's vision of the deployed submarine.

COMPREHENSION VERIFICATION CV17.1

Question: *JWST* will be a 6.5 m aperture telescope, which is larger than the 2.4 m primary of the *HST*. How much greater light gathering capacity will *JWST* have compared to *HST*? If *HST* operates at a wavelength of 500 nm and *JWST* operates at 2 μ (2000 nm), which observatory will produce the greatest spatial resolution? *Hint*: Do not actually calculate the resolutions, but ratio one against the other.

Answer: *JWST* has approximately 7.34 times greater light gathering capacity.

$$\frac{(6.5)^2}{(2.4)^2} = 7.33507$$

HST has nearly twice the resolution of *JWST*.

$$\frac{\text{Resolution}_{HST}}{\text{Resolution}_{JWST}} = \frac{2.4/500}{6.5/2000} = \frac{0.0048}{0.00235} \approx 2.04$$

18

TELECOMMUNICATIONS

Telecommunication is the transportation of information over long distances by technological means, particularly using electromagnetic or photonic signals. Some of the earliest forms of telecommunication were sent via visual signals (e.g., signal flags, beacons, and smoke signals) or via sounds (e.g., drumbeats and lung-blown horns). Telephones and landline telegraphs came into being in the 1800s, including successful undersea cables being established across the Atlantic in 1866, Australia being connected in 1872, and finally encircling world in 1902. Wireless telegraphy, known as radiotelegraphy, technologies ushered in a revolution in telecommunication in the early 1900s, at first allowing ships a few tens of kilometers off shore to communicate with land stations via Morse Code. Subsequently, radiotelegraphy and radio (or radio frequency—RF) communications achieved global reach. All of these forms of communication were either not private (i.e., anyone with a receiver could listen) or were limited to a single connection at a time requiring a separate set of wires to transmit information. Early telephone poles, for example, required a large number of wires, each connected to a central switchboard, to permit a relatively few number of connections simultaneously. Today, the majority of all communications still rely on wires and fiber optic cables for at least a portion of the transmission path between the source and destination, despite the rapid growth in popularity of smart phones and other wireless devices in recent decades. Not only is it interesting in its own right, the physics of wire cables still remains important to the advancement of high-quality telecommunications.

The major thrust of modern telecommunication companies is the transmission of large volumes of information with minimal losses over long distances in an environment having interference signals and inherent system noise. Networking, multiplexing, and data compressing systems dramatically increased the world's capacity to communicate.

Modern Devices: The Simple Physics of Sophisticated Technology, First Edition.
Charles L. Joseph and Santiago Bernal.

Companion website: www.wiley.com/go/joseph/moderndevices

The conversion from analog to digital signals and the replacement of wires in critical segments with optical fibers also considerably enhanced the volume of communications. The effective capacity of the world to communicate as measured in optimally compressed bytes/day of information went from 281 petabytes/day (281×10^{15} bytes/day) in 1986, to 2.2 exabytes/day (2.2×10^{18} bytes/day) in 2000, and to 65 exabytes/day in 2007. The telecommunications sector of the world economy was estimated to be $4.7 trillion (USD) in 2012, and its collective service revenue in 2010 was estimated to be $1.5 trillion (USD) corresponding to about 2.4% of the world's gross domestic product.

There are three primary components in any telecommunication system: a transmitter, a transmission medium or physical channel, and a receiver. Most are duplex (two-way) systems, relying on transponders that both receive and send signals. A transmitter takes information, converts it into a signal, and sends the encoded signal into the medium. A transmission medium can be a free-space channel as used by radio broadcasting stations, or it can be a point-to-point communication such as telephone voice sent over wires or fiber optic cables. Multiplexing allows several telecommunication devices to share a single physical channel concurrently, providing major reductions in operating costs. Multiplexed systems are designed as networks with routing nodes that direct individual signal packets to the correct destinations.

INTERESTING TIDBIT TB18.1

Amos Emerson Dolbear, an American physicist and professor at Tufts University, was well ahead of his time. He invented the first permanent-magnet telephone receiver in 1865, 11 years before Alexander Graham Bell. He also pioneered a wireless telegraph in 1882, ahead of German scientist Heinrich Hertz and Italian inventor Guglielmo Marconi. Unfortunately for Dolebear, Mr. Bell is credited with the invention of the telephone and Guglielmo Marconi is considered the inventor of radio, sharing the Nobel Prize in physics for his contributions to wireless telegraphy. A 1881 article in *Scientific American* articulated his plight: "...had [Dolbear] been observant of patent office formalities, it is possible that the speaking telephone, now so widely credited to Mr. Bell would be garnered among his own laurels."

18.1 PHYSICAL CONNECTIONS: PHONE LINES, COAXIAL CABLE, AND FIBER OPTICS

We discuss in this section some of the physics of point-to-point communication, including landline wire telecommunications in particular. Both wire lines and fiber optic cables share the same attributes: the loss of signal strength as a function of distance, pulse broadening over long path lengths, cross talk between channels, and the susceptibility to external noise. Some of these characteristic properties are depicted graphically in Figure 18.1. The physics of optical fibers, which enable information to

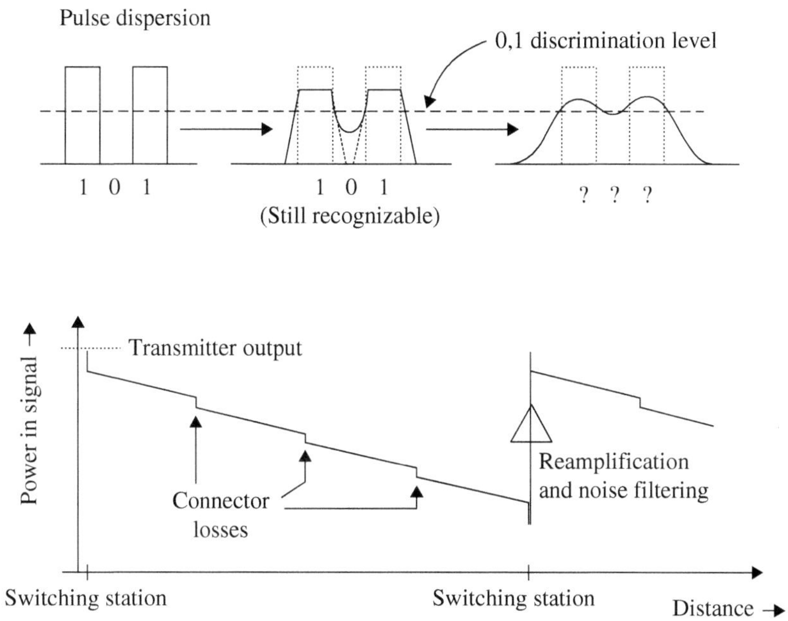

FIGURE 18.1 The losses that occur in long-distance wire cables. High-frequency pulses travel faster than low-frequency components, resulting in pulse dispersion (top), the spreading of the signal. Transmission losses (bottom) occur continuously along a pair of wires and small discrete losses arise at each connector that mates two cables. The signal has to be amplified periodically as well as have the external sources of noise reduced through filtering.

be transferred with 30 times greater efficiency compared to wires, have been discussed in Chapter 11. (Compare Fig. 18.1 with Figs. 11.4 and 11.10.)

Any two wires that run next to each other, especially over distances of many kilometers, have the physical properties of capacitance, C, resistance, R, and inductance, L. (See Chapter 6 for a discussion regarding techniques for simplifying the impedance of any complex circuitry to a simple equivalent LRC circuit.) The dominant component of impedance in long-range cables is capacitance. A conventional capacitor is simply two parallel conducting plates with a small separation. Two wires in a long cable are equivalent physically to a very-long, but very-narrow capacitor. Conductors with the exception of superconductors also have trace amounts of resistance, which are negligible over short distances, but are important over long path lengths. Moreover, a current carried by a wire produces a small magnetic field that is cylindrically symmetric about the axis of the wire. Any signal in one wire is affected by the magnetic field produced by the signal in another nearby wire, causing an inductive coupling between wires in the same cable. Figure 18.2 represents a circuit diagram of the impedance from each incremental length (Δx) of a two-wire cable. The resistance along the wire segment is $R\Delta x$, while the resistance between the two parallel wire segments, $G\Delta x$ in Figure 18.2, accounts for the tiny leakage current through the less-than-perfect electrical insulator. To obtain the total impedance from

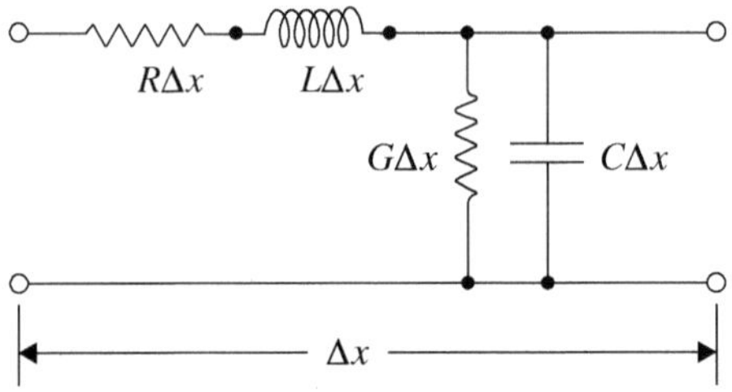

FIGURE 18.2 A schematic diagram of the electronic circuit for a long two-wire cable. The corresponding mathematical representation is the Telegrapher's equation.

a long cable, one has to integrate mathematically along all of the Δx segments. Two differential equations—one for voltage, $V(x,t)$, and the other for current, $I(x,t)$—are the mathematical counterparts to the circuit diagram. Each equation is known as a variant of the *Telegrapher's equation.*

The solution to the current, $I(x,t)$, from the Telegrapher's equation is

$$I(x,t) = \frac{f_1(\omega t - kx)}{Z_0} + \frac{f_2(\omega t + kx)}{Z_0}, \tag{18.1}$$

where f_1 and f_2 are arbitrarily shaped functions comprising multiple waves that make up the signal. This signal has an effective wavenumber, k, and an angular frequency, ω. (c/f the Intro Physics Flashback FB18.1.) Function f_1 travels left-to-right, while f_2, propagates right-to-left, both being modified by the impedance, Z_0, where

$$Z_0 = \sqrt{\frac{L}{C}}. \tag{18.2}$$

The velocity, v, in which information travels down the cable, is given by the following equation:

$$k = \omega\sqrt{LC} = \frac{\omega}{v} \Rightarrow v = \frac{1}{\sqrt{LC}} \tag{18.3}$$

Note: an unusually large inductance or capacitance will result in a slow transmission speed. One method to reduce C and L is to use small diameter wires with thicker insulations to increase the wire separation. Unfortunately, a thin wire has a larger R (resistance) than a thick one, increasing the signal dissipation per unit length. In addition, the signal strength decays with distance as S_0e^{-ax}, where $\alpha = R/2Z_0$, assuming the transmission line is only slightly lossy ($G \approx 0$ and the R coefficient is small). In other words, the signal will travel faster in a thinner wire, but will not travel as far before degrading into an unrecognizable signal. Tradeoffs have to be made between all of these parameters in the design of a cable.

Present-day wire cables contain hundreds or thousands of wire pairs, each with shunt capacitance, inductance, and resistance that also couple to the other pairs.

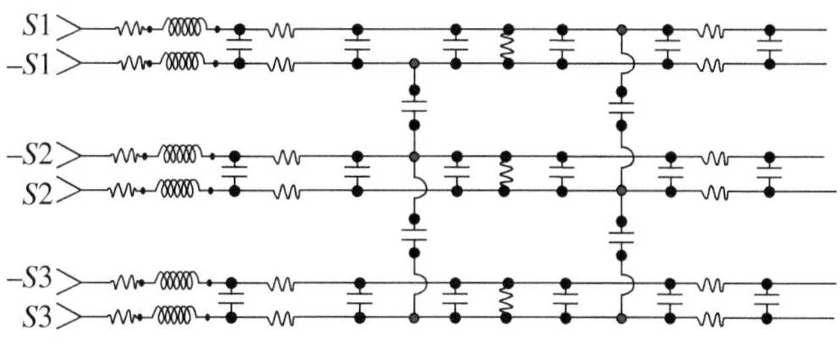

FIGURE 18.3 The coupling of signals within a cable containing three sets of wire pairs. The diagram indicates balanced pair operation, in which the difference signal is detected at the destination receiver.

Schematically, the circuit diagram for a cable with three pairs is shown in Figure 18.3. *Note*: the additional shunt capacitance between pairs. While resistance paths between signal pairs exist, these are negligible compared to the impedance coupling between a signal pair and the external environment so are usually ignored. *Differential-mode transmission*, using balanced pair operation, is often used to remove noise and spurious signals picked up by the cable. For balanced pair operation, the two wires carry equal and opposite (negative) signals, denoted (*S*1, −*S*1), (*S*2, −*S*2), and (*S*3, −*S*3) in Figure 18.3. Noise or erroneous signals picked up from extraneous electric or magnetic fields along the transmission path tend to impact both wires in the pair equally, which cancel out in a difference signal.

Several cable designs are available to improve signal isolation, inhibiting electromagnetic interference (EMI). Two such designs, twisted pairs and coaxial, are depicted in Figure 18.4 as cutouts. Unshielded twisted pairs (UTPs) is the least expensive and a commonly used cable in telephones and computer networks. Twenty-five color-coded UTP pairs is a standard indoor phone cable in the United States. Coiling each wire in the pair around each other and using balanced pair operation dramatically reduces the electric and magnetic fields emanating from the pair since the two signals are opposite. In other words, the magnetic field produced from say *S*1 signal is cancelled by the equal and opposing field from −*S*1. However, the benefit of differential mode transmission is partially negated if adjacent pairs have equal twist rates (twist pitch), especially in cables with a small number of twisted pairs. UTP cables such as that shown in Figure 18.4 (top) usually specify different twist pitches for this reason. Moreover, individual pairs or a subset of pairs are sometimes wrapped in foil to enhance EMI shielding, while the UTP cable as a whole might be enshrouded in a braided screen or braided foil for the same reason.

The coaxial cable or its shortened name: coax is generally the best EMI protected design, albeit a far more expensive one. Common applications include cable television distribution, RF, and microwave transmission. As its name suggests, the inner conductor, the insulating layers, and the grounded shield mesh in a coaxial cable, all share the same geometrical axis. *Note*: there are also triaxial cables that have a second layer of grounded mesh, used for measuring very weak signals with precision. While technically these too are coaxial, the term *triax* or *triaxial* is the preferred nomenclature, reserving coax for cables with only one layer of shielding.

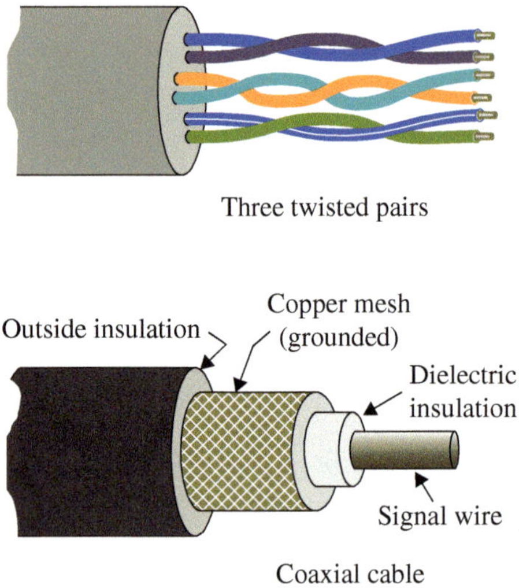

FIGURE 18.4 Two cutaways of cable designs, unshielded twisted pairs (top) and a coaxial cable (bottom), to minimize crosstalk and external EMI.

Coaxial cables such as the cutaway depicted in Figure 18.4 provide superior performance compared to other shielded cables, especially for RF signals up to a few gigahertz. These cables have tight tolerances on material quality and on the uniform diameter of each layer, driving costs. In addition, the conductors have large cross sections compared to twisted pairs. The size and dielectric properties of the insulator separating the signal wire from the grounded shield is also important. Physically, coaxial designs confine the electromagnetic fields from the signal wire to the dielectric insulator with very little leakage outside of the grounded shield. Conversely, external EMI are largely prevented from reaching the signal wire. In this sense, a coax may be viewed as a waveguide for RF signals, exhibiting the same physical properties as optical fibers, but for electromagnetic waves at RF frequencies.

INTRO PHYSICS FLASHBACK FB18.1

Wave Propagation

Modern telecommunications are based primarily on wave propagation physics and a deeper, richer comprehension of the former can be had through a basic understanding of waves. As demonstrated in many introductory physics courses, waves constructively and destructively interfere with each other. This phenomenon either results in standing waves (i.e., stationary bright and dark or loud and soft regions), or results in moving beat patterns. Here, we want to extend our treatment somewhat beyond two-wave interference normally presented in most introductory courses.

Consider the movements of a stereo speaker to render multi-note chords. The speaker displacement as a function of time for a C major chord is plotted at the bottom of Figure FB18.1, along with the three separate sinusoidal waves associated with American piano notes: C, E, and G (top plot). A simple point-by-point summation of the three input waves yields the bottom plot, which also ideally tracks the motion of the speaker.

The salient point is that there are inflection points on the plot, marking the times where the speaker coil changes its directional motion before going through its "zero" rest position. It is not necessary to have three separate speakers to handle a three-note chord. A single speaker renders the composite waveform, identical to the superposition of three speakers. This behavior is more evident in Figure FB18.2, showing the speaker displacement for the C major dominant thirteenth chord (*notes*: C, E, G, B^b, D, F, and A).

Insight from these more complex interference patterns enhances one's grasp of telecommunications. Sounds from multiple sources, whether these are from an orchestra or a group conversation, result in more complex, irregular waveforms than from musical chords. All of the informational content (i.e., the net electrical signal or the net speaker displacement) is contained in this relatively low-frequency, complex waveform. The ultimate receiver (e.g., ear plus brain) can disentangle simultaneous sounds from multiple sources from these composite waveforms. In telecommunications, these composite waveforms are used as an input variable to a simple carrier wave, which has a frequency that is approximately 1000 times higher. The carrier wave, for example, can be amplitude modulated by this signal (complex waveform) as seen in Figure 18.9.

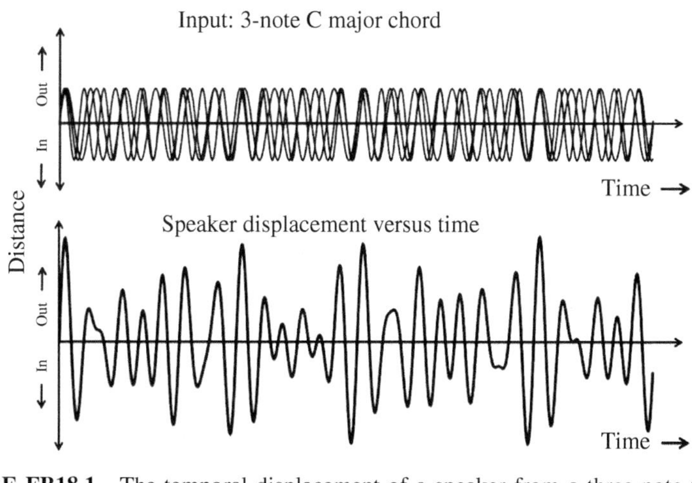

FIGURE FB18.1 The temporal displacement of a speaker from a three-note C major chord.

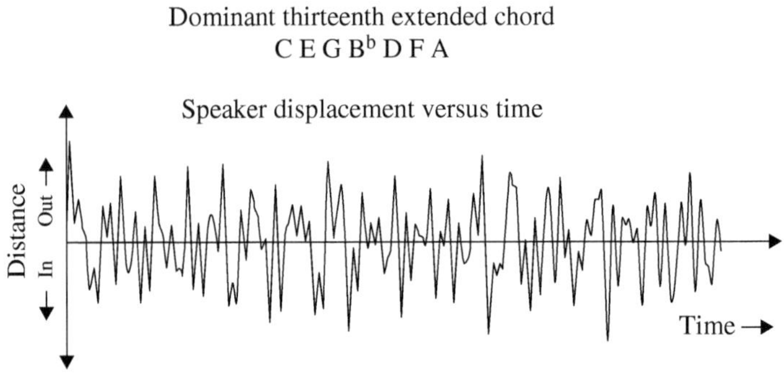

FIGURE FB18.2 The resulting speaker displacement from a dominant thirteenth extended chord. *Note*: the multiple times where the displacement does not cross zero over a few cycles.

18.2 ANALOG FREE-SPACE CHANNELS: TV, RADIO, MICROWAVE CONNECTIONS

Some of the most common forms of free-space telecommunications are radio broadcasting, shortwave radio, and television broadcasting. Mobile phones (cell phones) and Wi-Fi (wireless internet) are also examples that use free-space channels. All of these applications use electromagnetic radiation, propagating through the atmosphere at various radio wavelengths. We will limit our attention in this section to analog signals, which are still widely used and have the same wave propagation physics that free-space transmission of digital signals have. Moreover, while there are several methods for encoding and transmitting analog information, we will focus on amplitude modulation (AM) and frequency modulation (FM), the two earliest types and still the most commonly used in broadcasting. Table 18.1 lists some of the RF bands (or equivalently wavelength bands) along with familiar telecommunication applications using those bands. The opacity of Earth's atmosphere (see Fig. 17.10) from deep-space zenith to sea level is presented once again for the radio-wavelength portion in Figure 18.5 with the frequency bands of Table 18.1 overlaid for reference. Atmospheric absorption is strongest for low-frequency (long-wavelength) RF. The ionosphere causes RF refraction and reflection, especially for the shortest wavelength (highest frequency) RF radiation, reducing signal strength. *Note*: the two prominent opacity peaks around 10 cm in Figure 18.5 are associated with the two primary ionospheric layers.

While radio-wave signals travel in straight lines, RF communications can be sent all around the world, reflecting repeatedly off the uppermost parts of the atmosphere. UV light from the Sun establishes a layer of ions and electrons known as the ionosphere in Earth's tenuous upper atmosphere. Actually, the ionosphere comprises two major plasma layers that are always present plus an addition daylight layer, all of which vary in thicknesses and fluctuate in electron densities diurnally, seasonally, and at times unpredictably (e.g., with changes in solar activity). Note: we are using the physicist's

TABLE 18.1 A Few Radio Bands and Telecommunication Uses

Frequency Band	Frequency (Wavelength)	Example Applications
MF (medium)	300–3000 kHz (1 km to 100 m)	AM broadcast, amateur radio
HF (high)	3–30 MHz (100–10 m)	Ham and Citizens' band radio, RFID, first-generation mobile phones
VHF (very high)	30–300 MHz (10–1 m)	FM radio, television broadcasts, aircraft-to-aircraft communications
UHF (ultra high)	300 MHz to 3 GHz (1 m to 100 mm)	Television, mobile phones, wireless LAN, GPS
SHF (super high)	3–30 GHz (100–10 mm)	Communications satellites

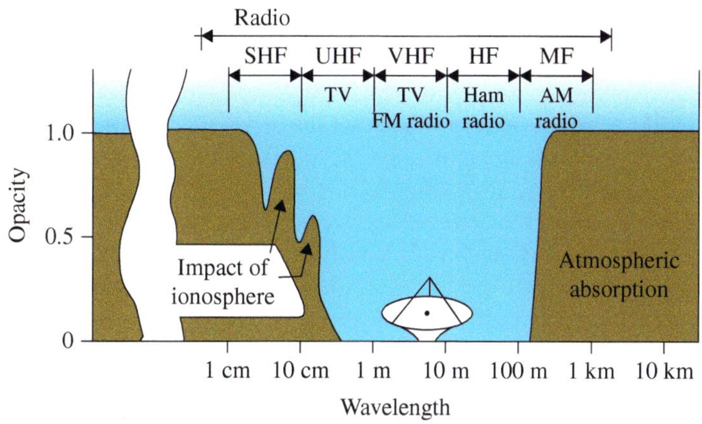

FIGURE 18.5 The RF portion of Figure 17.10 with broadcast ranges and frequency bands overlaid. Opacities only represent path lengths from deep-space zenith to sea level. Source: NASA.

definition of plasma: a partially or fully ionized gas. The ionospheric layers at the highest altitudes, however, always persist even during the nighttime hours because recombination rates are sufficiently slow because the atmosphere at those altitudes is very tenuous. Multiple plasma layers produce multimode RF reflections that can introduce radio reception problems. For simplicity, we will treat the ionosphere as a single plasma layer at a fixed altitude for the purposes of the present discussion.

When a radio wave from a ground antenna reaches the ionosphere, a tiny fraction of the energy is absorbed as a few electrons recombine with ions, some of the beam is refracted into deep space, but a portion of it is re-radiated and reflected back to the surface. This behavior is depicted in Figure 18.6 (left side), which is identical to a visible wavelength radiation encountering a surface between two materials with different indices of refraction. In addition, total internal reflection from the ionosphere occurs for sufficiently low-frequency RF signals, the same wavelength-dependent response as visible light incident on a surface. (See Chapter 10 and especially the Intro Physics Flashback 11.1.) Figure 18.6 (right)

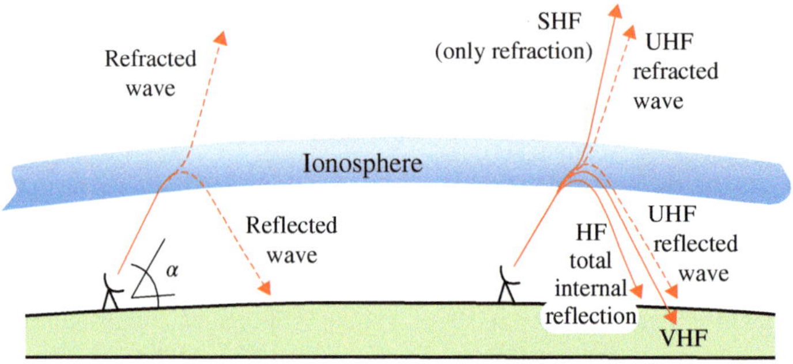

FIGURE 18.6 A schematic representation of an RF beam encountering the ionosphere (left). An example of the frequency-dependent interactions of RF waves refracting and reflecting in the ionosphere (right).

depicts an example condition where a small range of frequencies in the UHF is both refracted and reflected, while higher frequencies only refract and lower ones undergo total internal reflection. The exact range depends on ionospheric environment and more importantly on the angle of attack, α, of the transmitted beam with respect to the ground.

As noted previously, radio waves and visible light are both portions of the same electromagnetic spectrum, and hence have identical properties. In all cases, some of the electrons move in response to the oscillating electric and magnetic fields of the EM waves, refracting and reflecting EM waves of the same frequencies. (Ions also participate, but with masses that are several thousand times larger than electrons, the interaction with EM waves is tiny compared to that of the electrons.) For glass or plastic optical elements, the electrons move over limited distances within the dielectric material. Conventional mirrors, being simply glass or clear plastic with a silver coating, reflect light via moving electrons in the conduction band of the silver. It is free electrons in the ionosphere that respond to RF waves via cyclical motions in-phase with the fluctuating EM fields, provided the frequency is not too high. If the RF frequency is too great, the free electrons cannot respond fast enough and the wave simply refracts into deep space.

If the RF wavelength is sufficiently long (low enough frequency) or if the angle of incidence of the RF wave is shallow enough, then the wave will experience total internal reflection by the ionosphere. The ionosphere is an RF waveguide under these conditions with properties identical to those of a fiber optical cable, enabling multiple reflections off the upper atmosphere and over-the-horizon signal transmission. Figure 18.7 depicts the impact of the angles of attack, α_L and α_R, for two identical signals sent from the same location. The far-left antenna sends its signal more vertically where it is partially reflected and partially refracted. Signal losses are substantial at each reflection point, significantly limiting the reception range. In contrast, the adjacent antenna has a small angle of attack, α_R, producing sky wave or skip propagation, where the signal reflects completely back from the ionosphere repeatedly. Surface water and trace minerals in the soil are also reasonably good electrical conductors, enabling a sizeable fraction of the RF wave to be reflected, albeit somewhat diffusely off the surface and

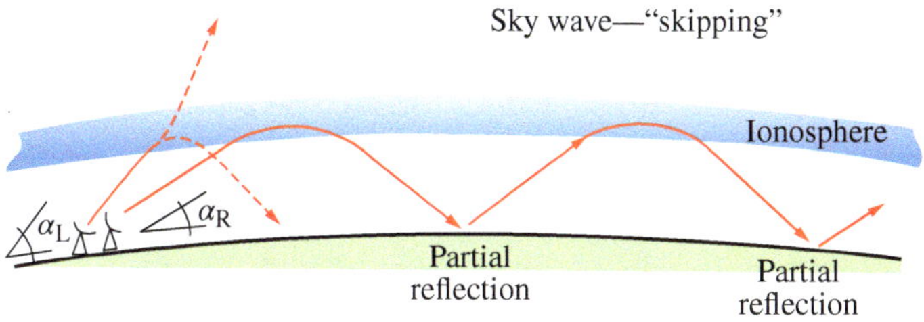

FIGURE 18.7 An RF signal can be sent to the other side of the Earth using a sufficiently small beam angle to sky wave (or skip).

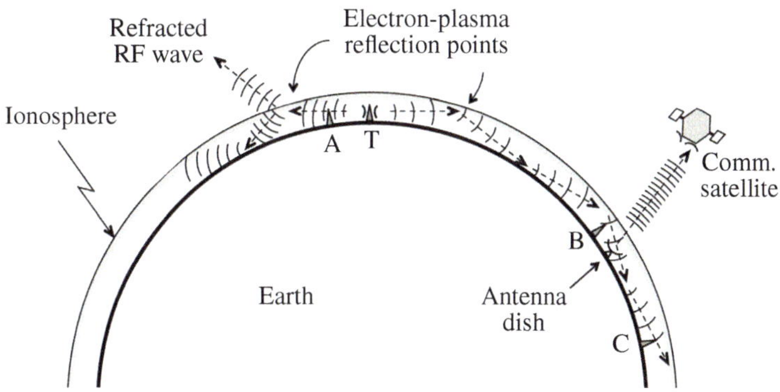

FIGURE 18.8 Schematic representation of how radio waves reflect off the ionosphere, allowing strong signals to be collected by antennas all over the world. The relative sizes are not to scale. See text for more details.

back toward the ionosphere. Path distances up to approximately 3500 km can be obtained with a single hop and transatlantic expanses in two or three hops.

Consider Figure 18.8 to broaden our understanding of over-the-horizon and ground-to-satellite RF telecommunications. (*Note*: the relative sizes are not to scale. The altitude of the ionosphere in actuality is 85–600 km, making its mean altitude about 1.05 times the Earth radii. In other words, the gap between Earth's surface and the bottom of the ionosphere is about one-fourth as large as that shown. The antennae towers would not be visible if displayed to scale. Nevertheless, Fig. 18.8 is instructive.) There are antennae located at A, B, C, and T in this example. The T transmitter has two radio dishes, each sending nearly horizontal RF signals, but in opposite directions and in different wavelength bands. A high-frequency (short-wavelength) signal is sent counterclockwise and can be received through direct line-of-sight transmission at location A. The signal continues past location A until it reaches the ionosphere, which refracts a portion and reflects the rest back toward the ground. (We are ignoring the tiny amount that is absorbed.) A significant fraction of the signal is lost at each boundary, severely limiting the transmission range of the RF wave. A second low-frequency (long-wavelength) signal is sent clockwise from T that can be received at locations B, C, and points further away. The atmosphere in this case is a waveguide with total internal reflections each time the wave encounters the ionosphere, enabling the signal to travel transcontinental distances, provided the initial signal is strong. Surface-to-satellite duplex telecommunications are also possible at location B via a dish antenna, which uses super high-frequency SHF (very short RF wavelength) signals that oscillate too rapidly for ionospheric electrons to follow or to reflect. *Note*: antenna dishes are only necessary to direct RF beams in particular directions; the tower antennae at A, B, and C can receive or transmit in all directions simultaneously. It is important to realize that reflections off the surface are always partial reflections with some of the beam being absorbed and some being scattered. Only strong signals can be sky wave propagated far over the horizon.

Up to this point, we have presented the wave physics associated with RF telecommunication, providing physical insights into the beam propagation of RF signals through the atmosphere. However, no informational content can be transmitted via a constant-amplitude single-frequency EM wave. While telecommunication protocols (i.e., encoding, sending, and receiving information) are really engineering rather than scientific tasks, significant wave physics is involved. The process requires a carrier wave that is much faster (usually thousands to millions of times higher frequency) than any rate of change in the signal strength. Then, the carrier wave must be modulated in some fashion to transmit sounds, video pictures, or other data.

Prior to a discussion of the modulation techniques, the reader should be made aware that graphical devices (i.e., plots on paper or computer screens) impose severe limitations on the presentation of these methods. Specifically, an actual carrier wave cannot be plotted to scale against the signal wave since the resolution of graphical screens or a sheet of paper are approximately 1000 times too coarse. A simple examination of some benchmark numbers quickly reveals the crux of the problem, the lack of sufficient resolution. The largest display systems are measured in megapixels (i.e., thousands of vertical columns). Standard HDTV, for example, has 1280×720 lines (~1 megapixel), while the planned QFHD has 3840×2160 lines or about 8 megapixel. (See Chapter 15.) A minimum of nine columns per wavelength are necessary to plot a distinguishable sine wave, indicating a maximum of approximately 100 crests and troughs that can be fit clearly on a single graph. Only a small segment (<1 half wavelength) of a signal wave can be displayed to scale, if the carrier wave frequency is to be depicted thousands of times faster than that of the signal wave. Further analysis of benchmark numbers reveals additional insights. The range of pitches that can be produce by choir members not including overtones is 80–1100 Hz, a more limited frequency range than the 20–20,000 Hz (maximum of 15 kHz for most adults) audible to the human ear. The range of broadcast channels is 530–1620 kHz for AM stations and 87.5–108 MHz for FM stations with the exact limits dependent on the regulations from individual nations. Thus, the carrier wave frequency compared to the highest pitch of a human voice is 480 times or greater for AM and 80,000 times or greater for FM broadcasting. Only one-fifth of a signal wave could be over plotted in a to-scale graphical depiction that included the carrier wave and this plot would represent the extreme end of favorable conditions. The reader should realize from this benchmark analysis that

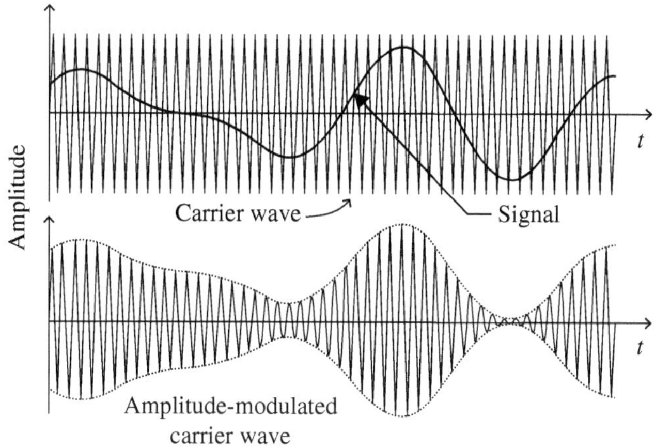

FIGURE 18.9 Plots of the amplitudes of the signal and carrier waves as a function of time (top) and the resulting amplitude modulated carrier wave plotted as a function of time (bottom).

all graphical depictions of telecommunication modulation found on the internet or in textbooks have to be grossly exaggerated to reveal basic properties.

The first method developed was AM, which is excellent for voice broadcast or conversation communications but is prone to noise interference from thunderstorms and EMI from other RF sources. Figure 18.9 (top) depicts an example of a varying signal level over-plotted against a carrier wave. The signal represents the superposition of multiple sounds, each with a different frequency or pitch. (See the Intro Physics Flashback FB18.1 to see how multiple waves combine constructively and destructively to produce a signal.) This multicomponent signal wave is used to alter continuously the amplitude of the carrier wave as plotted in Figure 18.9 (bottom). Observe the signal (thick line on top plot) has the exact same profile of the upper dotted line, which shows the continuously changing amplitude of the carrier wave. Also flipping or mirroring vertically the plotted signal forms the exact profile of the lower dotted line in the bottom plot.

Another common encoding method used in radio broadcasting is FM, which has its roots in the 1930s a couple of decades latter than AM radio. FM is preferred for music radio and for telemetry with spacecraft since it is less sensitive to noise pickup than AM transmissions. FM does require, however, more bandwidth than does AM, meaning there must be a greater frequency separation between adjacent channels. FM broadcasting occurs in the VHF range, specifically the 87.5–108 MHz portion of it, except in Japan (76–90 MHz) and in Russia (65.9–74 MHz). FM channels typically have 150–200 kHz separations with ±50 to ±75 kHz for signal modulation and 25–50 kHz gaps between channels. For example, FM stations in the United States broadcast at frequencies center on odd-numbered tenth decimal of megahertz (e.g., 99.1, 99.3, and 99.5 MHz), reserving 0.150 MHz (±0.075 MHz) for modulated changes to the carrier wave.

Encoding a signal via frequency modulation is depicted graphically in Figure 18.10, which uses the same signal that was plotted in Figure 18.9. An FM carrier wave

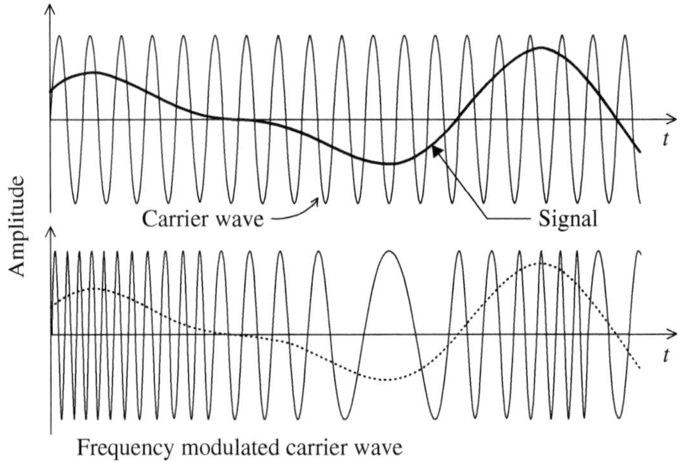

FIGURE 18.10 Frequency modulation. Carrier wave and signal wave are not to scale along the time axis, t. See text for more detail.

changes its frequency or equivalently its wavelength according to the instantaneous amplitude of the signal. The frequency of the carrier wave is highest (shortest wavelength) when the signal (bold line) is at its most positive values and the lowest frequency (longest wavelength) when the signal is at its most negative value. For easy reference, the signal strength has been over-plotted (weak dotted line) along with the frequency-modulated carrier wave at the bottom. Let us define f_0 as the frequency of the carrier wave depicted (top) and shown in regions where the amplitude of the signal is near zero (bottom). The modulated carrier wave as depicted at the bottom has a frequency, $f \sim 3f_0$ when the amplitude of the signal is at its most positive values and $f \sim 0.3f_0$ when the amplitude is most negative. As noted previously, the plotted carrier wave modulation of Figure 18.9 is only symbolic of the changes in wavelengths or frequencies, which have been grossly exaggerated for visualization purposes. The minima of the signal, which barely shows 2π (one cycle) of the carrier wave in the bottom plot of Figure 18.10, should in actuality contain approximately 20,000 cycles.

18.3 DIGITALLY MODULATED FREE-SPACE CHANNELS

Virtually, all of the physics associated with digital RF telecommunications is embodied in our discussion of the previous section on analog free space channels. Digital transmission and reception simply entail a series of data bits with two possible states, representing a "one" or a "zero" instead a continuous range of analog values. Usually, these bits are lumped together to form an 8-bit datum known as a byte that can take integer values from 0 to 255. A single pixel (picture element) in a digital television broadcast, for example, can have 256 shades of gray running from black (0) to white (255), or it can take various shades of hue (color) according to its byte value. See Chapter 15 on displays. Each subsequent pixel consists of eight bits until an entire line of picture elements has been sent. Periodically, there has to be a unique and unusual string of bits used to synchronize the receiver with sender electronics in case there is a data drop out. For broadcast television, each line (row) of display pixels is sandwiched between two overscan bit streams, containing a series of bits to indicate the exact start of each new row of display elements. In addition, the start of a new picture frame is indicated by another unique set of bits. Telecommunication of computer data proceeds pretty much the same way, except the bit stream has to include periodic error checking and "hand-shaking" signals returned to the sending source, indicating whether the latest portion of data was received correctly or has to be resent.

Digital encoding and decoding methods are known as keying schemes. There are three primary systems: amplitude-shift, frequency-shift, and phase-shift keying, as well as several hybrid methods. The three primary shift keying schemes are depicted in Figures 18.11, 18.12, and 18.13 for the transmission of a 0-1-1-0-1 bit sequence. For these three figures, the upper plot shows the voltage levels of the digital signal and the bottom plot shows the input carrier wave. These schemes are identical in nature to the analog systems already discussed, although the carrier wave only takes one of two states. The dashed vertical lines mark the boundaries between adjacent bits, which are determined by constant-period clock pulses. Carrier wave discontinuities occur at these boundaries whenever the binary state changes from a "0" to a "1" or visa versa. While the amplitude of the carrier wave remains fixed for frequency-shift and phase-shift keying, abrupt changes in frequency or in phase occur respectively during bit-state changes. In contrast, the carrier wave at the "1" to "1" bit transition is undisturbed in Figures 18.11, 18.12, and 18.13. *Note*: the amplitude of the carrier wave is never zero in the amplitude-shift keying mode (Fig. 18.11), maintaining the presence of the carrier wave even in the "0" state.

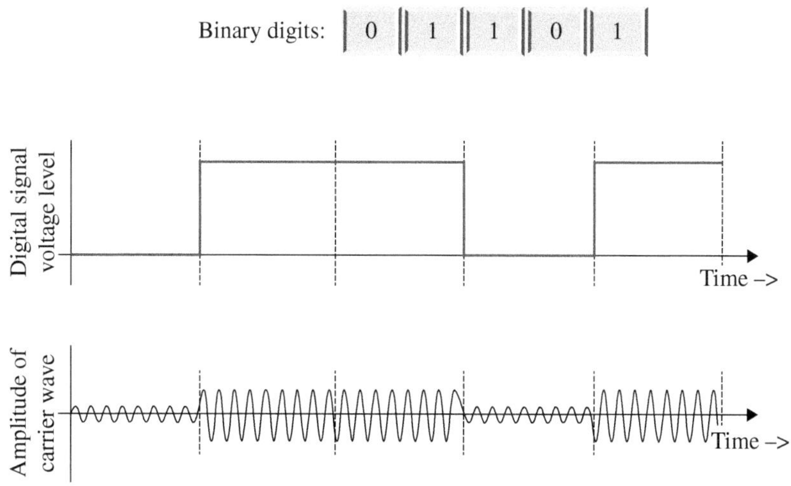

FIGURE 18.11 Digital transmissions through amplitude keying.

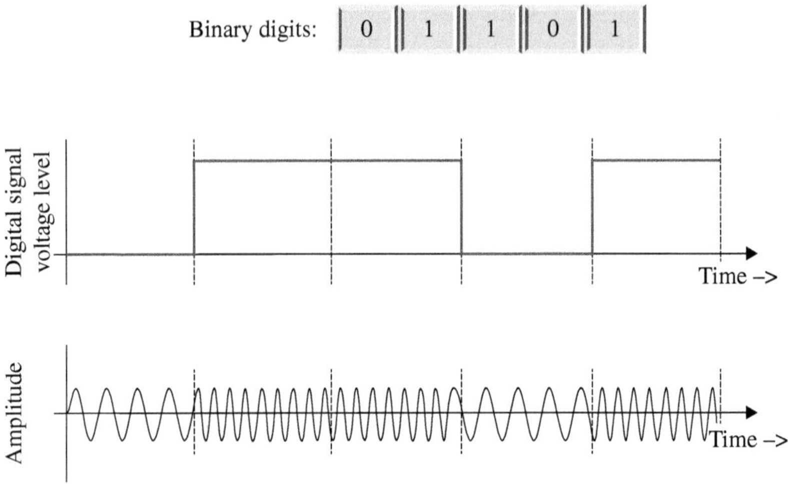

FIGURE 18.12 Digital transmissions through frequency keying.

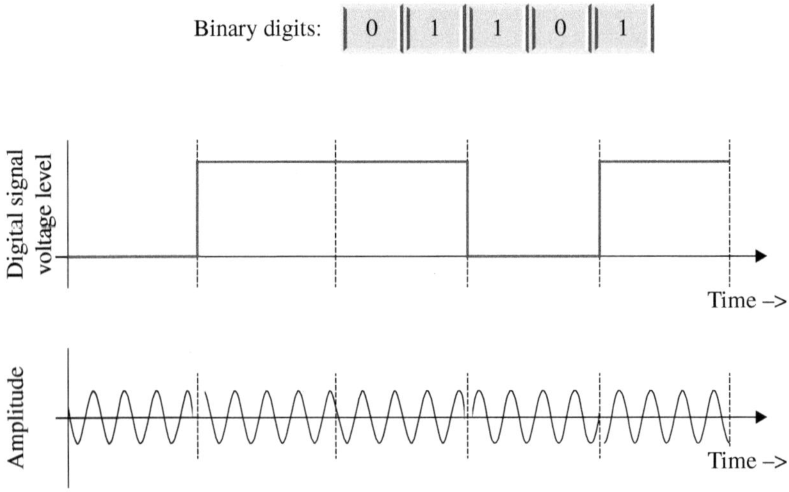

FIGURE 18.13 Digital transmissions through phase shift keying.

18.4 THE NETWORK, MULTIPLEXING, AND DATA COMPRESSION

It is instructive to know how computers communicate with each other and with peripheral devices prior to delving into a general discussion of telecommunication networks. An example of a small computer network that might be found in a home is shown in Figure 18.14. Central to the system are three multiwire cables known as buses, which every computer and peripheral device must be connected to share information. In wireless systems, each bus uses its own frequency transmitter/ receiver and a series of radio pulses are used instead of wires. A hardwire cable,

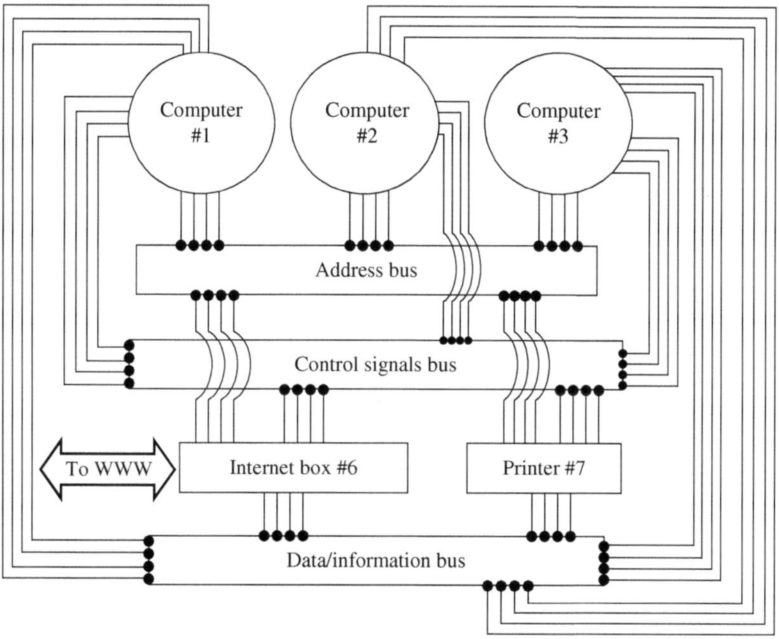

FIGURE 18.14 A typical configuration of a household network showing the three communication buses used to communicate between devices.

which is easier to visualize, will be used for clarity. The data bus is the actual cable where informational contents (e.g., text, pictures, and audible files) are exchanged, but it would be rendered useless if all computers attempted to use this bus simultaneously. Information transfer requires an address and a control bus. One device needs to be designated as the server, directing the digital bytes from and to various destinations. As an example, let us assign Computer 1 in Figure 18.14 as the server and observe the sequence of events necessary for Computer 3 to send a file to the printer. In addition to its normal functions, Computer 1 continually sequences through address 1–7, polling the status of each device via signals on the control bus. Most of the time, the various devices addressed simply return an "all clear" on the control bus. This time, however, when 3 is addressed and signaled for status, 3 returns on the control bus the bit pattern for "request printer access." The server then sets the address bus to 7 and sends on the control bus the bit pattern for it to receive input over the data bus. Once the printer is waiting for input, the server readdresses 3, telling it to output the file onto the data bus. These three buses are necessary to control the flow of information between multiple computers/devices, all sharing the same data bus. The details of these exchanges are more complicated than described here and other network configurations exist, but all rely on a three-bus system.

Generally, telecommunication networks are organized in a tree-branch system (Fig. 18.15) with individual communication trees being interconnected through central nodes. (Small networks or subnetworks are sometimes organized in loops.)

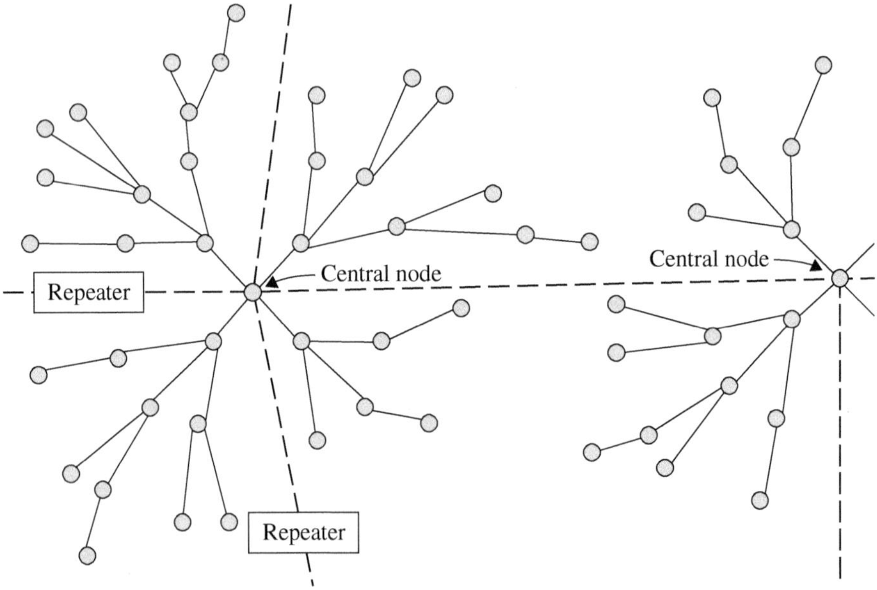

FIGURE 18.15 An example of a tree-branch network, showing two central nodes.

Each connection (solid or dashed line in Fig. 18.14) really consists of three separate and parallel connections known as planes or sometimes as buses for local networks. The data plane (also referred to as the user plane or bearer plane) is the set of wires, fiber optics, or free-space channels used to carry the informational content. A central processor enables the various devices on the local network to function together using the management plane. All devices are always listening to the management plane, consisting of two components: address bytes and control bytes. To enable one computer to send a picture to a second one, for example, the central processor has to address both sequentially, directing the second one to receive and the first to send the information.

Each computer or terminal (e.g., telephone, FAX machine, and printer) is known as a node and several nodes are often connected to a server, a dedicated processor with the sole purpose of routing packets of bytes to/from individual nodes. These servers are often combined with other servers to form a local area network (LAN), which usually interconnects many computers within a single office building. LAN servers and additional individual nodes in turn are linked to a more centralized server to form a wide area network (WAN) that normally covers a geographical area such as a city. (There are several other area network nomenclatures such as CAN (campus), MAN (municipal), and IAN (internet-cloud) as well as there many more telecommunication details such as protocol standards. We are, however, concerned in this text with the internal structure, the physics of networking.) Central nodes are usually connected to a few other central nodes via backbone channels, denoted by bold dashed lines in Figure 18.14. Multiple backbone channels enable telecommunications networks to reroute long-distance informational content along a different backbone, avoiding high-volume bottlenecks.

Many long-distance communications between two central nodes require one or more repeaters, devices that amplify and regenerate the signal stream. We observed in both Sections 11.1 and 18.1 how signals spread out (disperse), weaken (lose power in signal), pickup noise, and experience small drifts in relative arrival times while traveling over long distances. The signal processing done by a repeater circuitry for an example bit stream is shown in Figure 18.16. In this case, the signals of the bit stream has weakened and dispersed. Simple amplification of the incoming raw signal still leaves rounded edges caused by dispersion and the second 1-0-1 group has a small timing drift relative to the first 1-0-1 group. Pulse reshaping converts the rounded signals back to discrete (square) digital bits and retiming circuitry adjusts the clock intervals for each bit.

Today, digital transmission is preferred because it achieves higher reliability at lower costs than analog systems. Some analog signals such as voice communications, however, require analog-to-digital (abbreviated A/D or A to D) conversion and D/A back at the destination. Most modern television cameras and other image sensors have A/D circuitry embedded in the readout circuitry, avoiding this step in the network. Other advantages of digital communication include the ability to reduce the amount of redundant binary information (data compression) while simultaneously embedding a few error checking bits. An easily understood example of data compression is an astronomical picture of a sparse star field. Most of the image is black with a few bright point-like spots (the stars). Instead of transmitting very long series of eight-bit zeros followed by a few nonzero values, data compression algorithms might use two bytes to encode the number of pixels until the next pixel with a nonzero value. If there are on average 20 black pixels for every illuminated pixel, the compressed data uses two bytes instead of 20 bytes for the background, resulting in

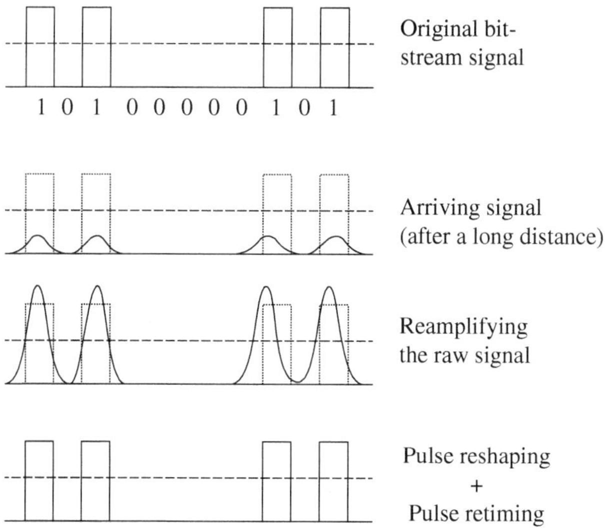

FIGURE 18.16 The process by which a repeater circuit restores a digital bit stream back to its original signal form.

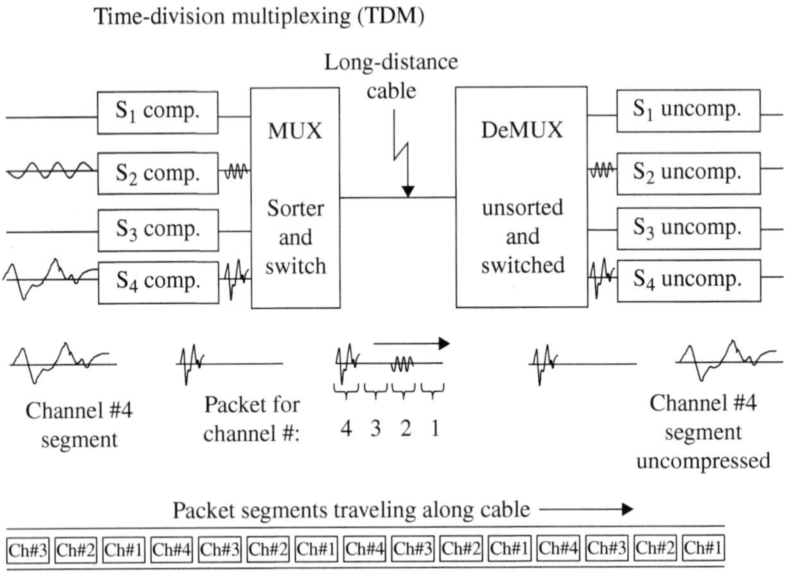

FIGURE 18.17 The method of time-division multiplexing (TDM).

nearly 10-fold image compression. Obviously, images and text documents have differing amounts of redundancy and the levels of data compression achieved will vary.

Multiplexing is another effective method of maximizing the volume of information that can be transmitted. Frequency-division and time-division multiplexing (TDM) are the two primary techniques. Frequency-division multiplexing was examined in somewhat in Chapter 11. Figure 18.17 depicts a four-channel TDM system that might be employed for telephone calls. Sound wave segments enter from the left and are compressed (converted) into high-frequency waves with identical waveforms but shortened time intervals. Only channels 2 and 4 have active inputs, a single tone and a voice segment, respectively. The other channels are idle. Next, a multiplexer (MUX) sequentially switches through each channel, sending each compressed segment (packet) down the long-distance cable in turn. While the first set of packets is being sent, the MUX is loading another complete set of segment packets. The chain of packet segments traveling along the long-distance cable is shown at the bottom of Figure 18.17. *Note*: there are small gaps between the segments, enabling a seamless switching between channels. Once the train of packet segments reaches their destination, the process is reversed. A de-MUX circuit sends the appropriate packets to each channel where each is uncompressed back to the original waveforms and outputted.

19

PHYSICS OF INSTRUMENTS FOR BIOLOGY AND MEDICINE

Common to most instrumentation, medical devices have uses in biological and nonbiological industrial applications as well as in environmental monitoring. For example, ultrasound pulses are used for several medical purposes, including to image a fedus durring gestation, to create echocardiograms—a sonogram of the heart, to beak up kidney stones and gallstones, or to heat and distroy diseased and cancerous tissue, all by focusing ultrasonic waves. Ultrasound pulses are also used for underwater range finding known as sonar as well for autofocusing by most cameras. Bats and dolphin among other animals use ultrasonic waves for navigation and silient dog whistles use it too. In addition, ultrasonic cleaners are the most effective way to remove residual contaminants from surfaces. Computed tomography (CT), the process of imaging by sections (or sectioning) using penatrating waves, is used extensively in medine as well as in archaeology, astrophysics, geophysics, materials science, quantum information, and industrial testing among others. In this chapter, we focus on the physics behind some familiar medical equipment. Similarly, the physics behind other medical or biological instruments can be found in various chapters throughout this text.

19.1 IMAGING INSTRUMENTS

Various types of medical imaging form an important set of noninvasive diagnostic tools. Two-dimensional images range from shaddows cast onto X-ray sensors to scintigraphy where radioactive materials are injected into the patient and the gamma rays detected. Three-dimensional images are formed using CT scans, which most commonly uses X-rays. *Note*: the computed axial tomography (CAT) has essentially been replaced by CT. Other forms of tomography include magnetic resonance imaging (MRI), single-photon

Modern Devices: The Simple Physics of Sophisticated Technology, First Edition.
Charles L. Joseph and Santiago Bernal.

Companion website: www.wiley.com/go/joseph/moderndevices

emission computer tomography (SPECT), and positron emission tomography (PET). SECT and PET devices use hard, ionizing radiation such as gamma rays, which are sometimes concentrated from external sources or from radioactive liquids that are injected directly into the patient. In addition, ultrasound tomography such as echocardiography and optical coherence tomography are also forms of 3D imaging. The contrast agents (dyes) are often used in all computed tomography to enhance the imaging of one organ or another or various tissues. We will not consider contrast agents further.

19.1.1 CT Scanners

Conventional CT scans have less resolution than an MRI, but are less expensive and scans can normally be scheduled more quickly than can an MRI. The patient lies on a sliding bed, remaining motionless for approximately 20 seconds while multiple image slices are made. The process can be repeated as necessary to extend the length of the three-dimensional (3D) image. These scanners commonly use X-rays, an ionizing radiation. CT devices have an advantage over simple 2D shaddow casting X-ray machines since superimposed shadows from various tissues are separated. However, CT scans generally require radiation doses that are approximately 1000 times greater than a conventional X-ray.

X-ray CT scanners are considered to be moderate- to high-radiation devices. Doses per organ typically range from 10 to 20 mGy and can be as much as 80 mGy. Actual doses vary significantly from one scan to the next and from one portion of the body to the next. The gray (Gy) is the SI derived unit of the absorbed dose per unit mass of tissue. It has units of Joules per kilogram (J/kg). See Chapter 20 on radiation for more details and comparisons to other units of radiation such as the Roentgen (R) and the Sievert (Sv). A good comparative benchmark number is 2.4 mGy/year, which is the typical dose of radiation received by an individual from all background sources (e.g. the Sun, trace amounts of radioactive elements in the soil, etc.).

A schematic representation of a CT scanner is shown in Figure 19.1. The individual lies on the table while an X-ray tube plus an array of detectors inside a large toroid revolve around him or her. A 2D single image slice is computer generated during a single revolution of the toroid. After each scan, the patient table is stepped and the next image slice is generated, creating a 3D image. (2D pictures generated with electronic sensors have picture elements called pixels, while a single 3D picture element is called a voxel.) Some CT devices use a continuous-motion table, running on a worm drive. In this case, the computer compensates for the lateral motion that occurs during a revolution of the toroid. Many CT devices use several sets of sensors, collecting several planes (slices) simultaneously to speed up the process. Other scanners use two pairs of X-ray tubes/detectors on opposite sides of the toroid, again for increased collection efficiency. Figure 19.1 shows the simplest configuration, a single image slice device.

19.1.2 Magnetic Resonance Imaging

To comprehend an MRI, one has to have a basic understanding of magnetic resonance, the interactions of spinning charges in a magnetic field. The mechanical analog is the physics of gyroscopic actions (i.e., spinning tops) in a gravitational field. (These concepts are reviewed in Intro Physics Flashback FB19.1, if needed.)

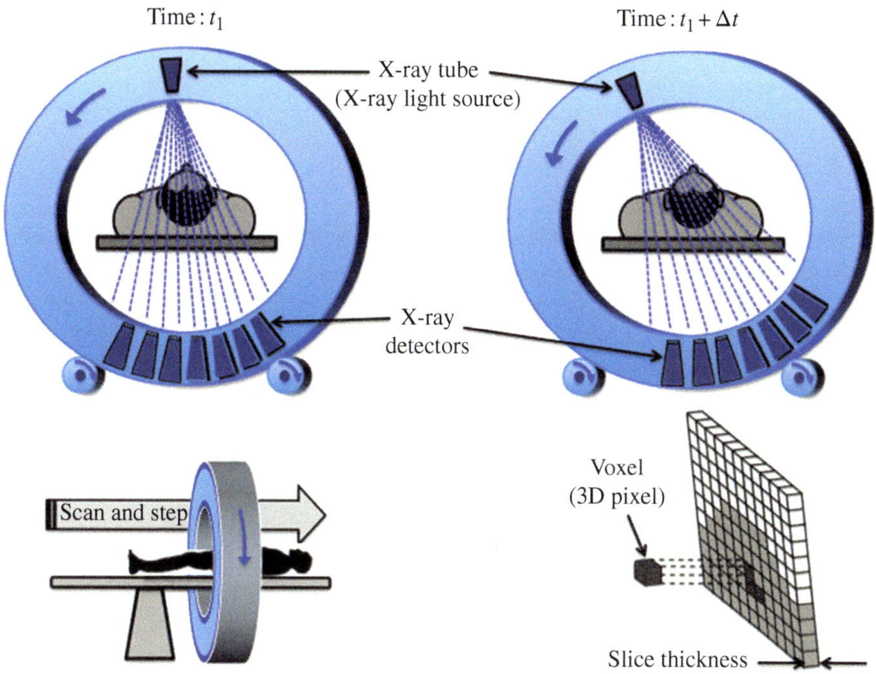

FIGURE 19.1 A schematic representation of an X-ray CT scanner.

One important distinction to keep in mind, however, is the quantum nature of subatomic particles, which can only exchange energy in discrete levels.

An MRI scanner is far more complex and sophisticated than a CT device. Most of the complexity of an MRI scanner results form the need to establish several strong magnetic fields. Fortunately, this provides an opportunity to demonstrate that a complex system usually consists of several simple concepts operating together. If one examines each component as a block function within a hierarchy and organize one's thoughts accordingly, the physics behind the instrument becomes evident and easily understood. Details and subtleties can be glossed over intially until you gain an appreication of the overall functioning of the device. It is important, however, to understand first what is to be measured and then evaluate how each block of instrumentation contributes an essential component of that measurement. The basic MRI messurement technique is encapsulated in Figures 19.2 and 19.3.

Fermions (protons, neutrons, electrons) are spin ½ particles and any nucleii with unpaired fermions (e.g. H and ^{13}C) have net magnetic moments that align with any strong magnetic field. These nucleons also experience torques in the presence of a second, nonparallel magnetic field. Figure 19.2 shows the spin orientations with and without a single, strong constant magnetic field, $\vec{B}_0$. When a second, high-frequency rotating $\vec{B}$ field is also switched ON, all of the poles of the spinning nucleons start recessing around the $\vec{B}_0$ direction with a Larmor angular frequency, ω_L, (top—Fig. 19.3). The phases of these poles initially are randomly oriented, becoming in-phase over time. These recessing poles emit Larmor radiation, $U(t)$ as a function of time

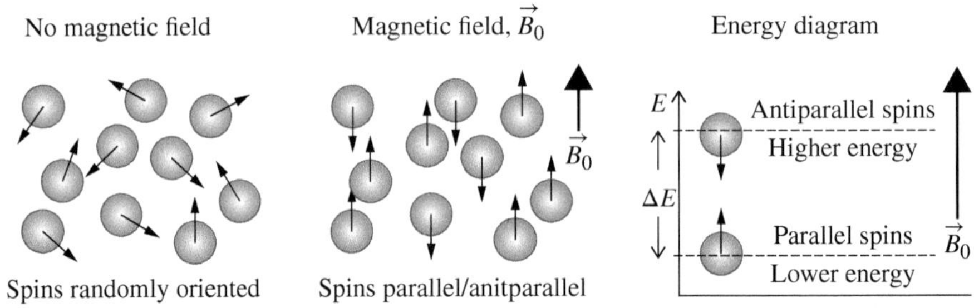

FIGURE 19.2 The proton spins in hydrogen atoms are randomly oriented in the absence of a strong magnetic field. These spins align parallel or antiparallel to a strong, static magnetic field, falling into two discrete energy levels.

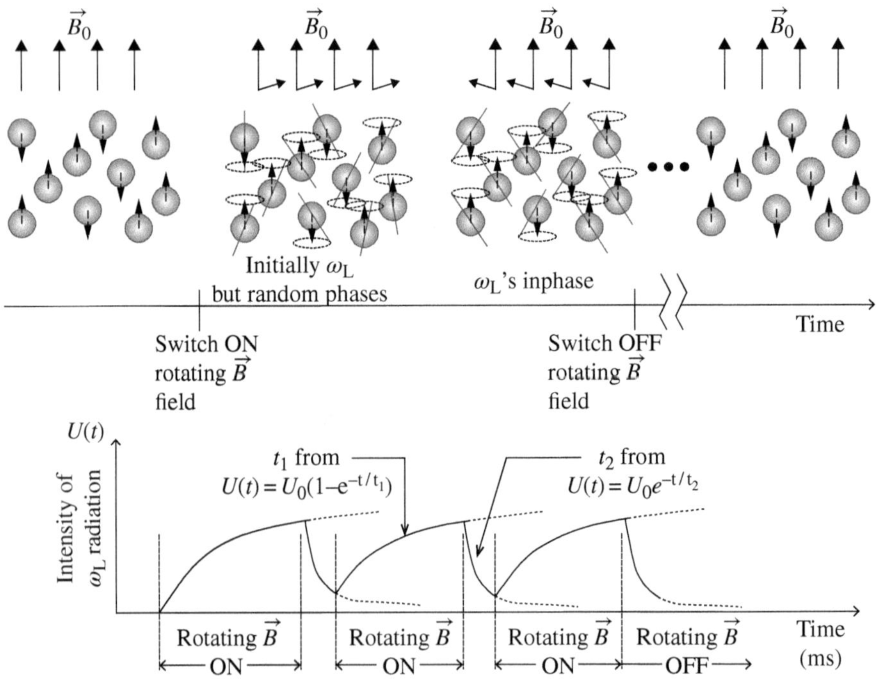

FIGURE 19.3 The basic measurement of an MRI machine. A strong, static field, $\vec{B}_0$, causes unpaired fermions to align to it (top). Switching ON a smaller, rapidly rotating $\vec{B}$ field, causes these fermions to precess, precessions that eventually become in-phase with each other. Disorder returns through relaxation to equalibrium when this field is switched OFF. The intensity of the associated radiation is plotted at the bottom. Several cycles are used to image various tissues, determining t_1 or t_2 that differ from one tissue to the next.

that is proportional to the percentage of the pole that are in-phase. (Randomly phased poles emit radiation as well, except the random phases results in destructive interference.) Each type of atom has a uniquely different Larmor frequency and each ω_L is proportional to the $\vec{B}_0$ field strength. The weaker, rotating $\vec{B}$ field has to be tuned to the resonance frequency, ω_L, of each tissue to "light up" that particular tissue. A series of ω_L radiation are applied to the patient. Each tissue has two distinct timescales: one for increasing $U(t)$ toward saturation (all in-phase) and the other for relaxation back to random equalibrium, t_1 and t_2, respectively. *Note*: these transitions between field ON and OFF or back each has an exponential mathematical form. The time-dependent intensity of ω_L for one type of tissue is plotted at the bottom of Figure 19.3. This $U(t)$ is the basic measurement made by an MRI device with a determination of either t_1 or t_2 indicating the types of tissues involved. For example, the hydrogen atoms in H_2O molecules are often the chosen resonators since the bulk of a human body consists of water. The contrast between two types of tissue is the result of different fractions of water in each.

A Swedish university MRI machine is shown in Figure 19.4. The depth of an MRI machine compared to a CT instrument is generally larger (i.e., the scanner housing obscures more of the patient at any given time). The added bulk is indicative of the sizeable number of components and *does not* signify the ability to collect extra image slices simultaneously.

We begin our detailed analysis of an MRI scanner starting with the outermost components and working toward the central axis, sequentially including those elements that are essential to the various modes of operation. The outermost portion of most MRI devices contains the main superconducting coil magnet, consisting of

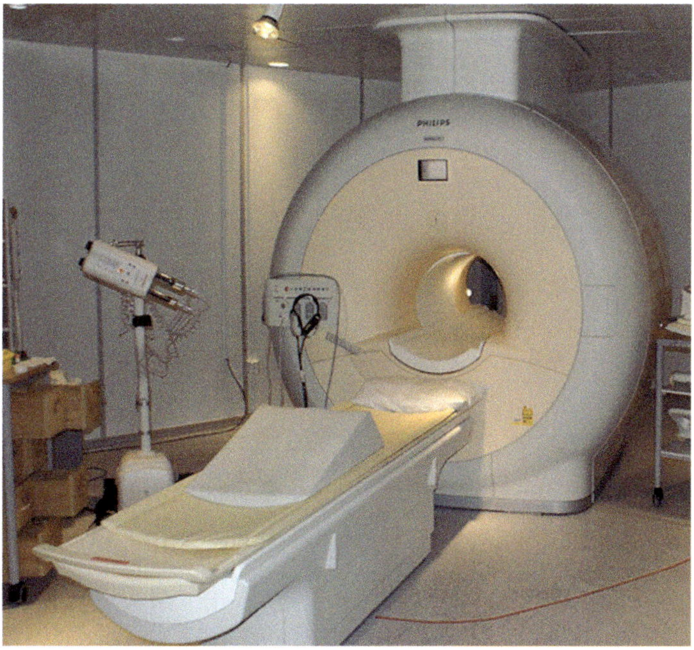

FIGURE 19.4 A magnetic resonance imaging (MRI) scanner manufactured by Philip Corporation used by Sahlgrenska Universitetsjukhuset, Gothenburg, Sweden. Source: Ainali, https://commons.wikimedia.org/wiki/File:MRI-Philips.JPG. Used under CC-BY 3.0 https://creativecommons.org/licenses/by/3.0/deed.en.

superconducting cables immersed in cryogenically cooled to liquid helium (LHe) temperatures (~4 K, −269°C). The container holding the cryogenic liquid is called a cryostat, employing three physical mechanisms to insulate it. These three techniques are identical to those used by an ordinary thermos bottle. Recall from Chapter 3 that a DC electrical current moving in a coil of wire creates a static magnetic field oriented along the cylindrical axis. This main magnet creates the constant, uniform magnetic field, $\vec{B}_0$, along the z axis (central axis) as depicted in Figure 19.5. $\vec{B}_0$ needs to be uniform to 1 ppm (part per million) over a 1 L volume, which is a much stricter requirment than can be met by production tolerances and the distortions introduced by feromagnetic materials in the walls of the building and in external electronic equipment. Consequentially, the basic $\vec{B}_0$ field needs to be magnetically shimmed, either by applying an adjustable amount of current through an extra set of resistive coils (not shown in Figs. 19.5, 19.6, 19.7, 19.8, and 19.9) or by the introduction of steel pieces with good magnetic properties near the main superconducting magnet.

The resolution of an MRI is proportional to the field strength of $\vec{B}_0$. While permanent magnets require less operating costs including less maintenance, it is only practical to manufacture physically large magnet systems with a maximum field strength of 0.4 T (Tesla). A simple solenoid of copper wire is an alternative, but again $\vec{B}$ field strengths are limited and stability is an issue. These simple resistive electromagnets are expensive to operate, requiring enormous amounts of energy and large a mounts of additional energy to cool the coils. Superconducting electromagnets can be fabricated with $\vec{B}$ fields up to approximately 10 T.

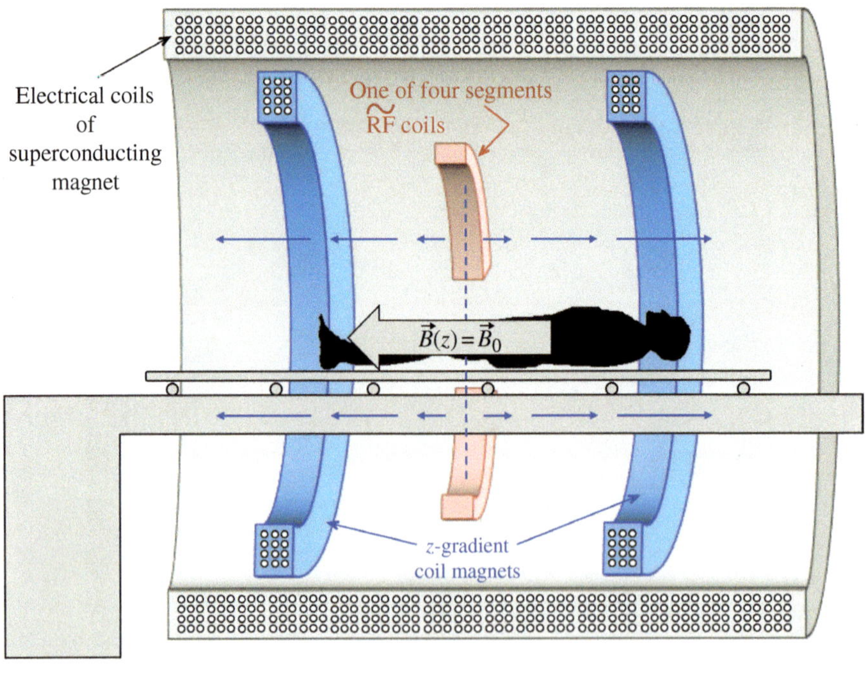

FIGURE 19.5 A schematic representation of the two outermost magnet structures. The coordinate system is shown at center point.

Many MRI scanners operate with a field strength $\vec{B}_0 = 1.5\,\text{T}$, although commercially available systems are available that operate between 0.2 and 7T. If the coil is made of superconducting materials, then the resistance is zero, the coil can be shorted, and the power supply shut off once the current is established. The current will continue to flow and the corresponding magnetic field maintained indefinitely without any further application of voltage or current. In reality, the cryostat cannot be completely thermally isolated from the outside world, resulting in minor losses of the LHe cryogen and the periodic need to refill. Some LHe cryostats are surrounded by a separate liquid nitrogen (LN2) cryostat, acting as a thermal buffer to extend the interval between He replenishments. Other LHe cryostats incorporate a mechanical cryocooler to reliquify the escaping cold gaseous He. A few MRI machines use the cryocooler to refrigerate the superconducting coils, but direct cooling requires a huge cyrocooler. All of these configurations are able to maintain the intense magnetic field for several months between servicing.

The next inward set of magnets establishes a z-gradient field. In physics, $\vec{B}_0$, the primary magnetic field, defines the z axis, the x axis is chosen, and the right-hand rule then defines the y axes. Recall, tomography requires image slices and a $\vec{B}$-field gradient is required to limit the volume where magnetic resonance occurs to a single slice of the body. This $z=0$ plane is denoted in Figure 19.5 by the vertical dashed blue line. $\vec{B}(z)$ is stronger than $\vec{B}_0$ to the left of the central plane and is weaker than $\vec{B}_0$ to the right. Only voxels (3D image elements) in the $z=0$ plane will be in resonance. This gradient field is created by a set of reverse Helmholtz coils, sending electrical currents in opposite directions through the coils to create a divergent supplemental B_z field. Additional slices can be made by repositioning a different portion of the patient's body at $z=0$.

A third set of electromagents produce the rotational, excitational $\vec{B}$ field that excites the Larmor recession of the nucleon within a portion of the body. Two of the four RF coils (red), surrounding $z=0$ location, are shown in Figure 19.5 and an end projection of all four coils is depicted in Figure 19.6 (red). Each coil contributes one component to a sinusoidal $\vec{B}$ field, each with an appropriate phase. These four coils emit radiowave radiation and also serve as radio receivers, using Faraday's Law, the exact same physics principals as metal detectors and radio frequency identification

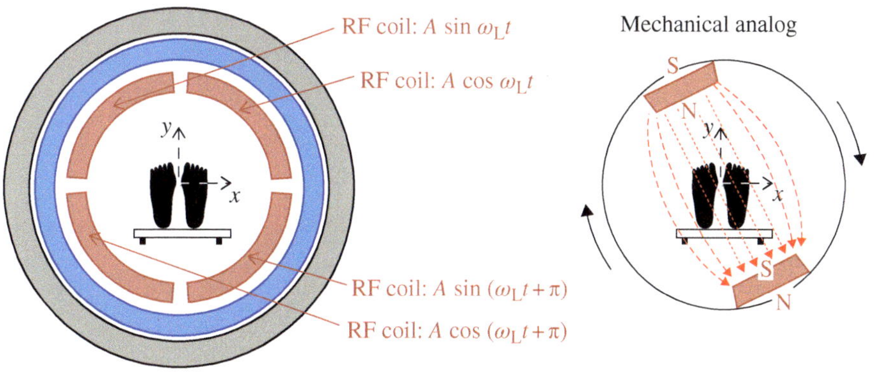

FIGURE 19.6 Schematic depiction (end cross section) of MRI showing four RF coils (red) that create a rotating magnetic field. The mechanical analog is only conceptal since it is impossible to rotate magnets sufficiently fast.

(RFID) encoders. (See Chapter 3.) Conceptually, the net alternating $\vec{B}$ field is equivalent to a pair of short permanent bar magnets, rotating around the patient as depicted by the mechanical analog (right). The resonance (Larmor) frequency of a proton in a 1.5T field is 63MHz. (*Note*: the mechanical analog pictured is only for conceptual purposes since the required rotational speeds are more than a 1000-fold faster than can be achieved mechanically. Compare the resonance frequency to benchmark numbers: 2000–3000rpm (33–50Hz) for automobile engines, 165,000rpm (2.75kHz) for the fastest gas turbine engine on jets, or 60,000–200,000rpm (1–3kHz) for flywheel energy storage.)

Some diagnostic MRI scans require tomographic image slices taken in the x or y direction. Pairs of electromagnetic coils, one set for each direction, form the fourth layer of concentric magnets. The reverse Helmoltz coils, schematically depicted in Figure 19.7 in orange, establish the y-gradient magnetic field that enables y-axis scanning. To offset the zero point (i.e., the y plane in which the nucleons are in resonance), the current strengths in top coils are adjusted relative to the bottom coils. Figure 19.8 shows the y resonance plane (dashed orange line) offset in the $+y$ direction. The electromagnets on top have larger electrical currents than those at the bottom in this example. x-axis scanning is accomplished with identical reverse Helmholtz coils, but located orthoganally to the y-gradient magnets.

In summary, Figure 19.9 schematically depicts all of the essential components of an MRI scanner from the end vantage point. There are two static $\vec{B}$ fields and one

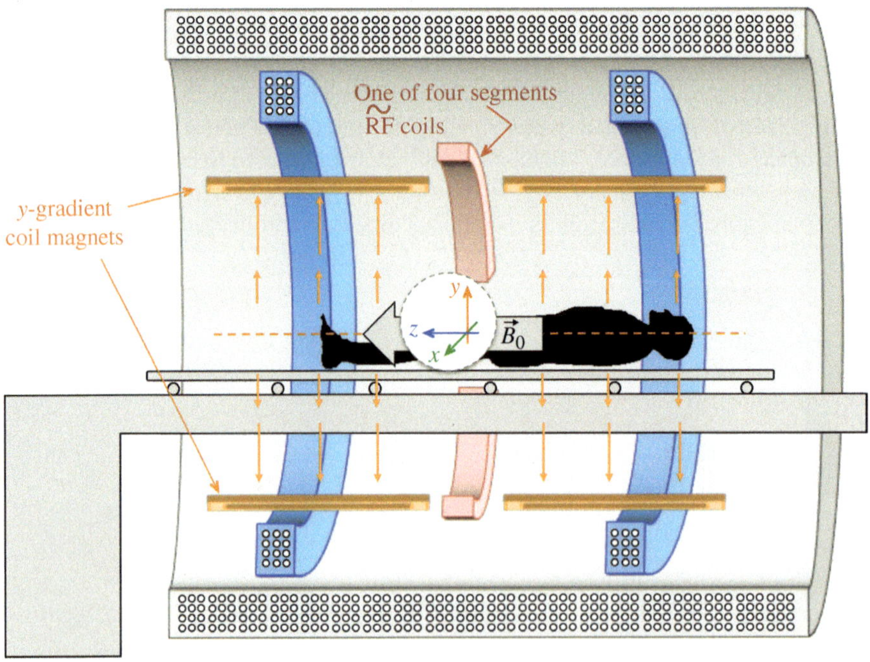

FIGURE 19.7 A y-gradient magnetic field can be established using the reverse Helmholtz coils pictured in yellow.

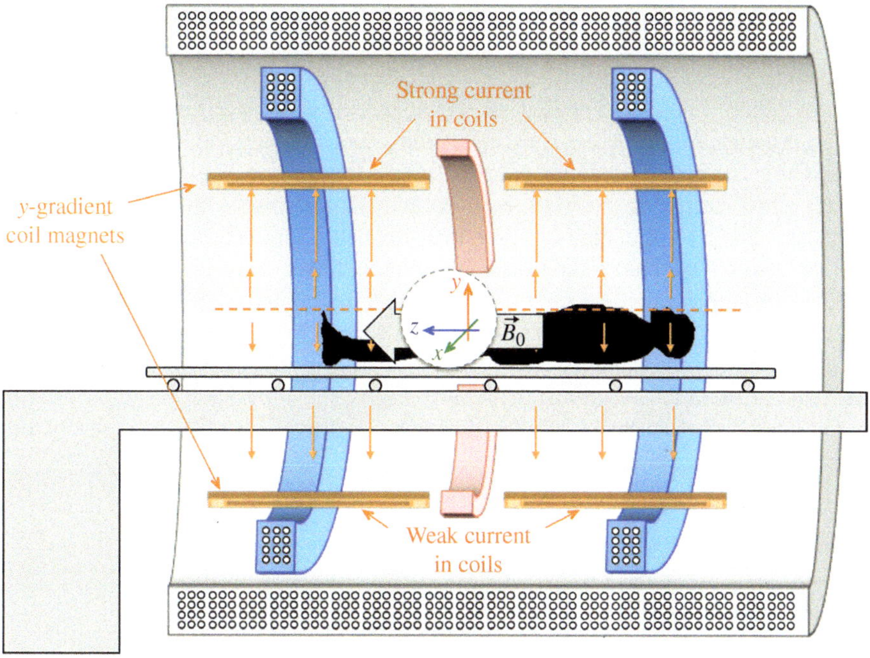

FIGURE 19.8 The resonance plane can be offset in $\pm y$ direction by altering the relative currents in the top y-gradient coil magnets compared to those on the bottom.

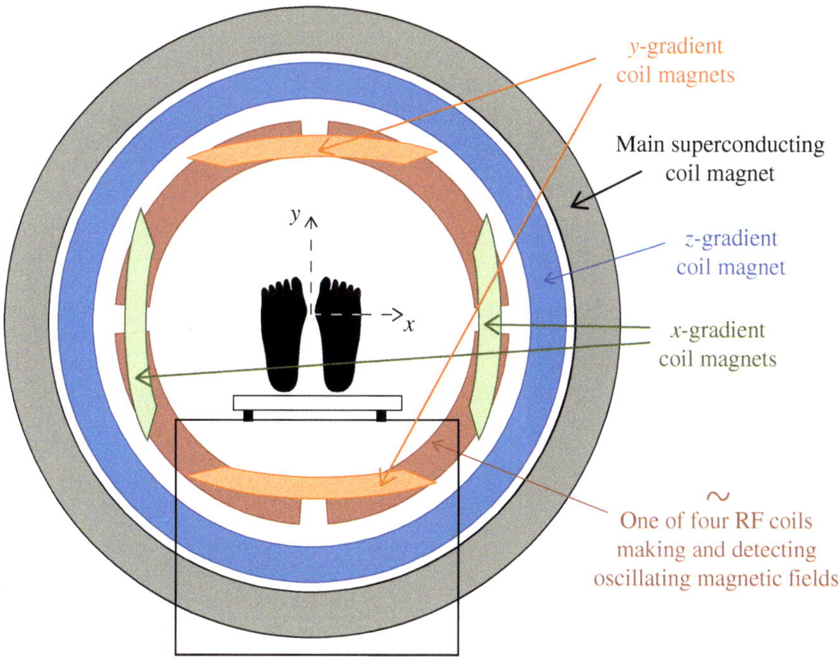

FIGURE 19.9 Schematic representation of the end view of the internal structure of an MRI scanner.

oscillating $\vec{B}$ field necessary to establish a resonance plane (image plane) where the Larmor precessions are in-phase. The superconducting main coil magnet establishes static, primary $\vec{B}$ field, which determines the z direction and the Larmor precession axis. Only one set of gradient magnets are used at a time, if the scan direction is purely along the x, y, or z axis. (In practice, an MRI scan can be obtained with any tip angle by using combinations of gradient magnets.) Gradient magnets repeatedly switch ON/OFF over several milliseconds, maintaining a constant-strength (static) $\vec{B}$ field for brief intervals. The variable (rotating) $\vec{B}$ field is established by four RF coils (red in Fig. 19.9), which emit and receive radio wave (1–100 MHz) radiation. The RF signal strength returned from any voxel (3D image element) is proportional to the number of resonating subatomic particles in that voxel. Different types of tissue (e.g., ordinary cells vs cancerous ones) will contain a dissimilar number of resonators, producing contrast in the imaging.

INTRO PHYSICS FLASHBACK FB19.1

Gyroscopic Effect

A sphere spinning exactly over its axis of rotation will continue to spin simply as depicted on the left side of Figure FB19.1. However, if the sphere is spun up with a tilt with respect to the gravitational force, it will also undergo precssion as it spins (right side of Fig. FB19.1). This gyroscopic effect is the continual cyclical change in the position of the axis of rotation. Precession is observed in gryroscopes, toy tops, and the Earth itself in the presence of gravitational torques. *Note*: gyroscopic precession can only occur when a force that is not parallel to the axis of rotation is applied to a rotating object.

The gyroscopic effect observed from different perspectives, including the recession of the Earth's axis, provides a more generalized understanding. If the Earth

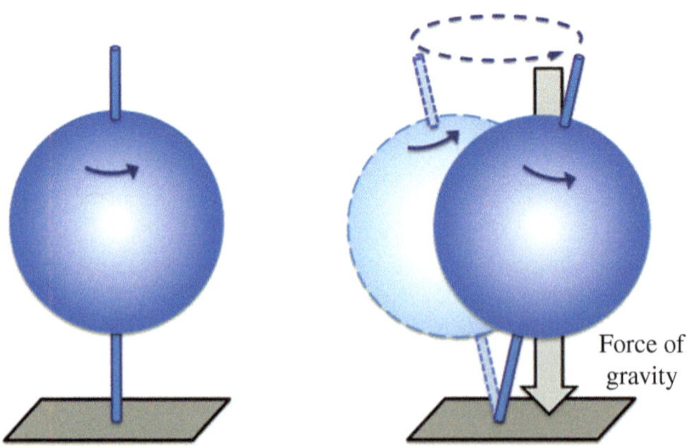

FIGURE FB19.1 The gyroscopic effect, the precession of the axis of rotation in the presence of a constant restoring force.

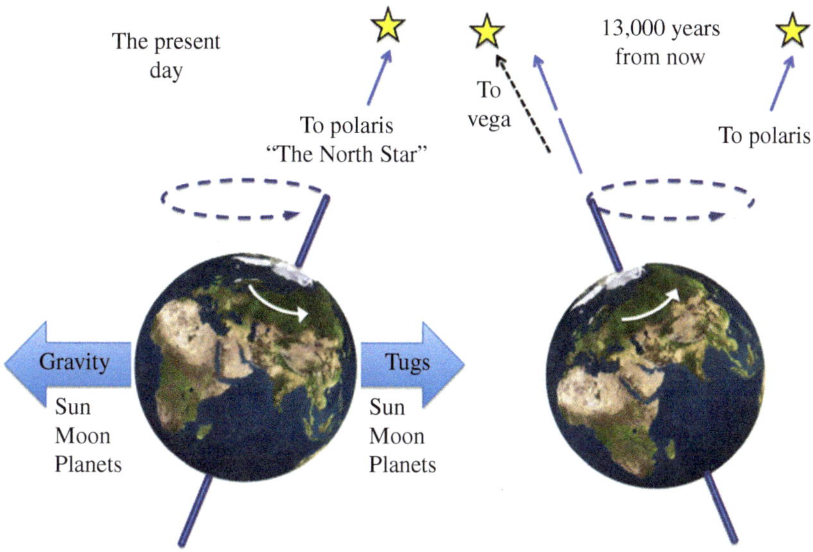

FIGURE FB19.2 Precession of the Earth's rotational axis due to the gyroscopic effect.

were perfectly spherical and its density was uniformly distributed around its axis of rotation, then there would be no precession. The North Pole, which currently points to a position in the sky close to Polaris as shown in Figure FB19.2, would remain that way indefinitely. However, the Earth has a small equatorial bulge and tidal bulges produced by the gravitational pull of the Moon and Sun. These minor distortions combined with the gravitational pull from the Sun, Moon, and planets produce forces (tugs) that cause the Earth to precess with a period of approximately 26,000 years to change slowly (~1°/year) the positions of all the stars with respect to their celestial longitudes and latitudes. Thirteen thousand years from now, the North Pole will point to a spot in the sky very close to Vega, while in 26,000 years, the Pole will be once again be close to Polaris, the North Star.

INTERESTING TIDBIT TB19.1

Precession of the Earth's rotational axis has made Astrology more than 2000 years out of date. The Sun now passes through 13 (not 12) constellations over the course of the year, and the Astrological start/stop dates of any sign are on average off by approximately 2 weeks. The time that the Sun's position spends in any one constellation now varies dramatically from one to the next. Moreover, the planets collectively pass through a combination of 23 constellations in the twenty-first century. The next time someone asks, "What's your sign?" inquire as to whether they want your astronomical or astrological sign since these are not the same.

INTERESTING TIDBIT TB19.2

The MRI was originally called a nuclear magnetic resonance imaging (NMRI) since the subatomic particles resonating are spinning nucleons. However, patient anxiety regarding anything "nuclear" led to the dropping of the "N" from all medial scanners. An NMRI is identical to an MRI and the former is still used by many nonmedical facilities.

19.1.3 Ultrasonography and Ultrasonic Lithotripsy

Diagnostic ultrasonography uses ultrasound, which is simply sound waves with frequencies (pitches) higher than can be detected by a human ear ($f > 20\,\text{kHz}$). Ultrasonographic scanners operate by sending pulses of ultrasound waves into a tissue and measuring the strength and round-trip travel times of the reflected waves. The depth into the tissue is determined by time interval for the reflection to occur, while the signal strength is indicative of the relative brightness (contrast) for each picture element. The resulting analog signals are captured via a standard frame grabber electronics and displayed. *Note*: the transmissions are short-duration pulses of specific carrier waves rather than clicks. There are several types of ultrasonic imaging. B-mode imaging is perhaps the best known, displaying a 2D cross section of tissue such as that used to image a developing fetus. Another type is *Doppler sonography*, which is used for realtime observations of blood flow or the relative stiffness of a tissue. The advantages of ultrasonography compared to other methods of medical imaging include the following: (i) images can be obtained in realtime rather than having to wait to schedule a time for the scan or having to wait for the results to be returned; (ii) the devices are low cost and portable, being able to be brought to the patient's bedside if necessary; and (iii) it does not use harmful ionizing radiation. Some shortcomings are that good imaging depends on a skilled operator and the field of view (FOV) is limited.

Typically, all medical ultrasonography begins and ends with with one or more piezioelectric transducers that first convert electrical signals into vibrations and then transform the returning reflected soundwaves back into electrical signals. Most crystals and some ceramic materials have the property that sharp accelerations or changes in pressure liberate free electrons. These piezoelectric materials also respond in turn to changing electrical voltages by vibrating. (Most microphones and stereo speakers operate via the peizioelectric effect, which is not effective for measuring static voltages or static mechanical stress. Also, the piezoelectric materials cannot be used for signal conversions in both directions simultaneously.)

There are three commonly used methods to focus ultrasonic waves, which are depicted in Figure 19.10. Acoustic lenses have essentially been replaced with electrically phased arrays in the twenty-first century, although acoustic lenses are used occasionally in theraputic applications to further improve the ultrasound focusing. As shown one-dimensionally in central diagram of Figure 19.10, the electronic

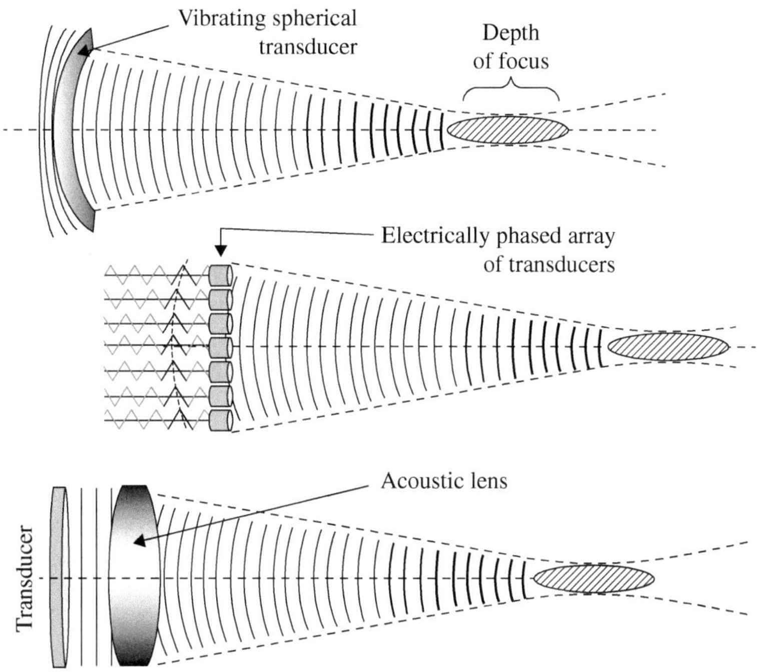

FIGURE 19.10 Three methods for concentrating (focusing) ultrasonic waves. In medicine, an accustic lens is only used in high-intensity focused ultrasound applications.

pulses sent to the array of transducers have various phase delays to produce ultrasound wavefronts equivalent to those of a spherical vibrating transducer. The phases are depicted by darkened portions of the input electronic signals as shown on the left side of the central figure. *Note*: the concentration of ultrasonic waves (i.e., the focus volume) is always elongated in the transmission–reflection direction, compared to the orthogonal axes.

The same ultrasound technique is used for theraputic lithotripsy by focusing the energy tightly into a small volume. Lithotripsy is a medical procedure to destroy kidney stones, bezoars, or gallstones by concentrating the ultrasonic waves sufficiently to "break or pulverize the stones." High-intensity focused ultrasound (HIFU) extends the technique even further, increasing the intensity of the ultrasound waves and focusing the waves more tightly into highly localized volumes. HIFU causes the targeted tissue to absorb the energy and to heat it locally. HIFU beams can usually be focused with the characteristic "cigar" shaped volume depicted in Figure 19.10 on the order of millimeters deep into the tissue. At sufficiently high intensities, HIFU can cause cavitation, the formation and growth of microbubbles, which can lead to very high temperatures within the bubble and which can subsequently collapse to form shockwaves that can cause localized mechanical damage to the tissue.

19.2 MINIMALLY INVASIVE PROBES AND SURGERY

One of the most common medical probes is the endoscope used to inspect interior regions such as hollow portions of an organ, intestines, or major arteries. A flexible endoscope, pictured in Figure 19.11, includes (i) an eye piece or miniature camera, (ii) a narrow cable containing the fiber bundles and control wires, and (iii) a probe head with an objective lens and minaturized tools. Endoscopes have a second flexible tube, which remains externral to the patent and provides a separate channel for suction or for dispensing gases or liquids. Also, the endoscope usually supports a channel for inserting a surgical instrument, running through a flexible hollow tube and ending at the head. The stiffness of the cable is uniform throughout its length except near the head where it is more flexible. This enables the head to bend side to side by changing the tension in two control wires.

A schematic representation of an endoscope head is shown in Figure 19.12. The objective lens focuses the image onto the ends of a multi-fiber bundle. This lens can be attached independently within the head or incorporated as an integral part of the fiber bundle as long as the ends of multi-fiber bundle are located at the focal distance below the lens. (The physics behind a fiber optic waveguide is provided in Chapter 11, and the detailed use of a fiber bundle to transfer an image is given in Chapter 17.) Two additional individual fibers deliver illumination to the organ undergoing a surgical procedure or being inspected. Typically, two hallow tube channels exist in

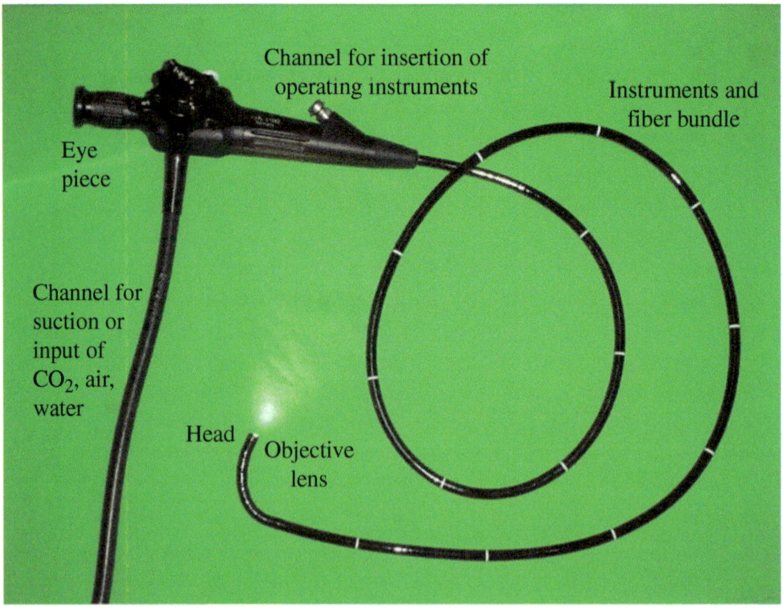

FIGURE 19.11 A flexible endoscope. Source: Adapted from de:Benutzer:Kalumet, https://commons.wikimedia.org/wiki/File:Flexibles_Endoskop.jpg. Used under CC-BY-SA 3.0 https://creativecommons.org/licenses/by-sa/3.0/deed.en

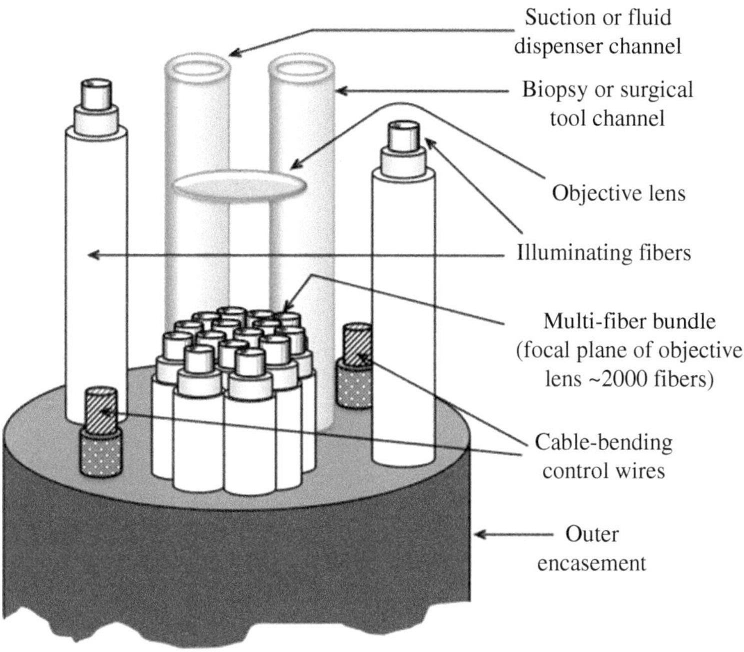

FIGURE 19.12 A schematic diagram of the head end of an endoscope, showing the interior components found inside the instrument and fiber bundle.

the flexible cable: one enabling suction or administration of gas or fluids and the other supporting the insertion of various small surgical tools. Finally, two wire cables that pull or push against the sides of the head permit the end of the cable head to be flexed left to right to follow, for example, a branch of a major blood vessel. While the rest of the insertion instrument and fiber bundle is stiffer than the head portion; it merely has to follow the head as it snakes to the problem area.

19.3 LASER TECHNOLOGIES

Lasers are used extensively in medicine, ranging from the removal of hair and minor vascular defects such as spider veins, to skin cancer eradication, to the reduction of benign thyroid nodules, to the distruction of malignant liver lesions, to periodontium surgery, to ophthalmology treatments, to laser-induced thermotherapy, and many other theraputic applications. Yttrium–aluminum–garnet (YAG) lasers in particular are widely used in medicine, especially in the treatment of eyes in the form of laser-assisted in situ keratomileusis (LASIK) —the reshaping of the cornea, laser capsulotomy—the removal of the after-cataract tissue, laser photocoagulation in diabetic retinopathy, and surgical iridectomy—the removal of part of the iris to treat closed-angle glaucoma and iris melanoma. (The physics principles behind laser cavities in general and a discription of a YAG laser are discussed in Chapter 9.)

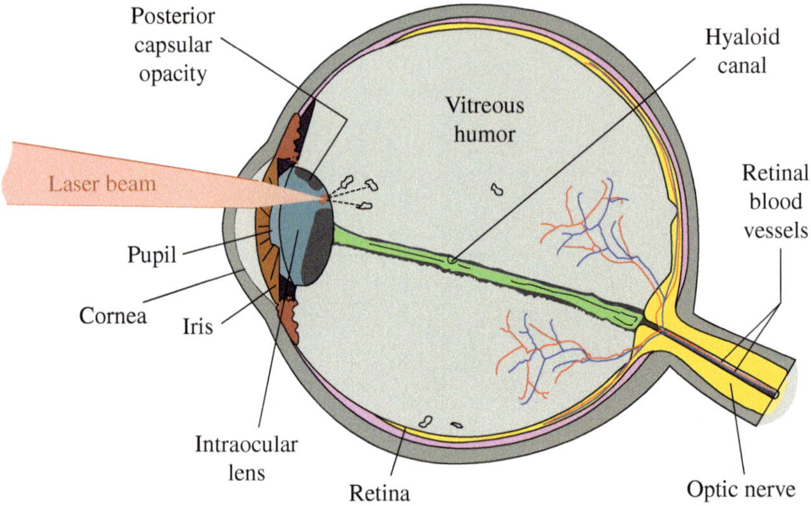

FIGURE 19.13 An application of a Nd:YAG laser for capsulotomy, the breaking up of posterior capsular opacity that sometimes forms after catarac surgery. Source: Adapted from Rhcastilhos, https://commons.wikimedia.org/wiki/File:Schematic_diagram_of_the_human_eye_horizontal_pt.svg. Used under CC0 1.0, https://creativecommons.org/publicdomain/zero/1.0/deed.en

We begin with the use of lasers to break up posterior capsular opacity that often forms in the eye after cataract surgery. This obscuration is the result of the regrowth of cells on the back wall of the capsular bag, clouding vision. Laser capsulotomy is an out-patient painless procedure that only takes a few minutes to perform. Figure 19.13 shows a Nd:YAG laser beam being focused onto a capsular opacity, breaking it into floaters. The resulting fragments float in the viterous humor for a few days until eventually settling permanently at the bottom of the eye. Patients rest their chins and foreheads on a visual field analyzer to remain stationary while the ophthalmologist inspects the interior of the dialated eye and directs the laser beam onto the obscuring regions.

Many ophthalmological surgical procedures require measurements of the size and shape of the cornea and intraocular lens, including any distortions that cause aberrations. The Shack–Hartmann wavefront sensor (SHWFS) is a popular tool for measuring optical imperfections. The basic principle of a SHWFS is to use a 2D array of lenslets to break an image into an array of small components, focusing each element into a tiny spot that is equidistance apart from its adjacent spots for a perfect distortion-free eye. If optical abberations are present, the lenslet spots are nolonger uniformly separated and a map of the spot locations is used to determine the geometrical imperfections of the optics. Conventional SHWFS are used extensively at astronomical observatories and by the military to restore images distorted through the atmosphere, which was discussed in detail in Chapter 17. The SHWFS instrumentation for the eye is shown one-dimensionally in Figure 19.14. *Note*: a beam splitter, which is a partially silvered mirror on flat glass held at a 45° angle, is used to deflect some light into the

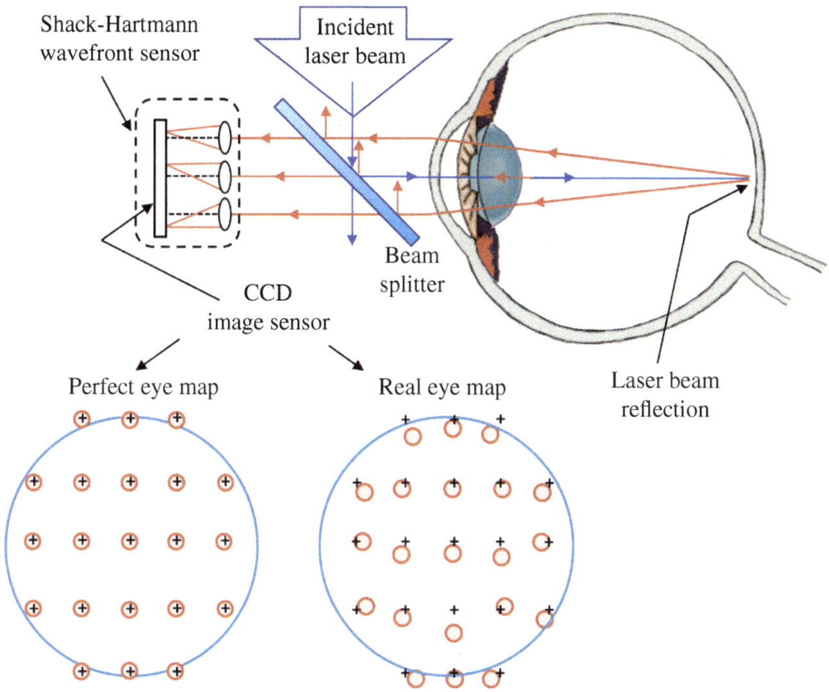

FIGURE 19.14 The use of a laser plus Shack–Hartmann wavefront sensor to map the lens aberration of an eye. Fiducial marks, denoted by the "+" symbol, indicate the locations of the Shack–Hartmann spots from a perfect lens. The differences between the spot centroids and the fiducials indicate the localized slope of the wavefront.

eye and transmit some of the return beam to the SHWFS. Each time the beam encounters the beam splitter, approximately half of the flux is transmitted and half is reflected 90° as shown. The array of spots from the SHWFS is presented at the bottom.

LASIK is a corneal reshaping technique and another application for lasers. The visual acuity of the patient is improved by a LASIK operation and is a permanent alternative to eyeglasses or contact lenses. The four basic steps of the LASIK procedure, portrayed in Figure 19.15, are (i) a hinged flap from the outer layers of the cornea is created either with a microkeratome (a small precission oscillating blade) or with a sequence of femtosecond laser pulses creating a series of closely arranged bubbles, (ii) the flap is pulled open exposing the underlying middle corneal tissue, (iii) an ultraviolet (193 nm) excimer laser is used to remodel the corneal stroma via ablation (vaporization) of stroma tissue, and (iv) the flap is carefully repositioned over the treatment area where it remains in position via natural adhesion until the eye has healed. Typical excimer laser pulses in step 3 are 10–20 ns long with an energy of about 1 mJ (millijoule). During step 3, an eye tracking system follows the patient's eye position approximately 4000 times/s, redirecting the precise placement of the excimer laser pulses within the treatment area. The lasers used in steps 1 and 3 deliver

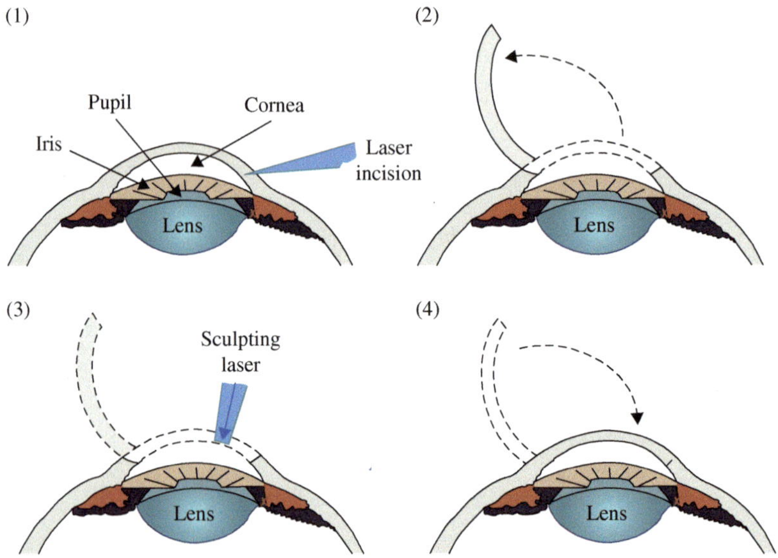

FIGURE 19.15 The four basic steps in a LASIK operation.

less energy to surrounding tissues and avoid permanently weakening the cornea compared to the use of a mechanical metal blade.

Two-thirds of the refractive power of the eye is provided by the cornea and one-third by the lens behind the iris. The focusing power of the cornea is fixed, however, and all of the distant-dependant focusing adjustments is done by the eye's lens. Laser remodeling of the stroma portion of the cornea results in a permanent alteration that cannot be reversed if a refractive error is made since the stroma consists of slow-growing cells. Laser ablation of tens of micrometers of stroma tissue in step 3 is accomplished without any burning or actual cutting. In constrast, the exceedingly thin surface layer of the cornea, the epithelial tissue layer, contains fast-growing easily regenerated cells, making it ideal for flap creation and healing in 2 or 3 days without further surgical intervention.

LASIK can only be performed on patients that have had stable eye prescriptions for a year preceding surgery. Prior to LASIK surgery, the thickness of the patient's cornea has to be examined using a pachymeter and the cornea surface contour (i.e, a topographic map of the cornea) made. Sometimes, these measurements counterindicate LASIK surgery. A pachymeter typically is a small ultrasonic device that captures a high-definition echogram of the cornea, while the topographic map, which reveals the basic correction needed plus any shape irregularities leading to astigmatism, is made with a wavefront sensor such as a Shack–Hartmann. Both instruments have accuracies measured in microns (micrometers). The ophthamologist uses these data to calculate the amounts and locations of corneal tissue to be removed. Some surgeries are now equipped with wavefront-guided LASIK, providing continual feedback to achieve a more optically perfect eye. The process, however, still depends on the physician's ability to predict the changes that will occur during the healing phase.

19.4 MISCELLANEOUS ELECTRONIC DEVICES

Artificial cardiac pacemakers are common medical devices that have been widely used for decades. Pacemakers send small electrical pulses to one or more electrodes attached to the heart muscles, causing a periodic contractions to regulate the heart beat. The reliability and the lifetime of the pacemaker improved significantly in the 1970s with the introduction of lithium-iodide battery cells. Today, most pacemakers are implanted in the patients.

One electrode is all that is needed for most cardiac pacemakers. However, the opposing walls do not contract simultaneously in 25–50% of heart failure patients. In these cases, a biventricular pacemaker (BVP) such as the one pictured in Figure 19.16 is used. BVP devices stimulate both the septum, the muscle tissue separating the left from right chamber (the two superior atria), and the ventricular septum, dividing the ventricals (lower chambers) from the upper chambers, respectively. A BCP improves overall cardiac function, reducing mortality rates and enhancing patient quality of life.

Artificial cardiac pacemakers continue to undergo enhancements as well as advancements in function. For instance, some pacemakers also incorporate a defibrillator into the same unit. Improved microprocessor control and RFID technologies are two areas undergoing upgrades. RFID technology presented in Chapter 3 is used to interrogate and reprogram the microprocessor. The cardiaologist can collect and

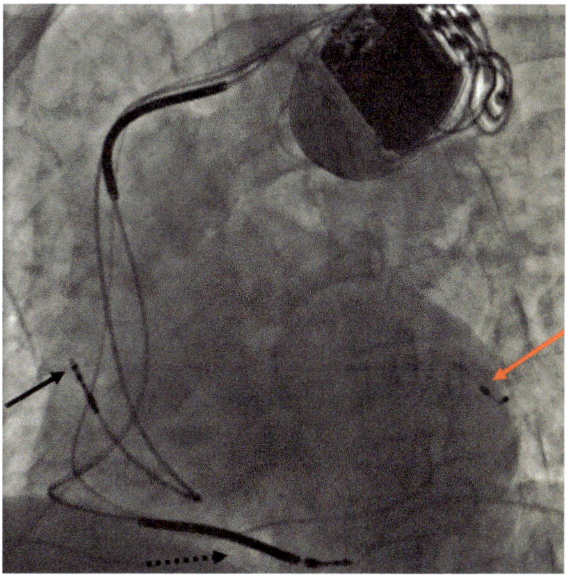

FIGURE 19.16 A biventricular pacemaker that can stimulate both the septal and lateral walls of the left ventricle. Three leads can be seen in this particular device: a right atrial lead (red arrow, left), a right ventricular lead (dashed arrow), and a coronary sinus lead (solid arrow, right). Source: Gregory Marcus, MD, MAS, UCSF School of Medicine, https://commons.wikimedia.org/wiki/File:Cardiac_resynchronisation_therapy.png. Used under CC-BY 3.0, https://creativecommons.org/licenses/by/3.0/deed.en

evaluate data on the device performance and the patient's health during each 6-month period between checkups. Stored data, for example, may indicate that one or more defibrillations occurred since the last checkup, for which the patient might not have been aware. The use of RFID technology in pacemakers, however, does have some drawbacks. Patients should avoid strong electromagnetic fields such as those from arc welding or MRIs, although one recently developed pacemaker has been approved for use in certain classes of MRI machines for specific wavelengths. Important progress is also being made on dynamic pacemaking technologies. Dynamic pacemakers adjust beat rates according to physical activities, which may be determined by an accelerometer, body temperature, adenosine triphosphate levels, or the presence of adrenaline.

Other types of artificial pacemakers include diphragmatic and brain pacemakers. A diaphragmatic pacemaker is a phrenic nerve stimulator, surgically implanted to help patients with spinal cord injuries breathe. A brain pacemaker is used in the treatment of patients with epilepsy, Parkinson's disease, major depression, or other brain disorters. It is implanted into the brain, sending electrical signals into the tissue. There are two types of treatment: deep brain stimulation and cortical stimulation.

We end this chapter with a brief discussion of two types of cautery tools: heating element and electrical current. Both types are depicted in Figure 19.17. The heating element is straight forward in which an electrical current is passed through a resistive wire at the tip or near the tip of the cautery tool. The hot surface can be extended by direct contact of a solid wire or of a small solid metal rod. This pencil shape tool is preferred for cauterization that are deep inside tissue with restricted access. Another type of cautery tool functions by passing electrical current through the tissue to be

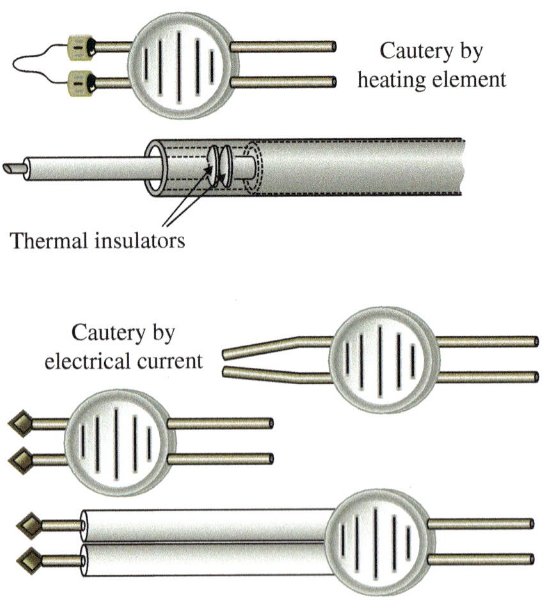

FIGURE 19.17 Two types of cautery instruments are heating element and electrical current.

cauterized. The AC current typically is 18 kHz (18,000 cycles/s), restricting all electrical burning to highly localized spots close to its two electrical tips. The significance of the high-frequency AC current from a cautery tool can be qualitatively understood by thinking of the tissue as a simple RC circuit, driven by the tool. (See Chapter 6.) The physics is all RC circuits have high-frequency roll-offs (attenuation proportional to frequency). In this case, the human tissue to be cauterized cannot respond electronically at 18 kHz over any significant distance away from the probes, localizing the volume of instantaneous heating.

INTERESTING TIDBIT TB19.3

One of the authors served on a jury for a medical malpractice case. The case hindged on whether a cautery tool, operating at 18 kHz, could have caused the extended damage claimed by the plaintif. None of the four expert witnesses, all physicians from prestegious medical teaching hospitals, understood the physics of this tool and the significance of the 18 kHz frequency. Consequently, the jury in a 5 to 1 vote awarded $1M USD to the plaintif for damages that could not have been caused by the surgeon or the cautery tool.

20

A-BOMBS, H-BOMBS, AND RADIOACTIVITY

Radioactivity is the term coined by Madame Marie Curie to describe the random spontaneous release of energy from unstable elements. Nuclear radioactive emissions include α particles, β particles, and γ rays. (Definitions and descriptions of each of these are given in Section 20.1.) Radioactivity was first discovered when photographic glass plates that had never been exposed to light became "cloudy" after being stored in the vicinity of certain mineral elements. Later, it was recognized that all elements in the periodic table with atomic numbers 83 and above are naturally unstable, decaying over time into lighter, more stable elements, and emitting radioactive particles and rays. Radiation has subsequently become a more encompassing term to include ordinary electromagnetic waves (e.g., radio waves, infrared and visible light, and ultraviolet light). There are numerous applications that use nuclear radioactive energy, ranging from medicine, industrial applications to weapons of mass destruction. There is significant folklore surrounding radioactivity and its associated radiations. We will attempt to shed light onto this esoteric topic and dispel some misconceptions.

Radioactive materials (i.e., trace amounts of radium, radon, uranium, and others) occur naturally in the soil, rocks, and in many building materials, continuously bombarding everyone with low-dose radiation. One such long-term exposure is radon gas released from the soil. Radon gas in low doses poses little risk, but is dangerous when it becomes concentrated, for example, by being trapped in the air of a basement. (An air exchanger is a simple, inexpensive solution to a home radon problem.) Another significant source of particle radiation comes directly from the Sun in the form of cosmic rays, neutrinos, and other subatomic particles. The Earth's magnetic field shields us from the most biologically damaging particle radiation from celestial sources. Additional shielding comes from the atmosphere, which reduces the number and degrades the energies of these particles. The atmosphere also blocks X-rays and

Modern Devices: The Simple Physics of Sophisticated Technology, First Edition.
Charles L. Joseph and Santiago Bernal.

Companion website: www.wiley.com/go/joseph/moderndevices

most ultraviolet photons. Our natural dose from the environment is small and is not a cause for alarm. A good comparison benchmark number is 2.4 mGy/year, an individual's typical dose from all natural background sources. Although we are surrounded by numerous naturally occurring sources of radiation, the nomenclature of "radioactive source" is generally reserved for the intentional concentration of radioactive material, or for any naturally occurring radioactive material with an abundance that is significantly above typical background levels.

The initiation of radioactive decay is completely random; it is a result of probability and the parameters set forth by quantum mechanics. The reaction must be energetically favorable, indicating it occurs because energy is released in the decay process; a reaction that is energetically unfavorable does not happen except via imparting external energy such as the bombardment with subatomic particles. For a given isotope's decay, the kinetic energy imparted to each escaping alpha particle is the same. If the reaction is energetically feasible, the potential for the decay to occur is there, just waiting for the proper internal configuration to permit the alpha decay. This is remarkable physics.

The rate of radioactive decay is exponentially governed, that is the probability of decay in a given time is constant for a particular isotope, and the probability of witnessing a single decay is proportional to the amount of that isotope present. Unique decay times for each isotope can be expressed in terms of half-life, the amount of time needed to reduce the isotope's abundance by 50%. Ra^{226} atoms, for example, have a half-life of 1700 years. Statistically speaking, if a radium sample contains one million Ra^{226} atoms in the year 2000, only 500 thousand will remain by year 3700. If at any time we know the quantity of a given isotope, N_0, and its half-life, T (or more commonly its *decay constant*, denoted λ), one can determine the amount, N, for all remaining times, t. This is mathematically expressed by relating the rate of decay (dN/dt) for a given number of atoms, N_0:

$$\frac{dN}{dt} = -\lambda N_0 \tag{20.1}$$

The solution to this differential equation is the exponential:

$$N = N_0 e^{-\lambda t} \tag{20.2}$$

We relate the half-life, T, to the decay constant λ:

$$T = \frac{\ln(2)}{\lambda} = \frac{0.693}{\lambda}$$

where ln(2) is the natural logarithmic value of 2, which equals 0.693. There are several useful variations on the equation to determine the remaining quantity, $N(t)$ of a decaying isotope for any give time from the reference quantity.

$$N(t) = N_0 e^{-t/\tau} = N_0 e^{-0.693t/T}$$

Another way to express the amount of time needed to reach one half of the initial count is as follows:

$$N(t) = N_0 \left(\frac{1}{2} \right)^{t/T} \tag{20.3}$$

Continuing the example decay rate noted before, there will be 500,000 Ra^{226} atoms in year 3700. Two hundred fifty thousand atoms will then decay in the following 1700-year interval, 125,000 in the 1700-year period after that, and 62,500 in the next half-life interval.

20.1 ALPHA, BETA, AND GAMMA RAY RADIATION

As noted, there are three primary types of radiation emitted from unstable nuclei: α, β, and γ. These forms of radiation were assigned Greek alphabetic letters to differentiate between each type since the discovering scientists did not immediately recognize any of these. Subsequently, an alpha particle (α) was found to be a fragment of a nucleus containing two protons and two neutrons, essentially a doubly ionized helium atom (He^{++}). The designation α particle remains due to its terminological legacy and for brevity reasons. Similarly, beta (β) particles are simply high-energy electrons (or in some cases positrons, the electron's antiparticle). In contrast with α and β particles, gamma rays (γ rays) are very energetic photons (very-high-frequency light waves) that travel at the speed of light in straight lines and behave exactly as all other electromagnetic radiation. Obviously, there is a plethora of other particles emitted directly from radioactive nuclei or via secondary processes. These extra particles, which are necessary to satisfy conservation rules (e.g., conservation of energy, momentum, and charge), are extremely weak interacting particles and very hard to detect.

Neutrons are ejected during many nuclear processes as well, but are rarely observed as a form of radiation except under specialized circumstances. Neutrons are particles with mass and are electrically neutral, making these also difficult to detect. Neutrons bound in a nucleus are generally stable, while "free" neutrons are not. Free neutrons do interact with materials, delivering a dose of radiation. These free particles have a characteristic lifetime of 881.5 ± 1.5 seconds or a corresponding half-life of 611.0 ± 1.0 seconds, spontaneously transforming via beta decay into a proton, electron, and a neutrino. Free neutrons are typically present at particle accelerators, nuclear reactors, and play a critical role in atomic bombs. A neutron can be captured readily by the nucleus of any atom since it is without an electrical charge and can freely approach any nuclei regardless of how many protons there are. A captured neutron produces an isotopic change or causes the nucleus to fission, the process of splitting of a nucleus into two lighter nuclei plus the release of a substantial quantity of energy.

The nucleons that form an alpha particle are donated by the parent nucleus and the remaining atom, known as the daughter, has two fewer protons and two less neutrons. Consider the alpha decay process of a radium atom. Radium is the 88th element, and as such has 88 protons. There are several isotopes of radium; the most naturally abundant is the isotope that contains 136 neutrons, for combined total of 226 nucleons (neutrons and protons). As radium-226 atoms alpha decay, each gives up two protons and two neutrons, leaving the daughter element, ${}^{222}_{86}\mathrm{Rn}$ (Radon-222). This decay process is symbolically written as follows:

$$ {}^{226}_{88}\mathrm{Ra} \rightarrow {}^{222}_{86}\mathrm{Rn} + \alpha \quad (2.35\,\mathrm{MeV}). $$

As stated previously, the decay of an unstable isotope such as radium-226 is a matter of chance, following the fundamental exponential law. *Note*: each spontaneous alpha decay of a radium-226 atom produces an alpha particle with the same 2.35 MeV of energy. Ra^{226} has a half-life of 1700 years, but its daughter Rn^{222} has a half-life of only 3.82 days.

Alpha particles are relatively large; consequently, these do not travel far prior to interacting with any target material, delivering its energy in a very short distance. As a result, it is easy to shield against alphas; a thick sheet of paper or even the dead layer of human skin will suffice. Alpha particles do not pose an external biological hazard to humans. Conversely, if an α-emitting material is ingested or inhaled, the alphas will repeatedly bombard a very small volume of tissue, delivering a large amount of damaging energy to it.

Next in our Greek alphabet soup are beta particles, denoted by the Greek letter, β. Beta particles are also very familiar to the reader as energetic electrons. Their origins require a little more sophisticated description than alpha particles, coming from the nucleus but being created at the moment of emission. When a neutron is ejected during a nucleus decay, that neutron itself almost instantaneously decays into a proton, an electron, and a neutrino. The proton is then immediately recaptured by the nucleus, while the energetic electron (β particle) and neutrino fly away. In this process, the nucleon count remains the same, since a neutron is exchanged for a proton. However, the proton count has increased by one, and atom has decayed "up" the periodic table. The beta decay process for tritium can be symbolically written as follows:

$$ {}^{3}_{1}\mathrm{H} \rightarrow {}^{3}_{2}\mathrm{He} + \beta + \nu. $$

Here, β is our high-velocity electron and ν is the neutrino, the elusive particle that rarely interacts with matter. Note that the parent is tritium, an isotope of hydrogen, having only one proton and two neutrons for a total of three nucleons. An isotope of helium is formed as the neutron converts into a proton and is recaptured, resulting in helium with only one neutron. The beta particle as well as the neutrino carries away significant portions of the energy.

INTERESTING TIDBIT TB20.1

Once an alpha particle is emitted, it interacts with the nearby matter, imparting its energy to the surrounding atoms and slowing down. After it comes to rest, its net positive charge (two protons) attracts two electrons, reaching electrical neutrality and ultimately resulting in a relaxed helium atom. Alpha particle radiation is the primary source of world's helium supply. The next time you see a child's helium balloon, tell that child they have a bag full of retired alpha particles.

INTERESTING TIDBIT TB20.2

If you go into your backyard, scoop up a cup of dirt, and sift through it to isolate all the radium and wait to detect an alpha decay, you will have witnessed a momentous event. All radium atoms on Earth were created at the same time as our solar system 4.5 billion years ago; one radium atom from your backyard has held that pent up energy for all those intervening years and then decayed into a new element right before your eyes at that particular moment.

20.2 A-BOMBS, H-BOMBS, AND DIRTY BOMBS

An atomic bomb, also more appropriately known as a nuclear bomb, is a weapon of mass destruction, primarily exploiting the force generated from the immense pressure wave and fireball in a fashion similar to a conventional bomb. As such, nuclear weapons are classified by their TNT equivalent. Radioactive fallout, the most feared aspect of a nuclear weapon, was just an added windfall in terms of its military effectiveness.

The first nuclear weapon was detonated at the Trinity Site in New Mexico, USA, on July 16, 1945. It had the equivalent explosive power of 16,000 tons of TNT. Racing to beat Nazi Germany, the United States committed itself to an unprecedented scientific effort to apply the academic discovery of fission to a functional weapon of war. In 1949, the Soviet Union was the next to detonate an A-bomb under test conditions. (Only two nuclear explosions have ever been used in combat, which occurred at the conclusion of World War II in 1945.) The USSR's entry into the nuclear club triggered an arms race between the United States and the Soviet Union, which included the development of thermonuclear weapons, the so-called H-bomb based on *fusion*. Each side detonated at least 200 test bombs in the atmosphere and many more underground. Collectively, over 2000 A-bomb and H-bomb tests have been conducted by the world's nuclear powers, including 1054 by USA, 715 by USSR, 45 by United Kingdom, 210 by France, and 45 by China. Thermonuclear bombs extend the energy yield into the equivalent of tens of megatons of TNT. Coupled with the parallel development of intercontinental ballistic missiles, our entire globe found itself in a precarious position of being capable of complete self-annihilation.

A critical mass of fissile material is needed to sustain the chain reaction. On average, a fission of a U^{235} atom creates 2.42 free neutrons. In a subcritical state, most the free neutrons simply escape to the outside environment where these rapidly decay. (In nuclear power plants, the rate that U^{235} atoms fission is carefully controlled by carbon rods that absorb an adequate percentage of the excess free neutrons to prevent a runaway condition.) If a significant amount of fissile material is in close proximity, the material is considered supercritical and those excess neutrons will strike other U^{235} atoms, leading to more and more fissions. If this continues, each generation of neutrons grows by a multiple of 2.42, quickly diverging to very large numbers in a chain reaction.

Generation 1 → $2.42^1 = 2.42$ neutrons
Generation 2 → $2.42*2.42 = 2.42^2 = 5.86$ neutrons
Generation 3 → $2.42*2.42*2.42 = 2.42^3 = 14.2$ neutrons

⋮

Generation 80 → $2.42^{80} = 5 \times 10^{30}$ neutrons

If each atom undergoing fission releases about 200 MeV of energy, the 80th generation is collectively creating 1×10^{33} MeV or 1.6×10^{20} J, which is the equivalent of 38,000 Megatons of TNT, a far greater amount of energy than the largest nuclear weapon ever fabricated. (*Note*: 1 ton of TNT = 4.184 GJ = 4.184×10^9 J. Good benchmark numbers to recall are 10 kilotons of TNT for a typical tactical atomic weapon and 20 Megatons of TNT for a thermonuclear bomb.) The potential of this storehouse of energy was realized almost from the beginning by the pioneers of nuclear physics; all that remained was to identify and isolate sufficient quantities of fissile material, which was the focus of the *Manhattan Project* in the United States during World War II. The explosive yields are limited (capped) by the number of fission generations that can be achieved before the fissile core disassembles itself. In other words, the chain reaction can only proceed so far before the rapidly increasing heat and pressure disrupts the fission process. Only a small fraction of the available fissile atoms actually participate in the chain reaction before the core expands beyond criticality. Most of the fissile material in the core simply goes unused. The first atomic bombs were very inefficient.

There are two basic A-bomb designs as depicted in Figure 20.1: the implosion type and the gun type. Uranium235 is usually the fissile material in the gun type (e.g., Little Boy dropped on Hiroshima, Japan), while plutonium is the material of choice for implosion (e.g., Fat Man dropped on Nagasaki, Japan). A fission bomb prior to detonation maintains its fissile material in a subcritical condition, usually using one of the two methods pictured in Figure 20.1. Two hemispheres of U^{235}, each subcritical, are kept separated in the gun-type assembly. The diameter of the sphere of plutonium in an implosion A-bomb is sufficiently large that density of plutonium is low enough to be subcritical. Both configurations then use chemical high explosives to compress the fissile material from its subcritical state into a supercritical one, leading to a self-sustaining chain reaction. In addition to the release a large amount of thermal energy, each atom that fissions also liberates excesses free neutrons, initiating more atoms to fission.

Fissile materials are limited to those that will fission under the influence neutrons; the list includes Th^{232}, U^{233}, U^{235}, and Pu^{239}. Only U^{235} and Pu^{239} have properties sufficient to sustain a chain reaction of catastrophic proportions necessary for an atomic bomb. More than 99% of the world's natural uranium is the U^{238} isotope, which itself is *not fissile*, but can be used in nuclear reactors to generate fissile

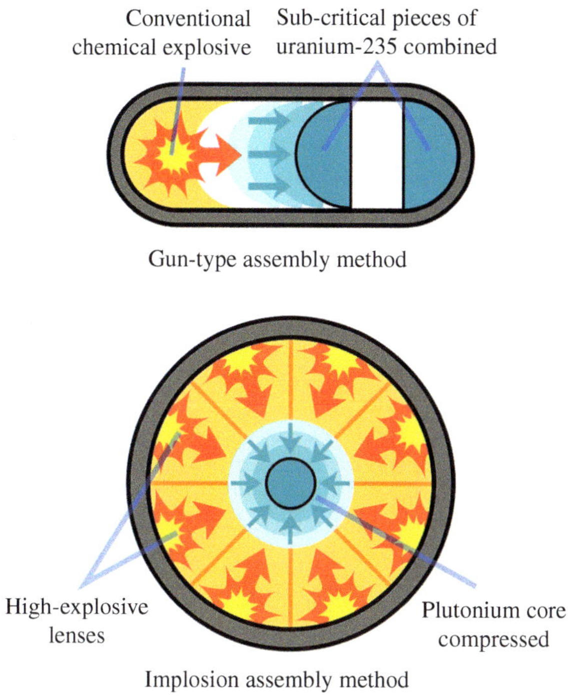

FIGURE 20.1 The two methods (A-bomb designs) using conventional explosions to transform the fissile material into a supercritical chain reaction. Source: Fastfission, https://commons.wikimedia.org/wiki/File:Fission_bomb_assembly_methods.svg. CC public domain.

plutonium-239. To obtain weapons-grade uranium, the tiny amounts (0.7%) of the U^{235} isotope have to be separated from the bulk of the naturally occurring uranium. Strong centrifuges are normally employed to separate this isotope in a process called enrichment. Weapons-grade uranium normally contains 80% or more U^{235}, although weak A-bombs can be constructed with as little as 20% concentrations of U^{235}. It is important to note that all fissile elements can also be used for the peaceful harnessing of atomic energy for the controlled power generation.

There are two phases of radiation resulting from an atomic blast. The first is the intense neutron and gamma flux at the moment of detonation. A person in the vicinity of a nuclear detonation would most likely succumb to the shockwave of the blast before he or she received a lethal dose of direct radiation poisoning. However, if adequate physical shielding protects the individual from the blast, fatal doses of radiation can still be delivered indirectly. The second phase of radiation associated from a nuclear detonation is the widespread nuclear "fallout." A ground-level blast with its intense neutron flux "radio-activates" soil, debris, ash, and dust, while the quickly rising fireball creates a strong updraft carrying these particulates into the upper atmosphere.

In a span of time ranging from minutes to days, these radioactive particulates will fall back to Earth. The intensity of this fallout is determined from the size and height of the blast, the wind direction, weather conditions, rain, etc. The fallout can be extremely radioactive, containing alpha, beta, and gamma emitters. A single blast can create fallout measuring 1000 R/h—delivering a fatal dose in 30 minutes. Fallout shelters, mostly a nicety of the 1960s, aimed to keep people isolated from the high dose rates of fallout, not the blast. For this reason, most shelters are centrally located in the basement of buildings, away from external walls and roofs; fallout shelters relied on heavy construction material to absorb the strong gamma rays. The rich variety of fission products and radionuclides created from a nuclear detonation have fairly short half-lives. Many above-ground nuclear tests of the 1950s indicated that shelter stays of only 3 weeks would be necessary to wait out the dangerous levels of fallout from a single bomb.

Hydrogen bombs, also known as thermonuclear or H-bombs, are *fusion* bombs utilizing the favorable energy conditions of combining lighter elements into heavier ones. (The fusion process is exothermic for all elements lighter than iron—all elements with atomic numbers <26 and is not exothermic above iron.) The tritium–deuterium reaction of an H-bomb, results in a helium nucleus, a neutron, and an excess of 14.1 MeV of energy.

$$^{3}_{1}\mathrm{H} + ^{2}_{1}\mathrm{H} \rightarrow ^{4}_{2}\mathrm{He} + \mathrm{n} + 14.1\ \mathrm{MeV}$$

Extreme conditions are required to initiate this and other similar fusion reactions. The needed pressures and temperatures are present in the smaller fission devices. The Teller–Ulam H-bomb design, depicted in Figure 20.2, is the basic architecture for many thermonuclear weapons and the first one test detonated. A conventional plutonium A-bomb is used as a triggering device to initiate hydrogen fusion. Unlike fission weapons, there is not an upper limit on the explosive yield and is ostensibly proportional to the quantity of fusion material available to "burn." Several countries have demonstrated energy yields into the tens of megatons regime. In 1961, the USSR's Tsar Bomba, a 50–58 Megatons thermonuclear weapon, was the most powerful weapon ever test detonated. Modern high-yield thermonuclear weapons are based on a three-stage hybrid fission–fusion–fission reaction. The initial fission explosion creates the necessary condition to ignite the thermonuclear components. Energetic neutrons are released in the fusion of the thermonuclear components, which in turn further drive fissionable components efficiently attaining very high explosive yields. Approximately half of the total explosive yield in large thermonuclear explosions comes from the third-stage reaction of plutonium fission.

Both nuclear and thermonuclear bombs are considered weapons of mass destruction. The distance the radioactive debris travels from a surface detonation is dependent on a number of factors, but correlates with the altitude of the mushroom cloud, which in turn is proportional to the blast strength as seen in Figure 20.3. The discovery and designs of nuclear weapons are technically challenging and fascinating. However, their introduction into the arsenal of the world has created a tenuous situation for the denizens of Earth. Despite drastic reductions in stockpiles reductions over the past

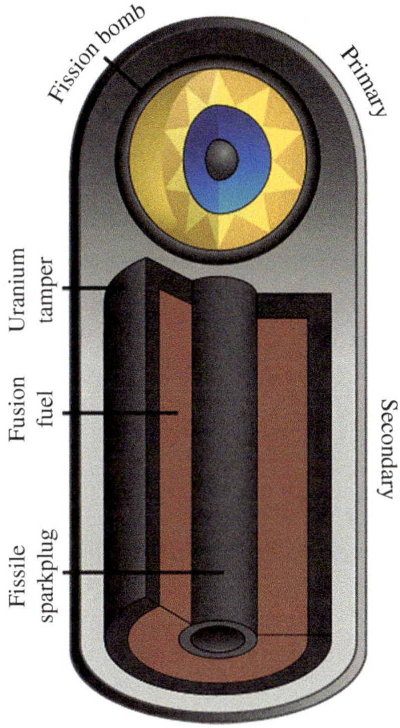

FIGURE 20.2 A Teller–Ulam design for a thermonuclear bomb. The top portion is a conventional plutonium A-bomb, used as a trigger to initiate the H-bomb. Source: Fastfission, https://commons.wikimedia.org/wiki/File:Teller-Ulam_device_3D.svg. CC public domain.

few decades, there is still sufficient explosive power to destroy most life on our planet, begging some to question whether or not the social maturity of Earth's human population has kept pace with its technical achievements.

Dirty bombs are not nuclear per se since these do not have fissile material supporting a chain reaction that results in mass destruction. Instead, dirty bombs contain small amounts of radioactive atoms (the dirty part) to be dispersed by an improvised explosive device (IED). Terrorists might use these weapons to induce mass hysteria, relying on the intense anxiety felt by the public over the minutest quantities of radioactivity. Obtaining some radioactive elements, strapping this material to an IED, and contaminating the immediate environment via an explosion constitutes a dirty bomb. The radiological threat is contained to local contamination surrounding the initial explosion, primarily a few city blocks. A successful dirty bomb temporarily creates an uninhabitable zone during the cleanup procedure, which can generally be handled by proper, but costly, decontamination procedures. Beyond creating a mess, a dirty bomb instills public fear and anxiety over areas much larger than those actually contaminated and well past the time the central area is no longer radioactive.

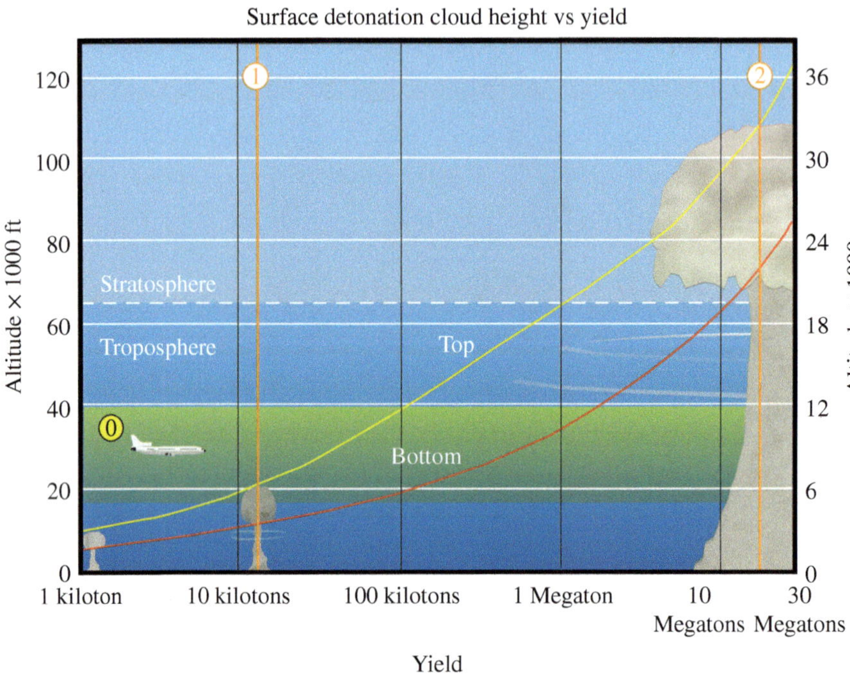

FIGURE 20.3 The height of the mushroom cloud as a function blast yield. The two lines labeled (1) and (2) represent the yields from Fat Boy, the U^{235} bomb dropped on Hiroshima in 1945, and Castle Brovo, a dry fuel, H-bomb detonated at the Bikini Atoll, Marshall Island in 1954, respectively. Source: Anybody, https://en.wikipedia.org/wiki/File:Nukecloud.png. Public domain.

20.3 RADIATION SAFETY, DETECTION, AND PROTECTION

A common question asked is as follows: How long does one have to wait for a radioactive source to decay to nothing? Mathematically, the answer is after an infinite amount of time (i.e., never). This conclusion is not very satisfactory to many people. A common standard to achieve an insignificant quantity is taken to be seven half-lives, resulting in less than 1% of the original amount of radioactive material.

$$N(t) = N_0\left(\frac{1}{2}\right)^7 = N_0\left(\frac{1}{2}\right)\left(\frac{1}{2}\right)\left(\frac{1}{2}\right)\left(\frac{1}{2}\right)\left(\frac{1}{2}\right)\left(\frac{1}{2}\right)\left(\frac{1}{2}\right) = N_0\frac{1}{128}$$

Perhaps, a better standard would be to define an insignificant quantity as the human exposure limit. For example, consider Cobalt-60, (^{60}Co) with a half-life of 5.27 years, a “medium-lived” radioactive isotope. If a fresh source of ^{60}Co has an initial activity of 100,000 Ci (100 kCi), the radioactivity of this source will be 800 Ci after the 7-half-life period, 37 years later. However, from a human safety perspective, a Co^{60} source of 800 Ci can easily deliver a fatal dose in a few seconds.

Unfortunately, there are many units of measure associated with radioactivity, some of these express the same information, but differ in scale. The quantities of radioactive sources have historically been measured in non-SI units of roentgens (R) or curies (Ci), and subsequently in SI units of becquerel (Bq). Biologists and medical professionals have in recent decades introduced the units of gray (Gy) and sievert (Sv), both of which are units of the radiation absorbed. As demonstrated, there are several types of radiation, each with varying energy ranges. Various biological tissues respond (absorb) each radiation type and each energy range differently. Consequently, the gray is used primarily in radiation doses for nonhuman tissue and the sievert specifically accounts for the absorption by humans. It is not important for the purposes of this book to learn or understand any of these parameters. Instead, the reader should realize there are disconnects among researchers in the choice of units; the physical scientists generally prefer quantities expressed in terms of radioactive sources, while those studying the effects of radiation poisoning prefer either the sievert or gray.

One must be careful when describing a samples' activity. As was noted earlier, the rather long-lived Ra^{226} decays into a short-lived Rn^{222}. On a coarse enough timescale, it can be stated the every Ra^{226} decay is promptly followed by an Rn^{222} decay. The presence of one Ci of radium also indicates there is one Ci of Rn^{222} after an equilibrium is achieved in a few weeks time. Moreover, we have not considered the activity contributed by the chain of radioactive daughters following the Rn^{222} decay products.

Absolute activity is used to assign prices when selling radioactive material and to define limits on licenses issued to users of radioactive materials. Concentration of the radioactivity is important too. Maximum theoretical-specific activity defines the maximum radioactivity per unit mass of a given isotope. Considered a carbon based biological tracer laden with some fraction of Carbon-14. If every carbon atom in a sample is ^{14}C, then it is said to be at the theoretical maximum specific activity. Concentrations are important in many applications, and are commonly found defining regulatory thresholds. For example the EPA sets the remediation action concentration of Radon to 4 pCi/L of air inside a home.

The buildup of radioactive atoms occurs deliberately in the manufacturing of radioactive sources and also as unwanted waste byproducts in the surrounding shield materials. Particle accelerators, nuclear reactors, and parent isotope generators create artificially manufactured radioisotopes. This manufacturing process presents a unique problem in that the "product" begins to decay as soon as the generation process occurs. To obtain practical quantities of the radioisotope, the creation rate must outpace the decay rate. In accelerators, typically a stable isotope is subjected to high-energy charged particles or neutrons that transform the target nucleus into the requested isotope. The rate of isotope production depends on the incident particle flux, its probability of absorption in the target nuclei, and the quantity of target material. We begin with v cm^3 of target material with an absorption probability of Σ_a that is subjected to a constant beam of particles with flux, ϕ. As soon as the irradiation begins to generate the new isotopes at a rate of $v\Sigma_a\phi$, these begin to decay at the rate of $-\lambda N$. The net isotope production rate, dN/dt, is equal to the formation rate less the decay rate.

Cyclotron beams have on-target incident fluxes on the order 1 to 5×10^{15} protons/(cm^2 s), but over very limited target areas (typically <1 cm^2). In contrast, nuclear reactors can achieve neutron fluxes on the order of 1 to 5×10^{14} neutrons/(cm^2 s) with enormous coverage areas measured in many square meters.

Regarding radiation safety, the as-low-as-reasonably achievable (ALARA) principle applies. (This principle is referred to as ALARP in the United Kingdom, where the word "practical" is used instead of "achievable.") Radiation in this context is energy that can do work, but that is also damaging to human tissue, electronics, and materials. The specialized field of radiation safety is known as health physics. There are many subtle aspects to radiological safety. First, the radiation worker has to evaluate the hazard and adequately prepare with appropriate personal protective equipment. Each radiation environment is different. Many texts are dedicated to the expanse of problems and challenges. However, there is a universal ideology that applies to all situations, in which radiation is present, this is the ALARA principle. In other words, while some radiation dose may be necessary, try to minimize it. This approach may seem obvious, but its implementation and exact dose levels may not be clear. Three factors contribute to a dose: time, distance, and shielding. The first is clear, the less time spent in a radiation field, the lower the received dose will be—this is a linear effect. Distance from a typical radiation source is quadratically favorable. Consider the radiation from a point source where the alpha, beta, or gamma radiation emanates in all direction from a radioactive source whose size dimensions are small compared to the distance from it. The intensity of the radiation field under these circumstances drops as $1/r^2$, where r is the distance to the source—this is known as the inverse square spreading effect. If the dose rate at 1 ft. from the source measures 400 mR/h, then backing up another foot drops the dose rate to 100 mR/h. Finally, shielding is the placement of a physical barrier between a radiation source and an individual or equipment to block (absorb) the radiation. Each type of radiation, its energy, and its intensity must be considered. We learned earlier that the most energetic α particles can be stopped by a thick sheet of paper. Wood or a thin metal sheet can easily shield beta particles. Gamma rays are much more penetrating and require high-density materials, such as lead or large quantities of concrete.

COMPREHENSION VERIFICATION CV20.1

Problem: Consider the example 100 kCi Co^{60} source. How many half-lives must pass before 1 μCi is all that remains? *Note*: typical safe, low-level instrument check sources that are exempt from licensing requirements have an activity of 1 μCi or less.

Solution:

$$1 \times 10^{-6} = 1 \times 10^{5} \left(\frac{1}{2}\right)^{T}$$

$$\frac{1 \times 10^{-6}}{1 \times 10^{5}} = 1 \times 10^{-11} = \left(\frac{1}{2}\right)^{T}$$

$$\text{Log}\left(1\times10^{-11}\right) = T\log\left(\frac{1}{2}\right)$$

$$\frac{\text{Log}\left(1\times10^{-11}\right)}{\log\left(\frac{1}{2}\right)} = T = 36.5$$

Thus, 36.5 half-lives, or 192 years, are required for the Co^{60} source to decay to level that is no longer a regulatory concern.

20.4 INDUSTRIAL AND MEDICAL APPLICATIONS

There are many uses of radioactive materials and ionizing radiation in medicine for both imaging diagnostics and for direct therapeutic treatments. One of the most common applications is the X-ray imaging found in most computed tomography (CT) scanners. Single-photon emission CT (SPECT), and positron emission tomography (PET) most often image γ-ray emitting solutions, injected directly into the tissues of humans or laboratory experimental animals. Nonsurgical cancer treatment is undertaken in some cases using proton therapy or neutron therapy. Cancer cells can be destroyed in highly localized regions of tissue, using 2–3 beams of protons or neutrons made to converge to a single spot. A few of these were discussed in the previous chapter.

Industry uses radiation in many applications, of which much goes unnoticed. Many food products are exposed to the strong γ rays of a Co^{60} irradiator for preservation, while many plastics are exposed to accelerator-produced beta particles to improve material strength. Every plastic grocery bag, high-end trash bag, and even garden hoses have all been irradiated as part of the manufacturing process. Ultraviolet radiation is effective at sterilizing liquids, gases, and the surfaces of solids. The United States Postal Service uses penetrating γ-ray radiation on letters and packages addressed to high-ranking government officials to destroy any harmful substances with DNA. Density gauges utilize radiation sources to measure the thickness and density of construction materials. Radioactive sources are used for on-site inspection of large pipe welds. Oil well logging companies use specialized sources that produce neutrons for detecting the presence of oil reserves. As we saw in Chapter 2, Americium oxide (AmO_2), an α-emitter radioactive source is a by-product of plutonium-241 decay from nuclear power plants and is used in home smoke detectors. There are numerous other commercial applications for radioactive sources.

21

POWER GENERATION

The terms *energy* and *power* are often used interchangeably in popular speech since electrical power is usually sold in units of kilowatt hour (i.e., energy per time multiplied by time). In addition, other energy requirements in a modern industrial society such as the heat for buildings and the mechanical power for transportation, both of which have relied primarily on fossil fuels. In physics, energy is the ability to do work, measured in joules (J), and power is rate of energy usage (J/s). Our focus in this chapter will be on electrical power with its unit of measure being the watt ($1\ W = 1\ J/s = 1\ kg \times m/s$). We choose to retain the distinct physics definitions of energy and power instead of the commercial definition (i.e., power × time, the kilowatt hour). This choice enables easy comparisons of the energy contents of various fuels and other raw energy sources. The other energy requirements will only be discussed comparatively. *Note*: hydrogen fuel cell architecture was discussed in Section 3.4 on transportation technologies. While the underlying technologies are the same between hydrogen-fueled automobiles and stationary hydrogen power plants the former may never be cost effective, and the latter currently is.

Electrical power generation is accomplished by converting kinetic energy into electrical energy. The mechanical energy to turn the rotor in major generators is normally obtained from one of several sources (e.g., steam-driven turbines and hydrodynamic flow of water from a damn). Most of the US electricity production in 2013 was driven by steam turbines from burning fossil fuels (coal 39.1% and natural gas 27.4%), and steam generated via nuclear fuels 19.4%. In contrast, 78.1% of the electricity consumed in France during 2013 was from nuclear powered steam. Electrical power can also be stored via a battery or other storage mechanism and then transferred back to the grid or used directly by localized equipment. *Note*: hydrogen fuel is generally considered to be power storage similar to batteries rather than a power

Modern Devices: The Simple Physics of Sophisticated Technology, First Edition.
Charles L. Joseph and Santiago Bernal.

Companion website: www.wiley.com/go/joseph/moderndevices

generation source since it takes more energy to separate the constituent hydrogen atoms from various molecules than can be recovered again from the hydrogen as a fuel.

Various power generation stations connected together via national or continental distribution grids, often simply referred to as "the grid," supply most the available electrical power. The varying amount of demand for electrical power must be matched equally by the quantity being generated. Traditionally, additional power stations are brought on line as needed to supplement any increased demand. Demand that exceeds the full generating capacity of the grid results in "brown outs" or possibly "black outs." To transition to a grid supplied primarily by renewable power sources (e.g., solar and wind), which are inherently variable, requires either the creation of numerous high-capacity energy storage facilities or the retention of significant number of conventional power plants. Current energy storage capacity is notably inadequate to support present-day power demands, and there is no technological solution available to increase it dramatically. It will become necessary, therefore, to alter human behavior in such a way as to use electrical power primarily when it is available. For example, we may have to schedule our use of clothes washer and driers, dishwashers, or even the times we drive our automobiles to correspond when green power is available.

21.1 PRINCIPLES OF ELECTRIC GENERATORS

We discussed the interplay of alternating electrical currents and magnetic fields (Faraday interactions) in Chapter 3. As noted, both AC electric motors and electric generators can use the exact same type of components, but the former converts AC currents into mechanical (kinetic) energy, while the latter takes mechanical energy to produce alternating electricity. Small electrical generators often have coils of wire inside a set of permanent magnets and must be forced to spin by an internal combustion engine. For very large commercial power generation, however, the strength of the fields that can be produced by permanent magnets is rather limited, and the magnets are particularly bulky and heavy. Instead, AC induction electrical generators as depicted in Figure 21.1 are used most by electric companies to produce substantially more power. The kinetic energy required to spin the rotor is often obtained from a turbine, powered by a renewable resource such as wind or a drop in elevation of water, or by steam pressure created from heating water.

Two stationary brushes coming from the left side of the figure send a single DC current through the rotating electromagnets, creating oscillating magnetic fields that Faraday couple to the stator coils to produce an AC electrical current. Once AC power is established, some of the output can be siphoned off to maintain the DC current being supplied to the rotor coils. The initial DC voltage at start up can be obtained by small trigger generators or simply pulled off the grid.

An induction generator can also be *black started* on its own via a process called self-excitation when widespread, complete outages occur. Magnetization remains in a metal after the external DC driving current is turned off. In fact, the method practiced in introductory physics courses to permanently magnetize an iron rod places

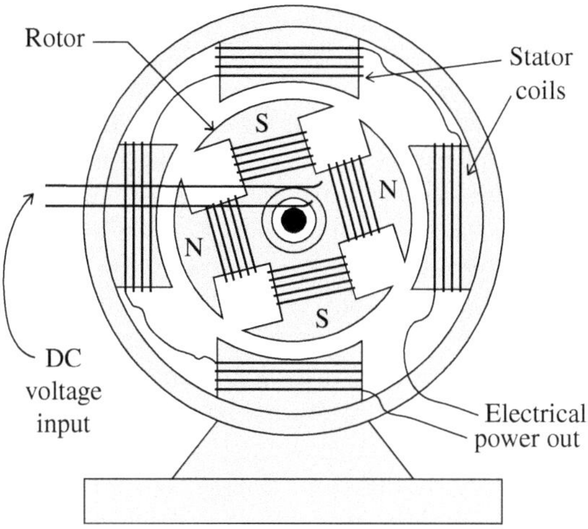

FIGURE 21.1 The design of an electrical generator used by power companies. Small start-up generators are used to the establish currents in coils, which can then be removed. The internal alternating-current electromagnets are maintained via Faraday interactions supplied by external mechanical energy. Figure based on a Siemens generator diagram.

that rod inside a coil with a DC electrical current. This ruminant magnetism initially produces a small amount AC output as the generator starts to turn. A portion of the output AC current is rectified (converted to DC) and sent back to the rotor coils, where it generates stronger magnetic fields. This bootstrap process continues until saturation occurs and a steady-state power output is achieved. A black start requires a substantially longer time than the use of small trigger generators to bring the power station back on line.

Large induction generators can have any even number of poles. *Note*: Figure 21.1 shows a four-pole design with identical poles (N–N or S–S) being on opposite sides, which is necessary to the reverse polarity at each passing. Power generators commonly have 8–40 poles. The shaft rotational speeds of both generators and motors are measured in revolutions per minute (RPM) and each electrical grid operates on a fixed oscillation frequency. For example, the North American grid operates at 60 Hz (60 oscillation/s), while Europe uses 50 Hz. All generators on the grid must output the same AC frequency, requiring each generator to spin at a synchronous speed according to its internal number of poles. For example, an eight-pole generator connected to a 60 Hz grid has to be spun at 900 RPM, while a 40-pole generator at 150 RPM to produce the same AC power.

21.2 POWER STORAGE AND POWER CONTENT OF FUELS

If electrical grids are ever to transition to a nearly complete reliance on renewable sources of energy, power storage will have to become a critical enabling technology. The main shortcoming of power storage is its economy of scale. One has to have a combined storage capacity that is at least 25% of the total daily usage if the grid has multiple, independent types of green generators supporting the grid and especially if smart-grid technologies are used by the vast majority of end users. (The storage capacity has to be at least 40% of daily use if the demand is largely at the discretion of private and corporate customers.) As shall be demonstrated shortly, it is impossible to build enough power storage capacity into any continental-sized grids using the available technologies and no technological breakthroughs appear to be on the near-term horizon to change this fact.

There are several methods of storing electrical energy. The most obvious one that comes to mind is the familiar battery, common in small portable devices. Large-scale energy storage is most often accomplished with pumped-storage hydroelectric (PSH) facilities, which consists of a reservoir that has an elevation of 100 m or more higher than a lake or second reservoir. Water flow from the first to the second basin can be used to generate electricity during peak demand usually during daylight hours. Energy is taken back from the grid at night to run the turbines in reverse, pumping the water back to the original reservoir. In this sense, a pump storage hydroelectric power station acts identically to a battery, storing power repeatedly and releasing it again back to the grid when it is most needed. A pumped storage (PS) facility attached to Nickajack Lake on the Tennessee River is shown schematically in Figure 21.2. The facility, called the Raccoon Mountain Pumped-Storage Plant, is owned and operated by the Tennessee Valley Authority (TVA).

This particular facility has a net generation capacity of 1652 MW, corresponding to an effective daily storage capacity of 14.3 GWh. PSH was first used in Italy and Switzerland in the 1890s. As of 2014, there were more than 50 PSH power stations

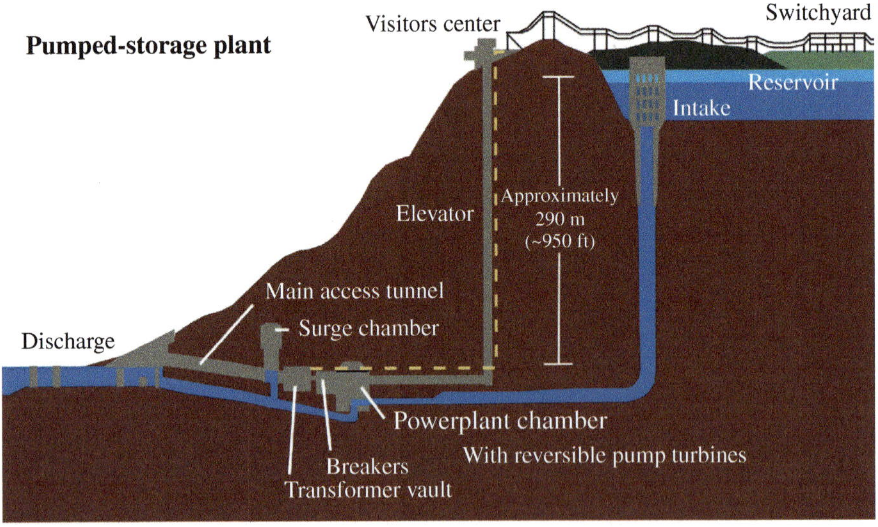

FIGURE 21.2 The cross-sectional diagram of the Raccoon Mountain Pumped Storage Plant in Southeastern Tennessee, USA. Source: Adapted from the United States Tennessee Valley Authority.

each with 1 GW or more capacity in 16 countries and more under development. The largest PS facilities are Bath County PS Station, USA; Guangdong PS Power Station, China; Huizhou PS Power Station, China; Okutataragi PS Power Station, Japan; and Ludington PS Power Plant, USA. Worldwide PSH generation was 127 GW in 2012, accounting for 95% of the total bulk storage capacity. While these are impressive numbers, the global demand for electricity is measured in thousands of terawatts-hours per year (e.g., 20,261 TWh/year in 2008). According to a 2013 US DOE survey the total large-scale storage (i.e., PSH) capacity is only approximately 2.5% of US generating capacity, a factor of a 10th of that needed to transition completely to renewable energies. Moreover, other power storage technologies are immature and require substantial prolonged investments to make these viable, especially for short-term and mid-level power ratings. *Note*: the distinction between PSH and conventional hydroelectric generation from dammed rivers and natural waterfalls. The latter produced 92% of the renewable generation or about 15% of the total electricity in 2008.

One of the most efficient and effective means of comparing the energy content of various fuels and energy storage devices is an energy density plot such as that of Figure 21.3, showing the volumetric and gravimetric densities. The total amount of available power is simply the product of volume times the volumetric density of the fuel or the product of the mass of fuel times its gravimetric density. For example, data points falling near the bottom of the graph (e.g., natural gas and pressurized H_2 gas) require relatively large storage volumes, while fuels such as diesel and gasoline (near the top) do well with relatively small holding tanks resulting in the widespread use of

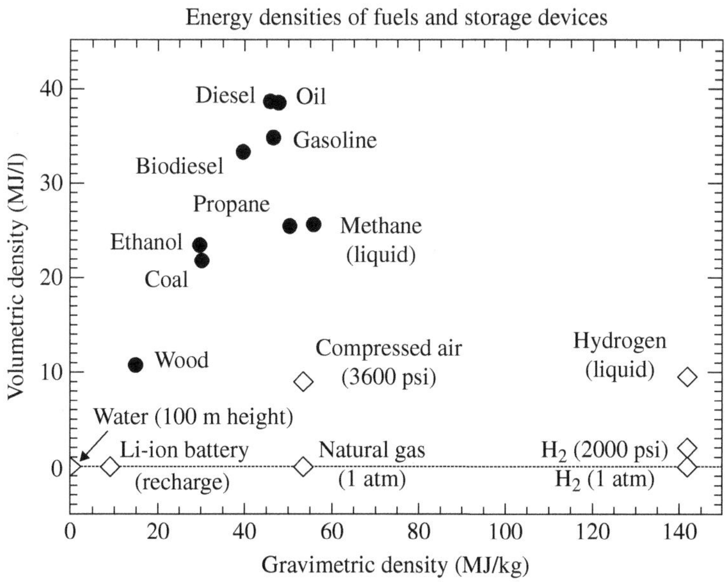

FIGURE 21.3 A plot of the energy densities contained in various fuels (solid dots) and energy storage mechanisms (open diamonds). The vertical axis denotes the energy content per volume, while the horizontal axis denotes the energy per weight.

these fuels for transportation vehicles. Moreover, the low volumetric energy densities of various forms of hydrogen makes stationary power plants feasible but not hydrogen-powered vehicles in a so-called "hydrogen economy." Fuels are plotted as solid dots and energy storage devices as open diamonds. *Note*: none of the data points take into account the amounts of energy that must be expended to obtain the various fuels or any inefficiency experienced during use. Thus, the figure only represents upper limits of the energy contents of the materials in hand and ready for use. Also *note*: the data point for water at a height of 100 m is (0.001, 0.001) not zero.

What generalities can be inferred from an energy diagram? First, fossil fuels are highly dense forms of energy, although these levels of concentration required very high pressures and millions of years to form. Second, compressed air, used to drive pneumatic tools, takes more energy to compress it to 3600 psi than 2400 psi, and it contains a larger stored energy than the lower pressure gas. Similarly, liquid hydrogen has a much higher energy density than H_2 gas compressed to 2000 psi, but it takes far more energy per kilogram to liquefy it than to compress it. Fuel sources such as natural gas and hydrogen fuels require substantial volume capacity, which are more suitable for pipeline delivery and stationary power plants than for transportation vehicles. (See Section 3.4.2 for more information on hydrogen-fueled automobiles.) Rechargeable batteries and elevated water have relatively low densities, requiring power storage of substantial mass and volumes. Lithium-ion batteries used in electric and hybrid automobiles are a substantial fraction of the volume and weight of the vehicle as well as are a significant fraction of the total life cost since these batteries must be replaced periodically. Similarly, the location of the water data point in Figure 21.3 reveals the reason that pump storage hydroelectric facilities require large bodies of water and a height differential of at least 100 m. (The water data point being so close to zero (0.001, 0.001) places a stringent constraint on suitable locations for a commercial PSH facility.) Finally, it is important to realize that the efficiencies of all of the power storage technologies are expected to improve by no more than a factor of two over the next few decades, severely inhibiting a complete transformation to renewable sources of electricity.

While valuable information on total power can be gleaned from an energy density diagram, there are additional requirements for power storage for large electrical grids. Specifically, the system power rating and the duration that that level of power can be delivered are also important pieces of the puzzle. The three control regimes of a well-functioning grid system are shown in Figure 21.4 along with the potential capability ranges for various power storage technologies. Bulk power management mitigates daily variations in the demand for an area served by one or a few power plants. Its capability, therefore, has to be 10 MW to a few gigawatts, functioning for several hours. The compressed air energy storage (CAES) in Figure 21.4 only refers to utility-scale storage, requiring an underground cavern or abandoned mine. Moderate levels of power (1–100 MW), functioning for a few minutes to 1–2 hours, are needed to transfer loads from one power plant to another or to respond to rapid changes in demand prior to bringing more generation capacity on line. Finally, power quality and uninterruptable power supply (UPS) are required locally to protect sensitive or essential electronic equipment that must remain operational during power fluctuations or temporary outages.

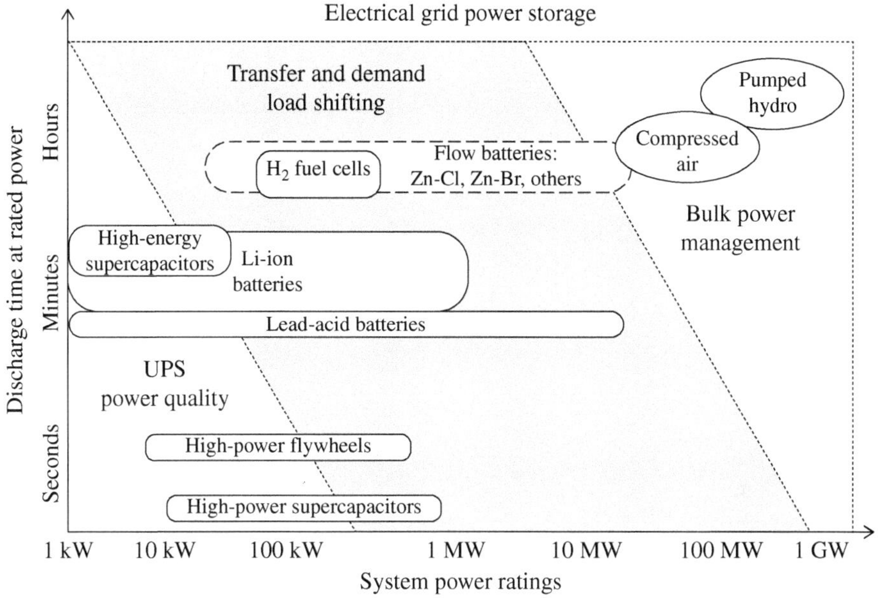

FIGURE 21.4 Various power storage technologies showing the duration that each can sustain its power levels. Overlaid are the three zones necessary for controlling the power.

21.3 THE POWER GRID

As already noted, most of the electrical power consumed in the industrialized societies is delivered via a countrywide or continent-wide distribution system called the grid. It is worth noting that the power transmitted along a wire is equal to current multiplied by voltage and that the current is equal to the voltage divided by resistance. Thus,

$$P = iV = \left(\frac{V}{R}\right) \times V = \frac{V^2}{R},$$

indicating a high-voltage (HV—the higher, the better) results in large amounts of power being transmitted (either AC or DC) over a fixed distance with a fixed resistance. In other words, the resistive losses are the same for both high and low voltages, but the amount of power delivered goes as the voltage squared. There are, however, other losses incurred in power transmission.

Grids usually consist of three components: the transmission grid that uses very high voltages over long distances, multiple distribution grids that use moderately high voltage to deliver power regionally, and numerous local supply hubs that step down the voltage to the consumer. The North American grid, a segment of it depicted schematically in Figure 21.5, primarily operates at 230 or 500 kV AC for transmission and 7.2 kV AC for distribution. (There are also subtransmission lines between two regions using 69 or 138 kV, which we include in the category of transmission.) A small fraction of the power is carried on high-voltage DC lines and subterranean

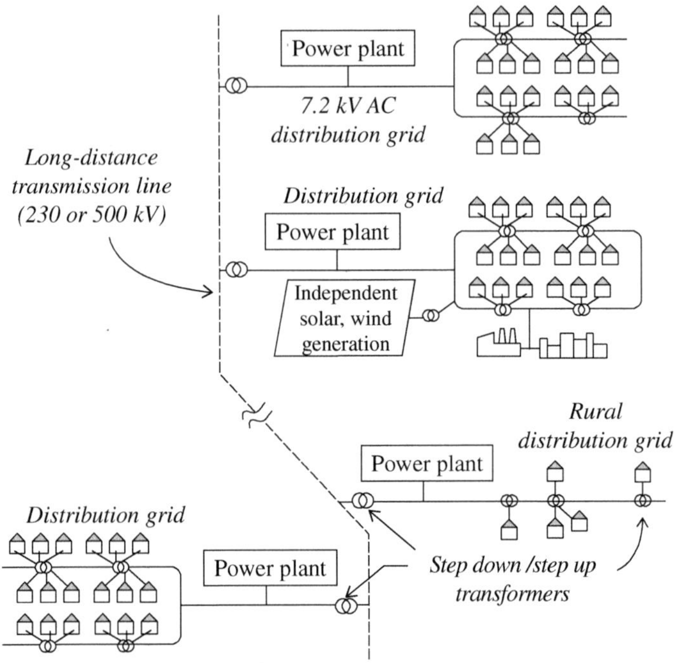

FIGURE 21.5 A schematic diagram of several distribution grids attached to a single, long-distance transmission line. Most distribution grids have multiple connections to the transmission grid.

lines. Grids are designed to share electrical energy seamlessly between many distribution grids, allowing a few power plants to be shut down for maintenance or to bring additional power stations online to cope with changes in demand.

In addition to the resistive losses already noted, capacitive coupling between adjacent power lines is important for AC transmission, as are the AC/DC losses from degrading HV insulators. Each step-down/step-up transformer incurs a further drop in efficiency. Total energy lost to various portions of the grid ranges from 20 to 40%, including losses in the regional subtransmission and local distribution networks. Some of the overall transmission losses are mitigated when electricity is generated onsite at the point of consumption such as a solar photovoltaic panel on the rooftop of a home. This component of the grid system is usually described as distributed generation.

Large portions of almost all grids transmit AC power due to the ease and efficiency of step-up/step-down transformers. (See Intro Physics Flashback FB3.1.) There are a few important constraints associated with an AC power grid. The most important requirement is all generators and every location on the grid has to have the exact same frequency and precisely the same phase of sinusoidal oscillation. A timing error of even 1 ms between any two points would result in huge amounts of power rushing back and forth between those two points. All nationwide and continental AC or DC grids must have safeguard control devices and procedures in place to prevent an

TABLE 21.1 Life Cycle Greenhouse Gas Emissions[a]

Generation Technology	CO_2 Equivalent Emissions[b] (g/kWh)	Range of CO_2 Emissions[c] (g/kWh)
Coal	1001	877–1130
Natural gas	469	422–548
Nuclear	16	8–45
Hydroelectric	4	3–7
Solar photovoltaic (PV)	46	29–80
Geothermal	45	45
Solar thermal	22	14–32
Wind (on shore)	12	8–20
Biomass	18	18–37

[a] Source: Moomaw et al. (2011).
[b] 50th percentile.
[c] 25th to 75th percentile.

electrical fault at one location from bring down the entire system through a cascading failure. AC grids also have to compensate for the inherent phase shifts that occur over long-distance cables. The nature of these phase shifts is identical to those in long-distance communications (see Chapter 18) or in all electronics (see Section 6.2). Moreover, large-scale grids act as giant antennas. Any sudden electromagnetic force (*emf*) pulses, especially one impacting a wide area, can bring the grid down. Two examples are as follows: the test detonation of thermonuclear weapons over South Pacific islands during the 1950s sometimes brought down the grids on the Hawaiian Islands; a single coronal mass ejection from the Sun, if the *emf* pulse is large and it impinges on the Earth, is capable of bringing down a continental grid. While the energy impacting any single power line is small, the collective energy imparted to the entire grid can be significant.

Finally, there are environmental concerns with electric power generation. Two important ones are the release of greenhouse gases resulting in climate change and the temperature increase to waterways from cooled discharges. Both of these issues can be mitigated with increased use of renewable energy generation. A considerable fraction of the generated electricity comes from steam turbines powered either by the burning of fossil fuels or by nuclear reactions. While these utilities are equipped with cooling towers or cooling ponds, the vast amounts of the water discharged back into the environment are usually still at somewhat elevated temperatures, adversely impacting aquatic ecosystems.

It is important to appreciate that all types of electricity generation result in greenhouse gas emissions over the lifecycle (the interval of operational time between replacement) of the generator. For example, energy is used in the manufacture and shipment of generators, turbines, and even in the construction and maintenance of buildings and auxiliary equipment. Consequently, greenhouse gases are created even to produce hydroelectric power. Table 21.1 lists the equivalent CO_2 greenhouse gas emissions per kilowatt-hour of electricity generated ($g_{eq}(CO_2)$/kWh) over the life cycle of various generation technologies. The Intergovernmental Panel on Climate Control (IPCC) compiled these data and presented their findings in a document entitled: "Annex II Methodology." The IPCC analyzed thousands of independent life cycle studies of greenhouse gas emissions, applied selection cuts according to quality of the data, and standardized the metrics of the report. The equivalent greenhouse gas emissions listed in Table 21.1 represents the 50th percentile for generation facilities using that technology. In other words, the 50th percentile represents the median value realized in practice for that particular form of distribution. We took the IPCC's 25th and 75th percentile values as a measure of the range equivalent CO_2 emissions/kWh. *Note*: geothermal energy produces electricity from large temperature differences found in unique locations of the Earth's core/mantle in such areas where molten lava is present near the surface such as found in hot springs. It is common but erroneous to classify heat pumps using long stretches of pipes buried in the soil as geothermal systems.

One additional method to reduce the net greenhouse gases released is to use the waste heat from power plants for other purposes. The process, called cogeneration takes the waste heat and uses it to manufacture small products or to heat nearby buildings. Several large US universities have an on-campus power plant and use the waste heat to keep sidewalks free of ice and snow in winter as well as heat buildings. There was 20,185 TWh of electricity generated worldwide in 2008 with an efficiency of 39 and 61% being waste heat. Only 3% of that energy was used for cogeneration.

INTERESTING TIDBIT TB21.1

In the late 1880s came the War of the Currents between proponents of DC power led by Thomas Edison and AC power advocated by George Westinghouse and by several European companies. Despite a publicity campaign by Edison challenging, the safety of AC HV lines, the lower cost of AC power distribution eventually prevailed. A key player in this War of the Currents was Nikola Tesla, who emigrated from Serbia to the United States in 1884 to work for Edison. Tesla soon struck out on his own, patenting the first practical AC motor and generator designs. Tesla's contributions to alternating current electricity were so important that the SI unit of magnetic flux density (magnetic inductivity) is named after him. Westinghouse Electric & Manufacturing licensed Tesla's patents and even hired him as a consultant. A tipping point in the War of the Currents occurred when generating facilities at Willamette Falls, Oregon, USA, and Niagara Falls, Canada/USA chose AC. A flood destroyed the Willamette Falls DC power station in 1890. That same year the Niagara Falls Power Company was organized and an international commission led by Lord Kelvin was formed to evaluate proposals to harness the power of the falls.

22

PARTICLE ACCELERATORS—ATOM AND PARTICLE SMASHERS

Electrically charged particles can be accelerated via electric fields not unlike the way gravity acts on massive objects. However, two major distinctions between gravity and electrical forces are apparent: first, electrical forces are stronger than gravitational forces by many orders of magnitude, and, second, electrical fields can be generated with time-varying magnetic fields and, conversely, magnetic fields can be induced with changing electric fields. These properties of electromagnetic forces enable the means to generate and manipulate charged-particle beams in ways not possible at present with gravity. Particle accelerators in particular have permitted the production of beams of particles with laboratory energies spanning some 12 orders of magnitude, from 1 MeV (mega-electron volt) to 100,000 TeV (tera-electron volt, i.e., 10^{12} eV or 10^{6} MeV), with a wealth of applications in basic science and technology. One electron volt is the energy that a particle of charge e acquires when accelerated through a potential difference or voltage of 1 V.

In Figure 22.1, we show a Livingston plot of the energy that accelerators have achieved over the past 70 years.

Due to the particle–wave duality of quantum physics, the energy can be related to a corresponding wavelength as was done in Chapter 9. The wavelengths range from 0.04 nm (1 keV electrons) to 10^{-4} fm (10 TeV protons—roughly the maximum center-of-mass energy at the Large Hadron Collider, or LHC). The first number corresponds to a fraction of the size of an atom, while the second is much smaller than the proton radius. Thus, electron microscopes, which operate with keV electrons, can peer at much finer details than optical microscopes (Chapter 13). High-energy particle accelerators like LHC, on the other hand, are the "ultimate microscopes" of modern technology and are capable of exploring the structure of "elementary particles."

Modern Devices: The Simple Physics of Sophisticated Technology, First Edition.
Charles L. Joseph and Santiago Bernal.

Companion website: www.wiley.com/go/joseph/moderndevices

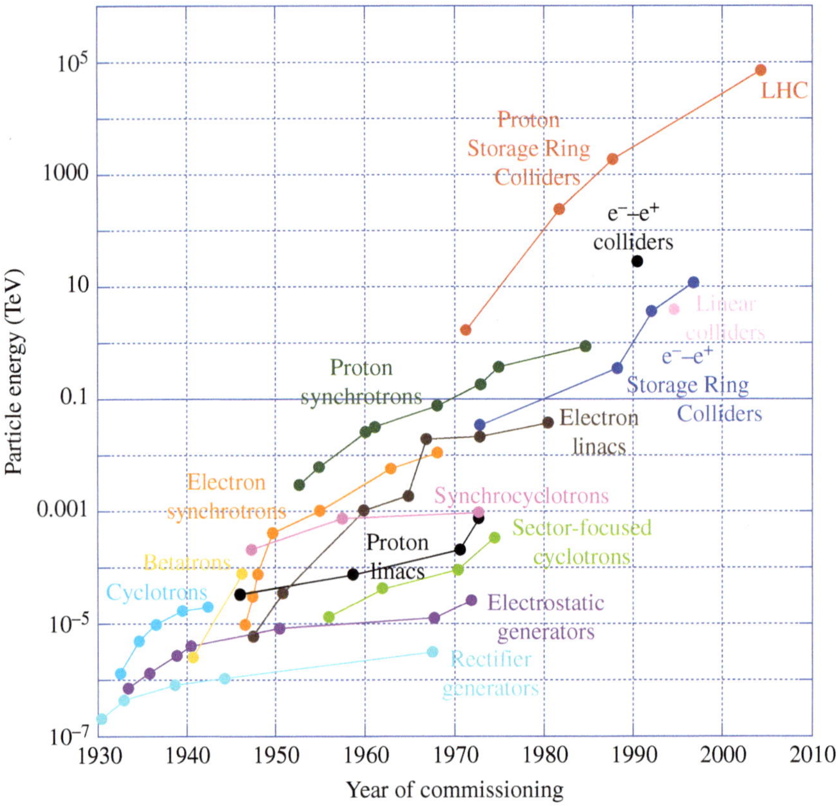

FIGURE 22.1 Evolution of energy of accelerators from 1930 to 2010: Livingston plot of particle energies in the laboratory. For colliders, the energies are normally quoted as "center-of-mass" energies. Source: Adapted from the 2001 Snowmass Accelerator R&D Report.

LHC represents but one application of accelerator technology; thousands of other accelerators operate on a daily basis in a large number of applications that affect us all. Accelerators come in all sizes, from desktop or room-sized to machines of tens of kilometers in length or circumference. The small- and medium-sized accelerators have applications in medicine (e.g., isotope production, sterilization, X-rays, and tumor therapy), materials processing (e.g., ion implantation, hardening of materials, and polymerization), detectors for home security and defense, and research (energy, materials science, nuclear physics, life sciences, and others). The largest accelerators are almost exclusively used for high-energy and nuclear physics research, although some are envisioned for radioactive waste processing and energy production.

The vast majority of particle accelerators operate on electrons to produce radiation from the infrared through ultraviolet and X-rays. Almost all of the rest use protons, but machines do exist that can accelerate ions of almost any element in the periodic table, including radioactive ones. The dental X-ray machines (Section 9.5) use an accelerator to smash electrons of tens of keV energy into a fixed target; the X-rays are produced by the decelerating electrons in a process called *bremsstrahlung.* Radiation can also be produced when the trajectory of a charged particle is bent in a magnetic field or in a series of magnets of alternating polarities (wigglers and undulators); this is the basis of *synchrotron* radiation (SR) of accelerators and free-electron lasers (FELs) which collectively are often called *light sources.* In a different process called *spallation*, copious neutrons can be produced when a very high energy proton beam strikes a mercury target; the neutrons are used in materials research, life sciences and other types of investigations. In yet another type of accelerator, beams of very high energy and circulating in opposite directions collide to produce new particles that shed light on the structure of matter and the origin of elementary particles.

The basic components of any accelerator include the source for ions or electrons, a pipe evacuated to a very low pressure (to eliminate scattering of beam particles with the background gases), accelerating structures, magnets for bending and focusing, injection/extraction systems to guide the particle beam in an out of the accelerator into a target or other accelerator, beam diagnostics to measure key beam characteristics, a cooling system (especially for superconducting components), and a control room where beam diagnostics and other components are monitored. In this chapter, we describe some of the listed components as well as the principles for a number of machines, from direct current (DC) accelerators to radio-frequency (RF) linear accelerators (linacs), cyclotrons and synchrotrons/light sources.

INTERESTING TIDBIT TB22.1

The Chocolate Bar and The H-Bomb—Calories Versus Megatons

The equivalence of mass and energy embodied in the famous equation $E=mc^2$ must be appreciated not only in the context of relativistic kinematics (motion). Every time energy is released in chemical or nuclear processes some mass is converted into energy, but the total energy is conserved. "Heat," for example, is a form of energy (thermal) with a mass equivalent. A 43 g Hershey chocolate bar is advertized as having 210 calories; these are actually 210 kcal using the physics "calorie" energy unit. This energy corresponds to $210\times4184\,\text{J}=8.8\times10^5\,\text{J}$; but the *total* energy content of the bar is $0.043\,\text{kg}\times(3\times10^8\,\text{m/s})^2=3.9\times10^{15}\,\text{J}$! Thus only a very minute fraction of the mass of the bar is converted into energy that the human body can use. On the other hand, an H-bomb with an advertised yield of 10 Megatons would release an energy equivalent to 10 million tons of dynamite, or $4.2\times10^{16}\,\text{J}$, only some 10 times more than the total energy of the chocolate bar! This example illustrates the enormous *total* energy content of even 1 g of mass *at rest.*

22.1 LORENTZ FORCE, DEFLECTION, AND FOCUSING

If gravity acted on masses in the same way that magnetism acts on moving charged particles, objects would not fall in straight lines but would instead spiral down (or even up!) without changing speed, or stay on horizontal circular orbits. Magnetic forces act "sideways" to both the velocity and field directions, following the right-hand rule (RHR): if the fingers of the right-hand curl in the direction connecting the velocity and the field vectors, then the thumb will point in the direction of the force on a positively charged particle. Figure 22.2 illustrates this by showing the effect of a dipole magnet for bending the trajectory of a moving positively charged particle. A similar effect was already shown in Figure FB2.3 to illustrate the action of a uniform magnetic field, or the combined actions of magnetic and electric fields. In the present example, the dipole magnet provides a magnetic field that is uniform only near its center, fading out at the edges. Thus the particle's trajectory is only circular inside and near the center of the dipole but straight outside; in the fringe-field regions, on the other hand, the trajectory's radius is increasing as the $\vec{B}$-field weakens.

The RHR provides also a way to understand the *vector product* of any two vectors $\vec{A}_1$ and $\vec{A}_2$: $\vec{A}_3 = \vec{A}_1 \times \vec{A}_2$, where $\vec{A}_3$ is perpendicular to both $\vec{A}_1$ and $\vec{A}_2$ and in the direction given by the RHR. The *Lorentz force* is a particular case: $\vec{F} = q\vec{v} \times \vec{B}$, where q is the electrical charge of a particle moving with velocity $\vec{v}$ and in the presence of a magnetic field $\vec{B}$.

If the goal is to bring a bundle of trajectories to a focal point, or else, to make the bundle diverge, dipole magnets can also accomplish this through effects from the fringe-fields (Section 22.4.4 and Interesting Tidbit TB22.4). But dipoles are not very effective at doing this; instead, a special configuration of four magnetic poles of alternating polarities, as in Figure 22.3, is employed in most accelerators. Application of the Lorentz force then shows that positively charge particles are deflected in the directions of the arrows if they move initially into the plane of the quadrupole. A bundle of particles moving in initially parallel trajectories would all be affected by the quadrupole field shown in Figure 22.3 in such a way that the trajectories would come closer together in the vertical plane but farther apart in the horizontal plane. Thus, the quadrupole shown in Figure 22.3 is focusing in the vertical plane but defocusing in the horizontal plane; in the language of optics, such a lens is said to be totally *astigmatic* (Section 10.5). A combination of quadrupoles of opposite polarities, therefore, provides

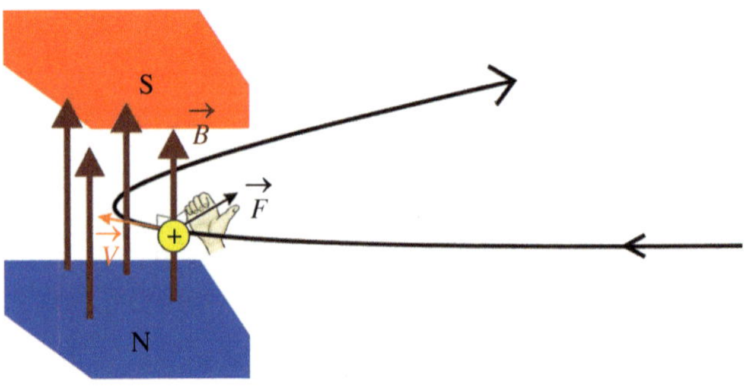

FIGURE 22.2 Right-hand rule and deflection of the trajectory of a positively charged particle by a dipole magnet.

net focusing in both transverse directions. The action of this *alternating gradient* (AG) focusing scheme can be illustrated with a paper straw representing a charged-particle beam: the straw is squeezed in perpendicular directions that alternate along its length; these positions correspond to the location of quadrupoles of opposite polarities. Figure 22.4 is a photo of the "AG" straw.

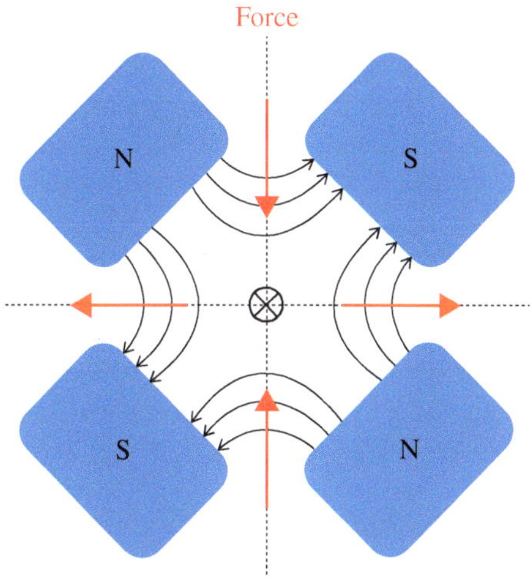

FIGURE 22.3 Quadrupole magnet for vertical focusing of beams of positively charged particles. The arrows indicate the magnetic forces (following the right-hand rule of vector multiplication, as illustrated in Fig. 22.2) acting on particles moving into the plane of the figure.

FIGURE 22.4 Photo of the "alternating gradient" (AG) paper straw. The straw is deformed to illustrate how a charged particle beam is focused/defocused by quadrupoles of alternating polarities.

22.2 BEAM GENERATION, MANIPULATION, AND CHARACTERIZATION

Electron beams are generated in many different ways: thermionic emission, field emission, photoemission, and others, some new ones even involving bundles of carbon nanotubes. The first method is the simplest one; it involves heating a cathode or filament made of a material like tungsten (as in a light bulb) to around 1000°C to boil electrons off its surface. The emitted electrons are then accelerated through an anode plate to a voltage of up to a few tens of kilovolt (naturally, higher voltages are possible as we will see in Section 22.3). This arrangement is called a *diode electron gun*. When a metal mesh or grid is added in front of the cathode to control the electron flow, the configuration is called a *triode gun*. A fourth electrode, called *Pierce electrode* after its inventor, is implemented in high current guns to counteract the effects of internal repulsion of the electrons (space charge effects). Figure 22.5 shows the schematics of a triode electron gun.

The currents produced from thermionic electron guns range from nA to a few 100A, and pulse durations from nanoseconds to DC. Electron guns can be found in cathode ray tubes (CRTs—see Chapter 15), electron microscopes (Chapter 13), klystrons and other microwave tubes (Chapter 2), dental X-ray machines, and many accelerators.

Thermionic cathodes are limited to maximum emitted current densities of around 1.0A/cm^2. In contrast, photoelectric emission from *RF photocathodes*, can yield more than 100A/cm^2. Pulsed RF photoguns are the preferred electron sources for many linear accelerators (linacs) because of their high current and other technical reasons. Linacs are described in Section 22.3.

In most accelerators, beams are deflected and focused with separate elements. Dipole magnets are used for deflection; while quadrupole lenses, based on magnetic or electric fields, are used for focusing. We presented in Section 22.1 the basics of magnetic forces for deflecting and focusing beams. Figure 22.6 shows photos of actual dipole and quadrupole magnets.

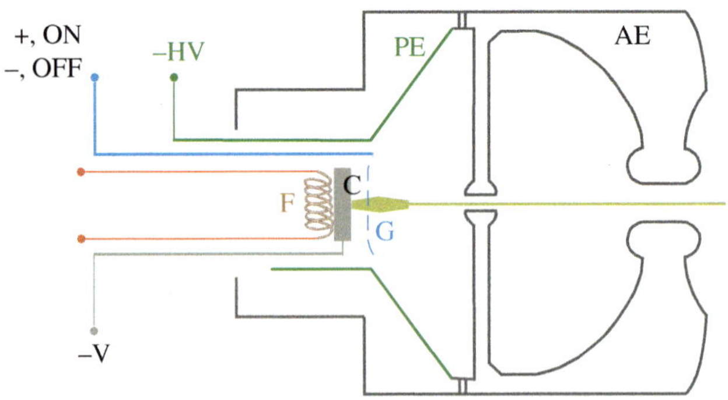

FIGURE 22.5 Thermionic electron source: cross-section schematics of a triode electron gun. AE, anode electrode; C, cathode; F, filament; G, grid; HV, high voltage; PE, Pierce electrode.

(a) (b)

FIGURE 22.6 Photos of accelerator magnets: (a) dipole, and (b) quadrupole. Source: (a) Florian DO, https://commons.wikimedia.org/wiki/File:Hetdipole.jpg. CC public domain. (b) O'Neill, https://commons.wikimedia.org/wiki/File:Aust.-Synchrotron,-Quadrupole-Focusing-Magnet,-14.06.2007.jpg. Used under CC-BY-SA 3.0, https://creativecommons.org/licenses/by-sa/3.0/deed.en

The deflection angle $\Delta\theta$ by a dipole field of field strength ΔB and length L is given by

$$\Delta\theta = \frac{\Delta B \cdot L}{X_m}, \tag{22.1}$$

with X_m giving the *magnetic rigidity* which can be shown to be equal to the ratio p/q, where p is the particle's linear momentum.

Focusing/defocusing of charged particle beams is realized with lenses whereby electrical or magnetic forces, or a combination of both, are used to bend the particle trajectories just as in standard optics. In an ideal linear lens, the force varies linearly from zero at the axis of the lens to a maximum value at the periphery. Typically, magnetic quadrupole lenses, based on either permanent magnets or electromagnets, are employed for focusing of relativistic particles, while electrostatic quadrupoles are often used at low energies for ion beams. The strength of a *magnetic* quadrupole is proportional to the *B-field gradient* "g_{M}" in Tesla/meter (T/m) and effective length L. This strength can be related to the focal length f by

$$\frac{1}{f} = \frac{g_{\mathrm{M}} \cdot L}{X_m}, \tag{22.2}$$

An electrical gradient can be similarly defined. Since the maximum electric and magnetic fields achievable in practice are of the order of 10^7 V/m and 2 T, respectively, the ratio of maximum magnetic to electric gradients can be shown to be proportional to $60v$, where v is the particle's speed. This is why magnetic focusing is preferred at high particle energies.

Focusing by quadrupoles is called *strong focusing*, as opposed to *weak focusing* in machines like cyclotrons (see Section 22.4.4). In the latter case, focusing arises from edge effects from the magnets, which normally provide combined deflection-focusing functions. As mentioned earlier and illustrated in Figure 22.4, strong focusing with quadrupoles arranged with alternating polarities is also called alternating gradient (AG) focusing.

To conclude this section, we mention the parameters typically employed to characterize beams in accelerators. A beam can be characterized by specifying its average or peak current, energy and energy spread, pulse duration and/or total charge, and its overall quality. For colliders, a common beam measure of interest is the *luminosity*, defined as the number of collision events per unit beam cross section. The transverse beam quality, on the other hand, is expressed as the transverse *emittance*. This latter quantity combines measures of the size of the beam cross section (coordinate space) and the spread in divergence angle (transverse velocity space); the combination of the two spaces is called *phase* or *trace space*. Thus, transverse emittance is essentially the product *beam dimension* (horizontal or vertical) × *beam divergence* (horizontal or vertical); the *beam brightness* (not to be confused with the brightness of SR—Section 22.6), is the particle density in *phase space*. Since emittance is a measure of the area in phase space, brightness is the number of particles divided by the emittance.

22.3 DC ACCELERATORS

A high-voltage power supply, which consists of an AC transformer and a rectifier circuit, can in principle provide the accelerating voltage for electrons or ions in an accelerator. However, electrical breakdown limits this simple approach to a few tens of keV. To overcome this restriction, a cascade of rectifier systems can be employed to multiply the voltage several times, as in the Cockcroft–Walton accelerator dating back to the 1930s. Figure 22.7 shows the 1.6 MV Cockcroft–Walton accelerator at Clarendon Lab, Oxford University.

Above 5 MeV or thereabouts, direct mechanical transfer of charge is used to achieve accelerating energies of tens of MeV. The best known example in this category is the Van de Graaf accelerator. DC accelerators are the most commonly used in industry. Examples of these machines are the ion-implantation accelerators of the semiconductor industry, small neutron generators for the oil industry, and electron systems for materials processing and sterilization of medical products.

22.4 RF LINEAR ACCELERATORS

22.4.1 Motivation and History

The final energy of a particle in a DC linear accelerator cannot exceed the product eV where V is the maximum DC voltage applied. RF accelerators, on the other hand, make possible the repeated application of electric fields to achieve voltage gains that greatly exceed the maximum applied voltage. Electromagnetic waves in vacuum, however,

FIGURE 22.7 Cockcroft–Walton accelerator at Clarendon Lab, Oxford University in 1948. Source: Chetvorno https://upload.wikimedia.org/wikipedia/commons/8/8e/Cockcroft-Walton_accelerator_Clarendon_Lab_Oxford.jpg. CC Public Domain.

cannot be tapped to accelerate particles because the electric fields are perpendicular to the wave propagation and are also continually changing polarity. To solve both problems, special metallic structures called *resonant cavities* are used to produce longitudinal electric fields that not only have the right polarity but also keep up with the motion of the accelerated particles. See Advanced Concept AC22.1 for additional explanation.

RF linear accelerators (RF linacs) can be as short as 1 m and as long as a few kilometer, ranging in cost from around \$1M to \$1B. The first successful RF linac was built by Wideröe in 1928; his design evolved soon into what is known today as the Alvarez *drift-tube linac* ("DTL" for short) or Alvarez DTL after his inventor. The RF technology before World War II (WWII) allowed the acceleration of relatively heavy ions like mercury or potassium to energies of tens of keV, but the acceleration of protons and electrons required much higher frequencies and power. The fast development of RF technology during WWII, especially in connection to *radar*, made possible the construction of proton and electron RF linacs. The first high-power microwave source, the *klystron*, was a major step in this direction. Invented in 1937 by the Varian brothers at Stanford University, the first klystron operated at 3000 MHz, a frequency still in use for electron linacs. For the heavier protons, the frequencies needed are around 200 MHz.

22.4.2 Linac Components and Operation

The main components of a linac are DC particle injector, RF power system (e.g., klystrons), accelerating cavities, focusing magnets, vacuum system, and cooling system. The latter is needed to remove the heat from the RF power dissipated on the cavities; water is used for cooling normal metal (typically copper) cavity structures. In the case of superconducting RF (SRF) cavities and magnets, liquid helium is needed to sustain the superconducting state. Figure 22.8 shows the schematics of an iris-loaded waveguide of the type used at the 3-km electron linac facility at the Stanford Linear Accelerator Center (SLAC). The RF power at SLAC is applied as a 2856 MHz *traveling wave* in such a way that the electrons are moving in synchronicity with the wave and very rapidly achieve a velocity very close to the speed of light. At the final energy of near 50 GeV (before the present-day modifications—see Section 22.6), the electrons move essentially at the speed of light. In contrast, the LANSCE proton linac at Los Alamos National Laboratory is based on a 200 MHz, 62-m DTL structure for initial acceleration to 100 MeV, and an 800 MHz, 731-m side-coupled linac (SCL) structure for final acceleration to 800 MeV (at this energy the protons move at 84% of the speed of light.) The SCL structures use a standing RF wave pattern that shortens the overall length of the accelerator by moving the field nulls outside the beamline. The same scheme is employed in many electron linacs for clinical use such as X-ray radiation therapy; these linacs operate in the energy range 4–35 MeV and are only a few meters, or shorter, in length.

Linacs require many RF cavities to achieve high energies because the field strengths per cavity are limited to a few MeV per meter. This fact explains why linacs designed for the highest energies are very long machines. To compound the cost, the power consumption can be very demanding for copper cavities, of the order of 100 MW, overall, for LANSCE, for example. SRF cavities, under development since the 1960s, allow much higher field strengths per cavity (up to about 30 MV/m) and also consume much less RF power. However, the technical

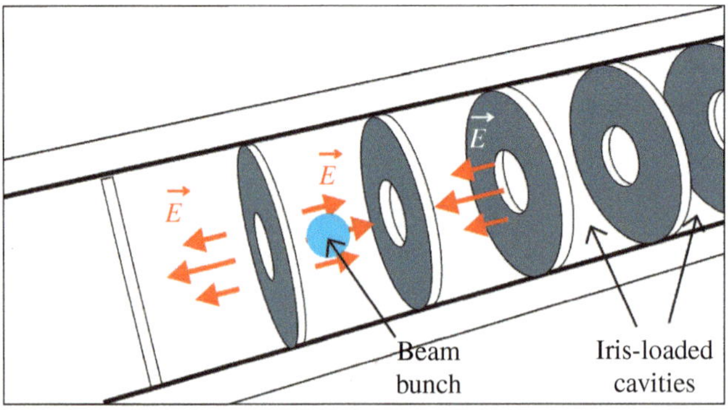

FIGURE 22.8 Schematics of iris-loaded structure for RF acceleration: The arrows indicate the direction of the electric field at an instant when the electron beam bunch (blob) is passing. The RF power source is not shown.

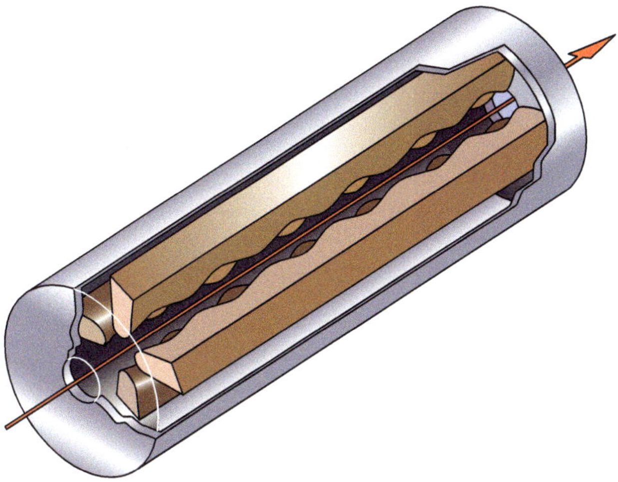

FIGURE 22.9 Schematics of radio-frequency quadrupole (RFQ). Source: Cover of Wangler (2008). Reproduced with permission of John Wiley & Sons, Inc.

benefits of superconducting technology have to be weighed against the economics of the additional hardware needed to maintain the low temperatures that SRF demands.

Unlike electron linacs, modern industrial ion linacs rely on the RF quadrupole (RFQ) and operate at frequencies 100–600 MHz. The RFQ, which became practical in the 1970s, permits the simultaneous acceleration and focusing of high current beams in a compact structure. The RFQ consists of four vanes in a cavity, as shown in Figure 22.9; the vanes provide quadrupole focusing (Section 22.1) and also acceleration through longitudinal electric fields induced by the special modulation of the vane tips.

In a typical electron linac, the electrons from a DC thermionic source or a photocathode gun (Section 22.1) are injected into the RF accelerating structure. The first few RF cavities capture the low-velocity electrons and accelerate them to velocities close to the speed of light; at the same time, the electrons tend to be grouped in *bunches*. The geometry of the main accelerating cavities that follow is designed for acceleration of relativistic electrons. In the *traveling-wave* linac, for example, the electrons ride a traveling EM wave (whose *phase velocity* is essentially equal to c) as a surfer rides a sea wave. The cavity structure, as in Figure 22.8, is periodic and is equivalent to a series of coupled oscillators that can sustain special field patterns called *modes* if the frequency of the RF is above a certain *cutoff frequency* that depends on the cell geometry. (See Advanced Concept AC22.1 for more information.) In the *stationary-wave* type of linac, the wavelength of a longitudinal E-field mode is equal to an integer multiple of the iris separation; this is shown in Figure 22.10 for the $2\pi/3$ mode where the wavelength is equal to three times the iris separation. In the traveling-wave linac, on the other hand, the *phase advance per cell* of the RF wave is specified (e.g., $2\pi/3$ for the linac at SLAC).

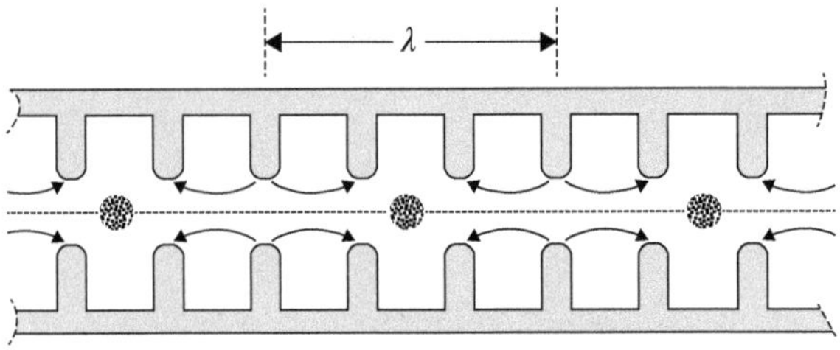

FIGURE 22.10 Field variation of the $2\pi/3$ mode.

22.4.3 Beam Bunch Stability and RF Bucket

A crucial issue for linac operation is the beam bunch *stability*. Typically, the electron bunch is injected slightly ahead of the peak RF field. Therefore, and as shown in Figure 22.11a, the particles see a quasi-linear restoring force if the bunch length is small compared to the RF wavelength: the particles that arrive early at the accelerating cavity are given a smaller RF kick than those that arrive late, but those that arrive with just the right phase ϕ_S, the so-called isochronous particles, are given the same kick every time. Therefore, particles inside the bunch undergo longitudinal oscillations not unlike a pendulum; these oscillations are called *synchrotron* oscillations. The analogy with the pendulum can be carried out further if we picture a biased pendulum, that is, a pendulum whose equilibrium position is off from the vertical by an angle ϕ_S. The pendulum will be stable for oscillation angles that do not depart too much from ϕ_S. In fact, the phase space of the biased pendulum which is formed by the coordinates $(\dot{\phi},\phi)$ is mathematically equivalent to the phase space $(\delta U,\phi)$ of the bunch particles, where δU is the energy error, that is, the energy deviation from the energy of the synchronous particles.

In Figure 22.11b, we have plotted δU versus ϕ for $\phi_S = 60°$ and five values of ϕ_0 (0°, 20°, 30°, 40°, and 50°). These values of the initial phase correspond to five values of particle's energy; the closer the initial phase is to $\phi_S = 60°$, the closer the oscillations are to pure harmonic oscillations. If the energy (initial phase) deviates too much, on the other hand, the particle motion will be unbound, as illustrated by the open curves in Figure 22.11b. The stability of particles in the bunch then depends on capturing particles with right velocities inside a region of the longitudinal RF field; the corresponding region in phase space is called the *RF bucket* and is bounded by the *separatrix* curve (the curve labeled $\phi_0 = 120°$ in Fig. 22.11b). The RF bucket contains a stable bunch of particles called a *micropulse*. If the RF is pulsed, then the group of beam bunches per pulse is called a *macropulse*.

22.4.4 Power Budget and Linac Applications

The RF power budget in a linac is as important as the stability just discussed. Ideally, all the RF power would be transmitted to the beam, but this is not possible as significant losses occur at the cavity walls. The power loss per unit length P_w is related to the peak accelerating field E_0 through a quantity called the *shunt impedance* per

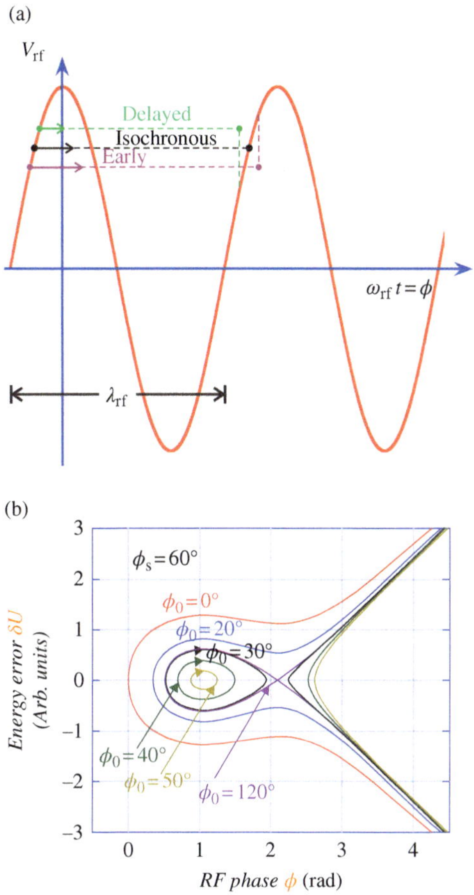

FIGURE 22.11 Phase stability and RF bucket: (a) principle of phase stability in a linac, (b) longitudinal phase space and RF bucket. Particles inside closed curves are stable. S. Bernal, 2016.

unit length, $r_s : P_w = E_0^2 / r_s$. As the shunt impedance increases with frequency for normal conducting cavities, higher operating RF frequencies are desired. But other important considerations that depend on the type of particle bunch desired may favor lower frequencies. Thus, the main design parameter in a linac is the frequency and is normally chosen as a compromise between power budget considerations and desired beam characteristics.

RF linacs have several advantages over circular accelerators: (i) there are no energy losses to *SR* (see Section 22.6), except possibly at special devices called "chicanes" where the beams are manipulated to compress the bunch duration; (ii) linacs can also operate with beam bunches of any duration, from DC (i.e., CW operation) to pulsed, so the average beam current can be much higher than in circular machines of comparable energy; (iii) the tolerances of magnets, RF cavities, and other devices are not as stringent as in circular machines because linacs are *single-pass* devices; (iv) because of their single-pass nature, linacs do not suffer as severely as circular accelerators from destructive phenomena such as beam resonances; and, finally, (v) linacs do not require complex optics for injection or extraction as in circular machines.

INTERESTING TIDBIT TB22.2

SLAC and the Discovery of Quarks

The particles called quarks were first proposed in 1964 by Gell-Mann and Zweig to model the structure of hadrons, that is, particles such as protons and neutrons (nucleons) and others that are subject to the "strong" force, the force that holds atomic nuclei together. A series of experiments between 1967 and 1973 at the SLAC in Menlo Park, California (now SLAC National Accelerator Laboratory) played a major role in the discovery of quarks. Studies by Richard Taylor and coworkers at SLAC and MIT led them to conclude that the protons and neutrons were composite particles. The inelastic scattering of giga-electron volt electrons off protons and bound neutrons on a fixed target revealed that scattering was greater than expected at large angles, strongly suggesting the presence of a hard core inside the nucleons. The situation paralleled the discovery of the nucleus by Rutherford, many years before, in scattering experiments of alpha particles off gold atoms in a thin foil.

Electron and ion linacs are used in many areas of basic science and technology: for basic research in high energy and particle physics (see Interesting Tidbit TB22.2 on the discovery of the quark at SLAC); to provide high- quality beams for free-electron lasers (see Section 22.6); as the acceleration stage for beams that are injected into accumulator rings; and to provide MeV electron beams for X-ray production for radiation therapy.

22.5 CYCLOTRONS

All circular accelerators have descended from the cyclotron. It was invented in 1929 by Professor Ernest O. Lawrence of the University of California, Berkeley, and the first working model was constructed in 1931 by his graduate student, M. Stanley Livingston. The internal cavity of the first cyclotron measured a mere 4 inches in diameter. Protons were injected to the center, spiraled in the cylindrical cavity between two magnet plates, and achieved ion energies of only 80 keV. Figure 22.12 shows a 1939 photograph of a 60-inch cyclotron at Berkeley, CA. Work with the machine led to the discoveries of the elements neptunium (McMillan) and plutonium (Seaborg). Seaborg and McMillan shared the Nobel Prize in 1951.

The cyclotron was the first machine to utilize a magnetic field to iteratively return ions to the accelerating gap, thereby simplifying the high voltage system necessary for accelerating to high energies. The cyclotron gains its success from two fundamental

FIGURE 22.12 Photograph of the 60-inch cyclotron at the University of California Lawrence Radiation Laboratory, Berkeley, in August, 1939. Source: Department of Energy. Office of Public Affairs, https://commons.wikimedia.org/wiki/File:Berkeley_60-inch_cyclotron.jpg. CC public domain.

physics principles. The first principle states that there can be no electric fields inside hollow conductors, such as the region deep inside a metallic pipe or can. Second, an ion traveling in a static magnetic field will execute a circular path orthogonal to the field lines, the diameter of the circular path being proportional to the particle's speed.

As a young Berkeley professor, Lawrence was flipping through a German physics journal and unable to read German, he simply looked at the figures and diagrams. One article by Rolf Wideröe, schematically described an idea for a *linear* accelerator with a simple operation. An ion would start from rest and be accelerated toward the end of a hollow metallic tube held at a negative voltage. Upon entering the tube, the ion will have gained energy qV, where q is the charge and V is the tube's voltage. Once the ion has traveled the length of the tube segment, the voltage is quickly reversed such that the exiting ion is now accelerated away, gaining another qV and doubling the final energy to $2qV$. For a practical accelerator, Wideröe's first tube would be followed by a string of many such tubes, synchronously flipping polarity to bring the ion to very high voltages. Ernest Lawrence immediately understood this from a single diagram in the *Archive for Electrotechnique*. He also understood the technical limitation of his day, and that for this type of accelerator to be useful it would have to be impractically long. Lawrence envisioned using a strong magnetic field to coil up the ion path into a compact accelerator. He equated the confining magnetic Lorentz force (see Section 22.1) to the centripetal force of an ion with mass m, charge q, and velocity υ circulating in a magnetic field B: $\vec{F}_{\text{magnetic}} = \vec{F}_{\text{centripetal}}$. Since the force is orthogonal to the direction of travel, we drop the vector notation and write $|\vec{B}| = B$

$$q\upsilon B = \frac{m\upsilon^2}{r}, \tag{22.3}$$

which can be simplified to

$$qB = \frac{m\upsilon}{r}. \tag{22.4}$$

Since the velocity υ can be expressed in terms of an angular frequency, ω (rad/s), by $\upsilon = \omega r$, we can now write

$$\omega = \frac{qB}{m}. \tag{22.5}$$

This is known as the *cyclotron frequency*, which remarkably is independent of radius. An ion of higher velocity maps out a larger radius, thus covering the greater path length with a faster velocity such that the circulation time is always the same. Lawrence correctly surmised that he could insert two hollow electrodes, each the shape of the letter D back to back, with alternating voltage across them into a magnetic field. (Each D-shaped hollow electrode is known as a DEE.) Adjusting the magnetic field and frequency of the alternating voltage to satisfy the cyclotron equation a resonant acceleration would occur. Every cycle of the applied voltage would extract a cluster of ions from the central ion source and iteratively accelerate them through an ever-increasing spiral path until they exited the chamber or hit an internal target. This accelerator would produce a continuous stream of bursts of particles at the revolution frequency. Lawrence named this technique magnetic resonance acceleration, but laboratory banter prevailed and "cyclotron" became the universally accepted name.

Confident from their success, Lawrence and Livingston proceeded with a larger cyclotron that aimed for the 1 million volt barrier (1 MeV), the voltage believed necessary to penetrate and peer into the atom's nucleus. The following year, the pair had succeeded with their 11-inch cyclotron! Cyclotron development continued at a rapid pace in Lawrence's Radiation Lab: achieving higher and higher energies with ever more massive cyclotrons. In 1937, a 27-inch cyclotron produced 5.5 MeV Deuterons; by 1938 a 37-inch cyclotron produced 8.0 MeV Deuterons. The year that Ernest Lawrence won the Nobel Prize, 1939, his current cyclotron, which utilized a 60-inch magnet (Fig. 22.12), produced 16 MeV deuterons. Construction was underway of a monstrous 184-inch magnet with the hopes of achieving ion energies in excess of 100 MeV. However, theoretical physicist Hans Bethe had predicted that the upper limit that a cyclotron could achieve was 16 MeV for protons, and a little higher for Deuterons. Bethe's prediction stemmed from Einstein theory of special relativity, in that as the particles gained energy, their effective mass would increase, which would affect the cyclotron condition, namely that the ions' revolution frequency, $\omega = qB/m$, would fall out of step with the alternating voltage applied to the DEEs. In other words, the ions' revolution time would develop a dependence on radius. If Bethe was correct, the 184-inch cyclotron would be a colossal failure.

Simultaneously, the 184-inch cyclotron's construction was diverted for use in WWII's Manhattan Project for the separation of U^{235}. During the intervening war years, a solution to the relativistic mass increase surfaced: adjust the applied frequency to keep in step with the increasing particle mass. This relativistic cyclotron became known as the frequency modulated cyclotron or the synchrocyclotron. While this was a successful solution to relativity, this would require a technical compromise in the reduction of the beam intensity: instead of every cycle containing ions, only the RF cycle that was in step with the magnetic field would successfully carry beam to the periphery. Consequently, the beam now exited the accelerator at the modulation frequency, at most a couple hundred pulses per second, as compared to the conventional cyclotron's millions of pulses per second.

Another solution to the synchronization problem was proposed by L. Thomas in 1938, but not fully appreciated until the 1950s. Thomas suggested azimuthally varying the magnetic field, with alternating high and low regions in such a way that the ions orbital path length would change with radius, allowing the revolution time to remain constant, or maintained isochronism. This method regained the continuous beam structure and returned to the simpler RF accelerating system, but it required very sophisticated precise magnetic field shaping. Today, all three types of cyclotron are in operation, the Thomas cyclotron being the workhorse.

Another important aspect of the physics of cyclotrons is the transverse confinement of the beam. An ideal vertical B-field, perfectly uniform across the pole faces of the cyclotron magnet (Fig. 22.13) will not work because the beam will not be

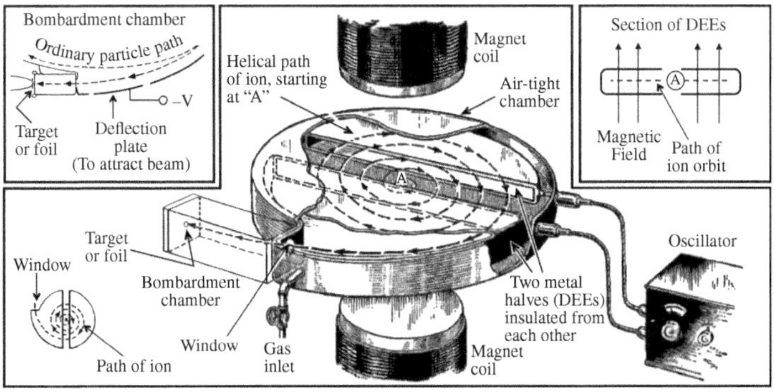

FIGURE 22.13 Basic components of a cyclotron accelerator: magnet poles, accelerating cavities (DEEs), RF voltage (oscillator), ion source (A) and extraction. Source: https://commons.wikimedia.org/wiki/File:Cyclotron_diagram.png. CC public domain.

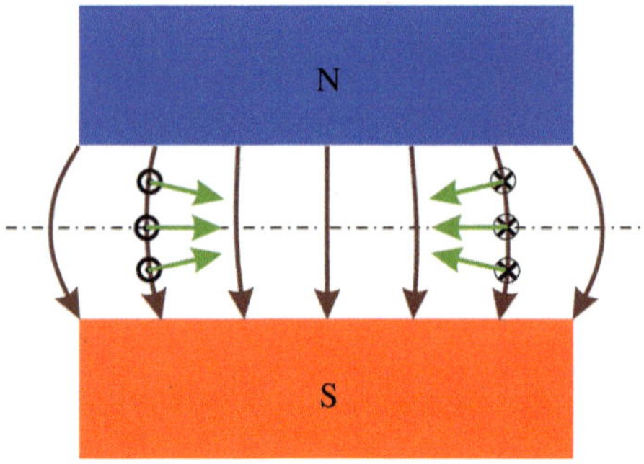

FIGURE 22.14 Radial confinement of a beam in a cyclotron: positively charged particles move into the plane of the paper on the right and out of the plane of the paper on the left; the forces are indicated at three vertical positions on both sides.

confined. This was soon realized during testing of the first cyclotrons. (See Interesting Tidbit TB22.3.) In fact, the vertical component of the B-field must decrease along the radial direction of the machine, but such that this variation does not exceed B/r, where B, r enter the definition of cyclotron frequency (Eqs. 22.4 and 22.5). When these conditions are met, the beam particles follow stable orbits along both radial and vertical directions. Figure 22.14 shows a cross section of a possible field configuration. The resulting transverse focusing of the beam corresponds to what is known as *weak focusing*, as opposed to the *strong focusing* of quadrupoles described in Section 22.1.

Cyclotrons have many applications. They were originally the exploratory machines of high energy physics, but they have been succeeded by synchrotrons, such as the Tevatron and the LHC. Presently, cyclotrons are used in the healing arts: in generating isotopes for diagnostics and therapy, and for proton and ion-beam therapy. The latter application relies in the way high energy protons or ions are absorbed by living tissue: contrary to photons, such as hard X-rays, which are absorbed more strongly near the tissue surface, protons or ions deliver their energy deep inside and in a single strong "kick," the so-called Bragg peak. Figure 22.15 illustrates the relative dose of photons, protons and ions versus the penetration depth in tissue for the hypothetical treatment of a deep tumor; the advantage of using protons or ions is obvious as they affect the surrounding tissue less than photons.

To conclude this section, we mention a close relative of the cyclotron that was invented in the 1950s and soon abandoned partly due to mid-twentieth-century technical limitations, but revived in the past few years. This "new" accelerator is the fixed-field AG accelerator (FFAG). These machines combine the advantages of the fast beam repetition rates of cyclotrons and the smaller magnets of synchrotrons, although FFAG magnets can be fairly complex. FFAGs are envisioned as accelerators for short-lived particles such as muons and also as machines for industrial irradiation, proton therapy and other applications.

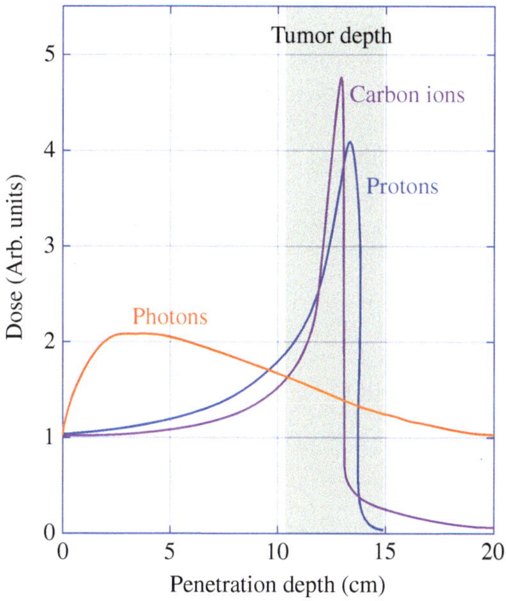

FIGURE 22.15 Photons versus protons and carbon ions for cancer treatment. The Bragg peak displayed by the absorption of energetic particles (protons or carbon ions) is the basis for using cyclotron-based therapy.

INTERESTING TIDBIT TB22.3

Theory Said It Should Work!

Milton Stanley Livingston meticulously soldered fine wires to span the open face off the electrode he called a DEE because of its letter D shape, in an attempt to smooth out the electric field. Still, there was no measurable beam. He then painstakingly placed thin slivers of soft iron between the magnet's pole faces and the vacuum chamber lid, shimming the magnet into creating a perfectly uniform field. Weeks of this tedious work left him frustrated and without beam—would his thesis project be a failure? He complained to Ernest Lawrence, the inventor of the cyclotron idea and Livingston's thesis advisor, but he insisted that it must work, and that all that Livingston needed to do was to do a better job shimming the magnetic field and tailoring a smooth electric field—just be patient. Ernest Lawrence left for vacation, but Livingston dutifully continued working in the lab until one evening, out of complete frustration he ripped all the fine wires off the face of the DEE and pulled out all the iron shims. He replaced the chamber between the magnet poles, turned on the vacuum pumps, the high-voltage oscillator, and ion source as done so many times before. This time, however, there instantly was a measurable beam! On December 1, 1930, Livingston made the cyclotron work for the first time. Very proud of his discovery, Livingston excitedly reported his perplexing findings to Lawrence. Lawrence immediately said "It is obvious what is going on: removing the fine wires causes a distortion to the electric field that focuses the ion back to the mid-plane. Similarly the removal of the iron shims causes a focusing force to the circulating ions." By removing the wires and shims, Livingston had introduced elements of focusing that would push ions that wandered away back to the median place of acceleration (see Fig. 22.14). These elements of focusing are perhaps the most importantly studied features of today's accelerators.

22.6 SYNCHROTRON RADIATION AND LIGHT SOURCES

22.6.1 Dipole Radiation and Larmor's Formula

Charged particles radiate electromagnetic waves when accelerated. *Dipole radiation*, for example, results from the back-and-forth motion of electrons in an antenna driven by an applied harmonic voltage $V(t) = \sin(\omega t)$. Figure 22.16 illustrates the progression of the radiation field pattern from an oscillating charge over one period. Note from Figure 22.16 that the maximum intensity of the radiation field occurs in the equatorial plane, that is, the plane perpendicular to the oscillating current, where the electric field lines are more densely concentrated. Another example is the microwave radiation produced by the very rapid and complicated oscillatory motion of electrons in the magnetron tubes of microwave ovens. (See Chapter 2.)

The force that accelerates a charged particle can be resolved into two components: one along and another one perpendicular to the instantaneous velocity direction. The total power radiated by a charged particle is greater from transverse than from longitudinal acceleration by a factor $\gamma^2 = 1/(1-\beta^2)$, where β is the particle's speed in units of the speed of light c. Electrons with a kinetic energy of 10 MeV, for example, have $\gamma^2 = 423$. Thus, to generate copious EM radiation, it is more efficient to bend the trajectory of an energetic charged particle than to simply push it along a straight line. The radiation generated from transverse acceleration is called *synchrotron radiation*

The total instantaneous power (in watts) radiated by an electron (electrical charge $-e$) with acceleration $a(t)$ is given by *Larmor's formula*:

$$P(t) = \frac{e^2 a^2(t)\gamma^4}{6\pi\varepsilon_0 c^3}. \tag{22.6}$$

If an electron of total energy E moves on a circular orbit of radius ρ under the action of a uniform magnetic field B, Equation 22.6 can be written in other useful forms:

$$P_R(t) = \frac{e^2 c}{6\pi\varepsilon_0}\frac{\beta^4\gamma^4}{\rho^2} = \frac{e^4}{6\pi\varepsilon_0 m_e^4 c^5}E^2B^2. \tag{22.7}$$

Equation 22.7 displays the scaling of SR power with bending radius ($P \propto 1/\rho^2$), rest mass ($P \propto 1/m^4$), and magnetic field of bending dipole ($P \propto B^2$).

E-field

B-field

Current i

0	$T/4$	$T/2$	$3T/4$	T
0	E_{max}	0	$-E_{max}$	0
B_{max}	0	$-B_{max}$	0	B_{max}
i_{max}	0	$-i_{max}$	0	i_{max}

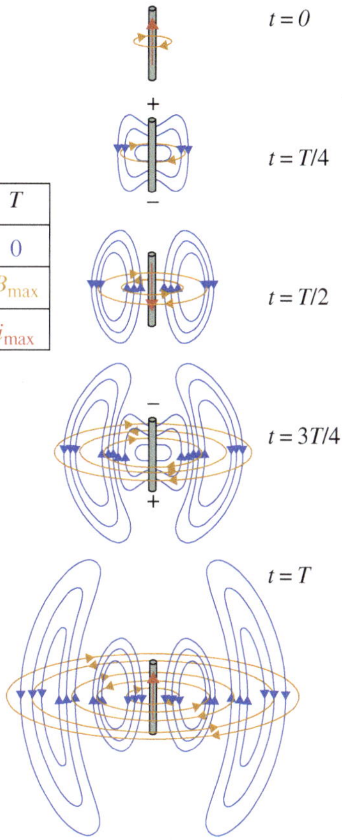

FIGURE 22.16 Radiation pattern from an oscillating electric dipole: the electric and magnetic near-fields are shown at various instances over one period "T" of the oscillation. The latter is driven by an external (not shown) AC power supply. The table illustrates the field magnitudes and current near the dipole. Source: ChemgaPedia, Elektromagnetische Schwingungen und Wellen, http://www.chemgapedia.de/vsengine/vlu/vsc/de/ph/14/ep/einfuehrung/emwellen/alles.vlu/Page/vsc/de/ph/14/ep/einfuehrung/emwellen/dipol3_abstrahlung.vscml.html. 2014 Copyright Wiley-VCH Verlag GmbH & Co. KGaA. Reproduced with permission.

The energy lost to SR per turn for electrons of total energy E (in GeV) in a field B (in Tesla) can be found from Equation 22.7:

$$\frac{\Delta E}{\text{turn}}(\text{keV}) = 88.5 \times \frac{E^4}{\rho} = 26.6 \times E^3 B, \tag{22.8}$$

where the bending radius ρ is given in meters. This energy has to be replenished by RF accelerating cavities to keep the electrons circulating with the same radius ρ. This shows the limitations of electron machines for achieving high energies. However, SR is in itself the reason for building electron machines as "light sources," as described in more detail later.

(a)

(b)

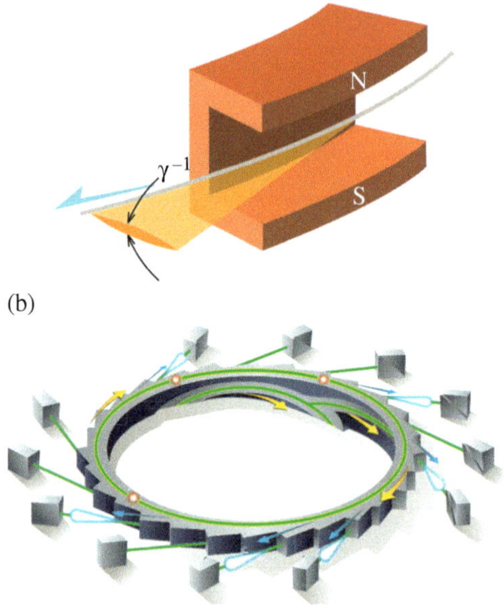

FIGURE 22.17 Angular distribution of synchrotron radiation power: (a) "searchlight" cone with vertical opening angle γ^{-1}; (b) schematics of beam stations around synchrotron accelerator. Source: Malamud (2013). Courtesy of American Physical Society, Division of Physics of Beams.

The radiation pattern of SR from ultra-relativistic particles forms a narrow beam cone with a vertical opening angle given by γ^{-1}, as shown in Figure 22.17. As an example, for electrons at 511 MeV the angle is 1 mrad, or 3.4 minutes of arc.

22.6.2 Wigglers and Undulators

SR can be obtained from bending by single magnets in circular machines as illustrated in Figure 22.17, or by using a combination of magnets of alternating polarities in insertion devices called *wigglers* (Fig. 22.18) and *undulators*. The bend angles from individual magnets in wigglers are large compared to γ^{-1}. In contrast, bend angles in undulators are of the same order as γ^{-1}. For bending magnet and wiggler sources, the spectrum of SR is continuous; half the power is radiated above and half is radiated below a *critical photon energy* E_C. We show an example of a continuous SR spectrum and the location of the critical photon energy in Figure 22.19. In undulators, interference effects yield a discrete spectrum, that is, a spectrum formed by a series of sharp peaks at certain wavelengths.

The figure of merit in many applications of SR is the *spectral brightness* or *brightness* for short (also called *brilliance* in Europe). Brightness is defined as the number of photons emitted by the SR source per unit time, per unit solid angle, per unit area at the source, and per unit bandwidth around a given frequency:

$$\text{Brightness} = \text{Photons} / \left(\text{s} \cdot \text{mm}^2 \cdot \text{mrad}^2 \cdot \text{BW}\right) \tag{22.9}$$

FIGURE 22.18 Schematics of a wiggler magnet.

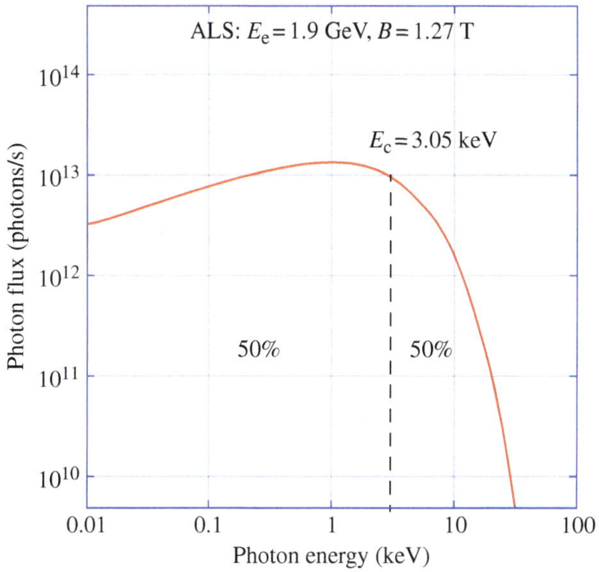

FIGURE 22.19 Spectrum of synchrotron radiation at the Advanced Light Source (Berkeley Lab): the spectrum is continuous, with critical photon energy of 3.05 keV. Note the log–log scale.

(The concept of solid angle and its units is presented in Advanced Concept AC10.3 near the end of Chapter 10.)

Another important property of SR radiation is the degree of *coherence*. The SR emitted by individual electrons in single magnets and wigglers adds *incoherently*: if the bunch contains N particles, the total intensity (per unit frequency or *spectral bandwidth*) is simply N times the intensity from an individual particle. In contrast, if the electron trajectories at a given pole are all in phase, the resulting intensity of SR will be proportional to N^2. This is the type of radiation generated from undulators; it is called *coherent synchrotron radiation* (CSR). Although SR from wigglers is incoherent, they yield more intense SR than single-bend magnets because they are longer and involve multiple bends. Further, since the number of particles in a typical bunch can be very large (10^{10} for a 1.6 nC bunch, for example), the intensity of a coherent source can be many orders of magnitude larger than the intensity of an incoherent one. In reality, the SR intensity always has incoherent and coherent components; their ratio is intimately related to the *structure* of the beam bunch and the geometry of the sources.

22.6.3 First-to-Fourth Generations of Light Sources and Applications of SR

SR sources, now called more generally "light sources," are based on storage rings and linear structures that can provide much more intense UV- and X-rays than conventional lamps and tubes; their dimensions range from a few meters in diameter for sources with 100–300 MeV electron beams to machines with kilometers dimensions and GeV beams. The *first-generation* light sources were built with high energy physics research in mind, but the parallel use of many of these machines as SR sources developed over time. *Second-generation* light sources, by contrast, were dedicated SR sources from the start; their use for multi-disciplinary research often involving multiple beam lines as shown in Figure 22.17b. An example of a second-generation light source is the 2 GeV SRS machine at Daresbury, UK. *Third-generation* light sources are storage rings optimized for high brightness (see Eq. 22.9) and high beam quality (low emittance—see end of Section 22.2) through the use of insertion devices like undulators. Examples are the 1.5 GeV Advanced Light Source (ALS) in Berkeley, CA, and the 3.0 GeV Diamond in Daresbury, UK. The smallest machines of this type produce SR with photon energies below 2 keV (soft X-rays), while the largest ones are designed for hard X-rays.

Finally, *fourth-generation* light sources can be defined as those that exceed third-generation light sources by one order of magnitude or more in a parameter like brightness, coherence, or (shorter) pulse duration. Fourth-generation light sources include storage rings with ultra-low emittance and photoinjector, linac-based free electron lasers (FELs) that produce VUV- to hard X-rays in ultra-short bursts (picosecond and femtosecond). Examples of these advanced SR sources are the FLASH FEL facility at DESY in Germany, the European X-ray FEL facility (XFEL) in Hamburg, and the LCLS at SLAC in the United States.

LCLS utilizes one-third (1-km) of the SLAC linac to accelerate electrons to 14.3 GeV; the 112-m-long undulator produces coherent X-rays at photon energies between 0.8 and 8.0 keV corresponding to wavelengths between 1.5 and 0.15 nm. The average brightness is 4.5×10^{22} photons/(s mrad^2 mm^2 0.1% bandwidth).

SR has broad application to basic and applied sciences. In basic science, the broad spectrum (from microwaves to hard X-rays), high brightness, flux, stability, polarization, and other properties of SR make it ideal for studies in materials science, chemistry, and biology via diffraction, spectroscopy, scattering, fluorescence, and other processes. The largest industrial use of SR is protein crystallography through X-ray diffraction for the development of drugs. Other important industrial uses are metallo-organic chemical vapor deposition (MOCVD) and lithography for the fabrication of semiconductor (e.g., sensors and lasers) and micromechanical (MEM) parts. Another interesting application of SR X-rays is in archeology where fluorescence imaging can be used to reveal original manuscripts that were hidden by other works. (See Interesting Tidbit TB22.5.)

Very recently, ultra-fast X-rays snapshots are being used to study protein folding and unfolding, an area of fundamental importance in molecular biology for the understanding of cell mutation and disease. X-ray FELs, to be described in some detail in Section 22.6.4, are envisioned as tools for controlling and real-time imaging of chemical reactions. Finally, SR from fourth-generation light sources is being employed for investigations of materials under extreme conditions, the "warm dense matter" (WDM) regime, with implications for plasmas in astrophysics and nuclear fusion.

INTERESTING TIDBIT TB22.4

From Horses to Molecules and the Advent of Ultra-Fast Imaging

In 1877, Eadweard Muybridge, renowned English photographer, was asked to settle the question of whether or not all four hooves of a racehorse left the ground while running. He not only answered the question with a single photograph but also went on to study the horse motion in more detail by setting a number of cameras along the race track. The resulting movie can be seen at http://www.electricalfun.com/ElectronCafe/LCLS_x-ray_laser.aspx. Fast photography developed quickly to the point that not just bullets but even light itself could be "stopped" in midair. Now with the advent of ultra-fast, ultra-bright coherent X-rays as in the LCLS, at SLAC, it is becoming possible to "stop" chemical reactions and other processes.

INTERESTING TIDBIT TB22.5

Another EUREKA Moment: Reading Archimedes with SR X-Rays

Researchers from RB Toth Associates, Rochester Institute of Technology, John Hopkins University, Conoco-Phillips and Rutgers University have employed SR X-rays to detect the traces of ink from Archimedes' Palimpsest. The Palimpsest is a 1000 year old parchment containing some of Archimedes' work as copied by a scribe, but also other work that was painted over much later. To distinguish the original writings from the new ones, 7.1 keV photons excite iron atoms in the ink and induce fluorescence at 6.4 keV photons. A detector tuned at the fluorescent signal in conjunction with image processing and optical character recognition has shown that the parchment contains Archimedes studies on floating bodies and the equilibrium of planes. The original parchment and its X-ray image can be seen at http://www2.slac.stanford.edu/tip/2005/may20/archimedes.htm. The intense and highly collimated X-ray beam comes from the SPEAR storage ring at the Stanford Synchrotron Radiation Lightsource (SSRL) facility.

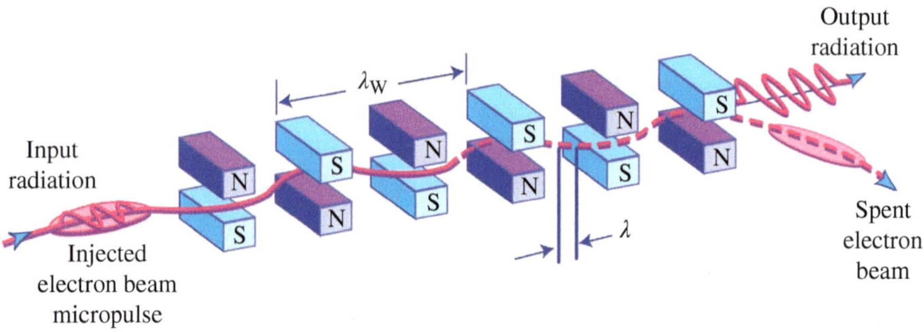

FIGURE 22.20 Free-electron laser concept: an electron beam interacts with a planar wiggler to produce radiation of wavelength λ. Source: O'Shea and Freund (2001). Reproduced with permission of IEEE.

22.6.4 Free-Electron Lasers

In FELs, the electron trajectories are modulated by the magnetic field of a wiggler, radiating photons that interact back with the electrons to produce *coherent bunching* accompanied with coherent radiation. Figure 22.20 shows the schematics of a basic FEL. To understand the process in some detail, we can describe the wiggler field as a traveling EM wave in the reference frame of the electron beam. This wiggler wave adds up to the SR wave to produce a beat wave with the same frequency of the SR wave but smaller phase velocity. Because the beat wave speed is less than c, it can get in synchronization with the electron beam (of axial speed v) and generate coherent bunches: essentially, the beat wave is an interference pattern that traps and synchronizes electron motion and radiation leading to a process akin to stimulated emission in atomic systems. If k is the light wave number and k_w the wiggler wave number, the phase-matching or resonance condition is $\omega/(k+k_w) \approx v$. Combining this expression with the basic formula for the light wave, $\omega = ck$, we obtain the following basic relation among FEL wavelength λ, wiggler spacing λ_w, and electron beam energy (as given by the gamma factor γ) for highly relativistic electrons:

$$\lambda \approx \frac{\lambda_w}{2\gamma^2}. \tag{22.10}$$

A more detailed derivation shows that a factor $(1+K^2)/2$ is needed in Equation 22.10. "K," called the FEL parameter, is of order the 1. From Equation 22.10, it is clear that the FEL wavelength is *tunable* through the electron energy. For example, at electron energies of 4.54 and 14.35 GeV, and undulator period $\lambda_w = 3$ cm, and $K = 3.71$ (LCLS parameters) we get $\lambda = 1.5$ nm, 0.15 nm.

FELs can be implemented as either *oscillators* or *amplifiers*. In the first case, the radiation is bounced back by mirrors like in a regular laser cavity until the coherent radiation builds up and escapes through a semitransparent mirror. These oscillator FELs, which can be implemented in storage rings, are low-gain devices that use high-energy, low-current electron beams. In the amplifier scheme, on the other hand, laser radiation of the desired wavelength passes once through the wiggler and is amplified by the generated coherent electron bunches; this is the type of FEL called "seeded."

Alternatively, coherent radiation can build up from noise alone in a process called self-amplified spontaneous emission (SASE). The SASE FELs (e.g., FLASH at DESY and LCLS at SLAC) are high-gain devices that require electron beams with very high peak current (kA) and low transverse emittance.

ADVANCED CONCEPT AC22.1

Radio-Frequency Electromagnetic Waves and Resonant Cavities

Electromagnetic waves in vacuum are always transverse, meaning that the electric and magnetic field vectors are perpendicular to the propagation direction. In transmission lines such as coaxial cables, the EM waves are also transverse and are referred to as "transverse electromagnetic" (TEM). In contrast, if the waves are confined in metallic hollow structures, no TEM waves are possible and, instead, waves can exist with *longitudinal* components of either the electric or magnetic fields (but not both). When the longitudinal component of the electric field is zero, the waves are called TE waves, for "transverse electric"; if the longitudinal component of the magnetic field is zero, the waves are called TM waves, for "transverse magnetic." These properties of EM waves in conducting enclosures are a consequence of the fact that no fields can exist inside the walls of (perfect) conductors, and the corresponding solutions of Maxwell's equations. It also follows that EM waves in enclosures, as in acoustical resonators, can have only wavelengths that relate simply to the enclosure dimensions. For example, in a *rectangular waveguide* of height "a" (along x), width "b" (along y), and "infinite" length (along z), the TE *modes* are labeled TE_{mn} where m and n are integer numbers such that the dimensions a and b are equal to n *half-wavelengths* and m *half-wavelengths*, respectively. The modes correspond to *standing waves* in the x–y plane, but a traveling wave in the z-direction. Furthermore, the modes exist only for frequencies above a *cutoff frequency* that depends on m, n, a, and b. The lowest cutoff frequency corresponds to the TE_{10} mode: $\omega_{10} = c\pi/a$. Figure AC22.1 shows the geometry and field pattern of a TE_{10} mode in a rectangular waveguide. As an example, the rectangular waveguide designated as WR-90 has dimensions $a = 2.286$ cm and $b = 1.016$ cm, so the cutoff frequency is 6.56 GHz. Thus, only frequencies above 6.56 GHz are possible in WR-90; in addition, in the so-called X-band of the microwave spectrum (8.20–12.50 GHz) only *single-mode* operation at TE_{10} will occur. At higher frequencies (e.g., 15 GHz), though, other modes and combinations can exist. Single-mode operation, however, is preferred.

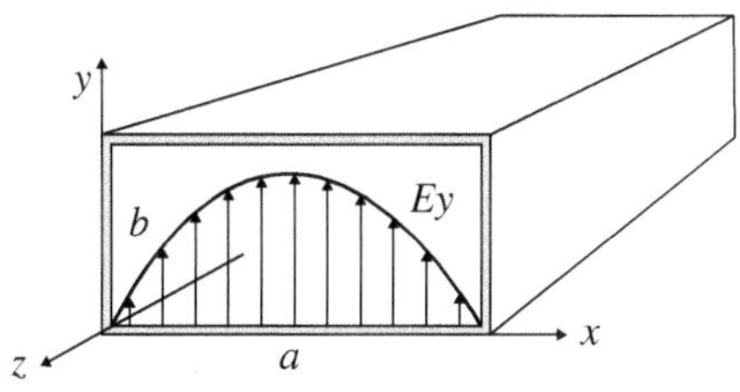

FIGURE AC22.1 Geometry of rectangular waveguide and pattern of electric field in the mode TE_{10}.

23

JET ENGINES, STRATOSPHERIC BALLOONS, AND AIRSHIPS

Jet engines operate on the same physics principle as rockets. The basic difference is rockets carry onboard a reservoir of an oxidizing agent, while jet engines, known as air breathers, obtain it from the atmosphere. In this chapter, we discuss jet engines, airships, and stratospheric balloons, many of which are able to operate at the edge of space in highly tenuous air above 99% of the atmosphere. To gain a quantitative perspective of the meaning "the edge of space," the reader is encouraged as always to absorb a few benchmark numbers. First, an appreciation of "the edge" can be seen in Figure 23.1, which is a picture taken from the cockpit of a U-2 spy plane while at an altitude of 23 km. Additional benchmark numbers are listed in Table 23.1, showing the extreme range of pressures ($1–10^{-3}$ atm) over which jet engines have to operate. These physical comparisons are approximate since actual values depend somewhat on a number of parameters (some variable), including latitude, temperature, and the phase of the solar activity cycle. Moreover, it is advantageous to round off numbers to at most two decimal places when establishing mental benchmarks to make these easier to recall. For instance, the standard atmospheric pressure at sea level is 1013 mbar, but we list 1000 mbar as the benchmark number. Note that most spacecrafts in low Earth orbits (LEOs) such as the *International Space Station* (ISO) and the *Hubble Space Telescope* (*HST*) still encounter trace amounts of Earth's atmosphere, causing tiny amounts of drag. Spacecraft in most LEOs must be boosted every few years back to its original orbit due to atmospheric drag. The rate of orbital decay depends on the phase of the solar activity cycle, which causes to atmosphere to expand and contract over an 11-year period.

Another useful benchmark number is the speed of sound. Aircraft, especially supersonic ones, moving through the atmosphere are often quoted in terms of Mach number, which is the ratio of its speed to that of the speed of sound. While the speed of sound varies inversely proportional to the square root of the air density ($v_s \propto 1/\sqrt{\rho}$),

---

Modern Devices: The Simple Physics of Sophisticated Technology, First Edition.
Charles L. Joseph and Santiago Bernal.

Companion website: www.wiley.com/go/joseph/moderndevices

FIGURE 23.1 Photo taken outside the window of a U-2 spy plane, flying at an altitude of 23 km (~75,000 ft). Source: USAF/CIA.

TABLE 23.1 Approximate Benchmark Numbers for the Atmosphere

Benchmark Position	Altitude (km)	Pressure (mbar)	Atmosphere Beneath Observer (%)
Sea level	0	1000	0
Mountain top	5	600	50
Commercial aircraft	12	200	80
Near-space (jets, stratospheric balloons)	20–40	40–4.5	99.0–99.6
Space station—low Earth orbit	100	0.001	99.999

the generally accepted value for the speed of sound at $T = 20°C$ is taken to be 331 m/s or equivalently about 1200 km/h (~745 miles/h). In this case, establish benchmark numbers for yourself by comparing typical highway speed in automobiles and the fact that commercial aircraft fly just below Mach 1. Other benchmarks to consider are: a common walking speed and perhaps spacecraft in low Earth orbit, which typically have speeds of 7 km/s.

We have already discussed the physics behind a rocket. Prior to our treatment of jet engines, we briefly address missiles, a special class of rockets used primarily by numerous militaries around the world. (*Note*: our definition of a rocket is somewhat broader than that commonly used by the armed forces since we include civilian exploratory uses throughout the solar system. In military parlance, an unguided missile is called a rocket.) If a rocket provides thrust only in its initial stage, it is a ballistic missile. One example is an Inter-Continental Ballistic Missile (ICBM), which may contain a single or multiple warheads. The latter is labeled a Multiple Independently Targetable Reentry Vehicles (MIRVs). An ICBM is considered a ballistic missile because a relatively short burn of its rocket engines determines its trajectory without any further active guidance. If the warhead is an MIRV, independent warheads separate toward the end of the flight, each having its own active guidance. The type of guidance and targeting systems, which we shall not discuss, further categorizes missiles. Cruise missiles, developed separately by a few governments, are distinctive in that these missiles usually fly at high speeds and low to the ground over long distances. These are essentially unmanned, rocket-propelled aircraft that are difficult for enemy targeting and destruction. Some examples of military missiles are shown in Figure 23.2.

FIGURE 23.2 Top: Aegis missile part of the anti-missile defense system, launched from the USS Shiloh cruiser; Middle left: an F-22 Raptor Jet fires an AIM-120 AMRAAM; Middle right: land-based Exocet missile launch. The missile is an anti-ship for small- to medium-sized ships and is built by MBDA, a European corporation; Bottom: Tomahawk cruise missile. Source: (Top) US Navy; (Middle left) US Air Force; (Middle right) US Navy/Pao; (Bottom) US Navy and Wikipedia.

23.1 RAMJETS, TURBOJETS, AND TURBOFAN JETS

The simplest and earliest type of jet engine is the ramjet depicted schematically in Figure 23.3. The four essential components are air intake, compression, ignition in the combustion chamber, and high-speed expulsion of gas. The engine does not receive its thrust from pushing against the ambient atmosphere. Rather, its thrust is the result of conservation of momentum similar to a rocket. (See Intro Physics Flashback FB16.2 as well as Figs. FB16.3 and FB16.4.) If a container full of air or oxygen where attached to the front end of a ramjet, it would function just as well in the vacuum of space. Ramjets, however, are highly inefficient and provide minimal thrust below Mach 0.5, half the speed of sound, since insufficient ram pressure does not result in adequately compressed air. Above Mach 0.5, a ramjet will be self sustaining. Normally, the fuel flow must be reduced at high speeds to prevent a runaway increase in speed, which could compromise the structural integrity of the engine and airframe.

A turbojet engine as represented in Figure 23.4 provides superior performance compared to a ramjet. The air intake and compression are controlled by turbine

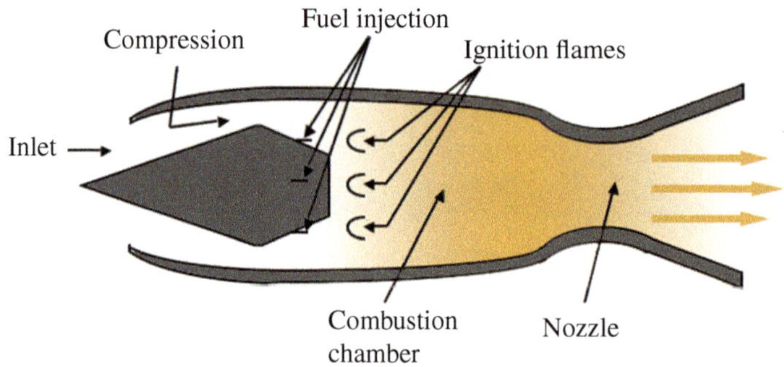

FIGURE 23.3 A schematic representation of the four elements of a ramjet: air intake, compression, combustion, and high-speed expulsion of gas.

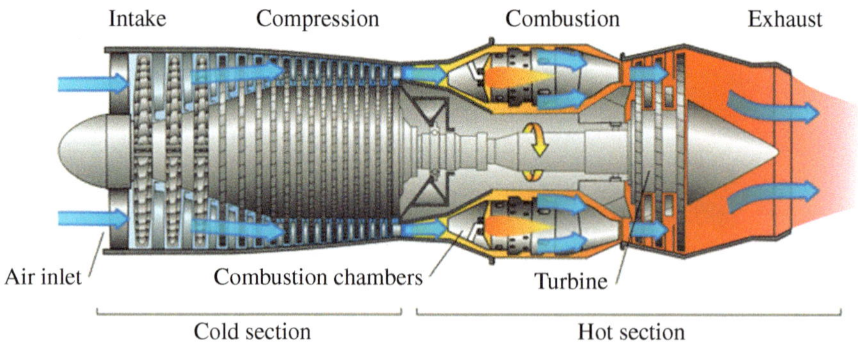

FIGURE 23.4 A diagram of a turbojet engine. Credit: Jeff Dahl/Wikipedia under the GNU Free Documentation License.

blades, which are attached via a shaft to a second set of turbine blades that are driven by the hot gas expelled from the combustion chambers. Although some of the energy produced by the engine is used to turn the compressor turbine, the enhanced compression of air leads to far greater combustion efficiency and higher velocity of the gas ejection, more than compensating for the energy used to drive the compressor turbine.

While turbojets out perform ramjets, both are particularly noisy and neither is as fuel efficient as turbofan jets (also known as turbofans or fan jets) used by commercial and military aircraft. A turbofan jet, schematically pictured in Figure 23.5, adds a second, outermost skin, forcing additional air (known as the bypass flow) exterior to the core of the engine. The bypass flow creates additional jet action thrust, albeit one with a slower ejection speed. It also assists the cooling of the high-pressure engine core, dampens noise, and improves overall fuel efficiency. Turbofan jets are wider than turbojets, causing these to have relatively large amounts of drag.

One important parameter of a turbofan jet is its bypass ratio, the rate of air mass flowing through the outer areas compared to that in the hot core. Commercial airlines fly turbofan jets with a high (5 : 1 to 6 : 1) bypass ratio, meaning 5–6 kg of air bypasses the core for every kilogram of air flowing through the core. High bypass turbofans produce more thrust and are more fuel efficient at subsonic speeds compared to those engines with low bypass ratios. In contrast, military turbofan jets use low bypass ratio of 2 : 1 or less. Fighter jets produce their maximum thrust at supersonic speeds, Mach 1 and above, but are less fuel efficient than the high-bypass engines used by commercial airlines.

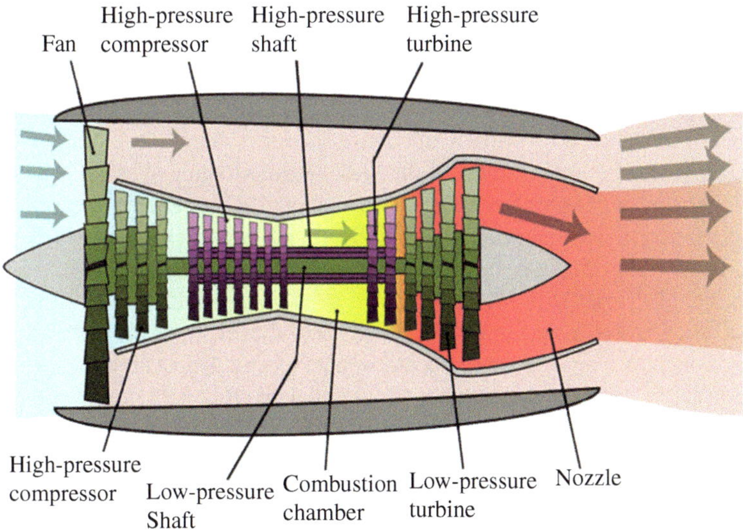

FIGURE 23.5 A turbofan jet design with a bypass flow of air surrounding the core turbojet. Image credit: K. Aainsqatsi, Wikipedia under the GNU Free Documentation License. The authors have modified outer engine profile.

INTERESTING TIDBIT TB23.1

One well-known commercial aircraft, the Concorde, in service from 1976 to 2003 flew at supersonic speeds, cutting transatlantic crossing times in half. Turbojet engines were used instead of turbofans to minimize drag, but the large noise from Concorde jets caused many governments to limit overflight routes as well as landing/takeoff destinations. High manufacturing and high fuel costs led to it being decommissioned. Concorde's engines were so inefficient at slow speeds, it used almost 2% of its maximum fuel load capacity taxiing to the runway.

23.2 STRATOSPHERIC BALLOONS

Stratospheric balloons have been used for more than 60 years in scientific studies of Earth's upper atmosphere as well as solar and astronomical research over a wide range of energies, including cosmic rays, hard X-rays, ultraviolet, visible, and IR wavelengths. Most stratospheric balloons have operated at the edge of space with floatation elevations of 20–40 km, depending on volume of the balloon used and the weight of the gondola, the frame holding the payload. While these floatation altitudes position the gondola above 99% or more of the atmosphere, the astronomical sky is still opaque in some wavelength bands. In particular, most ultraviolet and a large portion of X-ray wavelengths ($0.1\,\text{nm} < \lambda < 180\,\text{nm}$) cannot be observed. The opacity over these wavelengths also result in strong Sun–Earth coupling in the upper atmosphere with practical implications for humanity, one being the interference with telecommunications illustrated in Interesting Tidbit TB23.2.

Figure 23.6 shows the initial and floatation sizes of a standard zero-pressure balloon, using helium as the lighter-than-air lift gas. (Zero pressure means there is an opening at the bottom of the balloon, which allows gas exchanges between the balloon interior and the outside atmosphere.) At the time of release, the top of the balloon is approximately 230 m high, about 1.5 times has tall as the Washington Monument or about 80% as tall as the Eiffel Tower. Most of the balloon is not inflated at the time of release with only 25 m out of nearly 200 m having any significant horizontal extension. However, the balloon expands enormously by the time it arrives at floatation altitude. Maximum payload including ballast is 3000 kg (the weight of three small automobiles). Ballast weight is released twice during assent, about 5% of the total payload mass to drive the balloon through the coldest part of the atmosphere (normally just above the tropopause for stratospheric altitudes) and another 8% to maintain altitude at sunset.

Historically, many zero-pressure balloons have had flight times of approximately 23 hours with 3 hours for ascent, 8–20 hours at floatation, and a 45-minute descent. These experiments have been launched from selected sights during particular seasons when the prevailing winds of the lower atmosphere and stratosphere cause a balloon to drift in one direction for half of the flight and the opposite direction for the second half. Thus, the balloon with payload makes more or less a round trip, returning back close to the launch site. Figure 23.7 pictures the high altitude student platform (HASP) payload,

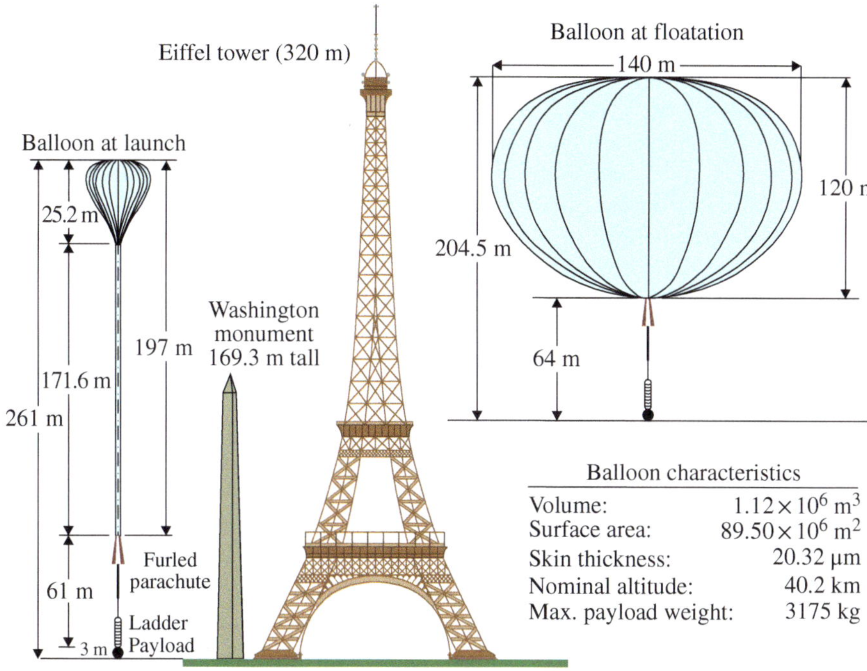

FIGURE 23.6 The relative heights and volumes of a zero-pressure stratospheric balloon. Source: NASA/Columbia National Balloon Facility.

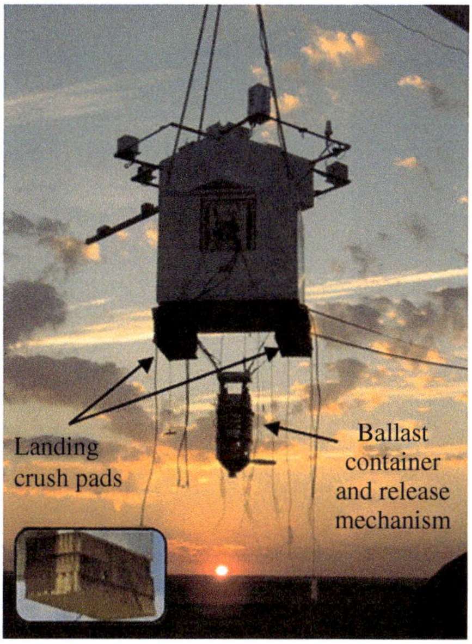

FIGURE 23.7 The HASP payload suspended from a crane at a launch site. Source: NASA & the Louisiana Space Consortium.

suspended from a crane at a launch site. Note the ballast container and release mechanism suspended below the gondola. Also most gondolas have crush pads made of corrugated cardboard attached on the underside to absorb the shock during landing. The inset reveals the structure of one of these pads. HASP is a joint venture between the NASA Balloon Program Office and the Louisiana Space Consortium. It is designed to carry as many as 12 student experiments for 15–20 hours at an altitude of 36 km.

Up to this point, we have been discussing exclusively zero-pressure balloons. The NASA and manufacturing vendors have been developing new balloon technologies since 1990. New materials technology have enabled balloons with much larger volumes and overpressurized balloons, enabling balloons to stay aloft for multiple days, weeks or months without having to shed ballast daily. (A zero-pressure balloon will gain some altitude during daylight hours, causing a small loss of helium since the fully inflated balloon experiences a constant volume but a lower external pressure. After sunset, the floatation altitude will be less than it was during the previous night. To maintain altitude over each diurnal cycle, a zero-pressure balloon has to release ballast. In contrast, an overpressurized balloon inherently maintains altitude.) Scientific balloons that stay aloft for 1–3 weeks are known as LDBs (long-duration balloons), while those that remain at the edge of space for a month or longer are known as ULDBs (ultra-long-duration balloons). New balloon technology is a remarkable feat of materials engineering. The balloon skin is only 20 μm thick, a fifth of the thickness of a 14-inch legal sheet of paper. Despite its ultra-thin skin, the balloon is so large that its mass is comparable to that of the payload and one could suspend a dozen jumbo jets inside when fully inflated. Great care must be taken to spool the balloon for shipment to the launch site and then to unspool it onsite so as to not damage the thin material. Two versions of an over-pressurized balloon are pictured in Figure 23.8, a long-duration balloon and a superpressurized ultra-long duration balloon. (The ULDB is referred to as a pumpkin skin.) Both of these balloon types are sealed and have over pressures of only 0.0363 PSI (2.50 mbar) when

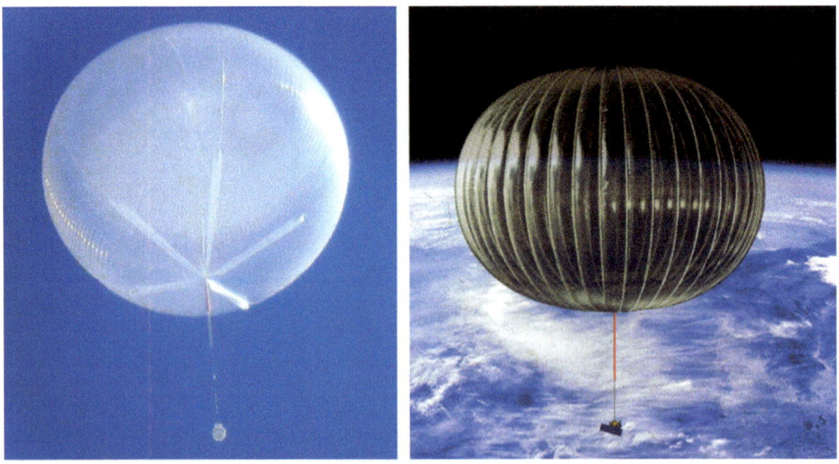

FIGURE 23.8 An over-pressurized LDB (left) and a superpressurized ULDB. Source: NASA/Columbia National Balloon Facility.

at floatation altitudes. Similar to vacuum tanks, all balloons have tiny, microscopic leaks. The primary difference between an LDB and an ULDB is the latter can hold its helium for longer periods, indicating it is more tightly sealed.

Stratospheric flow patterns enable balloon payloads to circumnavigate the Earth at specific latitudes but only during particular seasons. One flight path from Sweden to Canada, depicted in Figure 23.9, takes about 7 days. The stratospheric flows that enable this transatlantic path only occur during summer months, conditions where the payload spends most of its time in twilight. The oscillatory projection of the flight path is indicative of the diurnal light-dark period and the associated changes in latitude and small variations in the floatation altitude. An ULDB, launched from Alaska along a somewhat higher latitude flight path, is possible if over flight agreements can be reached between the US and Russia. Nations are reluctant to grant permission if the flight path goes over any populated areas. Flight path restrictions have been a significant impediment to scientific LDBs and ULDBs at low latitudes.

Another popular location to circumnavigate the Earth with a balloon is over the Antarctica continent, again during summer time for the southern hemisphere. All LDB and ULDB flights around the South Pole have been conducted in campaign mode with funding being shared between the NASA and the National Science Foundation (NSF) for US investigators. Figure 23.10 shows launch preparation at Willy Field, McMurdo Station in Antarctica. The balloon has been filled with He at the right; the unrolled,

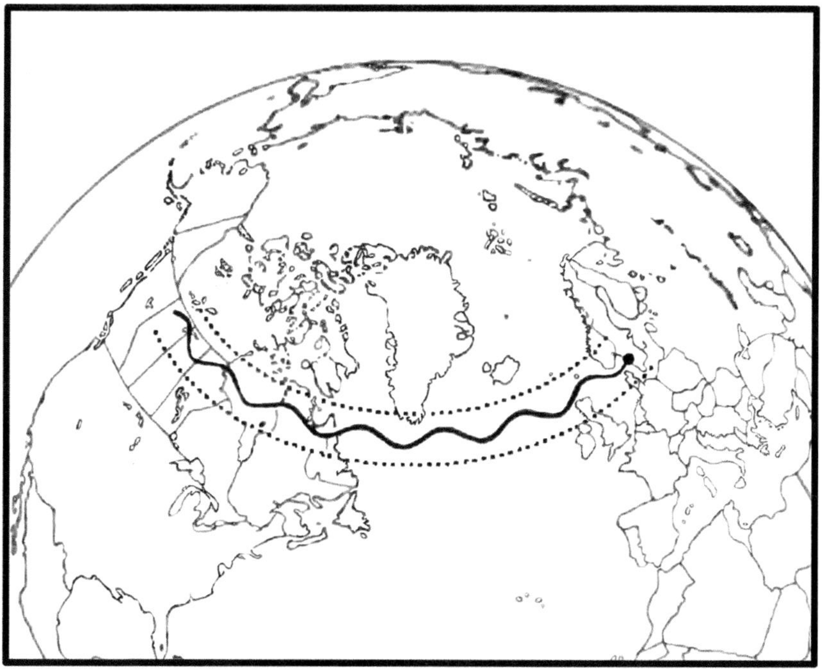

FIGURE 23.9 High-latitude stratospheric balloon flight path from Sweden to Northern Canada. Source: NASA.

FIGURE 23.10 Flight preparation at the Antarctica launch site. Source: NASA/Columbia National Balloon Facility.

uninflated portion extends to the unfurled parachute near a group of people (portion at center left). The payload and gondola are suspended on a mobile crane at the far left. Once the balloon is released from its tether, the mobile crane moves to position the payload directly underneath the ascending balloon as pictured in Figure 23.11. Compare this picture to the left diagram in Figure 23.6. The furled parachute can be seen near the middle along with the lengthy uninflated portion of the balloon. This crane positioning prior to release minimizes the pendulum swing of payload under the balloon.

Plotted in Figure 23.12 are the floatation altitudes achieved for various sizes of balloon as a function of suspended weight, the weight of the experiment and ballast. Balloon sizes are listed in million cubic meters (mcm). Million cubic feet (MCF) is sometimes used in the United States, which is the only industrial nation yet to go metric. A common practice in listing specifications such as those in Figure 23.6 is to quote the maximum values for each variable. The reader should note that all specs (specifications) cannot be met simultaneously. For example, Figure 23.6 indicates the maximum payload weight is 3175 kg and the nominal altitude is 40.2 km for a 1.12 mcm balloon. From Figure 23.12, we see that 40 km can only be achieved if the suspended weight is 500 kg or less. The maximum payload weight spec. is just the maximum that a 1.12 mcm balloon can be lifted to the stratosphere. In this case, an altitude of 37 km is indicated based on an extrapolation of the curve. In practice, a somewhat smaller balloon of 0.97H mcm provides better performance for heavy payloads, albeit at a somewhat lower range of elevations. The lifting advantage of the 0.97 mcm compared to a 1.11 mcm balloon is the lower weight of the balloon material itself.

Once the science mission is complete or the lift capability of the balloon has been compromised, the decent sequence is initiated. This phase of the flight has the highest risk. The payload plus parachute are detached from the balloon, an operation that is also designed to tear a large hole in the balloon so that it returns to Earth about as quickly as does the payload. At parachute deployment, the payload experiences a 5–8 g jerk. When the gondola impacts the ground, it typically receives another 6–8 g jolt, reduced from the 10 to 12 g that it would have had without crush pads. If there has been a significant cross wind and the parachute has not detached properly upon landing, a few payloads have been dragged along the ground, being damaged on large rocks.

There are several important benefits of using a stratospheric balloon to get above 99% of the atmosphere, compared to going to space. The primary advantage is cost. A mid- to large-sized balloon-borne payload can be launched, operated for weeks or months, and recovered for a few million US dollars. Normally, it can then be

FIGURE 23.11 The mobile crane moves the payload directly beneath the balloon. Source: NASA/Columbia National Balloon Facility.

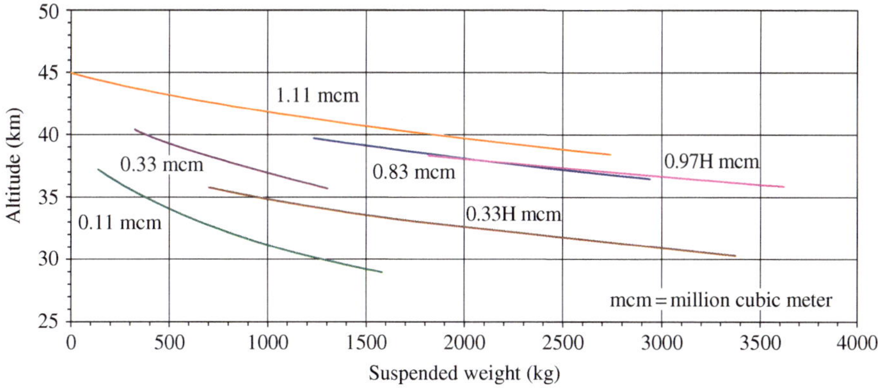

FIGURE 23.12 Plots of the maximum floatation altitude versus suspended weight for balloons of various sizes. Numerical values of the volumes are listed in million cubic meters (mcm). Balloons designed to carry heavy suspended weights have the "H" designation. For example, there are two plots having the same volume of 0.33 mcm. Source: NASA/Columbia National Balloon Facility.

refurbished, launched, and flown again inexpensively the following year. In comparison, the launch cost to place a mid-sized instrument in low Earth orbit is \$70M–\$150M (USD) with another several million USD per year to operate it. Moreover, most of the flight risk for a balloon occurs at the end of the mission, during decent and after the data have been collected. A balloon release is only a 1.1 g event, so there is little danger that optical components will be jerked out of alignment. In contrast, the greatest risk for a rocket payload occurs during launch before any observations are made. Rocket payloads receive sustained 12–20 g random accelerations in all three axes. For visualization, a 12 g impact shock is obtained by dropping an electronic box from chest high. Nevertheless, both rocket and balloon experiments have an approximate 1 in 20 chance of incurring a catastrophic failure.

Many countries have active scientific research programs using stratospheric balloons. The Sanriku Balloon Center in Japan, for example, has launched more than 400 balloons (10/year) since 1971 when it was founded. ESRANGE, originally operated by the Swedish Space Corporation and now by the European Space Agency (ESA), has 550 launches as of October 2012 to its credit. ESRANGE, located in northern Sweden, is an ideal location to carry out circumpolar balloon flight to study the atmosphere in the Arctic region and is used by the international scientific community. The NASA's Columbia National Balloon facility has also launched experiments for investigators from 12 countries.

COMPREHENSION VERIFICATION CV23.1

Question: What is the fraction of a zero-pressure balloon that is filled with helium, and why not put more He into the balloon to provide greater lift? (*Hint*: use Fig. 23.6 in the estimate.)

Answer: The fraction of the balloon filled with He can be estimated from the ratio of the volume at release to the volume at floatation. Here, we approximate each as a sphere even though the shape of the balloon departs significantly from a sphere at all times. We begin our estimate of the sizes by drawing two sets of circles on a copy of Figure 23.6. For convenience, we have cut and pasted two portions of Figure 23.6 into Figure FB23.1. The size of the balloon at release is approximated by the solid circle at the left side of the figure, which approximately has the same area as the drawing. A second, dashed circle is drawn against a convenient measured scale, 25.2 m in this case. Using a ruler, the ratio of diameters between the solid and dashed circles is about 82% so the diameter of the balloon at release is $0.82 \times 25.2\,\mathrm{m} \approx 21\,\mathrm{m}$. Recalling the volume of a sphere is $4/3\,\pi\,r^3$, the volume at launch is about $4800\,\mathrm{m}^3$. (Greater precision in our estimate is not justified by the inherent errors.) We use two similar sets of circles to estimate the diameter of the balloon at floatation. (See the right side of Figure FB23.1.) We estimate the volume to be $1.5 \times 10^6\,\mathrm{m}^3$. Compare this to the listed volume of $1.12 \times 10^6\,\mathrm{m}^3$, again underscoring more than two decimal places is not justified. From our estimate, the fraction of the balloon filled with He initially is

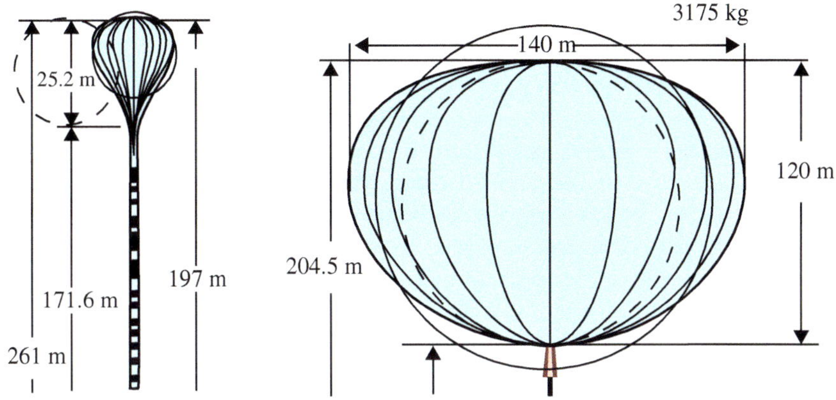

FIGURE FB23.1 How to estimate the volume of a high-altitude balloon at different elevations.

$4800/1.5\times10^6$ or about 0.0032 of the balloon's capacity. (It is 0.0043 if the more accurate numerical floatation value is used.)

Filling the balloon initially with additional He gas is only marginally useful and wasteful of a valuable resource since a zero-pressure balloon remains open to the atmosphere so that any excess He gas will simply overflow the balloon volume and be vented to the atmosphere. The volume of the initial interior gas will expand inversely proportional to the change in exterior pressure. If the balloon lifted off near sea level and went to a floatation altitude of 40 km, the pressure would go from 1000 to 4 mbar based on Table 23.1 and the interior volume of gas would expand by a factor of $1000/4=250$. Said another way, a full balloon at floatation under these particular circumstances would be only $4/1000=0.004$ at lift off, in general agreement with our estimates. Always be aware that these are very crude estimates, but performing these estimates for yourself enhances your insight into the physics and expands your estimating prowess.

INTERESTING TIDBIT TB23.2

The opacity of the upper atmosphere over soft X-rays or hard UV has important consequences. The primary energy deposition in the upper atmosphere is the hard-UV photons from our Sun, even though it emits orders of magnitude more visible photons. One important impact is the solar UV flux ionizes portions of the upper atmosphere, allowing radio waves to reflect back to ground and enabling radio communications around the world beyond a direct line-of-sight transmission. Early radio operators used to complain about "atmospherics" messing up their communications whenever the high-altitude weather disturbed this ionization layer sufficiently.

23.3 FUTURE AIRSHIPS

Airships also known as dirigibles are classified as aerostats, being able to remain airborne via static lift provided by the buoyancy of lighter-than-air gases. (*Note*: the technical definition of an aerostat is a tethered or moored balloon, which is narrower than our common characterization.) In contrast, fixed-wing planes and helicopters are both aerodynamic aircraft that produce lift by moving wings through air, which requires large expenditures of energy to remain aloft. Airships can be rigid, semi-rigid, or nonrigid. In general, aerostats provide large lift capacity and are able to stay aloft for extended periods. Dirigibles, which were the first airborne craft developed with a means of locomotion and steering, are slow and have huge drag coefficients (i.e., present a large cross section to winds), causing many to not maneuver well in winds above 35 km/h. Airplane performances began outpacing that of dirigibles in the 1930s and in the wake of high-profile disasters such as the Hindenburg fire, airships fell out of favor. Dirigibles were subsequently relegated to niche markets such as soft-skinned blimps, which today provide platforms for advertisements and for bird's-eye television cameras over U.S. football games.

In 1987, the U.S. Navy funded a demonstrator research program, designed to investigate whether blimps could be used to detect sea-skimming missiles such as the Exocet. While that program was eventually canceled, numerous commercial ventures and various governments around the world had acquired a renewed interest in the development of new airships by the twenty-first century. These aircraft do not require long runways for takeoffs and landings, enabling supplies and equipment to be transported easily from forward bases to any remote and relatively inaccessible areas. Commercial applications include the transport of materials for major construction projects to remote areas without the costs and added delay to improve access roads. Mining or oil drilling operations, for example, in Northern Canada, can be accomplished in areas where the cost of creating a road is prohibitive. Airships are also a complementary asset to today's unmanned aerial vehicles (UVAs), capable of loitering for longer periods with less fuel, operating at higher elevations for a wide field of vision, and carrying large amounts of surveillance equipment, ordinance, and additional special purpose winged drones.

Some examples of airships under development or created in the past decade are shown in Figures 23.13, 23.14, and 23.15. Lockheed Martin along with its partners, Stratcom International and others, developed the High Altitude Airship (HAA) shown in Figure 23.13 for the U.S. Army Space and Missile Defense Command. HAA is designed as a telecommunications relay or as a peacekeeper from its over-the-horizon geostationary position, operating well above the jet stream and any severe weather. (In this case, HAA is carried along partially by the Earth's upper atmosphere and maintains a geostationary position using its internal propulsion system to prevent it from drifting away from its ground location. This usage of geostationary should not be confused with a geostationary orbit, which is well above the

FIGURE 23.13 The High Altitude Airship (HAA) developed for the U.S. Army Space and Missile Defense Command. Source: U.S. Army.

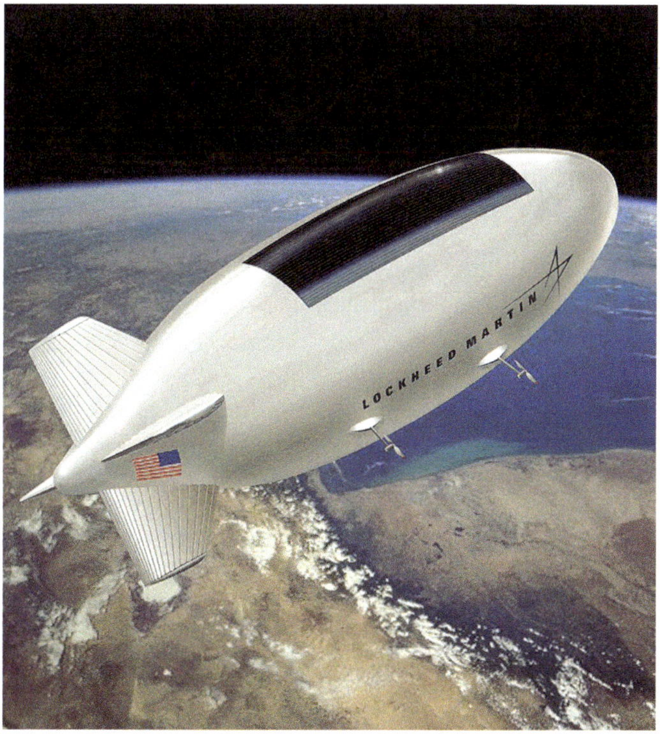

FIGURE 23.14 The solar-powered High-Altitude Long-Endurance Demonstrator (HALE-D). Source: U.S. Air Force.

FIGURE 23.15 The SkyFreighter, a very large capacity hybrid air vehicle, capable of landing and takeoff from the sea or land. Source: Millennium Airship, Inc. Reproduced with permission.

edge of space.) A follow-up to HAA also being developed by Lockheed Martin is the HALE-D, High-Altitude Long-Endurance Demonstrator, funded by the U.S. Air Force. HALE-D, pictured in Figure 23.14 is a subscale unmanned airship with a radar system that is capable of detecting an automobile hidden under a canopy of trees from an altitude of 300 km. Ultimately, the U.S. Air Force intends to replace its Boeing E-3 airborne warning and control system along with its Northrop Grumman E-8C JSTARS airborne ground surveillance aircraft with a fleet of stratospheric-roaming HALE airships. The HALE-D prototype is powered by a 15 kW thin-film solar cell array and rechargeable lithium-ion polymer batteries, providing 40 kWh energy storage, while providing 500 W of power to the payload and radar system. The airship is driven by two 2-kW electric propulsion motors. HALE-D is recoverable and reusable.

In Europe, the aerospace firms BAE Systems, EADS, and Finmeccania have formed a joint venture to design and build an unmanned airship. These military UAVs are expected to become available around the year 2030 as descendants of current Predator and Reaper drones and are intended to carry a suite of drones being developed by weapons manufacturer MBDA. Four sets of MBDA drones can be attached underneath the airship along with 16 even smaller drones. All of the drones have spring-loaded wings that deploy once dropped from the airship. The smaller of the air-dropped UAVs will serve as a scout with a maximum range of 30 km. It might seek out and laser illuminate a target for a larger laser-guided UAV or might carry a 1-kg warhead. The larger UAV will also carry a 1-kg warhead and able to remain airborne for much longer time (up to 2 hours). Both of these winged drones can fly at altitudes below any adverse weather conditions and seek out difficult to find targets.

Several airship designers have begun enhancing the capabilities of their vehicles by adding supplemental propulsion systems, effectively creating hybrid vehicles with increased speed and lift capacity, but with somewhat reduced operating times due to the limited amounts of fuel carried. In designing a hybrid airship, one has to deal with the fact that the center of mass (or equivalently the center of gravity) of conventional dirigibles is near the edge of the bulky lighter-than-air gas containment volume. Any attempt to apply significant thrust along a single axis, especially in variable wind conditions will tend to be unstable unless computer-controlled variable thrust can be applied rapidly by fly-by-wire systems. (To envision the basic problem, imagine driving your car down a highway in moderate winds with a boom and a very large sail attached to its top.) One particularly advanced hybrid airship design is the SkyFreighter being developed by Millennium Airship, Inc., and pictured in Figure 23.15. It incorporates the sophisticated Integrated Thrust and Maneuvering Management System (ITAMMS), a propulsion system being patented by Millennium Airship Corporation. The ITAMMS consists of four ultra-high-efficiency turbofans, mounted as small thrust wings on the sides of the airship's body. Each turbofan has a variable-pitch assembly and a separate engine room for both protection and the ability to service it in flight if need be. All four can be oriented perpendicular for maximum takeoff lift (as seen in the inset image) and then pitched to provide horizontal thrust with the small wings providing added lift during horizontal flight. The articulating portions of these thrust wings can be rotated 360° and can provide downward thrust whenever the vehicle is buoyantly light. The ITAMMS incorporates fly-by-wire control to simplify the pilot's control and maintain stability by rapidly adjusting amounts of thrust from each turbofan, its pitch angle, and redirecting thrust perpendicular airship body to compensate for changing wind flows, allowing takeoff and landings in high crosswinds.

SkyFreighter Canada Ltd. is expected to begin mining, oil drilling, and other operations in Northern Canada. SkyFreighter is designed to carry a 50-ton cargo over a range of 3000 km with a cruse speed of 80 knots (150 km/h). The cargo volume is 14 TEU (20-ft equivalent unit), enabling the shipment of structures much larger than can be shipped by roads. (A TEU is an international measure of a standard shipping container, 6.1 m long by 2.44 m wide and typically 2.59 m high.) Other hybrid airship designs include the Australian SkyLifter and Boeing's SkyHook. The air crane, SkyLifter, has the shape of a flying saucer with a hook on a cable extending down. SkyLifter will be capable of lifting up to 150 tons. SkyHook JHL-40 aircraft will be able to transport a 40-ton cargo up to 200 miles. Some future designs envisage the use of hydrogen instead of helium for maximum lift capabilities. While H_2 is combustible, new advanced materials provide strong, durable enclosures that can, for example, more readily dissipate electrical charge from a lightening strike, substantially increasing safety.

APPENDIX A

STATISTICS AND ERROR ANALYSIS

Statistical evaluations and analyses of uncertainties are becoming increasingly important as societies progressively rely on more complex technologies. The need to raise the overall sophistication and very basic knowledge of statistical concepts in the general population is perhaps the most important and most effective means we have to deal with many of the issues created by technology. Numerous software packages exist for the analysis of statistical data, enabling mathematically illiterate individuals to produce professional-looking presentations and to draw influential conclusions, but ones that are not supported by the data. It is important, therefore, that all science and engineering students as well as professionals in general give adequate attention to these subjects so they can reliably evaluate and interpret data. Most individuals who have become proficient scientists and engineers have already developed the sophisticated statistical skills well beyond those presented here. We provide this appendix as a learning tool aimed primarily at students still in their formative years.

One crucial point is that statistics cannot prove that a mathematical relationship exists; statistics can only be used to reject or invalidate causal connections. That is, if an equation or formula does not fit the data sufficiently well, that theory can be ruled out statistically. In contrast, the positive "a statistical proof" of any relationship is not actually a proof, but merely an increased confidence that alternative interpretations can be excluded. Also, a set of measurements can exhibit the appearance of trends and relationships simply by chance when none exists. This statement is especially true when the number of data points is relatively small. Moreover, statistics cannot be used to show a well-fit formula is unique. There may be other parameterizations that match the data equally well, statistically speaking. Only for a sufficiently large sampling of data with a suitably high correlation does the confidence level become high, but never perfect.

Modern Devices: The Simple Physics of Sophisticated Technology, First Edition.
Charles L. Joseph and Santiago Bernal.

Companion website: www.wiley.com/go/joseph/moderndevices

A comprehensive introduction either to statistics or error analysis is beyond the scope of this book. Instead, we present brief, very cursory treatments of a few of the essential topics that should be mastered elsewhere. In particular, we attempt to highlight a few of the most common mistakes that we have seen in presentations by even highly capable scientists with strong backgrounds in statistics. The student is referred to an excellent, best-selling freshman-level textbook titled *An Introduction to Error Analysis* by John R. Taylor [University Science Books, Sausalito, California], if he or she has not already mastered these most rudimentary skills. The student, for example, should insure that he/she understands the *propagation of errors*, the collective uncertainty when two or more numbers, each with its own level of uncertainty, are combined mathematically (added, subtracted, multiplied, or divided). Similarly, the student should understand the methods to calculate various statistics (e.g., the mean, the median, the standard deviation, and the variance) along with the distinction and proper use of each. While most scientists and engineers have already acquired these necessary skills, we highly recommend all students obtain formal training in statistics above the college freshman level.

All measurements have some level of uncertainty. There are two separate types of uncertainty: random and systematic. Potential systematical errors that are present in data are often missed or simply ignored by experimenters. Random and systematic uncertainties are also measures of the precision and accuracy, respectively. These two types of uncertainty are easily visualized from a target on a shooting range. The center of the bull's eye is the goal of the marksman. The x–y location of each bullet hole characterizes its error from a single shot. Figure A.1 shows the results from two separate marksmen. The one on the left was the more accurate of the two outcomes in that the average (mean) location of the holes is closest to the center of the target. Despite its larger random scatter, the average is closest to the center. If the same marksman fires additional rounds, the mean value becomes increasingly close to the correct (0,0) point, although the scatter is expected to remain constant. In contrast, the right target has its entire set of bullet holes tightly clustered to the lower right of the "bull's eye." This outcome has a higher precision in that hole locations are consistently repeatable, albeit a set of holes with a systematic error. This cause and effect

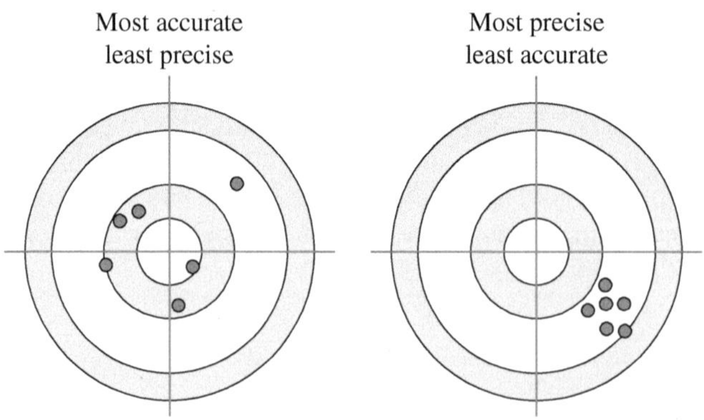

FIGURE A.1 The accuracy and precision of shots on a target.

might be the consequence of a better marksman using a rifle with a scope that is misaligned with the gun barrel. If this marksman fired additional rounds, he could compensate for the systematic error either by adjusting the crosshairs of his scope or simply aiming for the upper left of the target, producing an accurate and precise distribution. The precision versus accuracy visualization can be generalized to any theoretical prediction. In this case, the center of the target represents the theoretical value for the x and y parameters. Repeated, simultaneous measurements of x and y always produce a scatter of data points, occasionally having a measurable systematic component. While the random or systematic errors of the measurements can be diminishingly small, these are never zero.

The target illustration applies only to data sets that can be compared to theoretical predictions, where the expected result is known in advance. In many applications, there is no predicted value(s) and a statistical analysis must rely on the data itself to estimate its precision. Nothing can be said about any systematic error, except in those circumstances where it can be assessed independently. For example, if an investigator was using an old uncalibrated voltmeter, his measurements might have been only 90% of the values he would have obtained with a new meter. Once the old meter is calibrated against a new one, it is possible to compensate the original results without having to collect a whole new set of data. In other cases, there may be some observational limitation that causes a selection effect, biasing the results over a portion of the data. For example, the detection of a specific microbial species might be missed in certain water samples having pH levels that interfere with the functioning of the sensor. If the investigator is aware of these instrument limitations, sometimes a model based on other experiments can be use to mitigate its systematic impact.

The uncertainties from many measurements of a parameter often form what is known as a normal distribution, fit well with a Gaussian profile. That is, most measurements are very close to the average and very few are far from the average. While not all large sets of measurements form a normal distribution, most are approximately normal and the researcher can proceed safely with the analysis. If an experiment produces a small excess number of values at both extremes, then the Gaussian shaped distribution can be said to have extended wings. Occasionally, it is prudent for the investigator to simply exclude a few data points that are gross outliers, but any application of a mathematical rejection filter should be applied vary sparingly to avoid biasing the results. A large percentage of data falling far from the mean usually indicates a missing underlying phenomenon that is not included in the analysis or indicates the existence of unrecognized systematic errors.

Error bars on graphs or uncertainties listed in tables are usually expressed in terms of the *standard deviation*, often denoted as 1σ. Two-thirds (actually 68%) of all measurements in a normal distribution will fall within plus or minus of 1σ, while 95.4% will fall within $\pm 2\sigma$. *Note*: approximately one out of every 20 experimental data points will fall quite far (two or more standard deviations) from any curve fit to the data set. The standard deviation should not be confused with the full width at half maximum (FWHM), which is measured further down the vertical axis. For normal distributions, the FWHM $= 2.35\sigma$. The calculation of the standard deviation is independent of the distribution, even ones that are not normal.

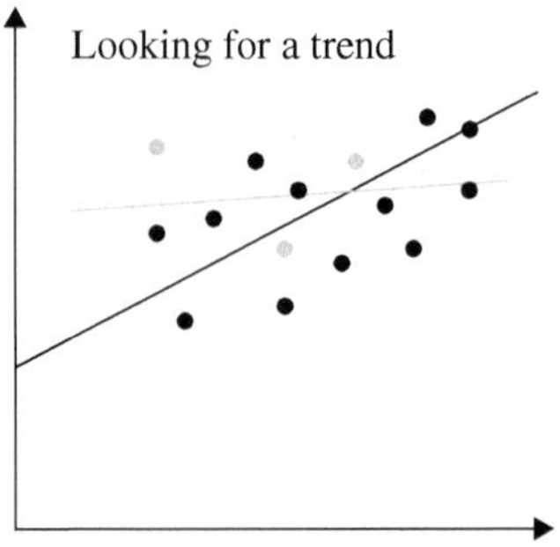

FIGURE A.2 A small number of data points, especially ones without error bars, can produce false appearances of a trend.

One of the simplest approaches to uniquely new data sets is to look for a trend. The researcher often plots one parameter against another, hoping to be able to draw a line through the data and infer a cause-and-effect relationship between two parameters. The black data points plotted in Figure A.2 show the appearance of such a trend with a black line reasonably positioned through the points. There are, however, two weaknesses in this example analysis. First, there are less than 30 data points spanning a very limited range of parameter space compared to the intrinsic scatter. If a few additional data points could be obtained, a dramatically different curve might be indicated as shown by the inclusion of the light-gray points and gray line. The larger data set is even consistent with no correlation between these two parameters. The second weakness is the lack of error bars assigned to individual data points. The lack of error assessments for individual measurements is often the case with totally new data sets since the uncertainties are often poorly understood. In these circumstances, one can only estimate an uncertainty from the observed scatter. If the observed scatter is eventually found to be significantly larger or smaller than would be expected from the intrinsic uncertainties, then one or more additional parameters may be impacting the relation or there are cross-correlated errors.

Once an adequate data set including an assessment of the uncertainties is in hand, the researcher might seek a best-fit linear relationship between the two parameters. The investigator can then go further and ask the following question: Is a fit justified? Is there a real physical relationship between the two parameters plotted, or did this particular set of observations just happen to produce an apparent relationship? To answer this question, the investigator calculates a correlation coefficient from a data set and then compare this calculated number against tabulated data for the probability that a particular number of measurements, N, of two uncorrelated variables would produce this value simply by chance. More formally, the investigator is calculating the linear correlation coefficient, r, given by

$$r = \frac{\Sigma(x_i - \bar{x})(y_i - \bar{y})}{\sqrt{\Sigma(x_i - \bar{x})^2 \, \Sigma(y_i - \bar{y})^2}}, \qquad \text{(A.1)}$$

where $x_i - \bar{x}$ and $y_i - \bar{y}$ are the apparent uncertainties (i.e., the values of each data point minus its distance to the fit line). The value of r measured from the data is then compared to tabulated values of the probability, $P_c(r,N)$, of obtaining that same value from a random sample of N observations taken from a parent population that is not correlated. For example, $N=10$ observations of two uncorrelated parameters will randomly produce a correlation coefficient, r, up to 0.549, one out of 10 times (i.e., $P=0.1$). $N=30$ observations will only exhibit an r as large as 0.463 with a probability, $P=0.01$. Said another way, a correlation coefficient of approximately 0.5 will occur about 10% of the time for a 10-sample size and about 1% of the time for a 30-sample size data of two uncorrelated parameters. As a rule of thumb, relationships inferred from data of 30 or more measurements exhibiting correlation coefficients of at least 50% are probably on very solid ground.

If there are a sufficient number of data points and there appears to be a reasonably strong correlation, the next step is to perform a *least square fit* calculation to determine a slope and intercept for a straight line, along with an estimate of the uncertainties of each. Unfortunately, if the investigator calculates a slope and intercept for parameters x and y with x as the independent variable, then repeats the least square calculation with y as the independent variable, the two sets of coefficients give different straight-line results as depicted in Figure A.3. This is known as *reciprocity in fitting* x versus y, and it is due to the fact that all uncertainties in both x and y are assigned to the dependent variable and none to independent variable in a least square fit. Generally, both parameters have associated error bars, leading to the ambiguity.

Another concern is when a single parameter is embedded in both plotted quantities to facilitate comparison trends. For example, astronomers examining the chemical evolution of a galaxy might be interested in oxygen abundances compared to those of iron. A direct O to Fe comparison for various locations within a galaxy might be meaningless unless these values are normalized to the corresponding

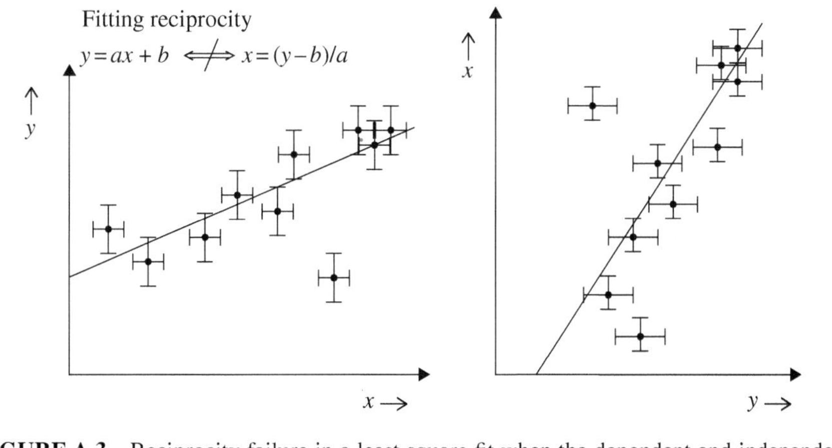

FIGURE A.3 Reciprocity failure in a least square fit when the dependent and independent variables are switched.

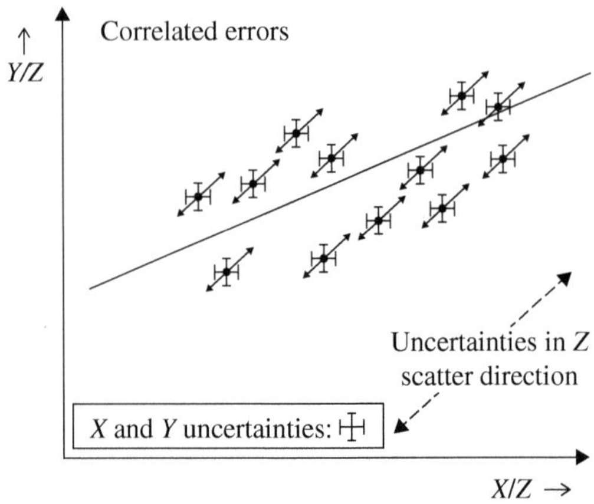

FIGURE A.4 Uncertainties in Z always increase any apparent correlation.

hydrogen abundances at those locations. Plotted in Figure A.4 are Y/Z versus X/Z. One inherent risk with such an approach is that any large uncertainties in Z (or in the hydrogen abundances noted before), if these exist, will introduce correlated errors. In effect, uncertainties in Z in Figure A.4 are equivalent to scattering all of the data points randomly along a diagonal. These Z uncertainties enhance the apparent correlation of the data or can create an apparent one when none exist. (It is not always easy to spot correlated errors since one or both parameters might be part of a complex formula.)

The robustness of a statistical conclusion is also important, especially if there are relatively few data points far from the main cluster. In many analyses of early results, much of the collected data is clustered in one portion of a scatter diagram with a few data points, representing the most observationally difficult to obtain and largest error bars, located some distance away. Whenever, a plot similar to Figure A.5 is presented, you can easily test the robustness of the conclusion by mentally moving one or two data points by two standard deviations (2σ) from their measured value(s). Recall, each measurement randomly differs one-third of the time from its actual value by 1σ and by 2σ about 1/20th of the time. This mental technique is demonstrated in Figure A.5 with the left-most black data point being moved to the location of the gray point. Permissible corresponding linear relationships (gray dashed lines) that would be consistent if the real value of this data point was 2σ above its initial value are shown, indicating the inferred result is not robust. More measurements are warranted.

Finally, a nonparametric test is a simple method for assessing the statistical validity of fitted data without going to the trouble of extensive calculations. As its name suggest, a nonparametric test does not use any parameters; rather, it relies completely on counting statistics and the likelihood of various outcomes, equivalent to the repeated flipping of a coin or the throwing of dice. If a coin is tossed into the air a limited number of times (say ten times), we are not at all surprised if it lands head

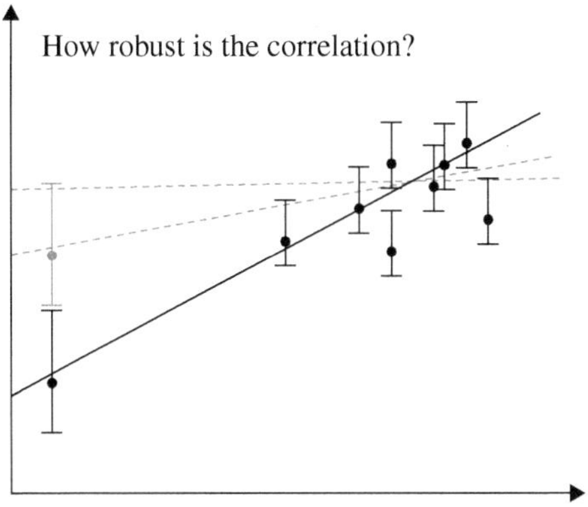

FIGURE A.5 Mentally moving one or two data points with the strongest leaver arms by two standard deviations easily tests the robustness of a result.

side up four, five, or six times. We are only mildly surprised if "heads" occurs seven times or three only times since this is a less probable outcome. If the tosses are even more lopsided (say 8 : 2), is this simply a lucky (or unlucky) accident or has the coin been modified somehow to produce skewed results? Regardless, the confidence in our conclusion is low due to the small numbers statistics of only 10 throws. For a much larger sampling of coin tosses, we expect "heads" to occur with a frequency very close to 50%, but not exactly 50%. Multiple tosses of ordinary coins form a binomial distribution given mathematically by

$$P(m,N) = \frac{N!}{m!(N-m)!}\left(\frac{1}{2}\right)^N \qquad \text{(A.2)}$$

where the probability, $P(m,N)$, of "heads" (or "tails") happening m times out of N coin tosses. Equation A.2 is more general than just for coin tosses, being applicable for any statistic where data points are expected to fall equally on either side of a fitted curve. If too many data points fall on one side or the other, then this is unlikely and the curve does not fit with a certain level of confidence. It is important for the individual performing a nonparametric test to state his/her chosen significance levels for rejecting the fit. Typically, a 5% likelihood outcome is taken as an improbable fit, while 1% likelihood is highly improbable. Let us demonstrate a nonparametric test on an example scatter diagram.

Figure A.6 shows a theoretical model fit to some observational data. There were scientific reasons for matching the two extreme ends with horizontal asymptotic lines with intermediate values of the curve being assigned continuously mixed (blended) values of these two extremes. In this example, the theoretically predicted curve only applies to the central region where the curve represents hybrid mixtures. We, therefore, exclude the six (three on each end) data points at the extremes, leaving 13 points for the nonparametric test. Thus, $m = 11$ and $N = 13$

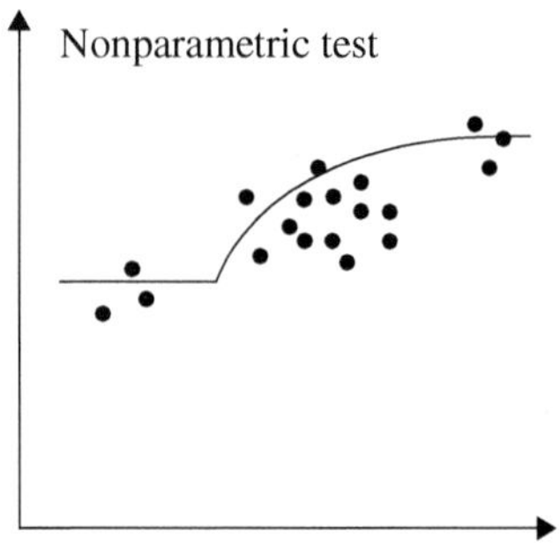

FIGURE A.6 A nonparametric test simply counts the net number of data points above (or below) the fit curve then compares the outcome to the probability this could happen by chance. There are 11 of 13 data points below the curve in this particular example.

for the data plotted in Figure A.6. Plugging these values into Equation A.1 gives $P(11,13) = 0.0095$ (0.95%), a highly significant reason to reject since the probability these observations fit the curve is less than a 1%.

BIBLIOGRAPHY

M. Altarelli, From Third- to Fourth-Generation Light Sources: Free-Electron Lasers in the UV and X-ray Range. In *Proceedings of the 11th International Conference on X-Ray Lasers*, [C. L. S. Lewis, D. Riley (eds)], Springer, Heidelberg, August 2008.

B. Berkowitz and A. Cuadra, Inside Your Technology: Behind the Screens, *The Washington Post*, p. E1, January 24, 2012.

S. Bernal, *A Practical Introduction to Beam Physics and Particle Accelerators*, Morgan & Claypool Publishers, IOP Publishing, 2016: http://iopscience.iop.org/book/978-1-6817-4076-8.

M. Bove Jr., Display Holography's Digital Second Act, *Proceedings of IEEE*, 100, 4, p. 918, April (2012).

P. J. Bryant and K. Johnsen, *The Principles of Circular Accelerators and Storage Rings*, Cambridge University Press, Cambridge/New York, 1993.

D. Carnoy and D. Katzmaier, CNET Reviews—LED TVs: 10 Things You Need to Know, June 4, 2010: http://reviews.cnet.com/led-tvs-review-10-things-you-need-to-know.

A. W. Chao and M. Tigner Ed., *Handbook of Accelerator Physics and Engineering*, 3rd Printing, World Scientific, Singapore/River Edge, NJ, 1999.

V. Coffey, *Photonics Spectra*, Laurin Publishing, vol. 46, December 2012a.

V. Coffey, Plastic Optics Provide Precision, *Photonics Spectra*, 46, 12, p. 50 December (2012b).

M. Conte and W. W. MacKay, *An Introduction to the Physics of Particle Accelerators*, World Scientific Publishing Co., Singapore/River Edge, NJ, 1991.

C. Cunningham and A. Russell, Precision engineering for astronomy: historical origins and the future revolution in ground-based astronomy. *Philosophical Transactions A of the Royal Society*, 370, p. 3852 (2012).

Modern Devices: The Simple Physics of Sophisticated Technology, First Edition.
Charles L. Joseph and Santiago Bernal.

Companion website: www.wiley.com/go/joseph/moderndevices

A. Davis and F. Kühnlenz, Optical Design Using Fresnel Lenses, *Optik & Photonik*, 4, pp. 52–55, December (2007).

M. De Graef, *Introduction to Conventional Transmission Electron Microscopy*, Cambridge University Press, Cambridge, 2003.

Digital Holographic Microscopy: http://en.wikipedia.org/wiki/Digital_holographic_microscopy.

D. Drosdoff and A. Widom, Snell's Law From an Elementary Particle Viewpoint, *American Journal of Physics*, 73, 10, pp. 973–975 (2005).

R. F. Egerton, *Physical Principles of Electron Microscopy: An Introduction to TEM, SEM, and AEM*, 1st Ed, Springer, Boston, MA, 2005.

D. Falk, D. Brill, and D. Stork, *Seeing the Light—Optics in Nature, Color, Vision, and Holography*, John Wiley & Sons, Inc., New York, 1986.

P. Falloon, "Multiple Slit Diffraction Pattern" From the Wolfram Demonstrations Project: http://demonstrations.wolfram.com/MultipleSlitDiffractionPattern/.

H. P. Freund and T. M. Antonsen, *Principles of Free-Electron Lasers*, 2nd Ed, Chapman & Hall, London, 1996.

J. M. Gibson, Reading and Writing with Electron Beams, *Physics Today*, 50, p. 56 October (1997).

R. L. Gregory, *Eye and Brain, the psychology of seeing*, World University Library, McGraw-Hill Book Company, New York, 1966.

R. Guenther, *Modern Optics*, John Wiley & Sons, Inc., New York, 1990.

S. Guttenberg, Sony Trinitron Television, in *Home Theater*, July 18, 2012: http://www.hometheater.com/content/sony-trinitron-television.

R. R. Hainich and O. Bimber, *Displays, Fundamentals & Applications*, CRC Press, Boca Raton, 2011.

R. W. Hamm and M. E. Hamm, The Beam Business: Accelerators in Industry, *Physics Today*, 64, p. 46 (2011).

P. K. Hansma, V. B. Elings, O. Marti, and C. E. Bracker, Scanning Tunneling Microscopy and Atomic Force Microscopy: Application to Biology and Technology, *Science*, 242, p. 209 (1988).

J. Hecht, Photonic Frontiers—Digital Holography: Digital Techniques Render Real-Time Response in Holography, *Laser Focus World*, 48 (7), pp. 53–58, July 1, 2012.

E. Hecht and A. Zajac, *Optics*, Addison-Wesley Publishing Co., Reading, MA, 1974, p. 234.

W. Henning and C. Shank, Accelerators for America's Future, Report from the US Department of Energy, June 2010.

C. Hernandez-Garcia, P. G. O'Shea, and M. L. Stutzman, Electron Sources for Accelerators, *Physics Today*, 61, p. 44 (2008).

L. Hornbeck, The 2009 National Inventors Hall of Fame Inductees: Digital Micromirror Device—DMD: http://invent.org/inductee-detail/?IID=397.

M. L. Huebschman, B. Munjuluri, and H. R. Garner, Dynamic Holographic 3-D Image Projection, *Optics Express* 11, 5, p. 437 (2003).

Š., Jakub, Magnet Types in Particle Accelerators, Wolfram Demonstrations Project and Contributors, 2012. Web. July 7, 2012: http://demonstrations.wolfram.com/MagnetTypesInParticleAccelerators/.

F.A. Jenkins and H.E. White, *Fundamentals of Optics*, 4th Ed, McGraw-Hill Kogakusha, Ltd, Tokyo, 1976.

C. Ji, Digital Micromirror Device, University of Buffalo Report, Electrical Engineering Department Course on Microelectromechanical Devices (MEMS), EE 541, May 4, 2001, http://www.ee.buffalo.edu/courses/ee541/reports/.

A. Jones, I. McDowall, H. Yamada, M. Bolas, and P. Debevec, Rendering for an Interactive 360° Light Field Display, USC Institute for Creative Technologies, 2007: http://gl.ict.usc.edu/Research/3DDisplay/3DDisplay_USCICT_SIGGRAPH2007.pdf.

S. Khan, TUTORIAL REVIEW, Free-Electron Lasers, *Journal of Modern Optics*, 55, 22, pp. 3469–3512, December (2008).

F. Krumeich, *Properties of Electrons, Their Interaction with Matter and Applications in Electron Microscopy*, Laboratory of Inorganic Chemistry, ETH Zurich, 2012. http://www.microscopy.ethz.ch/downloads/Interactions.pdf.

E.H. Land, Some Aspects of the Development of Sheet Polarizers, *Journal of the Optical Society of America*, 41, p. 957 (1951).

B. Lee, Three Dimensional Displays, Past and Present, *Physics Today*, 66, p. 36 April (2013).

J. R. Lewis, In the Eye of the Beholder, *IEEE Spectrum*, p. 24 May (2004). http://spectrum.ieee.org/biomedical/imaging/in-the-eye-of-the-beholder.

B. Mahon, *The Man Who Changed Everything—The Life of James Clerk Maxwell*, John Wiley & Sons, Ltd, Chichester, 2003.

E. Malamud Ed., *Accelerators and Beams—Tools for Discovery and Innovation*, 4th Ed, Division of Physics of Beams of the American Physical Society, College Park, March 2013.

A. T. McCann, Okay, but how do touch screens actually work? January 17, 2012: http://scienceline.org/2012/01/okay-but-how-do-touch-screens-actually-work/.

G. McMillan, Is Apple Planning a Holographic 3D TV? *Time Magazine*, December 8, 2010. http://techland.time.com/2010/12/28/is-apple-planning-a-holographic-3d-tv/.

F.E. Moolekamp and K.L. Stokes, "Polarization of an Optical Wave through Polarizers and Wave Plates" From the Wolfram Demonstrations Project (University of New Orleans): http://demonstrations.wolfram.com/PolarizationOfAnOpticalWaveThroughPolarizersAndWavePlates/.

W. Moomaw, P. Burgherr, G. Heath, M. Lenzen, J. Nyboer, and A. Verbruggen, Annex II: Methodology. In *IPCC Special Report on Renewable Energy Sources and Climate Change Mitigation* [O. Edenhofer, R. Pichs-Madruga, Y. Sokona, K. Seyboth, P. Matschoss, S. Kadner, T. Zwickel, P. Eickemeier, G. Hansen, S. Schlömer, C. von Stechow (eds)], Cambridge University Press, Cambridge/New York, United Kingdom and New York, NY, USA, 2011.

Next Generation Photon Sources for Grand Challenges in Science and Energy, A Report of a Subcommittee to the Basic Energy Sciences Advisory Committee, U.S. Department of Energy, May 2009. http://science.energy.gov/~/media/bes/pdf/reports/files/ngps_rpt.pdf.

Nikon, Microscopy U., The Source for Microscopy Education: http://www.microscopyu.com/articles/formulas/.

P. G. O'Shea and H. P. Freund, Free-Electron Lasers: Status and Applications, *Science*, 292, pp. 1853–1858 June (2001).

D. C. O'Shea, W. R. Callen, and W. T. Rhodes, *An Introduction to Lasers and Their Applications,* Addison-Wesley Publishing Co., Reading, MA, p. 186, 1978.

Online Dictionary of Crystallography: http://reference.iucr.org/dictionary/Main_Page.

W. K. H. Panofsky and M. Breidenbach, Accelerators and Detectors, *Reviews of Modern Physics*, 71, 2, p. 5121 (1999).

T. Parker, *Rules of Thumb—A Life Manual*, Workman Publishing, New York, 2008.

M. Perkins, Interlaced Video and Computer Monitors, February 17, 2011: http://www.cardinalpeak.com/blog/?p=834.

D. H. Pritchard, US Color Television Fundamentals—A Review, *IEEE Transactions on Consumer Electronics*, CE-23, 4, pp. 467–478, November (1977).

M. Riordan, The Discovery of Quarks, SLAC report SLAC PUB-5724, April 1992.

E. Ruska, The Development of the Electron Microscope and of Electron Microscopy, *Reviews of Modern Physics*, 59, 3, Part I, p. 627 July (1987).

P. Schatzkin, *The Boy Who Invented Television, A Story of Inspiration, Persistence and Quiet Passion*, TeamCom Books, Silver Spring, MD, 2002.

M. Sedlaček, *Electron Physics of Vacuum and Gaseous Devices*, John Wiley & Sons, Inc., New York, 1996.

A. Sessler and E. Wilson, *Engines of Discovery: A Century of Particle Accelerators*, World Scientific Publishing Co., Hackensack, 2007.

E. M. Slayter and H. S. Slayter, *Light and Electron Microscopy*, Cambridge University Press, Cambridge, 1992.

C. Smith, Looking at Life in Depth, *New Scientist*, 3, p. 21 January (1980).

A. Sygula, F. R. Fronczek, R. Sygula, P. W. Rabideau, and M. M. Olmstead, A Double Concave Hydrocarbon Buckycatcher, *Journal of the American Chemical Society*, 129, pp. 3842–3843 (2007).

The 3D DepthCube, Light-Space Technologies, October 5, 2010: http://www.youtube.com/watch?v=RAasdH10Irg.

The Nobel Prize in Physics 1991, Nobelprize.org. Nobel Media AB 2014. Web. December 21, 2015: http://www.nobelprize.org/nobel_prizes/physics/laureates/1991/.

J. Turley, The Two Percent Solution, EE Times, December (2002).

T. P. Wangler, *RF Linear Accelerators*, 2nd Ed, Wiley-VCH, Weinheim, 2008.

W.T. Welford, *Useful Optics*, The University of Chicago Press, Chicago, 1991.

H. Wiedemann, *Particle Accelerator Physics, Basic Principles and Linear Beam Dynamics,* Springer-Verlag, Berlin/New York, 1993.

K. Wille, *The Physics of Particle Accelerators, An Introduction*, Oxford University Press, New York, 2000.

D. B. Williams and C. Barry Carter, *Transmission Electron Microscopy, A Textbook for Materials Science*, 2nd Ed, Springer, New York, 2009.

E. Wilson, *An Introduction to Particle Accelerators*, Oxford University Press, Oxford/New York, 2001.

H. Winick Ed., *Synchrotron Radiation Sources—A Primer*, World Scientific, Singapore/London, 1994.

H. Winick, Fourth Generation Light Sources, Proceedings of the 17th IEEE Particle Accelerator Conference, Vancouver, BC, Canada, May 19, 1997, p. 37.

M. M. Woolfson, *The Fundamentals of Imaging—From Particles to Galaxies*, Imperial College Press, London, 2012.

L. A. Yoder, An Introduction to the Digital Light Processing (DLP™) Technology, Texas Instruments (no date): http://focus.ti.com/download/dlpdmd/119_Intro_Digital_Light_Processing.pdf.

S.-C. Yu, "CIE ChromaticityDiagram", from the Wolfram Demonstration Project: http://demonstrations.wolfram.com/CIEChromaticityDiagram/.

E. Zeleny, "Understanding Polarization with an Analogy" From the Wolfram Demonstrations Project: http://demonstrations.wolfram.com/UnderstandingPolarizationWithAnAnalogy/.

V. K. Zworykin, G. A. Morton, E. G. Ramberg, J. Hillier, and A. W. Vance, *Electron Optics and The Electron Microscope*, John Wiley & Sons, Inc., New York, 1945.

INDEX

aberrations, atmospheric 196, 362–364
aberrations, eye 412, 413
aberrations, lens 187, **205–211**, 236–237
  astigmatism 208
  chromatic 210–211
  coma 207–208, 354, 357
  distortion 209
  field curvature 208
  spherical 205–206
A-bombs 423–426, 428
absolute zero 96, 287, 368
accelerators, particle
  beam generation and manipulation 448–450
  cyclotron 456–461
  DC 450
  light sources 180, 462
  RF linear (Linac) 450–456
  synchrotron 462–469
active optics 362
adaptive optics (AO) 362–364
air conditioner (A/C) 10–15
air filters
  HEPA 89
  ULPA 89–90
airships 484–487
alpha particles 25, 26, 338, 419, **421–422**
alternating current (AC) **43–45**, 51, 172, 415, 434–435, 439–440
alternative fuels 57, **58–59**
altitude-azimuth (alt-az) mount 356–358
amplitude modulation (AM) 92, 384, 385, 389
application specific integrated circuit (ASIC) 91
attitude control system (ACS) 330
Auger electron spectrometer (AES) 69, 145–147
Auger process, Auger electrons 149–151, 271
augmented reality (AR) 318–319

balanced pair operation 381
benchmark numbers 6, 8, 75, 99, 140, 145, 153, 241, 388, 399, 406, 420, 424, 472
beta particles 421–422
bimetal 131–132
biological imaging *see* medical and biological imaging
biomimicry *see* materials science
birefringence, bi-refringent crystals 35–36, **190–194**, 308
  half-wave & quarter-wave plates 28, 35, 193

Modern Devices: The Simple Physics of Sophisticated Technology, First Edition.
Charles L. Joseph and Santiago Bernal.

Companion website: www.wiley.com/go/joseph/moderndevices

blackbody radiation (Planck distribution) **164–166**, 287
blackbody temperature and colors 164–167, 181
black box 92, 94, 104–105
blaze angle 199, 242–244
Blu-ray disc (BD) 27–30, 34
Bragg's diffraction 273–275
British Thermal Unit (BTU) 15

capacitance, capacitor 21–22, 91–93, 101–106, 118, 379–381, 440
carbon fiber composites 156, 340
carbon nanotube 158, 276, 448
Cassegrain focus, telescope 355–357
cathode 20–21, 61, 77, 84, 138, 170, 171, 268, 448 *see also* photocathode
cauterizing tools 416–417
Celsius *see* temperature scales
center of mass 18, 331, 487
charge-coupled device (CCD) 280–284
chromaticity diagram 298–299, 305
classical mechanics (CM) 3–5, 331–332, 346–347
cleanroom classification 87–88
CMOS *see* image sensors; transistor
coaxial cable or coax 381–382
coherence, temporal and spatial 179, 202–203, 215–216, 465
color computer file formats
  PNG, JPEG (jpg), TIFF (tif) 305
color displays
  HSB (hue-saturation-brightness) 296–297
  RGB (red, green, blue) 292, 297, 305–306
colorimetry 292, 296–300
commercially-available off the shelf (COTS) 9
compact disc (CD) 27–30, 34
complete internal reflection
computed tomography (CT) scan 177, **397–398**, 399–406
coronal mass ejection (CME) 327, 366
cosmic rays 284, 327, 341, 419, 476
Coudé focus 356–357
Coulomb's Law (electrostatic) 39–40
crossover pressure (CP) 71, 83
cryogen 283, 288, 403
cryostat 402, 403
crystal, crystalline 32, 91–101, 128, 136, 159, 179, 234, 273–274, 407
  defect class 93, 130–131
  lattice 111–113
  polycrystalline 95, 101, 127–128, 140–141
  wide-band-gap 98
crystal, light transmission/reflection 35–36, 190–193
crystals, liquid *see* image displays, LCD
cyclotron-based therapy 461
cyclotron frequency 458, 460

data compression 377–378, 395–396
deep space network (DSN) 342–344
depth of field 195, 265, 269
dichroism 190–191
diffraction 183, 194–200
diffraction grating *see* gratings
digital encoding/decoding (keying) 390–392
digital micro-mirror devices (DMD) 311–314
digital versatile disc (DVD) 27–35
diode 113–115
diopter (D) 186
dirty bomb 427
doping 98–99, 111–113, 116
Doppler effect **57–59**, 332–336
Doppler radar gun 62–63
Doppler sonography 408

electric power generator 433–434, **434–435**
electromagnetic force (*emf*) 45–46, 92, 441
electromagnetic induction 43–47, 51, 103, 434–435
electromagnetic radiation (light) 54, **161–166**, 183, 466–467
electromagnetism (EM) 3–5
electron gun (e-gun) 134–135, 138, 146, 268, 307–308, **448**
electron microscope
  SEM (scanning) 272–273
  STEM (scanning TEM) 273
  TEM (transmission) 266–270

electrons
  acceptor, donor 98, 111–114
  conduction band 95–98, 111–112, 114
  valence band 95, 97–100, 111–112
electrostatic 37, 39–41 *see also* Coulomb's law
electrostatic discharge (ESD) 123, 170
ellipsometer 147
endoscope 410–411
energy
  EM wavelength equivalent 161–166
  photon 161–166
  thermal equivalent (kT) 162, 283
  work function 97, 149–150
energy bands, semiconductor 95–98
energy density *see* power storage
energy efficiency rating (EER) 15
epilayer 128
epitaxial 128, 141, 157
equation of state 10
equatorial mount 356–357
e-reader, electrophoretic display 315–317
etch techniques 141–145
excimer laser 179, 413
expendable launch vehicle (ELV) 328, 348
eye, human 186, 293–295, 412–414

Fabry-Pérot (FP) 251–254
Faraday's law of induction **43–47**, 51–53, 403, 434–435
Fermi-Dirac statistics 95
Fermi level 96, 97, 111, 112
Fermion 5, 399–400
Fermi probability 95–98
fiber optical coupler 217, 222–224, 228
fiber optic gyro (FOG) 333–337
fiber optics 22, 218–222, 222–226
  signal degradation 223–227
field effect transistor (FET) 119–123
field of view (FOV) 249, 251, 293, 317, 364, 408
filters, electronic
  frequency selector 66, 105, 204
  high-pass 102–103
  low-pass 102–103
  notch suppression 105
filters, optical
  bandpass 250, 254, 279
  blocking 235, 240
  long pass 240
  neutral density 279
  notch suppression 256
  tunable interference *see* Fabry-Pérot
fissile material 423–427
fission 421, 423–426
frequency modulation (FM) 384–385, 388–389
Fresnel lens 197
fusion, nuclear 423, 426–427

gamma rays 162–163, **421–423**
geothermal 441, 442
global position system (GPS) **53–56**, 385
gradient $\vec{B}$ fields 403–406, 447, 449
grating, diffraction 198–200, 232–237, **241–246**
  blazed 241–243
  groove density/groove frequency 243–245, 247
  holographic 245–246, 255
  ruled 241–246
grating equation 243
greenhouse gas 60–61, 166, 441–442
gyroscopes or gyros 330–337 *see also* reaction wheel
  fiber optic guidance (FOG) 333–337
  mechanical 330
  ring laser gyroscope (RLG) 332
gyroscopic effect – for MRI tomography 399–401, 406–407

half-life 26, **420–422**, 428
Hall effect 148
H-bomb *see* thermonuclear
Helmholtz coils 403–405
high definition TV (HDTV) 292, 302–304, 307, 312
high-efficiency particulate air (HEPA) 89
high-speed rail 67–68
hollow-core fiber optic 221–222
holographic 3D displays *see* stereoscopic & holographic
holography 200–204
hydrogen economy 58, 438
hydrogen fuel cells 59–60, 433

ideal gas *see* equation of state
image displays
  augmented/virtual reality (AR/VR) 318
  computer monitor formats 303
  CRT (cathode ray tube) 306–308
  digital micro-mirror 311–313
  electrophoretic 315–317
  LCD (liquid crystal display) 292, 308–310
  near-eye displays (NED) 317–319
  plasma based 310–311
  raster scanning 292, 300, 308, 317
  stereoscopic & holographic displays 319–323
  touch screen 314–315
  TV (television) formats 301–302
  vector 292, 308
image intensifier 278, 285–286
image sensors *see* photoelectric image sensor
impedance, electronic 21, 101–106, 379–381
index of refraction 184–185, 189, 205
induction generator 434–435
inductor-resistor-capacitor (LRC) 21–22, 101–106, 379–381
infrared light 162, 222
infrared "thermal" image sensors 287–289
integral field unit (IFU)
  dense-pack optical fibers 249
  Fabry-Pérot 251–254
  image slicer 249–250
  microlens arrays 250–252
integrated circuit (IC) 6, 25–26, 87, **91–93**, 101, 110
  ASIC 91
  device structures and packaging 152–153
  failure rates 93
  VLSI (very large scale integration) 91–92
interference, interferometer
  amplitude *vs.* phase shift 337
  optical 33–36, 198–202, 337
  optical disks: CD, DVD, Blu-ray 27, 29–32
  Michelson 34
  Sagnac effect, optical gyros 332–336
Intergovernmental Panel on Climate Control (IPCC) 441–442
ionosphere 384–388
i-V diode curve 115

jet engines 474–475

Lagrangian points 329, 365
laminar flow 89–90
Large Hadron Collider (LHC) 443–444, 460
Larmor frequency/radiation 400, 405–406
Larmor's formula 462
laser **177–180**, 411–414
laser, free-electron 468–469
laser safety classification 180
LASIK 411–414
latent heat 17
lenslet/microlens arrays 250–252, 412
lens maker's formula 185
lens power *see* diopter
light *see* electromagnetic waves; photon
light emitting diode (LED) 26, **174–175**, 310
light sources *see also* radiation, electromagnetic
  continuum 161, 166, 169
  emission line 161, 170–171
  free-electron laser 468–469
  gas discharge 168–174
  germicidal 172
  laser **177–180**, 466
  synchrotron radiation (SR) 180, 462–467
  X-ray 175–177, 466
liquid crystal displays (LCD) 308–310
lithotripsy 19
Livingston plot 444
Lorentz force 446–447
low Earth orbit (LEO) 326–327, 365, 471–472
LRC or RLC circuits 21–22, 101–106, 379–381
luminous flux 166

Mach, unit of speed 471–472, 474
magnetic focusing *see* electromagnetic
magnetic resonance imaging (MRI) 398–406
magnetron **20–22**

magnification
  compound microscope 261–262
  single lens 186, 211, 260
  telescope 187
mass and energy equivalence
    443, 445
materials science 155–159
  biometry, biomimicry 155–156
  composite materials 156–157
  nanotechnology 158–159
materials, solid-state
  conducting 97
  crystalline 95, 97, 127–131, 140–141
  semiconducting 95–99, 111–114, 119
  silicon oxide insulator 92–93, 97
MCP (microchannel plate) intensifier
    145–146, 285–287, 288–289
medical & biological imaging
  CT scanner 397, 398–399
  MRI 399–406
  ultrasound 408–409
metal detector 43–45
metal-organic chemical vapor deposition
    (MOCVD) 132–133, 466
microscope
  compound 260–262
  electron *see* electron microscope
  optical 260–266
microwaves
  communications frequency bands
    341–342, 384–385
  cooking **18–22**
  part of EM radio waves 162
molecular beam epitaxy (MBE) 136–141
MOS *see* transistor
multi-layer insulator (MLI) 326, 339–340
multiple mirror telescope, 358, 361–362
multiplexing 226–227, 377, 396
MUX – multiplexer, deMux 225, 396

nanotechnology (nanotubes, nanoparticles
  nanowires, quantum dots) 158
near-eye displays (NED) 317–319
network, computer 392–396
neutron 421, 424
Newtonian focus telescope 355
Newton's laws of motion 346
night vision sensors 287–289
n-type material *see* semiconductor

optical coupler
  electronic isolation 228
  long-distance fiber connection 223–227
optical disc 27
optical fibers 218–222
optical registration and distortions
    144, 209
optics, reflecting/transmitting 184–187
orbits
  geosynchronous, HEO, LEO 327
  polar 328
  Solar 365–370

pacemaker 51, 415–416
parabolic reflective optics
  off-axis 233, 354–355
  telescope optics 354–355
particle accelerators 443–468
particle-wave duality 5, 443
Pauli Exclusion Principle 5
Periodic Table of Elements 98, 99
phase-lock loop (PLL) 66
phase shift 29–30, 333–337
phase transition 12–13, 16
photocathode 278, 284–289, 453
photoconductor 37
photocopier 37–39
photoelectric effect 147, 149–151
photoelectric image sensors 277–278
  CCD, CMOS 280–284
  detective quantum efficiency (DQE) 287
  infrared - night vision 288–289
  infrared – thermal solid-state 287–288
  photoconductive (visible) 280–284
  photoemissive (UV, X-ray) 284–287
  quantum efficiency (QE) 282,
    284–285, 287
photoelectrons 149–151, 14
photoemission spectroscopy (PES) 147
photolithography 141–144
photon, energy & momentum of 161–165
photonics 5–6, 217, 227, 377
piezoelectric transducer (PZT) 276,
    335–336, 362–364
planetary probes 372–376
PMT (photomultiplier tube) 138–140
p-n junction *see* diode; LED
polarization, polarized light 29, 35, 147,
    188–190, 309–310, 312

population inversion 178
positron emission tomography (PET) scan 398, 431
power cogeneration 442
power grid 439–440
power storage and power content of fuels 435–439
power *vs*. energy 433
pressure 10–16, **69–70**, 135–141, 471
see also vacuum systems
prism 232, 237–239, 240–241
programmable read only memory (PROM) 341
p-type material *see* semiconductor
pumped-storage hydroelectric 436–437

quantum efficiency (QE) 282, 284–285, 287
quantum mechanics (QM) 3–5, 95–96

radar – RAdio Detection and Ranging 60–67
radiance 166, 181
radiation
α, β, γ-ray 421–423, 426
dose, dose units 419–421, 425–426, **428–430**
electromagnetic 161–166, 183, 406, 419
industrial applications 431
Lamor (MRI) 400–401, 405–406
medical applications 398–399, 431
naturally occurring sources 419–420
safety 419, 428–431
synchrotron (SR) 443, 445, 450, 463, 466–467
radioactive/radioactivity **419–423**
decay – half life 26, **420–422**, 428
in smoke detectors 25
radio frequency identification (RFID) **51–53**, 385, 415–416
radioisotope heaters 339–340
radioisotope thermal generators (RTG) 337–339
radionuclides 426
Raman effect (Stokes/anti-Stokes) 256
raster scan 292, 300, 308, 317
Rayleigh scattering 256–257
RC circuit *see* LRC circuit and AC current
reaction wheel 332
reflective optics
adaptive 362–364
gratings 232–237, **241–246**
telescope 354–355, 358–362
refractive optics 184–187, 260, 261–263
refrigerant 10–17
renewable energy generation 435–436, 437–439, 441
resistance, electronic 98, 101, 106–107
resolution, optical
Airy disk 195–196, 263
Rayleigh criteria 237, 262–263
spectral 236–237, 241, 244
RF coils (of MRI scanners)
RF telecommunications
AM/FM broadcasting 384–390
digitally modulated free-space channels 390–392
sky wave – “skipping”, 385–388
RHEED probe 136–137, 138–141
rocket 345–348
rubber mirror 363–365

Sagnac effect 332–336
scanners, retail 47–49
scanning tunneling microscope (STM) 275–276
seasonally adjusted energy efficiency (SEER) 16
segmented mirror telescope 358, 359–362, 368
semiconductor
device architectures (CMOS, TTL, etc.) 119–123
energy bands 96–98
fabrication 132–145
failure rates 93, 94
majority/minority carriers 111–115
materials 95–99 *see also* materials
III-nitrate, III-V semiconductor 99, 284
n-type, p-type 111–115
substrate **127–130**, 133–139, 142
Shack-Hartmann wavefront sensor 363–364, 412–414
signal degradation 221, **223–227**, 379
simple harmonic oscillator (SHO) 104–105, 107–110
single-photon emission computer tomography (SPECT) 398, 431

Snell's Law of refractive optics **184–185**, 222, 241
special relativity 459
spacecraft subsystems
  attitude control 330–337
  command, control, & telemetry 341–344
  launch and propulsion 345–351
  power 337–339
  thermal & environmental control 339–341
spectral orders 235–236
  dispersion 236–240
  free spectral range 236
  spectral resolution 236
spectrographs
  dispersing element *see* gratings; prism
  Echelle 254–255
  Fabry-Pérot (FP) 251–254
  integral field *see* integral field unit 249–252
  long slit 248–249
  optical components 231–232, 236
  Raman 255–258
  Rowland circle 233–235
  Wadsworth-mount 232–233
Stokes/anti-Stokes Raman scattering 257
stratospheric balloons 476–482
superconducting 288, 401–402, 452
synchrotron radiation (SR) 443, 445, 450, 463, 466–467
synchrotron wigglers and undulators 464–466

TDRSS 342–344
telegrapher's equation 380
telescope designs 354–358
telescopes, ultra-large 358–362
temperature scale
  Celsius 8, 96
  Kelvin (absolute temperature) 96, 166
thermocouple (TC) 76
thermonuclear bomb (H bomb) 426–427, 428
thin films 127–131
  chemical vapor deposition (CVD) 133, 135–136
  defects, stress, and strain 130–131, 133
  grain boundary defects 128
  physical vapor deposition (PVD) 133–135, 136–137
  polycrystalline 128
  quality (uniformity, dense) 128–130
time-division multiplexing 6, **396**
total internal reflection 185, 218–219, 222, **228**, 385–388
transistor 115–123
  bipolar 115–118
  CMOS 122–123
  FET, JFET, MOSFET 119–122

ultrasonic lithotripsy 408
ultrasonography 408–409
ultraviolet (UV) light 162, 171, 466
  far-UV, near-UV, VUV 171
  germicidal 172
  UVA, UVB, UVC 165
unmanned aerial vehicle (UAV) 486

vacuum chambers 70–72
  ultrahigh vacuum (UHV) 72, 128, 138
vacuum gauges 72, 76–78
vacuum pumps 72, 74, 78–85
vacuum seals 72–74
voltage divider 125

waveguide 22, **218–219**, 221–223, 382, 386–387, 410, 452
wavelength 161–166
wave propagation 194, 382–383

X-ray photon spectrometer (XPS) 145–147
X-ray sources 146, 162, 175–177, 399
X-ray tubes/detectors 399

YAG laser 179, 218, 411–412